T0181536

Herbert Amann
Joachim Escher

Analysis II

Translated from the German
by Silvio Levy and Matthew Cargo

Birkhäuser
Basel · Boston · Berlin

Authors:

Herbert Amann
Institut für Mathematik
Universität Zürich
Winterthurerstr. 190
8057 Zürich
Switzerland
e-mail: herbert.amann@math.uzh.ch

Joachim Escher
Institut für Angewandte Mathematik
Universität Hannover
Welfengarten 1
30167 Hannover
Germany
e-mail: escher@ifam.uni-hannover.de

Originally published in German under the same title by Birkhäuser Verlag, Switzerland
© 1999 by Birkhäuser Verlag

2000 Mathematics Subject Classification: 26-01, 26A42, 26Bxx, 30-01

Library of Congress Control Number: 2008926303

Bibliographic information published by Die Deutsche Bibliothek
Die Deutsche Bibliothek lists this publication in the Deutsche Nationalbibliografie; detailed
bibliographic data is available in the Internet at <http://dnb.ddb.de>.

ISBN 978-3-7643-7472-3 Birkhäuser Verlag, Basel – Boston – Berlin

© 2008 Birkhäuser Verlag AG
Basel · Boston · Berlin
P.O. Box 133, CH-4010 Basel, Switzerland
Part of Springer Science+Business Media
Printed on acid-free paper produced of chlorine-free pulp. TCF ∞

ISBN 978-3-7643-7472-3
9 8 7 6 5 4 3 2 1

ISBN 978-3-7643-7478-5 (eBook)
www.birkhauser.ch

Foreword

As with the first, the second volume contains substantially more material than can be covered in a one-semester course. Such courses may omit many beautiful and well-grounded applications which connect broadly to many areas of mathematics. We of course hope that students will pursue this material independently; teachers may find it useful for undergraduate seminars.

For an overview of the material presented, consult the table of contents and the chapter introductions. As before, we stress that doing the numerous exercises is indispensable for understanding the subject matter, and they also round out and amplify the main text.

In writing this volume, we are indebted to the help of many. We especially thank our friends and colleagues Pavol Quittner and Gieri Simonett. They have not only meticulously reviewed the entire manuscript and assisted in weeding out errors but also, through their valuable suggestions for improvement, contributed essentially to the final version. We also extend great thanks to our staff for their careful perusal of the entire manuscript and for tracking errata and inaccuracies.

Our most heartfelt thank extends again to our "typesetting perfectionist", without whose tireless effort this book would not look nearly so nice.[1] We also thank Andreas for helping resolve hardware and software problems.

Finally, we extend thanks to Thomas Hintermann and to Birkhäuser for the good working relationship and their understanding of our desired deadlines.

Zürich and Kassel, March 1999   H. Amann and J. Escher

---

[1] The text was set in LaTeX, and the figures were created with CorelDRAW! and Maple.

Foreword to the second edition

In this version, we have corrected errors, resolved imprecisions, and simplified several proofs. These areas for improvement were brought to our attention by readers. To them and to our colleagues H. Crauel, A. Ilchmann and G. Prokert, we extend heartfelt thanks.

Zürich and Hannover, December 2003   H. Amann and J. Escher

Foreword to the English translation

It is a pleasure to express our gratitude to Silvio Levy and Matt Cargo for their careful and insightful translation of the original German text into English. Their effective and pleasant cooperation during the process of translation is highly appreciated.

Zürich and Hannover, March 2008   H. Amann and J. Escher

Contents

Chapter VII   Multivariable differential calculus

Chapter VIII   Line integrals

Chapter VI

Integral calculus in one variable

Integration was invented for finding the area of shapes. This, of course, is an ancient problem, and the basic strategy for solving it is equally old: divide the shape into rectangles and add up their areas.

A mathematically satisfactory realization of this clear, intuitive idea is amazingly subtle. We note in particular that is a vast number of ways a given shape can be approximated by a union of rectangles. It is not at all self-evident they all lead to the same result. For this reason, we will not develop the rigorous theory of measures until Volume III.

In this chapter, we will consider only the simpler case of determining the area between the graph of a sufficiently regular function of one variable and its axis. By laying the groundwork for approximating a function by a juxtaposed series of rectangles, we will see that this boils down to approaching the function by a series of staircase functions, that is, functions that are piecewise constant. We will show that this idea for approximations is extremely flexible and is independent of its original geometric motivation, and we will arrive at a concept of integration that applies to a large class of vector-valued functions of a real variable.

To determine precisely the class of functions to which we can assign an integral, we must examine which functions can be approximated by staircase functions. By studying the convergence under the supremum norm, that is, by asking if a given function can be approximated uniformly on the entire interval by staircase functions, we are led to the class of jump continuous functions. Section 1 is devoted to studying this class.

There, we will see that an integral is a linear map on the vector space of staircase functions. There is then the problem of extending integration to the space of jump continuous functions; the extension should preserve the elementary properties of this map, particularly linearity. This exercise turns out to be a special case of the general problem of uniquely extending continuous maps. Because the extension problem is of great importance and enters many areas of mathematics, we will discuss it at length in Section 2. From the fundamental extension theorem for uniformly continuous maps, we will derive the theorem of continuous extensions of continuous linear maps. This will give us an opportunity to introduce the important concepts of bounded linear operators and the operator norm, which play a fundamental role in modern analysis.

After this groundwork, we will introduce in Section 3 the integral of jump continuous functions. This, the Cauchy–Riemann integral, extends the elementary integral of staircase functions. In the sections following, we will derive its fundamental properties. Of great importance (and you can tell by the name) is the fundamental theorem of calculus, which, to oversimplify, says that integration reverses differentiation. Through this theorem, we will be able to explicitly calculate a great many integrals and develop a flexible technique for integration. This will happen in Section 5.

In the remaining sections — except for the eighth — we will explore applications of the so-far developed differential and integral calculus. Since these are not essential for the overall structure of analysis, they can be skipped or merely sampled on first reading. However, they do contain many of the beautiful results of classical mathematics, which are needed not only for one's general mathematical literacy but also for numerous applications, both inside and outside of mathematics.

Section 6 will explore the connection between integrals and sums. We derive the Euler–Maclaurin sum formula and point out some of its consequences. Special mention goes to the proof of the formulas of de Moivre and Sterling, which describe the asymptotic behavior of the factorial function, and also to the derivation of several fundamental properties of the famous Riemann ζ function. The latter is important in connection to the asymptotic behavior of the distribution of prime numbers, which, of course, we can go into only very briefly.

In Section 7, we will revive the problem — mentioned at the end of Chapter V — of representing periodic functions by trigonometric series. With help from the integral calculus, we can specify a complete solution of this problem for a large class of functions. We place the corresponding theory of Fourier series in the general framework of the theory of orthogonality and inner product spaces. Thereby we achieve not only clarity and simplicity but also lay the foundation for a number of concrete applications, many of which you can expect see elsewhere. Naturally, we will also calculate some classical Fourier series explicitly, leading to some surprising results. Among these is the formula of Euler, which gives an explicit expression for the ζ function at even arguments; another is an interesting expression for the sine as an infinite product.

Up to this point, we have will have concentrated on the integration of jump continuous functions on compact intervals. In Section 8, we will further extend the domain of integration to cover functions that are defined (and integrated) on infinite intervals or are not bounded. We content ourselves here with simple but important results which will be needed for other applications in this volume because, in Volume III, we will develop an even broader and more flexible type of integral, the Lebesgue integral.

Section 9 is devoted to the theory of the gamma function. This is one of the most important nonelementary functions, and it comes up in many areas of mathematics. Thus we have tried to collect all the essential results, and we hope you will find them of value later. This section will show in a particularly nice way the strength of the methods developed so far.

1 Jump continuous functions

In many concrete situations, particularly in the integral calculus, the constraint of continuity turns out to be too restrictive. Discontinuous functions emerge naturally in many applications, although the discontinuity is generally not very pathological. In this section, we will learn about a simple class of maps which contains the continuous functions and is especially useful in the integral calculus in one independent variable. However, we will see later that the space of jump continuous functions is still too restrictive for a flexible theory of integration, and, in the context of multidimensional integration, we will have to extend the theory into an even broader class containing the continuous functions.

In the following, suppose
- $E := (E, \|\cdot\|)$ is a Banach space;
  $I := [\alpha, \beta]$ is a compact perfect interval.

Staircase and jump continuous functions

We call $\mathfrak{Z} := (\alpha_0, \ldots, \alpha_n)$ a **partition** of I, if $n \in \mathbb{N}^\times$ and

$$\alpha = \alpha_0 < \alpha_1 < \cdots < \alpha_n = \beta \ .$$

If $\{\alpha_0, \ldots, \alpha_n\}$ is a subset of the partition $\overline{\mathfrak{Z}} := (\beta_0, \ldots, \beta_k)$, $\overline{\mathfrak{Z}}$ is called a **refinement** of $\mathfrak{Z}$, and we write $\mathfrak{Z} \le \overline{\mathfrak{Z}}$.

The function $f : I \to E$ is called a **staircase function** on I if I has a partition $\mathfrak{Z} := (\alpha_0, \ldots, \alpha_n)$ such that f is constant on every (open) interval (α_{j-1}, α_j). Then we say $\mathfrak{Z}$ is a partition for f, or we say f is a staircase function on the partition $\mathfrak{Z}$.

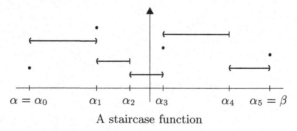

$$\alpha = \alpha_0 \qquad \alpha_1 \quad \alpha_2 \ \ \alpha_3 \qquad\quad \alpha_4 \quad \alpha_5 = \beta$$

A staircase function

If $f : I \to E$ is such that the limits $f(\alpha + 0)$, $f(\beta - 0)$, and

$$f(x \pm 0) := \lim_{\substack{y \to x \pm 0 \\ y \ne x}} f(y)$$

exist for all $x \in \mathring{I}$, we call f **jump continuous**.[1] A jump continuous function is **piecewise continuous** if it has only finitely many discontinuities ("jumps"). Finally,

---

[1] Note that, in general, $f(x + 0)$ and $f(x - 0)$ may differ from $f(x)$.

we denote by

$$\mathcal{T}(I,E), \qquad \mathcal{S}(I,E), \qquad \mathcal{SC}(I,E)$$

the sets of all functions $f \colon I \to E$ that are staircase, jump continuous, and piecewise continuous, respectively.[2]

$E = \mathbb{R}$

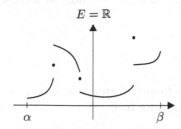

$E = \mathbb{R}$

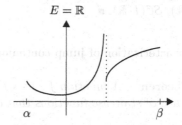

A piecewise continuous function   Not a jump continuous function

1.1 Remarks **(a)** Given partitions $\mathfrak{Z} := (\alpha_0, \ldots, \alpha_n)$ and $\overline{\mathfrak{Z}} := (\beta_0, \ldots, \beta_m)$ of I, the union $\{\alpha_0, \ldots, \alpha_n\} \cup \{\beta_0, \ldots, \beta_m\}$ will naturally define another partition $\mathfrak{Z} \vee \overline{\mathfrak{Z}}$ of I. Obviously, $\mathfrak{Z} \leq \mathfrak{Z} \vee \overline{\mathfrak{Z}}$ and $\overline{\mathfrak{Z}} \leq \mathfrak{Z} \vee \overline{\mathfrak{Z}}$. In fact, $\leq$ is an ordering on the set of partitions of I, and $\mathfrak{Z} \vee \overline{\mathfrak{Z}}$ is the largest from $\{\mathfrak{Z}, \overline{\mathfrak{Z}}\}$.

(b) If f is a staircase function on a partition $\mathfrak{Z}$, every refinement of $\mathfrak{Z}$ is also a partition for f.

(c) If $f \colon I \to E$ is jump continuous, neither $f(x+0)$ nor $f(x-0)$ need equal $f(x)$ for $x \in I$.

(d) $\mathcal{S}(I,E)$ is a vector subspace of $B(I,E)$.

Proof The linearity of one-sided limits implies immediately that $\mathcal{S}(I,E)$ is a vector space. If $f \in \mathcal{S}(I,E) \backslash B(I,E)$, we find a sequence (x_n) in I with

$$\|f(x_n)\| \geq n \quad \text{for } n \in \mathbb{N} . \tag{1.1}$$

Because I is compact, there is a subsequence (x_{n_k}) of (x_n) and $x \in I$ such that $x_{n_k} \to x$ as $k \to \infty$. By choosing a suitable subsequence of (x_{n_k}), we find a sequence (y_n), that converges monotonically to x.[3] If f is jump continuous, there is a $v \in E$ with $\lim f(y_n) = v$ and thus $\lim \|f(y_n)\| = \|v\|$ (compare with Example III.1.3(j)). Because every convergent sequence is bounded, we have contradicted (1.1). Therefore $\mathcal{S}(I,E) \subset B(I,E)$. ∎

(e) We have sequences of vector subspaces

$$\mathcal{T}(I,E) \subset \mathcal{SC}(I,E) \subset \mathcal{S}(I,E) \quad \text{and} \quad C(I,E) \subset \mathcal{SC}(I,E) .$$

(f) Every monotone function $f \colon I \to \mathbb{R}$ is jump continuous.

---

[2]We usually abbreviate $\mathcal{T}(I) := \mathcal{T}(I, \mathbb{K})$ etc, if the context makes clear which of the fields $\mathbb{R}$ or $\mathbb{C}$ we are dealing with.
[3]Compare with Exercise II.6.3.

Proof This follows from Proposition III.5.3. ∎

(g) If f belongs to $\mathcal{T}(I,E)$, $\mathcal{S}(I,E)$, or $\mathcal{SC}(I,E)$, and J is a compact perfect subinterval of I, then $f\,|\,J$ belongs to $\mathcal{T}(J,E)$, $\mathcal{S}(J,E)$, or $\mathcal{SC}(J,E)$.

(h) If f belongs to $\mathcal{T}(I,E)$, $\mathcal{S}(I,E)$, or $\mathcal{SC}(I,E)$, then $\|f\|$ belongs to $\mathcal{T}(I,\mathbb{R})$, $\mathcal{S}(I,\mathbb{R})$, $\mathcal{SC}(I,\mathbb{R})$. ∎

A characterization of jump continuous functions

1.2 Theorem *A function $f\colon I \to E$ is jump continuous if and only if there is a sequence of staircase functions that converges uniformly to it.*

Proof "⇒" Suppose $f \in \mathcal{S}(I,E)$ and $n \in \mathbb{N}^\times$. Then for every $x \in I$, there are numbers $\alpha(x)$ and $\beta(x)$ such that $\alpha(x) < x < \beta(x)$ and

$$\|f(s) - f(t)\| < 1/n \quad \text{for } s,t \in \big(\alpha(x),x\big) \cap I \quad \text{or} \quad s,t \in \big(x,\beta(x)\big) \cap I \ .$$

Because $\big\{\big(\alpha(x),\beta(x)\big) \ ; \ x \in I\big\}$ is an open cover of the compact interval I, we can find elements $x_0 < x_1 < \cdots < x_m$ in I such that $I \subset \bigcup_{j=0}^{m}\big(\alpha(x_j),\beta(x_j)\big)$. Letting $\eta_0 := \alpha$, $\eta_{j+1} := x_j$ for $j = 0,\ldots,m$, and $\eta_{m+2} := \beta$, we let $\mathfrak{Z}_0 = (\eta_0,\ldots,\eta_{m+2})$ be a partition of I. Now we select a refinement $\mathfrak{Z}_1 = (\xi_0,\ldots,\xi_k)$ of $\mathfrak{Z}_0$ with

$$\|f(s) - f(t)\| < 1/n \quad \text{for } s,t \in (\xi_{j-1},\xi_j) \text{ and } j = 1,\ldots,k \ ,$$

and set

$$f_n(x) := \begin{cases} f(x)\,, & x \in \{\xi_0,\ldots,\xi_k\}\,, \\ f\big((\xi_{j-1}+\xi_j)/2\big)\,, & x \in (\xi_{j-1},\xi_j)\,, \quad j=1,\ldots,k \ . \end{cases}$$

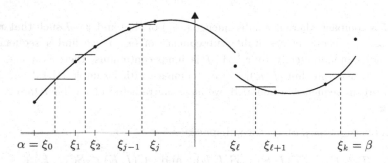

$$\alpha=\xi_0 \quad \xi_1 \ \xi_2 \quad \xi_{j-1} \ \xi_j \qquad\qquad \xi_\ell \quad \xi_{\ell+1} \qquad \xi_k=\beta$$

Then f_n is a staircase function, and by construction

$$\|f(x) - f_n(x)\| < 1/n \quad \text{for } x \in I \ .$$

Therefore $\|f - f_n\|_\infty < 1/n$.

"$\Leftarrow$" Suppose there is a sequence (f_n) in $\mathcal{T}(I, E)$ that converges uniformly to f. The sequence also converges to f in $B(I, E)$. Let $\varepsilon > 0$. Then there is an $n \in \mathbb{N}$ such that $\|f(x) - f_n(x)\| < \varepsilon/2$ for all $x \in I$. In addition, for every $x \in (\alpha, \beta]$ there is an $\alpha' \in [\alpha, x)$ such that $f_n(s) = f_n(t)$ for $s, t \in (\alpha', x)$. Consequently,

$$\|f(s) - f(t)\| \le \|f(s) - f_n(s)\| + \|f_n(s) - f_n(t)\| + \|f_n(t) - f(t)\| < \varepsilon \quad (1.2)$$

for $s, t \in (\alpha', x)$.

Suppose now (s_j) is a sequence in I that converges from the left to x. Then there is an $N \in \mathbb{N}$ such that $s_j \in (\alpha', x)$ for $j \ge N$, and (1.2) implies

$$\|f(s_j) - f(s_k)\| < \varepsilon \quad \text{for } j, k \ge N .$$

Therefore $\big(f(s_j)\big)_{j \in \mathbb{N}}$ is a Cauchy sequence in the Banach space E, and there is an $e \in E$ with $\lim_j f(s_j) = e$. If (t_k) is another sequence in I that converges from the left to x, then we can repeat the argument to show there is an $e' \in E$ such that $\lim_k f(t_k) = e'$. Also, there is an $M \ge N$ such that $t_k \in (\alpha', x)$ for $k \ge M$. Consequently, (1.2) gives

$$\|f(s_j) - f(t_k)\| < \varepsilon \quad \text{for } j, k \ge M .$$

After taking the limits $j \to \infty$ and $k \to \infty$, we find $\|e - e'\| \le \varepsilon$. Now e and e' agree, because $\varepsilon > 0$ was arbitrary. Therefore we have proved that $\lim_{y \to x-0} f(y)$ exists. By swapping left and right, we show that for $x \in [\alpha, \beta)$ the right-sided limits $\lim_{y \to x+0} f(y)$ exist as well. Consequently f is jump continuous. ∎

1.3 Remark If the function $f \in \mathcal{S}(I, \mathbb{R})$ is nonnegative, the first part of the above proof shows there is a sequence of nonnegative staircase functions that converges uniformly to f. ∎

The Banach space of jump continuous functions

1.4 Theorem *The set of jump continuous functions $\mathcal{S}(I, E)$ is a closed vector subspace of $B(I, E)$ and is itself a Banach space; $\mathcal{T}(I, E)$ is dense in $\mathcal{S}(I, E)$.*

Proof From Remark 1.1(d) and (e), we have the inclusions

$$\mathcal{T}(I, E) \subset \mathcal{S}(I, E) \subset B(I, E) .$$

According to Theorem 1.2, we have

$$\overline{\mathcal{T}(I, E)} = \mathcal{S}(I, E) , \quad (1.3)$$

when the closure is formed in $B(I, E)$. Therefore $\mathcal{S}(I, E)$ is closed in $B(I, E)$ by Proposition III.2.12. The last part of the theorem follows from (1.3). ∎

1.5 Corollary

(i) *Every (piecewise) continuous function is the uniform limit of a sequence of staircase functions.*

(ii) *The uniform limit of a sequence of jump continuous functions is jump continuous.*

(iii) *Every monotone function is the uniform limit of a sequence of staircase functions.*

Proof   The statement (i) follows directly from Theorem 1.2; (ii) follows from Theorem 1.4, and statement (iii) follows from Remark 1.1(f). ∎

Exercises

1   Verify that, with respect to pointwise multiplication, $\mathcal{S}(I,\mathbb{K})$ is a Banach algebra with unity.

2   Define $f : [-1,1] \to \mathbb{R}$ by

$$
f(x) := \begin{cases} \dfrac{1}{n+2}\,, & x \in \left[-\dfrac{1}{n}, -\dfrac{1}{n+1}\right) \cup \left(\dfrac{1}{n+1}, \dfrac{1}{n}\right], \\ 0\,, & x = 0\,. \end{cases}
$$

Prove or disprove that

(a) $f \in \mathcal{T}([-1,1],\mathbb{R})$;

(b) $f \in \mathcal{S}([-1,1],\mathbb{R})$.

3   Prove or disprove that $\mathcal{SC}(I,E)$ is a closed vector subspace of $\mathcal{S}(I,E)$.

4   Show these statements are equivalent for $f : I \to E$:

(i) $f \in \mathcal{S}(I,E)$;

(ii) $\exists\, (f_n)$ in $\mathcal{T}(I,E)$ such that $\sum_n \|f_n\|_\infty < \infty$ and $f = \sum_{n=0}^{\infty} f_n$.

5   Prove that every jump continuous function has at most a countable number of discontinuities.

6   Denote by $f : [0,1] \to \mathbb{R}$ the Dirichlet function on $[0,1]$. Does f belong to $\mathcal{S}([0,1],\mathbb{R})$?

7   Define $f : [0,1] \to \mathbb{R}$ by

$$
f(x) := \begin{cases} 1/n\,, & x \in \mathbb{Q} \text{ with } x \text{ in lowest terms } m/n\,, \\ 0\,, & \text{otherwise}\,. \end{cases}
$$

Prove or disprove that $f \in \mathcal{S}([0,1],\mathbb{R})$.

8   Define $f : [0,1] \to \mathbb{R}$ by

$$
f(x) := \begin{cases} \sin(1/x)\,, & x \in (0,1]\,, \\ 0\,, & x = 0\,. \end{cases}
$$

Is f jump continuous?

9 Suppose E_j, $j = 0, \ldots, n$ are normed vector spaces and

$$f = (f_0, \ldots, f_n) : I \to E := E_0 \times \cdots \times E_n \ .$$

Show

$$f \in \mathcal{S}(I, E) \iff f_j \in \mathcal{S}(I, E_j) \ , \quad j = 0, \ldots, n \ .$$

10 Suppose E and F are normed vector spaces and that $f \in \mathcal{S}(I, E)$ and $\varphi : E \to F$ are uniformly continuous. Show that $\varphi \circ f \in \mathcal{S}(I, F)$.

11 Suppose $f, g \in \mathcal{S}(I, \mathbb{R})$ and $\mathrm{im}(g) \subset I$. Prove or disprove that $f \circ g \in \mathcal{S}(I, \mathbb{R})$.

2 Continuous extensions

In this section, we study the problem of continuously extending a uniformly continuous map onto an appropriate superset of its domain. We confine ourselves here to when the domain is dense in that superset. In this situation, the continuous extension is uniquely determined and can be approximated to arbitrary precision by the original function. In the process, we will learn an approximation technique that recurs throughout analysis.

The early parts of this section are of fundamental importance for the overall continuous mathematics and have numerous applications, and so it is important to understand the results well.

The extension of uniformly continuous functions

2.1 Theorem (extension theorem) *Suppose Y and Z are metric spaces, and Z is complete. Also suppose X is a dense subset of Y, and $f : X \to Z$ is uniformly continuous.[1] Then f has a uniquely determined extension $\bar{f} : Y \to Z$ given by*

$$\bar{f}(y) = \lim_{\substack{x \to y \\ x \in X}} f(x) \quad \text{for } y \in Y \;,$$

and $\bar{f}$ is also uniformly continuous.

Proof (i) We first verify uniqueness. Assume $g, h \in C(Y, Z)$ are extensions of f. Because X is dense in Y, there is for every $y \in Y$ a sequence (x_n) in X such that $x_n \to y$ in Y. The continuity of g and h implies

$$g(y) = \lim_n g(x_n) = \lim_n f(x_n) = \lim_n h(x_n) = h(y) \;.$$

Consequently, $g = h$.

(ii) If f is uniformly continuous, there is, for every $\varepsilon > 0$, a $\delta = \delta(\varepsilon) > 0$ such that
$$d\big(f(x), f(x')\big) < \varepsilon \quad \text{for } x, x' \in X \text{ and } d(x, x') < \delta \;. \tag{2.1}$$

Suppose $y \in Y$ and (x_n) is a sequence in X such that $x_n \to y$ in Y. Then there is an $N \in \mathbb{N}$ such that
$$d(x_j, y) < \delta/2 \quad \text{for } j \geq N \;, \tag{2.2}$$

and it follows that

$$d(x_j, x_k) \leq d(x_j, y) + d(y, x_k) < \delta \quad \text{for } j, k \geq N \;.$$

From (2.1), we have

$$d\big(f(x_j), f(x_k)\big) < \varepsilon \quad \text{for } j, k \geq N \;.$$

---

[1] As usual, we endow X with the metric induced by Y.

Therefore $(f(x_j))$ is a Cauchy sequence in Z. Because Z is complete, we can find a $z \in Z$ such that $f(x_j) \to z$. If (x'_k) is another sequence in X such that $x'_k \to y$, we reason as before that there exists a $z' \in Z$ such that $f(x'_k) \to z'$. Moreover, we can find $M \geq N$ such that $d(x'_k, y) < \delta/2$ for $k \geq M$. This, together with (2.2), implies

$$d(x_j, x'_k) \leq d(x_j, y) + d(y, x'_k) < \delta \quad \text{for } j, k \geq M ,$$

and because of (2.1), we have

$$d\big(f(x_j), f(x'_k)\big) < \varepsilon \quad \text{for } j, k \geq M . \tag{2.3}$$

Taking the limits $j \to \infty$ and $k \to \infty$ in (2.3) results in $d(z, z') \leq \varepsilon$. This being true every positive ε, we have $z = z'$. These considerations show that the map

$$\bar{f} : Y \to Z , \quad y \mapsto \lim_{\substack{x \to y \\ x \in X}} f(x)$$

is well defined, that is, it is independent of the chosen sequence.

For $x \in X$, we set $x_j := x$ for $j \in \mathbb{N}$ and find $\bar{f}(x) = \lim_j f(x_j) = f(x)$. Therefore $\bar{f}$ is an extension of f.

(iii) It remains to show that $\bar{f}$ is uniformly continuous. Let $\varepsilon > 0$, and choose $\delta > 0$ satisfying (2.1). Also choose $y, z \in Y$ such that $d(y, z) < \delta/3$. Then there are series (y_n) and (z_n) in X such that $y_n \to y$ and $z_n \to z$. Therefore, there is an $N \in \mathbb{N}$ such that $d(y_n, y) < \delta/3$ and $d(z_n, z) < \delta/3$ for $n \geq N$. In particular, we get

$$d(y_N, z_N) \leq d(y_N, y) + d(y, z) + d(z, z_N) < \delta$$

and also

$$d(y_n, y_N) \leq d(y_n, y) + d(y, y_N) < \delta ,$$
$$d(z_n, z_N) \leq d(z_n, z) + d(z, z_N) < \delta$$

for $n \geq N$. From the definition of $\bar{f}$, Example III.1.3(l), and (2.1), we have

$$d\big(\bar{f}(y), \bar{f}(z)\big) \leq d\big(\bar{f}(y), f(y_N)\big) + d\big(f(y_N), f(z_N)\big) + d\big(f(z_N), \bar{f}(z)\big)$$
$$= \lim_n d\big(f(y_n), f(y_N)\big) + d\big(f(y_N), f(z_N)\big) + \lim_n d\big(f(z_N), f(z_n)\big)$$
$$< 3\varepsilon .$$

Therefore $\bar{f}$ is uniformly continuous. ∎

2.2 Application Suppose X is a bounded subset of $\mathbb{K}^n$. Then the restriction[2]

$$T : C(\overline{X}) \to BUC(X) , \quad u \mapsto u|X \tag{2.4}$$

---

[2] $BUC(X)$ denotes the Banach space of bounded and uniformly continuous functions on X, as established by the supremum norm. Compare to Exercise V.2.1.

is an isometric isomorphism.

Proof   (i) Suppose $u \in C(\overline{X})$. Then, because $\overline{X}$ is compact by the Heine–Borel theorem, Corollary III.3.7 and Theorem III.3.13 imply that u — and therefore also $Tu = u \,|\, X$ — is bounded and uniformly continuous. Therefore T is well defined. Obviously T is linear.

(ii) Suppose $v \in BUC(X)$. Because X is dense in $\overline{X}$, there is from Theorem 2.1 a uniquely determined $u \in C(\overline{X})$ such that $u \,|\, X = v$. Therefore $T \colon C(\overline{X}) \to BUC(X)$ is a vector space isomorphism.

(iii) For $u \in C(\overline{X})$, we have

$$\|Tu\|_\infty = \sup_{x \in X} |Tu(x)| = \sup_{x \in X} |u(x)| \leq \sup_{x \in \overline{X}} |u(x)| = \|u\|_\infty \;.$$

On the other hand, there is from Corollary III.3.8 a $y \in \overline{X}$ such that $\|u\|_\infty = |u(y)|$. We choose a sequence (x_n) in X such that $x_n \to y$ and find

$$\|u\|_\infty = |u(y)| = |\lim u(x_n)| = |\lim Tu(x_n)| \leq \sup_{x \in X} |Tu(x)| = \|Tu\|_\infty \;.$$

This shows that T is an isometry. ∎

Convention If X is a bounded open subset of $\mathbb{K}^n$, we always identify $BUC(X)$ with $C(\overline{X})$ through the isomorphism (2.4).

Bounded linear operators

Theorem 2.1 becomes particularly important for linear maps. We therefore compile first a few properties of linear operators.

Suppose E and F are normed vector spaces, and $A \colon E \to F$ is linear. We call A **bounded**[3] if there is an $\alpha \geq 0$ such that

$$\|Ax\| \leq \alpha \,\|x\| \quad \text{for } x \in E \;. \tag{2.5}$$

We define

$$\mathcal{L}(E, F) := \big\{\, A \in \operatorname{Hom}(E, F) \;;\; A \text{ is bounded} \,\big\} \;.$$

For every $A \in \mathcal{L}(E, F)$, there is an $\alpha \geq 0$ for which (2.5) holds. Therefore

$$\|A\| := \inf\{\, \alpha \geq 0 \;;\; \|Ax\| \leq \alpha \,\|x\| \,,\; x \in E \,\}$$

is well defined. We call $\|A\|_{\mathcal{L}(E,F)} := \|A\|$ the **operator norm** of A.

---

[3]For historical reasons, we accept here a certain inconsistency in the nomenclature: if F is not the null vector space, there is (except for the zero operator) *no* bounded linear operator that is a bounded map in the terms of Section II.3 (compare Exercise II.3.15). Here, a bounded linear operator maps bounded sets to bounded sets (compare Conclusion 2.4(c)).

2.3 Proposition *For $A \in \mathcal{L}(E, F)$ we have*[4]

$$\|A\| = \sup_{x \neq 0} \frac{\|Ax\|}{\|x\|} = \sup_{\|x\|=1} \|Ax\| = \sup_{\|x\| \leq 1} \|Ax\| = \sup_{x \in \mathbb{B}_E} \|Ax\| \ .$$

Proof The result follows from

$$\|A\| = \inf\{ \alpha \geq 0 \ ; \ \|Ax\| \leq \alpha \|x\|, \ x \in E \}$$

$$= \inf\left\{ \alpha \geq 0 \ ; \ \frac{\|Ax\|}{\|x\|} \leq \alpha, \ x \in E \setminus \{0\} \right\}$$

$$= \sup\left\{ \frac{\|Ax\|}{\|x\|} \ ; \ x \in E \setminus \{0\} \right\}$$

$$= \sup\left\{ \left\| A\left(\frac{x}{\|x\|}\right) \right\| \ ; \ x \in E \setminus \{0\} \right\}$$

$$= \sup_{\|y\|=1} \|Ay\| \leq \sup_{\|x\| \leq 1} \|Ax\| \ .$$

For every $x \in E$ such that $0 < \|x\| \leq 1$, we have the estimate

$$\|Ax\| \leq \frac{1}{\|x\|} \|Ax\| = \left\| A\left(\frac{x}{\|x\|}\right) \right\| \ ,$$

Therefore we find

$$\sup_{\|x\| \leq 1} \|Ax\| \leq \sup_{\|y\|=1} \|Ay\| \ .$$

Thus we have shown the theorem's first three equalities.

For the last, let $a := \sup_{x \in \mathbb{B}_E} \|Ax\|$ and $y \in \overline{\mathbb{B}} := \overline{\mathbb{B}}_E$. Then $0 < \lambda < 1$ means λy is in $\mathbb{B}$. Thus the estimate $\|Ay\| = \|A(\lambda y)\|/\lambda \leq a/\lambda$ holds for $0 < \lambda < 1$, as $\|Ay\| \leq a$ shows. Therefore

$$\sup_{\|x\| \leq 1} \|Ax\| \leq \sup_{x \in \mathbb{B}_E} \|Ax\| \leq \sup_{\|x\| \leq 1} \|Ax\| \ . \ \blacksquare$$

2.4 Conclusions (a) If $A \in \mathcal{L}(E, F)$, then

$$\|Ax\| \leq \|A\| \, \|x\| \quad \text{for } x \in E \ .$$

(b) Every $A \in \mathcal{L}(E, F)$ is Lipschitz continuous and therefore also uniformly continuous.

Proof If $x, y \in E$, then $\|Ax - Ay\| = \|A(x - y)\| \leq \|A\| \, \|x - y\|$. Thus A is Lipschitz continuous with Lipschitz constant $\|A\|$. $\blacksquare$

---

[4]Here and in similar settings, we will implicitly assume $E \neq \{0\}$.

(c) Let $A \in \mathrm{Hom}(E, F)$. Then A belongs to $\mathcal{L}(E, F)$ if and only if A maps bounded sets to bounded sets.

Proof "$\Rightarrow$" Suppose $A \in \mathcal{L}(E, F)$ and B is bounded in E. Then there is a $\beta > 0$ such that $\|x\| \leq \beta$ for all $x \in B$. It follows that

$$\|Ax\| \leq \|A\| \, \|x\| \leq \|A\| \, \beta \quad \text{for } x \in B \ .$$

Therefore the image of B by A is bounded in F.

"$\Leftarrow$" Since the closed unit ball $\bar{\mathbb{B}}_E$ is closed in E, there is by assumption an $\alpha > 0$ such that $\|Ax\| \leq \alpha$ for $x \in \bar{\mathbb{B}}_E$. Because $y/\|y\| \in \bar{\mathbb{B}}_E$ for all $y \in E \backslash \{0\}$, it follows that $\|Ay\| \leq \alpha \, \|y\|$ for all $y \in E$. ∎

(d) $\mathcal{L}(E, F)$ is a vector subspace of $\mathrm{Hom}(E, F)$.

Proof Let $A, B \in \mathcal{L}(E, F)$ and $\lambda \in \mathbb{K}$. For every $x \in E$, we have the estimates

$$\|(A + B)x\| = \|Ax + Bx\| \leq \|Ax\| + \|Bx\| \leq (\|A\| + \|B\|) \, \|x\| \tag{2.6}$$

and

$$\|\lambda Ax\| = |\lambda| \, \|Ax\| \leq |\lambda| \, \|A\| \, \|x\| \ . \tag{2.7}$$

Therefore $A + B$ and λA also belong to $\mathcal{L}(E, F)$. ∎

(e) The map

$$\mathcal{L}(E, F) \to \mathbb{R}^+ \ , \quad A \mapsto \|A\|$$

is a norm on $\mathcal{L}(E, F)$.

Proof Because of (d), it suffices to check directly from the definition of a norm. First, because $\|A\| = 0$, it follows from Proposition 2.3 that $A = 0$. Next choose $x \in E$ such that $\|x\| \leq 1$. Then it follows from (2.6) and (2.7) that

$$\|(A + B)x\| \leq \|A\| + \|B\| \quad \text{and} \quad \|\lambda Ax\| = |\lambda| \, \|Ax\| \ ,$$

and taking the supremum of x verifies the remaining parts of the definition. ∎

(f) Suppose G is a normed vector space and $B \in \mathcal{L}(E, F)$ and also $A \in \mathcal{L}(F, G)$. Then we have

$$AB \in \mathcal{L}(E, G) \quad \text{and} \quad \|AB\| \leq \|A\| \, \|B\| \ .$$

Proof This follows from

$$\|ABx\| \leq \|A\| \, \|Bx\| \leq \|A\| \, \|B\| \, \|x\| \quad \text{for } x \in E \ ,$$

and the definition of the operator norm. ∎

(g) $\mathcal{L}(E) := \mathcal{L}(E, E)$ is a **normed algebra**[5] with unity, that is, $\mathcal{L}(E)$ is an algebra and $\|1_E\| = 1$ and also

$$\|AB\| \leq \|A\| \, \|B\| \quad \text{for } A, B \in \mathcal{L}(E) \ .$$

Proof The claim follows easily from Example I.12.11(c) and (f). ∎

---

[5] Compare with the definition of a Banach algebra in Section V.4.

Convention In the following, we will always endow $\mathcal{L}(E, F)$ with the operator norm. Therefore

$$\mathcal{L}(E, F) := \big(\mathcal{L}(E, F), \|\cdot\|\big)$$

is a normed vector space with $\|\cdot\| := \|\cdot\|_{\mathcal{L}(E,F)}$.

The next theorem shows that a linear map is bounded if and only if it is continuous.

2.5 Theorem *Let $A \in \mathrm{Hom}(E, F)$. These statements are equivalent:*

(i) *A is continuous.*

(ii) *A is continuous at 0.*

(iii) *$A \in \mathcal{L}(E, F)$.*

Proof "(i)$\Rightarrow$(ii)" is obvious, and "(iii)$\Rightarrow$(i)" was shown in Conclusions 2.4(b).

"(ii)$\Rightarrow$(iii)" By assumption, there is a $\delta > 0$ such that $\|Ay\| = \|Ay - A0\| < 1$ for all $y \in \overline{\mathbb{B}}(0, \delta)$. From this, it follows that

$$\sup_{\|x\| \leq 1} \|Ax\| = \frac{1}{\delta} \sup_{\|x\| \leq 1} \|A(\delta x)\| = \frac{1}{\delta} \sup_{\|y\| \leq \delta} \|Ay\| \leq \frac{1}{\delta} \ ,$$

and therefore A is closed. ∎

The continuous extension of bounded linear operators

2.6 Theorem *Suppose E is a normed vector space, X is a dense vector subspace in E, and F is a Banach space. Then, for every $A \in \mathcal{L}(X, F)$ there is a uniquely determined extension $\overline{A} \in \mathcal{L}(E, F)$ defined through*

$$\overline{A}e = \lim_{\substack{x \to e \\ x \in X}} Ax \quad \text{for } e \in E \ , \tag{2.8}$$

and $\|\overline{A}\|_{\mathcal{L}(E,F)} = \|A\|_{\mathcal{L}(X,F)}$.

Proof (i) According to Conclusions 2.4(b), A is uniformly continuous. Defining $f := A$, $Y := E$, and $Z := F$, it follows from Theorem 2.1 that there exists a uniquely determined extension of $\overline{A} \in C(E, F)$ of A, which is given through (2.8).

(ii) We now show that $\overline{A}$ is linear. For that, suppose $e, e' \in E$, $\lambda \in \mathbb{K}$, and (x_n) and (x'_n) are sequences in X such that $x_n \to e$ and $x'_n \to e'$ in E. From the linearity of A and the linearity of limits, we have

$$\overline{A}(e + \lambda e') = \lim_n A(x_n + \lambda x'_n) = \lim_n Ax_n + \lambda \lim_n Ax'_n = \overline{A}e + \lambda \overline{A}e' \ .$$

Therefore $\overline{A} \colon E \to F$ is linear. That $\overline{A}$ is continuous follows from Theorem 2.5, because $\overline{A}$ belongs to $\mathcal{L}(E, F)$.

(iii) Finally we prove that $\bar{A}$ and A have the same operator norm. From the continuity of the norm (see Example III.1.3(j)) and from Conclusions 2.4(a), it follows that

$$\|\bar{A}e\| = \|\lim Ax_n\| = \lim \|Ax_n\| \leq \lim \|A\| \, \|x_n\| = \|A\| \, \|\lim x_n\| = \|A\| \, \|e\|$$

for every $e \in E$ and every sequence (x_n) in X such that $x_n \to e$ in E. Consequently, $\|\bar{A}\| \leq \|A\|$. Because $\bar{A}$ extends A, Proposition 2.3 implies

$$\|\bar{A}\| = \sup_{\|y\| < 1} \|\bar{A}y\| \geq \sup_{\substack{\|x\| < 1 \\ x \in X}} \|\bar{A}x\| = \sup_{\substack{\|x\| < 1 \\ x \in X}} \|Ax\| = \|A\| \; ,$$

and we find $\|\bar{A}\|_{\mathcal{L}(E,F)} = \|A\|_{\mathcal{L}(X,F)}$. ∎

Exercises

In the following exercises, E, E_j, F, and F_j are normed vector spaces.

1 Suppose $A \in \mathrm{Hom}(E, F)$ is surjective. Show that

$$A^{-1} \in \mathcal{L}(F, E) \iff \exists \alpha > 0 : \alpha \|x\| \leq \|Ax\| \; , \quad x \in E \; .$$

If A also belongs to $\mathcal{L}(E, F)$, show that $\|A^{-1}\| \geq \|A\|^{-1}$.

2 Suppose E and F are finite-dimensional and $A \in \mathcal{L}(E, F)$ is bijective with a continuous inverse[6] $A^{-1} \in \mathcal{L}(F, E)$. Show that if $B \in \mathcal{L}(E, F)$ satisfies

$$\|A - B\|_{\mathcal{L}(E,F)} < \|A^{-1}\|_{\mathcal{L}(F,E)}^{-1}$$

then it is invertible.

3 $A \in \mathrm{End}(\mathbb{K}^n)$ has in the standard basis the matrix elements $[a_{jk}]$. For $E_j := (\mathbb{K}^n, |\cdot|_j)$ with $j = 1, 2, \infty$, verify for $A \in \mathcal{L}(E_j)$ that

(i) $\|A\|_{\mathcal{L}(E_1)} = \max_k \sum_j |a_{jk}|$,

(ii) $\|A\|_{\mathcal{L}(E_2)} \leq \left(\sum_{j,k} |a_{jk}|^2 \right)^{1/2}$,

(iii) $\|A\|_{\mathcal{L}(E_\infty)} = \max_j \sum_k |a_{jk}|$.

4 Show that $\delta : B(\mathbb{R}^n) \to \mathbb{R}$, $f \mapsto f(0)$ belongs to $\mathcal{L}(B(\mathbb{R}^n), \mathbb{R})$, and find $\|\delta\|$.

5 Suppose $A_j \in \mathrm{Hom}(E_j, F_j)$ for $j = 0, 1$ and

$$A_0 \otimes A_1 : E_0 \times E_1 \to F_0 \times F_1 \; , \quad (e_0, e_1) \mapsto (A_0 e_0, A_1 e_1) \; .$$

Verify that

$$A_0 \otimes A_1 \in \mathcal{L}(E_0 \times E_1, F_0 \times F_1) \iff A_j \in \mathcal{L}(E_j, F_j) \; , \quad j = 0, 1 \; .$$

6 Suppose (A_n) is a sequence in $\mathcal{L}(E, F)$ converging to A, and suppose (x_n) is a sequence in E whose limit is x. Prove that $(A_n x_n)$ in F converges to Ax.

7 Show that $\ker(A)$ of $A \in \mathcal{L}(E, F)$ is a closed vector subspace of E.

---

[6]If E is a finite-dimensional normed vector space, then $\mathrm{Hom}(E, F) = \mathcal{L}(E, F)$ (see Theorem VII.1.6). Consequently, we need not assume that A, A^{-1}, and B are continuous.

3   The Cauchy–Riemann Integral

Determining the area of plane geometrical shapes is one of the oldest and most prominent projects in mathematics. To compute the areas under graphs of real functions, it is necessary to simplify and formalize the problem. It will help to gain an intuition for integrals. To that end, we will first explain, as simply as possible, how to integrate staircase functions; we will then extend the idea to jump continuous functions. These constructions are based essentially on the results in the first two sections of this chapter. These present the idea that the area of a graph is made up of the sum of areas of rectangles that are themselves determined by choosing the best approximation of the graph to a staircase function. By the "unlimited refinement" of the width of the rectangles, we anticipate the sum of their areas will converge to the area of the figure.

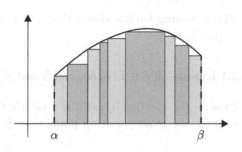

In the following we denote by

- $E := (E, |\cdot|)$, a Banach space;
  $I := [\alpha, \beta]$, a compact perfect interval.

The integral of staircase functions

Suppose $f \colon I \to E$ is a staircase function and $\mathfrak{Z} := (\alpha_0, \ldots, \alpha_n)$ is a partition of I for f. Define e_j through

$$f(x) = e_j \quad \text{for } x \in (\alpha_{j-1}, \alpha_j) \text{ and } j = 1, \ldots, n \ ,$$

that is, e_j is the value of f on the interval (α_{j-1}, α_j). We call

$$\int_{(\mathfrak{Z})} f := \sum_{j=1}^{n} e_j (\alpha_j - \alpha_{j-1})$$

the **integral of f with respect to the partition $\mathfrak{Z}$**. Obviously $\int_{(\mathfrak{Z})} f$ is an element of E. We note that the integral does not depend on the values of f at its jump discontinuities. In the case that $E = \mathbb{R}$, we can interpret $|e_j| (\alpha_j - \alpha_{j-1})$ as the area of a rectangle with sides of length $|e_j|$ and $(\alpha_j - \alpha_{j-1})$. Thus $\int_{(\mathfrak{Z})} f$ represents a weighted sum of rectangular areas, where the area $|e_j| (\alpha_j - \alpha_{j-1})$ is weighted by the sign of e_j. In other words, those rectangles that rise above the x-axis are weighted by 1, whereas those below are given -1.

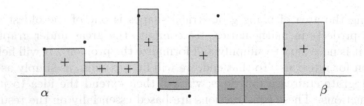

The following lemma shows that $\int_{(3)} f$ depends only on f and not on the choice of partition 3.

3.1 Lemma If $f \in \mathcal{T}(I, E)$, and 3 and $3'$ are partitions for f, then $\int_{(3)} f = \int_{(3')} f$.

Proof (i) We first treat the case that $3' := (\alpha_0, \ldots, \alpha_k, \gamma, \alpha_{k+1}, \ldots, \alpha_n)$ has exactly one more partition point than $3 := (\alpha_0, \ldots, \alpha_n)$. Then we compute

$$
\int_{(3)} f = \sum_{j=1}^{n} e_j(\alpha_j - \alpha_{j-1})
$$

$$
= \sum_{j=1}^{k} e_j(\alpha_j - \alpha_{j-1}) + e_{k+1}(\alpha_{k+1} - \alpha_k) + \sum_{j=k+2}^{n} e_j(\alpha_j - \alpha_{j-1})
$$

$$
= \sum_{j=1}^{k} e_j(\alpha_j - \alpha_{j-1}) + e_{k+1}(\alpha_{k+1} - \gamma) + e_{k+1}(\gamma - \alpha_k) + \sum_{j=k+2}^{n} e_j(\alpha_j - \alpha_{j-1})
$$

$$
= \int_{(3')} f .
$$

(ii) If $3'$ is an arbitrary refinement of 3, the claim follows by repeatedly applying the argument in (i).

(iii) Finally, if 3 and $3'$ are arbitrary partitions of f, then $3 \vee 3'$, according to Remark 1.1(a) and (b), is a refinement of both 3 and $3'$ and is therefore also a partition for f. Then it follows from (ii) that

$$
\int_{(3)} f = \int_{(3 \vee 3')} f = \int_{(3')} f . \blacksquare
$$

Using Lemma 3.1, we can define for $f \in \mathcal{T}(I, E)$ the **integral of f over I** by

$$
\int_I f := \int_\alpha^\beta f := \int_{(3)} f
$$

using an arbitrary partition $\mathfrak{Z}$ for f. Obviously, the integral induces a map from $\mathcal{T}(I, E)$ to E, namely,

$$\int_\alpha^\beta : \mathcal{T}(I, E) \to E , \quad f \mapsto \int_\alpha^\beta f ,$$

which we naturally also call the integral. We explain next the first simple properties of this map.

3.2 Lemma If $f \in \mathcal{T}(I, E)$, then $|f| \in \mathcal{T}(I, \mathbb{R})$ and

$$\left| \int_\alpha^\beta f \right| \leq \int_\alpha^\beta |f| \leq \|f\|_\infty (\beta - \alpha) .$$

Also $\int_\alpha^\beta$ belongs to $\mathcal{L}\big(\mathcal{T}(I, E), E\big)$, and $\| \int_\alpha^\beta \| = \beta - \alpha$.

Proof The first statement is clear. To prove the inequality, suppose $f \in \mathcal{T}(I, E)$ and $(\alpha_0, \ldots, \alpha_n)$ is a partition for f. Then

$$\left| \int_\alpha^\beta f \right| = \left| \sum_{j=1}^n e_j (\alpha_j - \alpha_{j-1}) \right| \leq \sum_{j=1}^n |e_j| (\alpha_j - \alpha_{j-1}) = \int_\alpha^\beta |f|$$

$$\leq \max_{1 \leq j \leq n} |e_j| \sum_{j=1}^n (\alpha_j - \alpha_{j-1}) \leq \sup_{x \in I} |f(x)| (\beta - \alpha) = (\beta - \alpha) \|f\|_\infty .$$

The linearity of $\int_\alpha^\beta$ follows from Remark 1.1.(e) and the definition of $\int_\alpha^\beta$. Consequently, $\int_\alpha^\beta$ belongs to $\mathcal{L}\big(\mathcal{T}(I, E), E\big)$, and we have $\| \int_\alpha^\beta \| \leq \beta - \alpha$. For the constant function $\mathbf{1} \in \mathcal{T}(I, \mathbb{R})$ with value 1, we have $\int_\alpha^\beta \mathbf{1} = \beta - \alpha$, and the last claim follows. ∎

The integral of jump continuous functions

From Theorem 1.4, we know that the space of jump continuous functions $\mathcal{S}(I, E)$ is a Banach space when it is endowed with the supremum norm and that $\mathcal{T}(I, E)$ is a dense vector subspace of $\mathcal{S}(I, E)$ (see Remark 1.1(e)). If follows from Lemma 3.2 that the integral $\int_\alpha^\beta$ is a continuous, linear map from $\mathcal{T}(I, E)$ to E. We can apply the extension Theorem 2.6 to get a unique continuous linear extension of $\int_\alpha^\beta$ into the Banach space $\mathcal{S}(I, E)$. We denote this extension using the same notation, so that

$$\int_\alpha^\beta \in \mathcal{L}\big(\mathcal{S}(I, E), E\big) .$$

From Theorem 2.6, it follows that

$$\int_\alpha^\beta f = \lim_n \int_\alpha^\beta f_n \text{ in } E \quad \text{for } f \in \mathcal{S}(I, E) ,$$

where (f_n) is an arbitrary sequence of staircase functions that converges uniformly
to f. The element $\int_\alpha^\beta f$ of E is called the (**Cauchy–Riemann**) **integral of f** or **the
integral of f over I** or **the integral of f from α to β.** We call f the **integrand**.

3.3 Remarks Suppose $f \in \mathcal{S}(I, E)$.

(a) According to Theorem 2.6 $\int_\alpha^\beta f$ is well defined, that is, this element of E is
independent of the approximating sequence of staircase functions. In the special
case $E = \mathbb{R}$, we interpret $\int_\alpha^\beta f_n$ as the weighted (or "oriented") area of the graph
of f_n in the interval I.

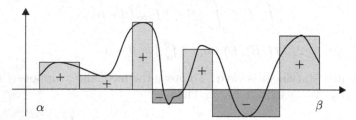

Because the graph of f_n approximates f, we can view $\int_\alpha^\beta f_n$ as an approximation
to the "oriented" area of the graph of f in the interval I, and we can therefore
identify $\int_\alpha^\beta f$ with this oriented area.

(b) There are a number of other notations for $\int_\alpha^\beta f$, namely,

$$\int_I f \;,\quad \int_I f \;,\quad \int_\alpha^\beta f(x)\,dx \;,\quad \int_I f(x)\,dx \;,\quad \int f(x)\,dx \;,\quad \int_I f\,dx \;.$$

(c) We have $\int_\alpha^\beta f \in E$, with $\left| \int_\alpha^\beta f\,dx \right| \le (\beta - \alpha)\|f\|_\infty$.

Proof This follows from Lemma 3.2 and Theorem 2.6. ∎

Riemann sums

If $\mathfrak{Z} := (\alpha_0, \ldots, \alpha_n)$ is a partition of I, we call

$$\triangle_{\mathfrak{Z}} := \max_{1 \le j \le n} (\alpha_j - \alpha_{j-1})$$

the **mesh** of $\mathfrak{Z}$, and we call any element $\xi_j \in [\alpha_{j-1}, \alpha_j]$ a **between point**. With this
terminology, we now prove an approximation result for $\int_\alpha^\beta f$.

3.4 Theorem *Suppose $f \in \mathcal{S}(I, E)$. There there is for every $\varepsilon > 0$ a $\delta > 0$ such
that*

$$\left| \int_\alpha^\beta f\,dx - \sum_{j=1}^n f(\xi_j)(\alpha_j - \alpha_{j-1}) \right| < \varepsilon$$

for every partition $\mathfrak{Z} := (\alpha_0, \ldots, \alpha_n)$ *of* I *of mesh* $\triangle_{\mathfrak{Z}} < \delta$ *and for every value of the between point* ξ_j.

Proof (i) We treat first the staircase functions. Suppose then that $f \in \mathcal{T}(I, E)$ and $\varepsilon > 0$. Also let $\widehat{\mathfrak{Z}} := (\widehat{\alpha_0}, \ldots, \widehat{\alpha_{\widehat{n}}})$ be a partition of f, and let $e_j = f \,|\, (\widehat{\alpha}_{j-1}, \widehat{\alpha}_j)$ for $1 \leq j \leq \widehat{n}$. We set $\delta := \varepsilon / 4\widehat{n} \, \|f\|_\infty$ and choose a partition $\mathfrak{Z} := (\alpha_0, \ldots, \alpha_n)$ of I such that $\triangle_{\mathfrak{Z}} < \delta$. We also choose between points $\xi_j \in [\alpha_{j-1}, \alpha_j]$ for $1 \leq j \leq n$. For $(\beta_0, \ldots, \beta_m) := \mathfrak{Z} \vee \widehat{\mathfrak{Z}}$ we have

$$
\int_\alpha^\beta f \, dx - \sum_{j=1}^n f(\xi_j)(\alpha_j - \alpha_{j-1}) = \sum_{i=1}^{\widehat{n}} e_i(\widehat{\alpha}_i - \widehat{\alpha}_{i-1}) - \sum_{j=1}^n f(\xi_j)(\alpha_j - \alpha_{j-1})
$$

$$
= \sum_{k=1}^m e_k'(\beta_k - \beta_{k-1}) - \sum_{k=1}^m e_k''(\beta_k - \beta_{k-1}) \quad (3.1)
$$

$$
= \sum_{k=1}^m (e_k' - e_k'')(\beta_k - \beta_{k-1}) \,,
$$

and we specify e_k' and e_k'' as

$$
\begin{aligned}
e_k' &:= f(\xi) \,, && \xi \in (\beta_{k-1}, \beta_k) \,, \\
e_k'' &:= f(\xi_j) \,, && \text{if } [\beta_{k-1}, \beta_k] \subset [\alpha_{j-1}, \alpha_j] \,.
\end{aligned}
$$

Obviously, $e_k' \neq e_k''$ can only hold on the partition points $\{\widehat{\alpha}_0, \ldots, \widehat{\alpha}_{\widehat{n}}\}$. Therefore the last sum in (3.1) has at most $2\widehat{n}$ terms that do not vanish. For each of these, we have

$$
|(e_k' - e_k'')(\beta_k - \beta_{k-1})| \leq 2 \|f\|_\infty \triangle_{\mathfrak{Z}} < 2 \|f\|_\infty \delta \,.
$$

From (3.1) and the value of δ, we therefore get

$$
\left| \int_\alpha^\beta f - \sum_{j=1}^n f(\xi_j)(\alpha_j - \alpha_{j-1}) \right| < 2\widehat{n} \cdot 2 \|f\|_\infty \, \delta = \varepsilon \,.
$$

(ii) Suppose now that $f \in \mathcal{S}(I, E)$ and $\varepsilon > 0$. Then, according to Theorem 1.4, there is a $g \in \mathcal{T}(I, E)$ such that

$$
\|f - g\|_\infty < \frac{\varepsilon}{3(\beta - \alpha)} \,. \tag{3.2}
$$

From Remark 3.3(c), we have

$$
\left| \int_\alpha^\beta (f - g) \right| \leq \|f - g\|_\infty \, (\beta - \alpha) < \varepsilon/3 \,. \tag{3.3}
$$

From (i), there is a $\delta > 0$ such that for every partition $\mathfrak{Z} := (\alpha_0, \ldots, \alpha_n)$ of I with $\triangle_{\mathfrak{Z}} < \delta$ and every value $\xi_j \in [\alpha_{j-1}, \alpha_j]$, we have

$$
\left| \int_\alpha^\beta g \, dx - \sum_{j=1}^n g(\xi_j)(\alpha_j - \alpha_{j-1}) \right| < \varepsilon/3 \,. \tag{3.4}
$$

The estimate

$$\left| \sum_{j=1}^{n} (g(\xi_j) - f(\xi_j))(\alpha_j - \alpha_{j-1}) \right| \leq \|f - g\|_{\infty} (\beta - \alpha) < \varepsilon/3 \qquad (3.5)$$

is obvious. Altogether, from (3.3)–(3.5), we get

$$\left| \int_{\alpha}^{\beta} f \, dx - \sum_{j=1}^{n} f(\xi_j)(\alpha_j - \alpha_{j-1}) \right|$$

$$\leq \left| \int_{\alpha}^{\beta} (f - g) \, dx \right| + \left| \int_{\alpha}^{\beta} g \, dx - \sum_{j=1}^{n} g(\xi_j)(\alpha_j - \alpha_{j-1}) \right|$$

$$+ \left| \sum_{j=1}^{n} (g(\xi_j) - f(\xi_j))(\alpha_j - \alpha_{j-1}) \right| < \varepsilon ,$$

and the claim is proved. ∎

3.5 Remarks   (a) A function $f : I \to E$ is said to **Riemann integrable** if there is
an $e \in E$ with the property that for every $\varepsilon > 0$ there is a $\delta > 0$, such that

$$\left| e - \sum_{j=1}^{n} f(\xi_j)(\alpha_j - \alpha_{j-1}) \right| < \varepsilon$$

for every partition $\mathfrak{Z} := (\alpha_0, \ldots, \alpha_n)$ of I with $\triangle_{\mathfrak{Z}} < \delta$ and for every set of between
points $\xi_j \in [\alpha_{j-1}, \alpha_j]$.

If f is Riemann integrable, the value e is called the **Riemann integral** of f
over I, and we will again denote it using the notation $\int_{\alpha}^{\beta} f \, dx$. Consequently, Theorem 3.4 may be stated as, *Every jump continuous function is Riemann integrable,
and the Cauchy–Riemann integral coincides with the Riemann integral.*

(b) One can show $\mathcal{S}(I, E)$ is a proper subset of the Riemann integrable functions.[1]
Consequently, the Riemann integral is a generalization of the Cauchy–Riemann
integral.

We content ourselves here with the Cauchy–Riemann integral and will not
seek to augment the set of integrable functions, as there is no particular need for
it now. However, in Volume III, we will introduce an even more general integral,
the Lebesgue integral, which satisfies all the needs of rigorous analysis.

(c) Suppose $f : I \to E$, and $\mathfrak{Z} := (\alpha_0, \ldots, \alpha_n)$ is a partition of I with between
points $\xi_j \in [\alpha_{j-1}, \alpha_j]$. We call

$$\sum_{j=1}^{n} f(\xi_j)(\alpha_j - \alpha_{j-1}) \in E$$

---

[1] The follows for example from that a bounded function is Riemann integrable if and only if
the set of its discontinuities vanishes in the Lebesgue measure (see Theorem X.5.6 in Volume III).

the **Riemann sum**. If f is Riemann integrable, then

$$\int_\alpha^\beta f\,dx = \lim_{\triangle_3 \to 0} \sum_{j=1}^n f(\xi_j)(\alpha_j - \alpha_{j-1})\ ,$$

expresses its integral symbolically. ∎

Exercises

1 Define $[\cdot]$ to be the floor function. For $n \in \mathbb{N}^\times$, compute the integrals

(i) $\displaystyle\int_0^1 \frac{[nx]}{n}\,dx$, (ii) $\displaystyle\int_0^1 \frac{[nx^2]}{n^2}\,dx$, (iii) $\displaystyle\int_0^1 \frac{[nx^2]}{n}\,dx$, (iv) $\displaystyle\int_\alpha^\beta \operatorname{sign} x\,dx$.

2 Compute $\int_{-1}^1 f$ for the function f of Exercise 1.2 and also $\int_0^1 f$ for the f of Exercise 1.7.

3 Suppose F is a Banach space and $A \in \mathcal{L}(E,F)$. Then show for $f \in \mathcal{S}(I,E)$ that

$$Af := \big(x \mapsto A(f(x))\big) \in \mathcal{S}(I,F)$$

and $A\int_\alpha^\beta f = \int_\alpha^\beta Af$.

4 For $f \in \mathcal{S}(I,E)$, there is, according to Exercise 1.4, a sequence (f_n) of jump continuous functions such that $\sum_n \|f_n\|_\infty < \infty$ and $f = \sum_n f_n$. Show that $\int_I f = \sum_n \int_I f_n$.

5 Prove that for every $f \in \mathcal{S}(I,\mathbb{R})$, f^+ and f^- also belong to $\mathcal{S}(I,\mathbb{R})$, and

$$\int_I f \le \int_I f^+\ ,\quad -\int_I f \le \int_I f^-\ .$$

6 Show that if $f : I \to E$ is Riemann integrable, then $f \in B(I,E)$, that is, *every Riemann integrable function is bounded*.

7 Suppose $f \in B(I,\mathbb{R})$ and $3 = (\alpha_0, \dots, \alpha_n)$ is a partition of I. Define

$$\bar{S}(f,I,3) := \sum_{j=1}^n \sup\big\{\, f(\xi)\ ;\ \xi \in [\alpha_{j-1}, \alpha_j]\,\big\}(\alpha_j - \alpha_{j-1})$$

and

$$\underline{S}(f,I,3) := \sum_{j=1}^n \inf\big\{\, f(\xi)\ ;\ \xi \in [\alpha_{j-1}, \alpha_j]\,\big\}(\alpha_j - \alpha_{j-1})$$

and call them the upper and lower sum of f over I with respect to the partition 3. Prove that

(i) if $3'$ is a refinement of 3, then

$$\bar{S}(f,I,3') \le \bar{S}(f,I,3)\ ,\quad \underline{S}(f,I,3) \le \underline{S}(f,I,3')\ ;$$

(ii) if 3 and $3'$ are partitions of I, then

$$\underline{S}(f,I,3) \le \bar{S}(f,I,3')\ .$$

8 Suppose $f \in B(I, \mathbb{R})$ and $\mathfrak{Z}' := (\beta_0, \ldots, \beta_m)$ is a refinement of $\mathfrak{Z} := (\alpha_0, \ldots, \alpha_n)$. Show that
$$\bar{S}(f, I, \mathfrak{Z}) - \bar{S}(f, I, \mathfrak{Z}') \le 2(m - n) \, \|f\|_\infty \, \Delta_\mathfrak{Z} \, ,$$
$$\underline{S}(f, I, \mathfrak{Z}') - \underline{S}(f, I, \mathfrak{Z}) \le 2(m - n) \, \|f\|_\infty \, \Delta_\mathfrak{Z} \, .$$

9 Let $f \in B(I, \mathbb{R})$. From Exercise 7(ii), we know the following exist in $\mathbb{R}$:
$$\overline{\int_I} f := \inf\{ \, \bar{S}(f, I, \mathfrak{Z}) \, ; \, \mathfrak{Z} \text{ is a partition of } I \, \}$$

and
$$\underline{\int_I} f := \sup\{ \, \underline{S}(f, I, \mathfrak{Z}) \, ; \, \mathfrak{Z} \text{ is a partition of } I \, \} \, .$$

We call $\overline{\int_I} f$ the **over Riemann(–Darboux) integral** of f over I; likewise we call $\underline{\int_I} f$ the **under Riemann integral**.

Prove that

(i) $\underline{\int_I} f \le \overline{\int_I} f$;

(ii) for every $\varepsilon > 0$ there is a $\delta > 0$ such that for every partition $\mathfrak{Z}$ of I with $\Delta_\mathfrak{Z} < \delta$, we have the inequalities
$$0 \le \bar{S}(f, I, \mathfrak{Z}) - \overline{\int_I} f < \varepsilon \quad \text{and} \quad 0 \le \underline{\int_I} f - \underline{S}(f, I, \mathfrak{Z}) < \varepsilon \, .$$

(Hint: There is a partition $\mathfrak{Z}'$ of I such that $\bar{S}(f, I, \mathfrak{Z}') < \overline{\int_I} f + \varepsilon/2$. Let $\mathfrak{Z}$ be an arbitrary partition of I. From Exercise 8, it follows that
$$\bar{S}(f, I, \mathfrak{Z}) \le \bar{S}(f, I, \mathfrak{Z} \vee \mathfrak{Z}') + 2m \, \|f\|_\infty \, \Delta_\mathfrak{Z} \, .$$
In addition, Exercise 7(i) gives $\bar{S}(f, I, \mathfrak{Z} \vee \mathfrak{Z}') \le \bar{S}(f, I, \mathfrak{Z}')$.)

10 Show these statements are equivalent for $f \in B(I, \mathbb{R})$:

(i) f is Riemann integrable.

(ii) $\overline{\int_I} f = \underline{\int_I} f$.

(iii) For every $\varepsilon > 0$ there is a partition $\mathfrak{Z}$ of I such that $\bar{S}(f, I, \mathfrak{Z}) - \underline{S}(f, I, \mathfrak{Z}) < \varepsilon$.

Show that $\overline{\int_I} f = \underline{\int_I} f = \int_I f$.

(Hint: "(ii)$\Longleftrightarrow$(iii)" follows from Exercise 9.
"(i)$\Rightarrow$(iii)" Suppose $\varepsilon > 0$ and $e := \int_I f$. Then there is a $\delta > 0$ such that

$$e - \frac{\varepsilon}{2} < \sum_{j=1}^{n} f(\xi_j)(\alpha_j - \alpha_{j-1}) < e + \frac{\varepsilon}{2}$$

for every partition $\mathfrak{Z} = (\alpha_0, \ldots, \alpha_n)$ of I with $\Delta_\mathfrak{Z} < \delta$. Consequently, $\bar{S}(f, I, \mathfrak{Z}) \le e + \varepsilon/2$ and $\underline{S}(f, I, \mathfrak{Z}) \ge e - \varepsilon/2$.)

4 Properties of integrals

In this section, we delve into the most important properties of integrals and, in particular, prove the fundamental theorem of calculus. This theorem says that every continuous function on an interval has a primitive function (called the anti-derivative) which gives another form of its integral. This antiderivative is known in many cases and, when so, allows for the easy, explicit evaluation of the integral.

As in the preceding sections, we suppose

- $E = (E, |\cdot|)$ is a Banach space; $I = [\alpha, \beta]$ is a compact perfect interval.

Integration of sequences of functions

The definition of integrals leads easily to the following convergence theorem, whose theoretical and practical meaning will become clear.

4.1 Proposition (of the integration of sequences and sums of functions) *Suppose (f_n) is a sequence of jump continuous functions.*

(i) *If (f_n) converges uniformly to f, then f is jump continuous and*

$$\lim_n \int_\alpha^\beta f_n = \int_\alpha^\beta \lim_n f_n = \int_\alpha^\beta f \ .$$

(ii) *If $\sum f_n$ converges uniformly, then $\sum_{n=0}^\infty f_n$ is jump continuous and*

$$\int_\alpha^\beta \left(\sum_{n=0}^\infty f_n \right) = \sum_{n=0}^\infty \int_\alpha^\beta f_n \ .$$

Proof From Theorem 1.4 we know that the space $\mathcal{S}(I, E)$ is complete when endowed with the supremum norm. Both parts of the theorem then follow from the facts that $\int_\alpha^\beta$ is a linear map from $\mathcal{S}(I, E)$ to E and that the uniform convergence agrees with the convergence in $\mathcal{S}(I, E)$. ∎

4.2 Remark The statement of Proposition 4.1 is false when the sequence (f_n) is only pointwise convergent.
Proof We set $I = [0, 1]$, $E := \mathbb{R}$, and

$$f_n(x) := \begin{cases} 0, & x = 0, \\ n, & x \in (0, 1/n), \\ 0, & x \in [1/n, 1], \end{cases}$$

for $n \in \mathbb{N}^\times$. Then f_n is in $\mathcal{T}(I, \mathbb{R})$, and (f_n) converges pointwise, but not uniformly, to 0. Because $\int_I f_n = 1$ for $n \in \mathbb{N}^\times$ and $\int_I 0 = 0$, the claim follows. ∎

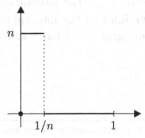

For the first application of Proposition 4.1, we prove that the important statement of Lemma 3.2 — about interchanging the norm with the integral — is also correct for jump continuous functions.

4.3 Proposition For $f \in S(I, E)$ we have $|f| \in S(I, \mathbb{R})$ and

$$\left| \int_\alpha^\beta f \, dx \right| \leq \int_\alpha^\beta |f| \, dx \leq (\beta - \alpha) \|f\|_\infty .$$

Proof According to Theorem 1.4, there is a sequence (f_n) in $\mathcal{T}(I, E)$ that converges uniformly to f. Further, it follows from the inverse triangle inequality that

$$\left| |f_n(x)|_E - |f(x)|_E \right| \leq |f_n(x) - f(x)|_E \leq \|f_n - f\|_\infty \quad \text{for } x \in I .$$

Therefore $|f_n|$ converges uniformly to $|f|$. Because every $|f_n|$ belongs to $\mathcal{T}(I, \mathbb{R})$, it follows again from Theorem 1.4 that $|f| \in S(I, \mathbb{R})$.

From Lemma 3.2, we get

$$\left| \int_\alpha^\beta f_n \, dx \right| \leq \int_\alpha^\beta |f_n| \, dx \leq (\beta - \alpha) \|f_n\|_\infty . \tag{4.1}$$

Because (f_n) converges uniformly to f and $(|f_n|)$ converges likewise to $|f|$, we can, with the help of Proposition 4.1, pass the limit $n \to \infty$ in (4.1). The desired inequality follows. ∎

The oriented integral

If $f \in S(I, E)$ and $\gamma, \delta \in I$, we define[1]

$$\int_\gamma^\delta f := \int_\gamma^\delta f(x) \, dx := \begin{cases} \int_{[\gamma, \delta]} f , & \gamma < \delta , \\ 0 , & \gamma = \delta , \\ -\int_{[\delta, \gamma]} f , & \delta < \gamma , \end{cases} \tag{4.2}$$

and call $\int_\gamma^\delta f$ "the integral of f from γ to δ". We call γ the **lower limit** and δ the **upper limit** of the integral of f, even when $\gamma > \delta$. According to Remark 1.1(g), the integral of f from γ to δ is well defined through (4.2), and

$$\int_\gamma^\delta f = - \int_\delta^\gamma f . \tag{4.3}$$

---

[1] Here and hence, we write simply $\int_J f$ for $\int_J f \,|\, J$, if J is a compact perfect subinterval of I.

4.4 Proposition (of the additivity of integrals) *For $f \in \mathcal{S}(I, E)$ and $a, b, c \in I$ we have*

$$\int_a^b f = \int_a^c f + \int_c^b f .$$

Proof It suffices to check this for $a \leq b \leq c$. If (f_n) is a sequence of staircase functions that converge uniformly to f and J is a compact perfect subinterval of I, then

$$f_n \,|\, J \in \mathcal{T}(J, E) \quad \text{and} \quad f_n \,|\, J \xrightarrow[\text{uniformly}]{} f \,|\, J . \tag{4.4}$$

The definition of the integral of staircase functions gives at once that

$$\int_a^c f_n = \int_a^b f_n + \int_b^c f_n . \tag{4.5}$$

Using (4.4) and Proposition 4.1, we pass the limit $n \to \infty$ and find

$$\int_a^c f = \int_a^b f + \int_b^c f .$$

Then we have

$$\int_a^b f = \int_a^c f - \int_b^c f = \int_a^c f + \int_c^b f . \blacksquare$$

Positivity and monotony of integrals

Until now, we have considered the integral of jump continuous functions taking values in arbitrary Banach spaces. For the following theorem and its corollary, we will restrict to the real-valued case, where the ordering of the reals implies some additional properties of the integral.

4.5 Theorem *For $f \in \mathcal{S}(I, \mathbb{R})$ such that $f(x) \geq 0$ for all $x \in I$, we have $\int_\alpha^\beta f \geq 0$.*

Proof According to Remark 1.3, there is a sequence (f_n) of nonnegative staircase functions that converges uniformly to f. Obviously then, $\int f_n \geq 0$, and it follows that $\int f = \lim_n \int f_n \geq 0$. $\blacksquare$

4.6 Corollary *Suppose $f, g \in \mathcal{S}(I, \mathbb{R})$ satisfy $f(x) \leq g(x)$ for all $x \in I$. Then we have $\int_\alpha^\beta f \leq \int_\alpha^\beta g$, that is, integration is monotone on real-valued functions.*

Proof This follows from the linearity of integrals and from Proposition 4.5. $\blacksquare$

4.7 Remarks   (a) Suppose V is a vector space. We call any linear map from V to its field is a **linear form** or **linear functional**.

Therefore in the scalar case, that is, $E = \mathbb{K}$, the integral $\int_\alpha^\beta$ is a continuous linear form on $\mathcal{S}(I)$.

(b) Suppose V is a real vector space and P is a nonempty set with

  (i) $x, y \in P \Rightarrow x + y \in P$,

  (ii) $(x \in P,\ \lambda \in \mathbb{R}^+) \Rightarrow \lambda x \in P$,

  (iii) $x, -x \in P \Rightarrow x = 0$.

In other words, P satisfies $P + P \subset P$, $\mathbb{R}^+ P \subset P$ and $P \cap (-P) = \{0\}$. We then call P a (in fact convex) **cone** in V (with point at 0). This designation is justified because P is convex and, for every x, it contains the "half ray" $\mathbb{R}^+ x$. In addition, P contains "straight lines" $\mathbb{R}x$ only when $x \neq 0$.

We next define $\leq$ through

$$x \leq y :\Longleftrightarrow y - x \in P\ , \tag{4.6}$$

and thus get an ordering on V, which is **linear on V** (or **compatible with the vector space structure of V**), that is, for $x, y, z \in V$,

$$x \leq y \Rightarrow x + z \leq y + z$$

and

$$(x \leq y,\ \lambda \in \mathbb{R}^+) \Rightarrow \lambda x \leq \lambda y\ .$$

Further, we call $(V, \leq)$ an **ordered vector space**, and $\leq$ is the **ordering induced by P**. Obviously, we have

$$P = \{\, x \in V\ ;\ x \geq 0 \,\}\ . \tag{4.7}$$

Therefore we call P the **positive cone** of $(V, \leq)$.

If, conversely, a linear ordering $\leq$ is imposed on V, then (4.7) defines convex cone in V, and this cone induces the given ordering. Hence there is a bijection between the set of convex cones on V and the set of linear orderings on V. Thus we can write either $(V, \leq)$ or (V, P) whenever P is a positive cone in V.

(c) When (V, P) is an ordered vector space and ℓ is a linear form on V, we call ℓ **positive** if $\ell(x) \geq 0$ for all $x \in P$.[2] A linear form on V is then positive if and only if it is increasing[3].

Proof   Suppose $\ell : V \to \mathbb{R}$ is linear. Because $\ell(x-y) = \ell(x) - \ell(y)$, the claim is obvious. ∎

(d) Suppose E is a normed vector space (or a Banach space) and $\leq$ is a linear ordering on E. Then we say $E := (E, \leq)$ is an **ordered normed vector space** (or an **ordered Banach space**) if its positive cone is closed.

---

[2] Note that the null form is positive under this definition.

[3] We may also say increasing forms are **monotone** linear forms.

Suppose E is an ordered normed vector space and (x_j) and (y_j) are sequences in E with $x_j \to x$ and $y_j \to y$. Then $x_j \le y_j$ for almost all $j \in \mathbb{N}$, and it follows that $x \le y$.

Proof If P is a positive cone in E, then $y_j - x_j \in P$ for almost all $j \in \mathbb{N}$. Then it follows from Proposition II.2.2 and Remark II.3.1(b) that

$$y - x = \lim y_j - \lim x_j = \lim(y_j - x_j) \in P \ ,$$

and hence P is closed. ∎

(e) Suppose X is a nonempty set. The pointwise ordering of Example I.4.4(c), namely,

$$f \le g :\Longleftrightarrow f(x) \le g(x) \quad \text{for } x \in X$$

and $f, g \in \mathbb{R}^X$ makes $\mathbb{R}^X$ an ordered vector space. We call this ordering the **natural ordering** on $\mathbb{R}^X$. In turn, it induces on every subset M of $\mathbb{R}^X$ the **natural ordering** on M (see Example I.4.4(a)). Unless stated otherwise, we shall henceforth always provide $\mathbb{R}^X$ and any of its vector subspaces the natural ordering. In particular, $B(X, \mathbb{R})$ is an ordered Banach space with the positive cone

$$B^+(X) := B(X, \mathbb{R}^+) := B(X, \mathbb{R}) \cap (\mathbb{R}^+)^X \ .$$

Therefore every closed vector subspace $\mathfrak{F}(X, \mathbb{R})$ of $B(X, \mathbb{R})$ is an ordered Banach space whose positive cone is given through

$$\mathfrak{F}^+(X) := \mathfrak{F}(X, \mathbb{R}) \cap B^+(X) = \mathfrak{F}(X, \mathbb{R}) \cap (\mathbb{R}^+)^X \ .$$

Proof It is obvious that $B^+(X)$ is closed in $B(X, \mathbb{R})$. ∎

(f) From (e) it follows that $\mathcal{S}(I, \mathbb{R})$ is an ordered Banach space with positive cone $\mathcal{S}^+(I)$, and Proposition 4.5 says *the integral $\int_\alpha^\beta$ is a continuous positive (and therefore monotone) linear form on $\mathcal{S}(I, \mathbb{R})$.* ∎

The staircase function $f : [0, 2] \to \mathbb{R}$ defined by

$$f(x) := \begin{cases} 1 \,, & x = 1 \,, \\ 0 \,, & \text{otherwise} \end{cases}$$

obviously satisfies $\int f = 0$. Hence there is a nonnegative function that does not identically vanish but whose integral does vanish. The next theorem gives a criterion for the strict positivity of integrals of nonnegative functions.

4.8 Proposition Suppose $f \in \mathcal{S}^+(I)$ and $a \in I$. If f is continuous at a with $f(a) > 0$, then

$$\int_\alpha^\beta f > 0 \ .$$

Proof   (i) Suppose a belongs to $\overset{\circ}{I}$. Then continuity of f at a and $f(a) > 0$ implies there exists a $\delta > 0$ such that $[a - \delta, a + \delta] \subset I$ and

$$|f(x) - f(a)| \le \frac{1}{2}f(a) \quad \text{for } x \in [a - \delta, a + \delta] .$$

Therefore we have

$$f(x) \ge \frac{1}{2}f(a) \quad \text{for } x \in [a - \delta, a + \delta] .$$

From $f \ge 0$ and Proposition 4.5, we get $\int_\alpha^{a-\delta} f \ge 0$ and $\int_{a+\delta}^\beta f \ge 0$. Now Proposition 4.4 and Corollary 4.6 imply

$$\int_\alpha^\beta f = \int_\alpha^{a-\delta} f + \int_{a-\delta}^{a+\delta} f + \int_{a+\delta}^\beta f \ge \int_{a-\delta}^{a+\delta} f \ge \frac{1}{2}f(a) \int_{a-\delta}^{a+\delta} 1 = \delta f(a) > 0 .$$

(ii) When $a \in \partial I$, the same conclusion follows an analogous argument. ∎

Componentwise integration

4.9 Proposition   *The map $f = (f^1, \ldots, f^n) \colon I \to \mathbb{K}^n$ is jump continuous if and only if it is so for each component function f^j. Also,*

$$\int_\alpha^\beta f = \left(\int_\alpha^\beta f^1, \ldots, \int_\alpha^\beta f^n \right) .$$

Proof   From Theorem 1.2, f is jump continuous if and only if there is a sequence (f_k) of staircase function that uniformly converge to it. It is easy to see that the last holds if and only if for every $j \in \{1, \ldots, n\}$, the sequence $(f_k^j)_{k \in \mathbb{N}}$ converges uniformly to f^j. Because f^j belongs to $\mathcal{S}(I, \mathbb{K})$ for every j, Theorem 1.2 implies the first claim. The second part is obvious for staircase functions and follows here because Proposition 4.1 also holds when taking limits for $f \in \mathcal{S}(I, \mathbb{K}^n)$. ∎

4.10 Corollary   *Suppose $g, h \in \mathbb{R}^I$ and $f := g + ih$. Then f is jump continuous if and only if both g and h are. In this case,*

$$\int_\alpha^\beta f = \int_\alpha^\beta g + i \int_\alpha^\beta h .$$

The first fundamental theorem of calculus

For $f \in \mathcal{S}(I, E)$ we use the integral to define the map

$$F \colon I \to E , \quad x \mapsto \int_\alpha^x f(\xi) \, d\xi , \tag{4.8}$$

whose properties we now investigate.

4.11 Theorem   *For $f \in \mathcal{S}(I, E)$, we have*

$$|F(x) - F(y)| \le \|f\|_\infty |x - y| \quad \text{for } x, y \in I .$$

Therefore the integral is a Lipschitz continuous function on the entire interval.

Proof   From Proposition 4.4, we have

$$F(x) - F(y) = \int_\alpha^x f(\xi)\, d\xi - \int_\alpha^y f(\xi)\, d\xi = \int_y^x f(\xi)\, d\xi \quad \text{for } x, y \in I .$$

The claim now follows immediately from Proposition 4.3. ∎

Our next theorem says F is differentiable where f is continuous.

4.12 Theorem (of the differentiability in the entire interval)   *Suppose $f \in \mathcal{S}(I, E)$ is continuous at all $a \in I$. Then F is differentiable at a and $F'(a) = f(a)$.*

Proof   For $h \in \mathbb{R}^\times$ such that $a + h \in I$, we have

$$\frac{F(a+h) - F(a)}{h} = \frac{1}{h}\Big(\int_\alpha^{a+h} f(\xi)\, d\xi - \int_\alpha^a f(\xi)\, d\xi \Big) = \frac{1}{h} \int_a^{a+h} f(\xi)\, d\xi .$$

We note $hf(a) = \int_a^{a+h} f(a)\, d\xi$, so that

$$\frac{F(a+h) - F(a) - f(a)h}{h} = \frac{1}{h} \int_a^{a+h} \big(f(\xi) - f(a) \big)\, d\xi .$$

Because f is continuous at a, there is for every $\varepsilon > 0$ a $\delta > 0$ such that

$$|f(\xi) - f(a)|_E < \varepsilon \quad \text{for } \xi \in I \cap (a - \delta, a + \delta) .$$

Using Proposition 4.3, we have the estimate

$$\left| \frac{F(a+h) - F(a) - f(a)h}{h} \right|_E \le \frac{1}{|h|} \left| \int_a^{a+h} |f(\xi) - f(a)|_E \, d\xi \right| < \varepsilon$$

for h such that $0 < |h| < \delta$ and $a + h \in I$, Therefore F is differentiable at a and $F'(a) = f(a)$. ∎

From the last result, we get a simpler, but also important conclusion.

4.13 The second fundamental theorem of calculus   *Every $f \in C(I, E)$ has an antiderivative, and if F is such an antiderivative,*

$$F(x) = F(\alpha) + \int_\alpha^x f(\xi)\, d\xi \quad \text{for } x \in I .$$

Proof   We set $G(x) := \int_\alpha^x f(\xi)\,d\xi$ for $x \in I$. According to Theorem 4.12, G is differentiable with $G' = f$. Therefore G is an antiderivative of f.

Suppose F is an antiderivative of f. From Remark V.3.7(a) we know that there is a $c \in E$ such that $F = G + c$. Because $G(\alpha) = 0$, we have $c = F(\alpha)$. ∎

4.14 Corollary   *For every antiderivative F of $f \in C(I, E)$, we have*

$$\int_\alpha^\beta f(x)\,dx = F(\beta) - F(\alpha) =: F|_\alpha^\beta \ .$$

We say $F|_\alpha^\beta$ is "F evaluated between α and β".

The indefinite integral

If the antiderivative of f is known, Corollary 4.14 reduces the problem of calculating the integral $\int_\alpha^\beta f$ to the trivial difference $F(\beta) - F(\alpha)$. The corollary also shows that, in some sense, integration is the inverse of differentiation.

Although the fundamental theorems of calculus guarantee an antiderivative for every $f \in C(I, E)$, it is not possible in the general case to find it explicitly.

To calculate an integral according to Corollary 4.14, we must provide, by some method, its antiderivative. Obviously, we can get a basic supply of antiderivatives by differentiating known differentiable functions. Using the results of Chapter IV and V, we will build a list of important antiderivatives, which we will give following Examples 4.15. In the next sections, we will see how using some simple rules and transformations yields an even larger class of antiderivatives. It should also be pointed out that there are extensive tables listing thousands of antiderivatives (see for example [Ape96], [GR81] or [BBM86]).

Instead of painstakingly working out integrals on paper, it is much simpler to use software packages such as MAPLE or MATHEMATICA, which can do "symbolic" integration. These programs "know" many antiderivatives and can solve many integrals by manipulating their integrands using a set of elementary rules.

For $f \in C(I, E)$ we call

$$\int f\,dx + c,$$

in the interval of f's antiderivative, the **indefinite integral** of f. This is a symbolic notation, in that it omits the interval I (although the context should make it clear), and it suggests that the antiderivative is only determined up to an additive constant $c \in E$. In other words: $\int f\,dx + c$ is the equivalence class of antiderivatives of f, where two antiderivatives are equivalent if they only differ by a constant. In the following, when we only want to talk about this equivalence class, we will drop the constant and write simply $\int f\,dx$.

4.15 Examples (a) In the following list of basic indefinite integrals, a and c are complex numbers and x is a real variable in I. Here, I must lie in the domain of definition of f but is otherwise arbitrary.

f	$\int f\,dx$	f	$\int f\,dx$
x^a, $\quad a \neq -1$,	$x^{a+1}/(a+1)$	$\sin x$	$-\cos x$
$1/x$	$\log|x|$	$1/\cos^2 x$	$\tan x$
a^x, $\quad a>0$, $\quad a\neq 1$,	$a^x/\log a$	$-1/\sin^2 x$	$\cot x$
e^{ax}, $\quad a\in\mathbb{C}^\times$,	e^{ax}/a	$1/\sqrt{1-x^2}$	$\arcsin x$
$\cos x$	$\sin x$	$1/(1+x^2)$	$\arctan x$

Proof The validity of this list follows from Examples IV.1.13 and the use of IV.2.10. ∎

(b) Suppose $a = \sum a_k X^k \in \mathbb{K}[\![X]\!]$ with radius of convergence $\rho > 0$. Then we have

$$\int\left(\sum_{k=0}^{\infty} a_k x^k\right) dx = \sum_{k=0}^{\infty} \frac{a_k}{k+1}\, x^{k+1} \quad \text{for } -\rho < x < \rho\,.$$

Proof This follows from Remark V.3.7(b). ∎

(c) Suppose $f \in C^1(I,\mathbb{R})$ such that $f(x) \neq 0$ for all $x \in I$. Then we have

$$\int \frac{f'}{f}\, dx = \log|f|\,.$$

Proof Suppose $f(x) > 0$ for all $x \in I$. From the chain rule it follows that[4]

$$(\log|f|)' = (\log f)' = f'/f\,.$$

If $f(x) < 0$ for $x \in I$, we have in analogous fashion that

$$(\log|f|)' = \big[\log(-f)\big]' = (-f)'/(-f) = f'/f\,.$$

Therefore $\log|f|$ is an antiderivative of f'/f. ∎

The mean value theorem for integrals

Here, we determine a mean value theorem for the integrals of real-valued continuous functions.

---

[4]We call f'/f the **logarithmic derivative** of f.

4.16 Proposition *Suppose $f, \varphi \in C(I, \mathbb{R})$ and $\varphi \geq 0$. Then there is an $\xi \in I$ such that*

$$\int_\alpha^\beta f(x)\varphi(x)\,dx = f(\xi) \int_\alpha^\beta \varphi(x)\,dx \ . \tag{4.9}$$

Proof If $\varphi(x) = 0$ for all $x \in I$, then (4.9) obviously holds for every $\xi \in I$. Thus we can assume $\varphi(x) > 0$ for some $x \in I$. Then Proposition 4.8 implies the inequality $\int_\alpha^\beta \varphi(x)\,dx > 0$.

Letting $m := \min_I f$ and $M := \max_I f$, we have $m\varphi \leq f\varphi \leq M\varphi$ because $\varphi \geq 0$. Then the linearity and monotony of integrals implies the inequalities

$$m \int_\alpha^\beta \varphi\,dx \leq \int_\alpha^\beta f\varphi\,dx \leq M \int_\alpha^\beta \varphi\,dx \ .$$

Therefore we have

$$m \leq \frac{\int_\alpha^\beta f\varphi}{\int_\alpha^\beta \varphi} \leq M \ .$$

The choice of m and M and the intermediate value theorem (Theorem III.5.1) immediately proves the theorem. ∎

4.17 Corollary *For $f \in C(I, \mathbb{R})$ there is a $\xi \in I$ such that $\int_\alpha^\beta f\,dx = f(\xi)(\beta - \alpha)$.*

Proof Set $\varphi := 1$ in Proposition 4.16. ∎

We note that — in contrast to the mean value theorem of derivatives (Theorem IV.2.4) — the point ξ need not lie in the interior of the interval.

We illustrate Corollary 4.17 with the following figures:

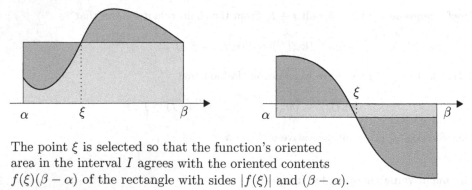

The point ξ is selected so that the function's oriented area in the interval I agrees with the oriented contents $f(\xi)(\beta - \alpha)$ of the rectangle with sides $|f(\xi)|$ and $(\beta - \alpha)$.

Exercises

1 Prove the statement of Remark 4.7(b).

2 For $f \in \mathcal{S}(I)$, show $\overline{\int_\alpha^\beta f} = \int_\alpha^\beta \bar{f}$.

3 The two piecewise continuous functions $f_1, f_2 : I \to E$ differ only at their discontinuities. Show that $\int_\alpha^\beta f_1 = \int_\alpha^\beta f_2$.

4 For $f \in \mathcal{S}(I, \mathbb{K})$ and $p \in [1, \infty)$ suppose

$$\|f\|_p := \left(\int_\alpha^\beta |f(x)|^p \, dx \right)^{1/p},$$

and let $p' := p/(p-1)$ denote p's **dual exponent** (with $1/0 = \infty$).
Prove the following:

(i) For $f, g \in \mathcal{S}(I, \mathbb{K})$, the **Hölder inequality** holds, that is,

$$\left| \int_\alpha^\beta fg \, dx \right| \leq \int_\alpha^\beta |fg| \, dx \leq \|f\|_p \|g\|_{p'} \ .$$

(ii) For $f, g \in \mathcal{S}(I, \mathbb{K})$, the **Minkowski inequality** holds, that is,

$$\|f + g\|_p \leq \|f\|_p + \|g\|_p \ .$$

(iii) $\left(C(I, \mathbb{K}), \|\cdot\|_p\right)$ is a normed vector space.

(iv) $\left(C(I, \mathbb{K}), \|\cdot\|_2\right)$ is an inner product space.

(v) For $f \in C(I, \mathbb{K})$ and $1 \leq p \leq q \leq \infty$, we have

$$\|f\|_p \leq (\beta - \alpha)^{1/p - 1/q} \|f\|_q \ .$$

(Hint: (i) and (ii) Consult the proof of Applications IV.2.16(b) and (c). (v) Use (i).)

5 Show that if $q \geq p$, the norm $\|\cdot\|_q$ on $C(I, \mathbb{K})$ is stronger[5] than $\|\cdot\|_p$. These norms are not equivalent when $p \neq q$

6 Show that $\left(C(I, \mathbb{K}), \|\cdot\|_p\right)$ is not complete for $1 \leq p < \infty$. Therefore $\left(C(I, \mathbb{K}), \|\cdot\|_p\right)$ is not a Banach space for $1 \leq p < \infty$, and $\left(C(I, \mathbb{K}), \|\cdot\|_2\right)$ is not a Hilbert space.

7 Suppose $f \in C^1(I, \mathbb{R})$ has $f(\alpha) = f(\beta) = 0$. Show that

$$\|f\|_\infty^2 \leq \frac{1}{2} \int_\alpha^\beta \left(f^2 + (f')^2 \right) dx \ . \tag{4.10}$$

How must (4.10) be modified when $f \in C^1(I, \mathbb{R})$ but only $f(\alpha) = 0$?
(Hint: Suppose $x_0 \in I$ has $f^2(x_0) = \|f\|_\infty^2$. Show that $f^2(x_0) = \int_\alpha^{x_0} ff' \, dx - \int_{x_0}^\beta ff' \, dx$.
Then apply Exercise I.10.10.)

8 Suppose $f \in C^1(I, \mathbb{K})$ has $f(\alpha) = 0$. Show that

$$\int_\alpha^\beta |ff'| \, dx \leq \frac{\beta - \alpha}{2} \int_\alpha^\beta |f'|^2 \, dx \ .$$

---

[5]If $\|\cdot\|_1$ and $\|\cdot\|_2$ are norms on a vector space E, we say $\|\cdot\|_1$ is **stronger** than $\|\cdot\|_2$ if there is a constant $K \geq 1$ such that $\|x\|_2 \leq K \|x\|_1$ for all $x \in E$. We say **weaker** in the opposite case.

(Hint: If $F(x) := \int_\alpha^x |f'(\xi)| \, d\xi$ then $|f(x)| \leq F(x)$ and $\int_\alpha^\beta 2FF' \, dx = F^2(\beta)$.)

9 The function $f \in C^2(I, \mathbb{R})$ satisfies $f \leq f''$ and $f(\alpha) = f(\beta) = 0$. Show that

$$0 \leq \max_{x \in I} f'(x) \leq \frac{1}{\sqrt{2}} \int_\alpha^\beta \left(f^2 + (f')^2 \right) \, dx \; .$$

(Hint: Let $x_0 \in I$ such that $f'(x_0) = \|f'\|_\infty$. Then show $f'(x_0) \leq 0$, so that $f = 0$.

If $f'(x_0) > 0$, there exists a $\xi \in (x_0, \beta)$ such that $f'(x) > 0$ for $x \in [x_0, \xi)$ and $f'(\xi) = 0$ (see Theorem IV.2.3). Then consider $\int_{x_0}^\xi (ff' - f'f'') \, dx$ and apply Exercise 7.)

10 Prove the **second mean value theorem of integrals**: Suppose $f \in C(I, \mathbb{R})$ and that $g \in C^1(I, \mathbb{R})$ is monotone.[6] Then there is a $\xi \in I$ such that

$$\int_\alpha^\beta f(x)g(x) \, dx = g(\alpha) \int_\alpha^\xi f(x) \, dx + g(\beta) \int_\xi^\beta f(x) \, dx \; .$$

(Hint: Letting $F(x) := \int_\alpha^x f(s) \, ds$, show that

$$\int_\alpha^\beta f(x)g(x) \, dx = Fg \big|_\alpha^\beta - \int_\alpha^\beta F(x)g'(x) \, dx \; .$$

Proposition 4.16 can be used for the integral on the right side.)

11 Suppose $a > 0$ and $f \in C([-a, a], E)$. Prove that

(i) if f is odd, then $\int_{-a}^a f(x) \, dx = 0$;

(ii) if f is even, then $\int_{-a}^a f(x) \, dx = 2 \int_0^a f(x) \, dx$.

12 Define $F : [0, 1] \to \mathbb{R}$ through

$$F(x) := \begin{cases} x^2 \sin(1/x^2) \, , & x \in (0, 1] \, , \\ 0 \, , & x = 0 \, . \end{cases}$$

Verify that

(i) F is differentiable,

(ii) $F' \notin \mathcal{S}(I, \mathbb{R})$.

That is, there are functions that are not jump continuous but nevertheless have an antiderivative. (Hint for (ii): Remark 1.1(d).)

13 Suppose $f \in C([a, b], \mathbb{R})$ and $g, h \in C^1([\alpha, \beta], [a, b])$. Further suppose

$$F : [\alpha, \beta] \to \mathbb{R} \, , \quad x \mapsto \int_{g(x)}^{h(x)} f(\xi) \, d\xi \; .$$

Show that $F \in C^1([\alpha, \beta], \mathbb{R})$. How about F'?

---

[6] One can show that the second mean value theorem for integrals remains true when g is only assumed to be monotone and therefore generally not regular.

14 Suppose $n \in \mathbb{N}^\times$ and $a_1, \ldots, a_n \in \mathbb{R}$. Prove that there is an $\alpha \in (0, \pi)$ such that

$$\sum_{k=1}^{n} a_k \cos(k\alpha) = 0 .$$

(Hint: Use Corollary 4.17 with $f(x) := \sum_{k=1}^{n} a_k \cos(kx)$ and $I := [0, \pi]$.)

15 Prove that

(i) $\displaystyle \lim_{n \to \infty} n\left(\frac{1}{n^2} + \frac{1}{(n+1)^2} + \cdots + \frac{1}{(2n-1)^2}\right) = \frac{1}{2}$,

(ii) $\displaystyle \lim_{n \to \infty} \sum_{k=nq}^{np-1} \frac{1}{k} = \log \frac{p}{q}$ for $p, q \in \mathbb{N}$ and $q < p$.

(Hint: Consider the appropriate Riemann sum for $f(x) := 1/(1+x)^2$ on $[0, 1]$ or, for (ii), $f(x) := 1/x$ on $[q, p]$.)

16 Suppose $f \in C(I \times I, E)$ and

$$g : I \to E , \quad x \mapsto \int_I f(x, y) \, dy .$$

Show that $g \in C(I, E)$.

5   The technique of integration

Through the fundamental theorem of calculus, rules for taking derivatives become corresponding rules for integration. In this section, we will carry out this conversion for the chain rule and the product rule, and so attain the important substitution rule and the method of integration by parts.

In this section we denote by

- $I = [\alpha, \beta]$ a compact perfect interval.

Variable substitution

5.1 Theorem (substitution rule)   *Suppose E is a Banach space, $f \in C(I, E)$, and suppose $\varphi \in C^1([a, b], \mathbb{R})$ satisfies $-\infty < a < b < \infty$ and $\varphi([a, b]) \subset I$. Then*

$$\int_a^b f(\varphi(x))\varphi'(x)\,dx = \int_{\varphi(a)}^{\varphi(b)} f(y)\,dy \ .$$

Proof   The fundamental theorem of calculus says that f has an antiderivative $F \in C^1(I, E)$. From the chain rule (Theorem IV.1.7), there then exists a function $F \circ \varphi \in C^1([a, b], E)$ such that

$$(F \circ \varphi)'(x) = F'(\varphi(x))\varphi'(x) = f(\varphi(x))\varphi'(x) \quad \text{for } x \in [a, b] \ .$$

Then Corollary 4.14 gives

$$\int_a^b f(\varphi(x))\varphi'(x)\,dx = F \circ \varphi \big|_a^b = F(\varphi(b)) - F(\varphi(a))$$

$$= F \big|_{\varphi(a)}^{\varphi(b)} = \int_{\varphi(a)}^{\varphi(b)} f(y)\,dy \ . \blacksquare$$

At this point, we should explain the meaning of the symbol "dx" which, although present in the notation and obviously marks the integration variable, has otherwise not been clarified. In Section VIII.3, we will give a more formal definition, which we will then justify and make precise. Here we will be content with a heuristic observation.

5.2 Remarks   **(a)** Suppose $\varphi \colon I \to \mathbb{R}$ is differentiable. We call[1] $d\varphi := \varphi'\,dx$ the **differential** of φ. If we choose φ to be the identity function id_I, then $d\varphi = \mathbf{1}\,dx$.

---

[1]The notation $d\varphi = \varphi'\,dx$ is here only a formal expression in which the function is understood to be differentiable. In particular, $\varphi'\,dx$ is *not* a product of independent objects.

It is obvious that the symbol $1\,dx$ can also be written as dx. Therefore dx is the differential of the identity map $x \mapsto x$.

(b) In practice, the substitution rule is written in the compact form

$$\int_a^b f \circ \varphi \, d\varphi = \int_{\varphi(a)}^{\varphi(b)} f \, dy \ .$$

(c) Suppose $\varphi : I \to \mathbb{R}$ is differentiable and $x_0 \in I$. In the following heuristic approach, we will explore φ in an "infinitesimal" neighborhood of x_0. We then regard dx as an increment between x_0 and x, hence as a real variable in itself.

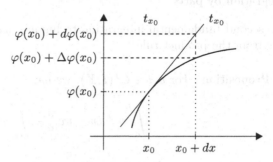

Suppose t_{x_0} is tangent to φ at the point $(x_0, \varphi(x_0))$:

$$t_{x_0} : \mathbb{R} \to \mathbb{R} \ , \qquad x \mapsto \varphi(x_0) + \varphi'(x_0)(x - x_0) \ .$$

In addition, define $\Delta\varphi(x_0) := \varphi(x_0 + dx) - \varphi(x_0)$ as the increment in φ, and define $d\varphi(x_0) := \varphi'(x_0)\,dx$ as the increment along the tangent t_{x_0}. From the differentiability of φ, we have

$$\Delta\varphi(x_0) = d\varphi(x_0) + o(dx) \quad \text{as} \quad dx \to 0 \ .$$

For small increments dx, $\Delta\varphi(x_0)$ can be approximated through the "linear approximation" $d\varphi(x_0) = \varphi'(x_0)\,dx$. ∎

5.3 Examples   (a) For $a \in \mathbb{R}^\times$ and $b \in \mathbb{R}$, we have

$$\int_\alpha^\beta \cos(ax + b)\,dx = \frac{1}{a}\big(\sin(a\beta + b) - \sin(a\alpha + b)\big) \ .$$

Proof   The variable substitution $y(x) := ax + b$ for $x \in \mathbb{R}$ gives $dy = a\,dx$. Therefore, it follows from Example 4.15(a) that

$$\int_\alpha^\beta \cos(ax + b)\,dx = \frac{1}{a}\int_{a\alpha + b}^{a\beta + b} \cos y \, dy = \frac{1}{a}\sin\Big|_{a\alpha + b}^{a\beta + b} \ . \ \blacksquare$$

(b) For $n \in \mathbb{N}^\times$, we have

$$\int_0^1 x^{n-1}\sin(x^n)\,dx = \frac{1}{n}(1 - \cos 1) \ .$$

Proof   Here we set $y(x) := x^n$. Then $dy = nx^{n-1}\, dx$ and thus

$$\int_0^1 x^{n-1} \sin(x^n)\, dx = \frac{1}{n} \int_0^1 \sin y\, dy = -\frac{\cos y}{n}\Big|_0^1 . \quad \blacksquare$$

Integration by parts

The second fundamental integration technique, namely, integration by parts, follows from the product rule.

5.4 Proposition   *For $u, v \in C^1(I, \mathbb{K})$, we have*

$$\int_\alpha^\beta uv'\, dx = uv\Big|_\alpha^\beta - \int_\alpha^\beta u'v\, dx . \tag{5.1}$$

Proof   The claim follows directly from the product rule $(uv)' = u'v + v'u$ and Corollary 4.14. $\blacksquare$

In differentials, (5.1) assumes the practical form

$$\int_\alpha^\beta u\, dv = uv\Big|_\alpha^\beta - \int_\alpha^\beta v\, du . \tag{5.2}$$

5.5 Examples   (a) $\int_\alpha^\beta x \sin x\, dx = \sin \beta - \sin \alpha - \beta \cos \beta + \alpha \cos \alpha$.

Proof   We set $u(x) := x$ and $v' := \sin$. Then $u' = \mathbf{1}$ and $v = -\cos$. Therefore

$$\int_\alpha^\beta x \sin x\, dx = -x \cos x\Big|_\alpha^\beta + \int_\alpha^\beta \cos x\, dx = (\sin x - x \cos x)\big|_\alpha^\beta . \quad \blacksquare$$

(b) (the area of a circle)   A circle of radius R has area πR^2.

Proof   We may place the origin of a rectangular coordinate system at the center. We can express points with radius at most R through the closed set

$$K_R = \big\{ (x, y) \in \mathbb{R}^2 \; ; \; |(x, y)| \leq R \big\} = \big\{ (x, y) \in \mathbb{R}^2 \; ; \; x^2 + y^2 \leq R^2 \big\} .$$

We can also express K_R as the union of the (upper) half disc

$$H_R := \big\{ (x, y) \in \mathbb{R}^2 \; ; \; x^2 + y^2 \leq R^2, \; y \geq 0 \big\}$$

with the corresponding lower half $-H_R$. They intersect at

$$H_R \cap (-H_R) = [-R, R] \times \{0\} = \big\{ (x, 0) \in \mathbb{R}^2 \; ; \; -R \leq x \leq R \big\} .$$

We can write the upper boundary of H_R as the graph of the function

$$[-R, R] \to \mathbb{R}, \quad x \mapsto \sqrt{R^2 - x^2},$$

and the area of H_R, according to our earlier interpretation, as

$$\int_{-R}^{R} \sqrt{R^2 - x^2}\, dx.$$

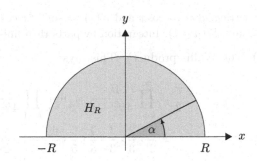

Here it is irrelevant whether the "lower boundary" $[-R, R] \times \{0\}$ is included in H_R, because the area of a rectangle with width 0 is itself 0 by definition. By symmetry[2], the area A_R of the disc K_R is simply double the area of the half disc H_R. Therefore

$$A_R = 2 \int_{-R}^{R} \sqrt{R^2 - x^2}\, dx.$$

To evaluate this integral, it is convenient to adopt polar coordinates. Therefore we set $x(\alpha) := R \cos \alpha$ for $\alpha \in [0, \pi]$. Then we have $dx = -R \sin \alpha\, d\alpha$, and the substitution rule gives

$$A_R = -2R \int_{\pi}^{0} \sqrt{R^2 - R^2 \cos^2 \alpha}\, \sin \alpha\, d\alpha$$

$$= 2R^2 \int_0^{\pi} \sqrt{1 - \cos^2 \alpha}\, \sin \alpha\, d\alpha = 2R^2 \int_0^{\pi} \sin^2 \alpha\, d\alpha.$$

We integrate $\int_0^{\pi} \sin^2 \alpha\, d\alpha$ by parts. Putting $u := \sin$ and $v' := \sin$, we have $u' = \cos$ and $v = -\cos$. Therefore

$$\int_0^{\pi} \sin^2 \alpha\, d\alpha = -\sin \alpha \cos \alpha \Big|_0^{\pi} + \int_0^{\pi} \cos^2 \alpha\, d\alpha$$

$$= \int_0^{\pi} (1 - \sin^2 \alpha)\, d\alpha = \pi - \int_0^{\pi} \sin^2 \alpha\, d\alpha,$$

and we find $\int_0^{\pi} \sin^2 \alpha\, d\alpha = \pi/2$. Altogether we have $A_R = R^2 \pi$. ∎

(c) For $n \in \mathbb{N}$ suppose $I_n := \int \sin^n x\, dx$. We then get the **recursion formula**[3]

$$n I_n = (n - 1) I_{n-2} - \cos x \sin^{n-1} x \quad \text{for } n \geq 2, \tag{5.3}$$

with $I_0 = X$ and $I_1 = -\cos$.

Proof Obviously, $I_0 = X + c$ and $I_1 = -\cos + c$. If $n \geq 2$, then

$$I_n = \int \sin^{n-2} x\, (1 - \cos^2 x)\, dx = I_{n-2} - \int \sin^{n-2} x \cos x \cos x\, dx.$$

---

[2]In this example we consider the absolute area and not the oriented area, and we forgo a precise definition of area, using instead heuristically evident arguments. In Volume III, we will delve into this concept and related issues and justify these heuristic arguments.

[3]Note the agreement with Example 4.15.

By setting $u(x) := \cos x$ and $v'(x) := \sin^{n-2}x \cos x$, we easily find that $u' = -\sin$ and $v = \sin^{n-1}/(n-1)$. Integration by parts then finishes the proof. ∎

(d) (the Wallis product) This says[4]

$$\frac{\pi}{2} = \prod_{k=1}^{\infty} \frac{4k^2}{4k^2 - 1} = \lim_{n \to \infty} \prod_{k=1}^{n} \frac{4k^2}{4k^2 - 1}$$

$$= \frac{2}{1} \cdot \frac{2}{3} \cdot \frac{4}{3} \cdot \frac{4}{5} \cdot \frac{6}{5} \cdot \frac{6}{7} \cdot \cdots \cdot \frac{2k}{2k-1} \cdot \frac{2k}{2k+1} \cdots .$$

Proof We get the claimed product by evaluating (5.3) on the interval $[0, \pi/2]$. We set

$$A_n := \int_0^{\pi/2} \sin^n x \, dx \quad \text{for } n \in \mathbb{N} .$$

From (5.3), it follows that

$$A_0 = \pi/2 , \quad A_1 = 1 , \quad A_n = \frac{n-1}{n} A_{n-2} \quad \text{for } n \geq 2 .$$

A simple induction argument gives

$$A_{2n} = \frac{(2n-1)(2n-3) \cdot \cdots \cdot 3 \cdot 1}{2n(2n-2) \cdot \cdots \cdot 4 \cdot 2} \cdot \frac{\pi}{2} , \quad A_{2n+1} = \frac{2n(2n-2) \cdot \cdots \cdot 4 \cdot 2}{(2n+1)(2n-1) \cdot \cdots \cdot 5 \cdot 3} .$$

From this follows the relations

$$\frac{A_{2n+1}}{A_{2n}} = \frac{2n \cdot 2n(2n-2)(2n-2) \cdot \cdots \cdot 4 \cdot 4 \cdot 2 \cdot 2}{\left[(2n+1)(2n-1)\right]\left[(2n-1)(2n-3)\right] \cdot \cdots \cdot [5 \cdot 3][3 \cdot 1]} \cdot \frac{2}{\pi}$$

$$= \frac{2}{\pi} \prod_{k=1}^{n} \frac{(2k)^2}{(2k)^2 - 1} \tag{5.4}$$

and

$$\lim_{n \to \infty} \frac{A_{2n+2}}{A_{2n}} = \lim_{n \to \infty} \frac{2n+1}{2n+2} = 1 . \tag{5.5}$$

Because $\sin^2 x \leq \sin x \leq 1$ for $x \in [0, \pi/2]$, we have $A_{2n+2} \leq A_{2n+1} \leq A_{2n}$. Therefore

$$\frac{A_{2n+2}}{A_{2n}} \leq \frac{A_{2n+1}}{A_{2n}} \leq 1 \quad \text{for } n \in \mathbb{N} ,$$

and the claim is a consequence of (5.4) and (5.5). ∎

---

[4]Suppose (a_k) is a sequence in $\mathbb{K}$ and $p_n := \prod_{k=0}^{n} a_k$ for $n \in \mathbb{N}$. If the sequence (p_n) converges, we call its limit the **infinite product** of a_k and write

$$\prod_{k=0}^{\infty} a_k := \lim_{n} \prod_{k=0}^{n} a_k .$$

The integrals of rational functions

By the **elementary functions** we mean all the maps generated by taking polynomials, exponential, sines and cosines and applying to them the four fundamental operations of addition, subtraction, multiplication, and division, together with functional composition and inversion. Altogether, these operations generate a vast set of functions.

5.6 Remarks (a) The class of elementary functions is "closed under differentiation", that is, the derivative of an elementary function is also elementary.

Proof This follows from Theorems IV.1.6–8 as well as from Examples IV.1.13, Application IV.2.10 and Exercise IV.2.5. ∎

(b) Antiderivatives of elementary functions are unfortunately not generally elementary. In other words, the class of elementary functions is not closed under integration.

Proof For every $a \in (0,1)$, the function

$$f : [0,a] \to \mathbb{R} , \quad x \mapsto 1/\sqrt{1 - x^4}$$

is continuous. Therefore

$$F(x) := \int_0^x \frac{dy}{\sqrt{1 - y^4}} \quad \text{for } 0 \le x < 1 ,$$

is a well-defined antiderivative of f. Because $f(x) > 0$ for $x \in (0,1)$, F is strictly increasing. Thus according to Theorem III.5.7, F has a well-defined inverse function G. It is known[5] that there is a subset M of $\mathbb{C}$ that is countable and has no limit points, such that G has a uniquely determined analytic continuation $\widetilde{G}$ on $\mathbb{C} \setminus M$. It is also known that $\widetilde{G}$ is **doubly periodic**, that is, there are two $\mathbb{R}$-linearly independent periods $\omega_1, \omega_2 \in \mathbb{C}$ such that

$$\widetilde{G}(z + \omega_1) = \widetilde{G}(z + \omega_2) = \widetilde{G}(z) \quad \text{for } z \in \mathbb{C} \setminus M .$$

Because elementary functions are at most "simply periodic", $\widetilde{G}$ cannot be elementary. Therefore F is also not elementary. Hence the elementary function f has no elementary antiderivative. ∎

In the following we will show that rational functions have elementary antiderivatives. We begin with simple examples, which can then be extended to the general case.

---

[5]Proving this and the following statement about the periodicity of $\widetilde{G}$ widely surpasses the scope of this text (see for example Chapter V of [FB95]).

5.7 Examples   (a) For $a \in \mathbb{C} \setminus I$, we have

$$\int \frac{dx}{X - a} = \begin{cases} \log|X - a| + c \,, & a \in \mathbb{R} \,, \\ \log(X - a) + c \,, & a \in \mathbb{C} \setminus \mathbb{R} \,. \end{cases}$$

Proof   This follows from Examples 4.15(c) and IV.1.13(e). ∎

(b) Suppose $a \in \mathbb{C} \setminus I$ and $k \in \mathbb{N}$ with $k \geq 2$. Then

$$\int \frac{dx}{(X - a)^k} = \frac{-1}{(k - 1)(X - a)^{k-1}} + c \,.$$

(c) Suppose $a, b \in \mathbb{R}$ with $D := a^2 - b < 0$. Then

$$\int \frac{dx}{X^2 + 2aX + b} = \frac{1}{\sqrt{-D}} \arctan\!\left(\frac{X + a}{\sqrt{-D}}\right) + c \,.$$

Proof   First,

$$q(X) := X^2 + 2aX + b = (X + a)^2 - D = |D|\left(1 + \left(\frac{X + a}{\sqrt{|D|}}\right)^2\right) \,.$$

Then, by defining $y := |D|^{-1/2}(x + a)$, the substitution rule gives

$$\int_\alpha^\beta \frac{dx}{q} = \frac{1}{\sqrt{-D}} \int_{y(\alpha)}^{y(\beta)} \frac{dy}{1 + y^2} = \frac{1}{\sqrt{-D}} \arctan\Big|_{y(\alpha)}^{y(\beta)} \,. \ ∎$$

(d) For $a, b \in \mathbb{R}$ with $D := a^2 - b < 0$, we have

$$\int \frac{X \, dx}{X^2 + 2aX + b} = \frac{1}{2} \log(X^2 + 2aX + b) - \frac{a}{\sqrt{-D}} \arctan\!\left(\frac{X + a}{\sqrt{-D}}\right) + c \,.$$

Proof   In the notation of the proof of (c), we find

$$\frac{X}{q} = \frac{1}{2} \frac{2X + 2a}{q} - \frac{a}{q} = \frac{1}{2} \frac{q'}{q} - \frac{a}{q} \,,$$

and the claim quickly follows from Example 4.15(c) and (c). ∎

(e) For $a, b \in \mathbb{R}$ with $D := a^2 - b = 0$ and $-a \notin I$, we have

$$\int \frac{dx}{X^2 + 2aX + b} = \frac{-1}{X + a} + c \,.$$

Proof   Because $q = (X + a)^2$, the claim follows from (b). ∎

(f) Suppose $a, b \in \mathbb{R}$ and $D := a^2 - b > 0$. For $-a \pm \sqrt{D} \notin I$, we have

$$\int \frac{dx}{X^2 + 2aX + b} = \frac{1}{2\sqrt{D}} \log \left| \frac{X + a - \sqrt{D}}{X + a + \sqrt{D}} \right| + c \,.$$

Proof The quadratic polynomial q has real zeros $z_1 := -a + \sqrt{D}$ and $z_2 := -a - \sqrt{D}$. Therefore $q = (X - z_1)(X - z_2)$. This suggests the ansatz

$$\frac{1}{q} = \frac{a_1}{X - z_1} + \frac{a_2}{X - z_2} \,. \tag{5.6}$$

By multiplying this equation by q, we find

$$1 = (a_1 + a_2)X - (a_1 z_2 + a_2 z_1) \,.$$

Therefore, from the identity theorem for polynomials (see Remark I.8.19(c)) and by comparing coefficients, we have $a_1 = -a_2 = 1/2\sqrt{D}$. The claim then follows from (a). ∎

In the last examples, we integrated the rational function $1/(X^2 + 2aX + b)$ using the decomposition (5.6), thereby reducing it to two simpler rational functions $1/(X - z_1)$ and $1/(X - z_2)$. Such a "partial fraction decomposition" is possible for every rational function, as we will now show.

A polynomial is said to be **normalized** if its highest coefficient is 1. By the fundamental theorem of algebra, every normalized polynomial q of positive degree has a factor decomposition

$$q = \prod_{j=1}^{n} (X - z_j)^{m_j} \,, \tag{5.7}$$

in which $z_1, \ldots, z_n$ are the zeros of q in $\mathbb{C}$ and $m_1, \ldots, m_n$ are the corresponding multiplicities. Therefore

$$\sum_{j=1}^{n} m_j = \deg(q)$$

(see Example III.3.9(b) and Remark I.8.19(b)).

Suppose now $p, q \in \mathbb{C}[X]$ with $q \neq 0$. Then according to Proposition I.8.15, it follows by polynomial division that there are uniquely determined $s, t \in \mathbb{C}[X]$ such that

$$\frac{p}{q} = s + \frac{t}{q} \,, \quad \text{where} \quad \deg(t) < \deg(q) \,. \tag{5.8}$$

Thus, in view of Example 4.15(b), we can restrict our proof to the elementary integrability of the rational function $r := p/q$ for $\deg(p) < \deg(q)$, and we can also assume that q is normalized. The basis of the proof is then the following theorem about the **partial fraction expansion**.

5.8 Proposition *Suppose $p, q \in \mathbb{C}[X]$ with $\deg(p) < \deg(q)$, and q is normalized. Then, with the notation of (5.7), we have*

$$\frac{p}{q} = \sum_{j=1}^{n} \sum_{k=1}^{m_j} \frac{a_{jk}}{(X - z_j)^k} \tag{5.9}$$

with uniquely determined $a_{jk} \in \mathbb{C}$.

Proof We make the ansatz

$$\frac{p}{q} = \frac{a}{(X - z_1)^{m_1}} + \frac{p_1}{q_1} , \tag{5.10}$$

where $a \in \mathbb{C}$, $p_1 \in \mathbb{C}[X]$, and

$$q_1 := \frac{q}{X - z_1} \in \mathbb{C}[X] .$$

By multiplying (5.10) by q, we get

$$p = a \prod_{j=2}^{n} (X - z_j)^{m_j} + (X - z_1) p_1 , \tag{5.11}$$

from which we read off[6]

$$a = p(z_1) \Big/ \prod_{j=2}^{n} (z_1 - z_j)^{m_j} . \tag{5.12}$$

Therefore $a_{1m_1} := a$ is uniquely determined. From (5.11) follows

$$\deg(p_1) < \Big(\sum_{j=2}^{n} m_j \Big) \vee \deg(p) \le \deg(q) - 1 = \deg(q_1) .$$

Next we can apply the above argument to p_1/q_1 and so get the claim after finitely many steps. ∎

5.9 Corollary *Every rational function has an elementary integral.*

Proof This follows immediately from (5.8), (5.9), and Examples 5.7(a) and (b). ∎

5.10 Remarks **(a)** If all the zeros of the denominator are simple, the partial fraction expansion (5.9) becomes

$$\frac{p}{q} = \sum_{j=1}^{n} \frac{p(z_j)}{q'(z_j)} \frac{1}{X - z_j} . \tag{5.13}$$

---

[6] As usual, the "empty product" for $n = 1$ is given the value 1.

In the general case, one makes an ansatz of the form (5.9) with undetermined coefficients a_{jk} and then determines the coefficients after multiplying through by q (see the proof of Example 5.7(f)).

Proof The statement (5.13) follows easily from (5.12). ∎

(b) Suppose $s \in \mathbb{R}[X]$. Every zero $z \in \mathbb{C}$ of s has a complex conjugate zero $\bar{z}$ with the same multiplicity.

Proof From Proposition I.11.3, $\overline{s(z)} = s(\bar{z})$. ∎

(c) Suppose $p, q \in \mathbb{R}[X]$ and $z \in \mathbb{C}\backslash\mathbb{R}$ is a zero of q with multiplicity m. Further suppose a_k and b_k for $1 \le k \le m$ are the coefficients of $(X - z)^{-k}$ and $(X - \bar{z})^{-k}$, respectively, in the expansion (5.9). Then $b_k = \bar{a}_k$.

Proof From (b) we know that $\bar{z}$ is also a zero of q with multiplicity m. Therefore b_k is uniquely determined. For $x \in \mathbb{C}\backslash\{z_1, \ldots, z_n\}$, it follows from (5.9) that

$$\sum_{j=1}^{n}\sum_{k=1}^{m_j} \frac{a_{jk}}{(\bar{x} - z_j)^k} = \frac{p(\bar{x})}{q(\bar{x})} = \overline{\frac{p}{q}}(x) = \sum_{j=1}^{n}\sum_{k=1}^{m_j} \frac{\bar{a}_{jk}}{(\bar{x} - \bar{z}_j)^k} \ .$$

Now the claim follows from the uniqueness result of Proposition 5.8. ∎

Suppose now that $r := p/q$ for $p, q \in \mathbb{R}[X]$ with $\deg(p) < \deg(q)$ is a real rational function. If $z \in \mathbb{C}\backslash\mathbb{R}$ is a zero of q with multiplicity m, then, by invoking the partial fraction expansion (5.9), we get, according to Remark 5.10(c), the summand

$$\frac{a}{X - z} + \frac{\bar{a}}{X - \bar{z}} = \frac{(a + \bar{a})X - (z\bar{a} + \bar{z}a)}{(X - z)(X - \bar{z})} = 2\frac{\operatorname{Re}(a)X - \operatorname{Re}(\bar{z}a)}{X^2 - 2\operatorname{Re}(z)X + |z|^2} \ ,$$

and we have $D := (\operatorname{Re} z)^2 - |z|^2 < 0$. Then Examples 5.7(c) and (d) show that

$$\int \left(\frac{a}{X - z} + \frac{\bar{a}}{X - \bar{z}}\right) dx$$
$$= A\log\left(x^2 - 2\operatorname{Re}(z)x + |z|^2\right) + B\arctan\left(\frac{x - \operatorname{Re} z}{\sqrt{-D}}\right) + c \tag{5.14}$$

for $x \in I$, where the coefficients A and B follow uniquely from $\operatorname{Re} a$, $\operatorname{Re} z$, $\operatorname{Re}(\bar{z}a)$ and $\sqrt{-D}$.

For $2 \le k \le m$, it follows from Remark 5.10(c) and Example 5.7(b) that

$$\int \left(\frac{b}{(x - z)^k} + \frac{\bar{b}}{(x - \bar{z})^k}\right) dx = \frac{-1}{k - 1}\left(\frac{b}{(x - z)^{k-1}} + \frac{\bar{b}}{(x - \bar{z})^{k-1}}\right) + c$$
$$= \frac{-2\operatorname{Re}\left(b(x - \bar{z})^{k-1}\right)}{(k - 1)\left(x^2 - 2\operatorname{Re}(z)x + |z|^2\right)^{k-1}} + c \tag{5.15}$$

for $x \in I$. Therefore from Proposition 5.8 as well as (5.14) and (5.15), we get the antiderivative in real form.

The exercises give concrete examples of integrating by using partial fraction expansions.

Exercises

1 Calculate the indefinite integrals

(i) $\displaystyle\int \frac{dx}{a^2x^2 + bx + c}$,   (ii) $\displaystyle\int \frac{x\,dx}{x^2 - 2x + 3}$,   (iii) $\displaystyle\int \frac{x^5\,dx}{x^2 + x + 1}$,

(iv) $\displaystyle\int \frac{dx}{x^4 + 1}$,   (v) $\displaystyle\int \frac{x^2\,dx}{x^4 + 1}$,   (vi) $\displaystyle\int \frac{dx}{(ax^2 + bx + c)^2}$.

2 By clever substitution, transform the following integrals

(i) $\displaystyle\int \frac{e^x - 1}{e^x + 1}\,dx$,   (ii) $\displaystyle\int \frac{dx}{1 + \sqrt{1 + x}}$,   (iii) $\displaystyle\int \frac{dx}{\sqrt{x}(1 + \sqrt[3]{x})}$

into integrals of rational functions. What are the corresponding antiderivatives?

3 Suppose $f \in C^1(I, \mathbb{R})$. Show[7] $\lim_{\lambda \to \infty} \int_I f(x) \sin(\lambda x)\,dx = 0$.

4 Suppose $f \in C([0, 1], \mathbb{R})$. Verify that

(i) $\displaystyle\int_0^\pi x f(\sin x)\,dx = \frac{\pi}{2} \int_0^\pi f(\sin x)\,dx$,

(ii) $\displaystyle\int_0^{\pi/2} f(\sin 2x) \cos x\,dx = \int_0^{\pi/2} f(\cos^2 x) \cos x\,dx$.

(Hints: (i) Substitute $t = \pi - x$.
(ii) Observe $\int_{\pi/4}^{\pi/2} f(\sin 2x) \cos x\,dx = \int_0^{\pi/4} f(\sin 2x) \sin x\,dx$, and then use the substitution $\sin 2x = \cos^2 t$.)

5 The **Legendre polynomials** P_n are defined by[8]

$$P_n(X) := \frac{1}{2^n n!} \partial^n [(X^2 - 1)^n] \quad \text{for } n \in \mathbb{N} ,$$

Prove

$$\int_{-1}^1 P_n P_m = \begin{cases} 0 , & n \neq m , \\ \dfrac{2}{2n + 1} , & n = m . \end{cases}$$

6 By using the substitution $y = \tan x$, prove that

$$\int_0^1 \frac{\log(1 + x)}{1 + x^2}\,dx = \frac{\pi}{8} \log 2 .$$

(Hint: $1 + \tan x = (\sqrt{2}\sin(x + \pi/4))/\cos x$.)

---

[7] See Remark 7.17(b).
[8] See Exercise IV.1.12.

7 Show
$$\int_0^{1/\sqrt{2}} \frac{4\sqrt{2} - 8x^3 - 4\sqrt{2}x^4 - 8x^5}{1 - x^8}\, dx = \pi \ .$$

(Hints: Substitute $x = y/\sqrt{2}$ and
$(y - 1)(y^8 - 16) = (y^2 - 2)(y^2 - 2y + 2)(y^5 + y^4 + 2y^3 - 4)$.)

8 Suppose $k, m \in \mathbb{N}^\times$ and $c \in (0,1)$. Show that
$$\int_0^c \frac{x^{k-1}}{1 - x^m} = \sum_{n=0}^{\infty} \frac{c^{nm+k}}{nm + k} \ .$$

(Hint: $\sum_{n=0}^{\infty} x^{nm} = 1/(1 - x^m)$.)

9 In Exercise 8, put $c = 1/\sqrt{2}$, $m = 8$, and $k = 1, 4, 5, 6$. Construct using Exercise 7 the following form of π:
$$\pi = \sum_{n=0}^{\infty} \frac{1}{16^n} \left(\frac{4}{8n + 1} - \frac{2}{8n + 4} - \frac{1}{8n + 5} - \frac{1}{8n + 6} \right) \ .$$

10 Supply recursion formulas for the integrals

(i) $I_n = \displaystyle\int x^n e^x\, dx$,   (ii) $I_n = \displaystyle\int x(\log x)^n\, dx$,

(iii) $I_n = \displaystyle\int \cos^n x\, dx$,   (iv) $I_n = \displaystyle\int \frac{dx}{(x^2 + 1)^n}$.

6 Sums and integrals

Suppose $m, n \in \mathbb{Z}$ with $m < n$ and $f : [m, n] \to E$ is continuous. Then

$$\sum_{k=m+1}^{n} f(k)$$

can be interpreted as a Riemann sum that approximates

$$\int_{m}^{n} f \, dx \ . \tag{6.1}$$

In the two previous sections, we have learned effective methods for calculating integrals, all of which followed from the fundamental theorem of calculus. In fact, in many cases, it is simpler to calculate the integral (6.1) than its corresponding sum. This brings up an interesting idea: "turn the tables" by using the known integral to approximate the sum. The technique for approximating sums will prove quite useful.

Naturally, the effectiveness of this idea will depend on how well the error

$$\left| \int_{m}^{n} f \, dx - \sum_{k=m+1}^{n} f(k) \right| \tag{6.2}$$

can be controlled. In this context, the Bernoulli numbers and polynomials are of interest, and we turn to them first.

The Bernoulli numbers

We begin with the proof that the function[1]

$$h : \mathbb{C} \to \mathbb{C} , \quad h(z) := \begin{cases} (e^z - 1)/z , & z \in \mathbb{C}^{\times} , \\ 1 , & z = 0 , \end{cases}$$

is analytic and has a disc around 0 on which h does not vanish.[2]

6.1 Lemma   *Suppose* $h \in C^{\omega}(\mathbb{C}, \mathbb{C})$ *such that* $h(z) \neq 0$ *for* $z \in 2\pi \mathbb{B}$.

Proof[3]   (i) By Proposition II.9.4, the power series $g(X) = \sum \left(1/(k+1)! \right) X^k$ has an infinite convergence radius, by Proposition V.3.5, it is analytic. We will now verify the equality of g and h. Obviously $g(0) = h(0)$, and therefore $z \in \mathbb{C}^{\times}$. Then $g(z) = h(z)$ follows from $z \cdot g(z) = e^z - 1$.

---

[1]For the following, note also Exercises V.3.2–4.
[2]The existence of an $\eta > 0$ such that $h(z) \neq 0$ for $z \in \mathbb{B}(0, \eta)$ follows, of course, from $h(0) = 1$ and the continuity of h.
[3]See also Exercise V.3.3.

(ii) For $z \in \mathbb{C} \setminus 2\pi i \mathbb{Z}$, we have

$$\frac{z}{e^z - 1} + \frac{z}{2} = \frac{z}{2} \frac{e^z + 1}{e^z - 1} = \frac{z}{2} \coth \frac{z}{2} . \tag{6.3}$$

Therefore, our theorem follows because $h(0) = 1$. ∎

From the last proof, it follows in particular that

$$f : 2\pi\mathbb{B} \to \mathbb{C} , \quad z \mapsto \frac{z}{e^z - 1}$$

is well defined.[4] Furthermore, the function f is analytic in a neighborhood of 0, as the next theorem shows.

6.2 Proposition   *There is a $\rho \in (0, 2\pi)$ such that $f \in C^\omega(\rho\mathbb{B}, \mathbb{C})$.*

Proof   The recursive algorithm of Exercise II.9.9 secures the existence of a power series $\sum b_k X^k$ with positive radius of convergence ρ_0 and the property

$$\left(\sum \frac{1}{(k+1)!} X^k \right) \left(\sum b_k X^k \right) = 1 \in \mathbb{C}[\![X]\!] .$$

With $\rho := \rho_0 \wedge 2\pi$, we set

$$\widetilde{f} : \rho\mathbb{B} \to \mathbb{C} , \quad z \mapsto \sum_{k=0}^{\infty} b_k z^k .$$

Then

$$\frac{e^z - 1}{z} \widetilde{f}(z) = \left(\sum_{k=0}^{\infty} \frac{z^k}{(k+1)!} \right) \left(\sum_{k=0}^{\infty} b_k z^k \right) = 1 \quad \text{for } z \in \rho\mathbb{B} .$$

Consequently, Lemma 6.1 and the choice of ρ give

$$\widetilde{f}(z) = f(z) = \frac{z}{e^z - 1} \quad \text{for } z \in \rho\mathbb{B} ,$$

and now the analyticity of f follows from Proposition V.3.5. ∎

In Remark VIII.5.13(c), we will show f is analytic without using recursive division of power series (Exercise II.9.9).

The **Bernoulli numbers** B_k are defined for $k \in \mathbb{N}$ through

$$\frac{z}{e^z - 1} = \sum_{k=0}^{\infty} \frac{B_k}{k!} z^k \quad \text{for } z \in \rho\mathbb{B} , \tag{6.4}$$

with properly chosen $\rho > 0$. From Proposition 6.2 and because of the identity theorem for power series (Corollary II.9.9), the B_k are uniquely determined through (6.4). The map f with $f(z) = z/(e^z - 1)$ is called the **generating function** of B_k.

---

[4]This means that we can interpret $z/(e^z - 1)$ as equaling 1 at $z = 0$.

Recursion formulas

From (6.4) we can use the Cauchy product of power series to easily derive the recursion formula for the Bernoulli numbers.

6.3 Proposition   *The Bernoulli numbers B_k satisfy*

(i) $\displaystyle\sum_{k=0}^{n}\binom{n+1}{k}B_k = \begin{cases} 1, & n=0, \\ 0, & n\in\mathbb{N}^\times; \end{cases}$

(ii) $B_{2k+1} = 0$ for $k\in\mathbb{N}^\times$.

Proof   By Proposition II.9.7, we have for $z\in\rho\mathbb{B}$

$$z = (e^z - 1)f(z) = z\Big(\sum_{k=0}^{\infty}\frac{z^k}{(k+1)!}\Big)\Big(\sum_{j=0}^{\infty}\frac{B_j}{j!}z^j\Big)$$

$$= z\sum_{n=0}^{\infty}\Big(\sum_{k=0}^{n}\frac{B_k}{k!}\frac{1}{(n+1-k)!}\Big)z^n .$$

The identity theorem for power series then implies

$$\sum_{k=0}^{n}\frac{B_k}{k!}\frac{1}{(n+1-k)!} = \begin{cases} 1, & n=0, \\ 0, & n\in\mathbb{N}^\times. \end{cases}$$

The theorem then follows from multiplying this identity by $(n+1)!$.

   (ii) On the one hand, we find

$$f(z) - f(-z) = z\Big(\frac{1}{e^z-1} + \frac{1}{e^{-z}-1}\Big) = z\frac{e^{-z}-1+e^z-1}{1-e^z-e^{-z}+1} = -z .$$

On the other, we have the power series expansion

$$f(z) - f(-z) = \sum_{k=0}^{\infty}\Big(\frac{B_k}{k!}z^k - \frac{B_k}{k!}(-z)^k\Big) = 2\sum_{k=0}^{\infty}\frac{B_{2k+1}}{(2k+1)!}z^{2k+1} .$$

Therefore, from the identity theorem for power series, we get $B_{2k+1} = 0$ for $k\in\mathbb{N}^\times$. ∎

6.4 Corollary   *For the Bernoulli numbers, we have*

$$B_0 = 1, \quad B_1 = -1/2, \quad B_2 = 1/6, \quad B_4 = -1/30, \quad B_6 = 1/42, \quad \cdots$$

   Using the Bernoulli numbers, we can find a series expansion for the cotangent, which will prove useful in the next section.

6.5 Application   For sufficiently small[5] $z \in \mathbb{C}$, we have

$$z \cot z = 1 + \sum_{n=1}^{\infty} (-1)^n \frac{4^n}{(2n)!} B_{2n} z^{2n} .$$

Proof   Using (6.3), (6.4), and Proposition 6.3(ii), we get from $B_1 = -1/2$ that

$$\frac{z}{2} \coth \frac{z}{2} = \sum_{n=0}^{\infty} B_{2n} \frac{z^{2n}}{(2n)!} \quad \text{for } z \in \rho \mathbb{B} .$$

By replacing z with $2iz$, the theorem follows. ∎

The Bernoulli polynomials

For every $x \in \mathbb{C}$ the function

$$F_x : \rho \mathbb{B} \to \mathbb{C} , \quad z \mapsto \frac{z e^{xz}}{e^z - 1}$$

is analytic. In analogy to the Bernoulli numbers, we will define the **Bernoulli polynomials** $B_k(X)$ through

$$\frac{z e^{xz}}{e^z - 1} = \sum_{k=0}^{\infty} \frac{B_k(x)}{k!} z^k \quad \text{for } z \in \rho \mathbb{B} \text{ and } x \in \mathbb{C} . \tag{6.5}$$

By the identity theorem for power series, the functions $B_k(X)$ on $\mathbb{C}$ are uniquely determined through (6.5), and the next theorem shows that they are indeed polynomials.

6.6 Proposition   *For $n \in \mathbb{N}$, we have*
   (i) $B_n(X) = \sum_{k=0}^{n} \binom{n}{k} B_k X^{n-k}$,
   (ii) $B_n(0) = B_n$,
   (iii) $B_{n+1}'(X) = (n+1)B_n(X)$,
   (iv) $B_n(X+1) - B_n(X) = nX^{n-1}$,
   (v) $B_n(1-X) = (-1)^n B_n(X)$.

Proof   The first statement comes, as in the proof of Proposition 6.3, from using the Cauchy product of power series and comparing coefficients. On one hand, we

---

[5] From the proof of Application 7.23(a), it will follow that this holds for $|z| < 1$.

have

$$F_x(z) = e^{xz}f(z) = \Big(\sum_{k=0}^{\infty} \frac{x^k z^k}{k!}\Big)\Big(\sum_{j=0}^{\infty} \frac{B_j}{j!}z^j\Big) = \sum_{n=0}^{\infty}\Big(\sum_{k=0}^{n} \frac{x^{n-k}}{(n-k)!}\frac{B_k}{k!}\Big)z^n$$

$$= \sum_{n=0}^{\infty}\Big(\sum_{k=0}^{n}\binom{n}{k}B_k x^{n-k}\Big)\frac{z^n}{n!}$$

and, alternately, $F_x(z) = \sum_n B_n(x)z^n/n!$.

The statement (ii) follows immediately from (i). Likewise from (i) we get (iii) because

$$B'_{n+1}(X) = \sum_{k=0}^{n}\binom{n+1}{k}B_k X^{n-k}(n+1-k)$$

$$= (n+1)\sum_{k=0}^{n}\binom{n}{k}B_k X^{n-k} = (n+1)B_n(X) .$$

Finally, (iv) and (v) follow from $F_{x+1}(z) - F_x(z) = ze^{xz}$ and $F_{1-x}(z) = F_x(-z)$ by comparing coefficients. ∎

6.7 Corollary *The first four Bernoulli polynomials read*

$$B_0(X) = 1 , \qquad\qquad B_1(X) = X - 1/2 ,$$
$$B_2(X) = X^2 - X + 1/6 , \qquad B_3(X) = X^3 - 3X^2/2 + X/2 .$$

The Euler–Maclaurin sum formula

The Bernoulli polynomials will aid in evaluating the error in (6.2). But first, let us prove a helpful result.

6.8 Lemma *Suppose $m \in \mathbb{N}^\times$ and $f \in C^{2m+1}[0,1]$. Then we have*

$$\int_0^1 f(x)\,dx = \frac{1}{2}\big[f(0) + f(1)\big] - \sum_{k=1}^{m} \frac{B_{2k}}{(2k)!}f^{(2k-1)}(x)\Big|_0^1$$

$$- \frac{1}{(2m+1)!}\int_0^1 B_{2m+1}(x)f^{(2m+1)}(x)\,dx .$$

Proof We apply the functional equation $B'_{n+1}(X) = (n+1)B_n(X)$ and continue to integrate by parts. So define $u := f$ and $v' := \mathbf{1}$. Then $u' = f'$ and $v =$

$B_1(X) = X - 1/2$. Consequently,

$$\int_0^1 f(x)\, dx = B_1(x) f(x) \Big|_0^1 - \int_0^1 B_1(x) f'(x)\, dx$$

$$= \frac{1}{2}[f(0) + f(1)] - \int_0^1 B_1(x) f'(x)\, dx \; . \tag{6.6}$$

Next we set $u := f'$ and $v' := B_1(X)$. Because $u' = f''$ and $v = B_2(X)/2$, we find

$$\int_0^1 B_1(x) f'(x)\, dx = \frac{1}{2} B_2(x) f'(x) \Big|_0^1 - \frac{1}{2} \int_0^1 B_2(x) f''(x)\, dx \; .$$

From this, it follows with $u := f''$ and $v' := B_2(X)/2$ (and therefore with $u' = f'''$ and $v = B_3(X)/3!$) the equation

$$\int_0^1 B_1(x) f'(x)\, dx = \left[\frac{1}{2} B_2(x) f'(x) - \frac{1}{3!} B_3(x) f''(x) \right] \Big|_0^1 + \frac{1}{3!} \int_0^1 B_3(x) f'''(x)\, dx \; .$$

A simple induction now gives

$$\int_0^1 B_1(x) f'(x)\, dx = \sum_{j=2}^{2m+1} \frac{(-1)^j}{j!} B_j(x) f^{(j-1)}(x) \Big|_0^1$$

$$+ \frac{1}{(2m+1)!} \int_0^1 B_{2m+1}(x) f^{(2m+1)}(x)\, dx \; .$$

From Proposition 6.6, we have $B_n(0) = B_n$, $B_n(1) = (-1)^n B_n$ and $B_{2n+1} = 0$ for $n \in \mathbb{N}^\times$, and the theorem follows. ∎

In the following, we denote by $\widetilde{B}_n$ the 1-periodic continuation of the function $B_n(X) \,|\, [0,1)$ on $\mathbb{R}$, that is,

$$\widetilde{B}_n(x) := B_n(x - [x]) \quad \text{for } x \in \mathbb{R} \; .$$

Obviously the $\widetilde{B}_n$ are jump continuous on $\mathbb{R}$ (that is, on every compact interval). For $n \geq 2$, the $\widetilde{B}_n$ are actually continuous (see Exercise 4).

6.9 Theorem (Euler–Maclaurin sum formula) *Suppose $a, b \in \mathbb{Z}$ with $a < b$ and f belongs to $C^{2m+1}[a, b]$ for some $m \in \mathbb{N}^\times$. Then*

$$\sum_{k=a}^{b} f(k) = \int_a^b f(x)\, dx + \frac{1}{2}[f(a) + f(b)] + \sum_{k=1}^{m} \frac{B_{2k}}{(2k)!} f^{(2k-1)}(x) \Big|_a^b$$

$$+ \frac{1}{(2m+1)!} \int_a^b \widetilde{B}_{2m+1}(x) f^{(2m+1)}(x)\, dx \; .$$

Proof We decompose the integral $\int_a^b f\,dx$ into a sum of terms $\int_{a+k}^{a+k+1} f\,dx$, to which we apply Lemma 6.8. Therefore

$$\int_a^b f(x)\,dx = \sum_{k=0}^{b-a-1} \int_{a+k}^{a+k+1} f(x)\,dx = \sum_{k=0}^{b-a-1} \int_0^1 f_k(y)\,dy$$

with $f_k(y) := f(a+k+y)$ for $y \in [0,1]$. Then

$$\frac{1}{2} \sum_{k=0}^{b-a-1} \big[f_k(0) + f_k(1) \big] = \frac{1}{2} \sum_{k=0}^{b-a-1} \big[f(a+k) + f(a+k+1) \big]$$

$$= \sum_{k=a}^{b} f(k) - \frac{1}{2}\big[f(a) + f(b) \big]$$

and

$$\int_0^1 B_{2m+1}(y) f_k^{(2m+1)}(y)\,dy = \int_0^1 \widetilde{B}_{2m+1}(a+k+y) f^{(2m+1)}(a+k+y)\,dy$$

$$= \int_{a+k}^{a+k+1} \widetilde{B}_{2m+1}(x) f^{(2m+1)}(x)\,dx \ .$$

The theorem then follows from Lemma 6.8. ∎

Power sums

In our first application of Euler–Maclaurin sum formula, we choose f to be a monomial X^m. The formula then easily evaluates the power sums $\sum_{k=1}^n k^m$, which we have already determined for $m = 1, 2$ and 3 in Remark I.12.14(c).

6.10 Example For $m \in \mathbb{N}$ with $m \geq 2$, we have

$$\sum_{k=1}^n k^m = \frac{n^{m+1}}{m+1} + \frac{n^m}{2} + \sum_{2j<m+1} \binom{m}{2j-1} \frac{B_{2j}}{2j} n^{m-2j+1} \ .$$

In particular,

$$\sum_{k=1}^n k^2 = \frac{n^3}{3} + \frac{n^2}{2} + nB_2 = \frac{n(n+1)(2n+1)}{6} \ ,$$

$$\sum_{k=1}^n k^3 = \frac{n^4}{4} + \frac{n^3}{2} + \frac{3}{2}n^2 B_2 = \frac{n^2(n+1)^2}{4} \ .$$

Proof For $f := X^m$, we have

$$f^{(\ell)} = \begin{cases} \binom{m}{\ell} \ell!\, X^{m-\ell} \ , & \ell \leq m \ , \\ 0 \ , & \ell > m \ , \end{cases}$$

and, for $2j - 1 < m$,

$$\frac{B_{2j}}{(2j)!} f^{(2j-1)}(n) = \frac{B_{2j}}{(2j)!} \binom{m}{2j-1} (2j-1)! \, n^{m-2j+1} = \frac{B_{2j}}{2j} \binom{m}{2j-1} n^{m-2j+1} \ .$$

Because $f^{(m)}(x)\big|_0^n = 0$, the formulas follow from Theorem 6.9 with $a = 0$ and $b = n$. ∎

Asymptotic equivalence

The series (a_k) and (b_k) in $\mathbb{C}^\times$ are said to be **asymptotically equivalent** if

$$\lim_k (a_k/b_k) = 1 \ .$$

In this case, we write $a_k \sim b_k \ (k \to \infty)$ or $(a_k) \sim (b_k)$.

6.11 Remarks (a) It is not difficult to check that $\sim$ is an equivalence relation on $(\mathbb{C}^\times)^{\mathbb{N}}$.

(b) That (a_k) and (b_k) are asymptotically equivalent does not imply that (a_k) or (b_k) converges, nor that $(a_k - b_k)$ is a null sequence.

Proof Consider $a_k := k^2 + k$ and $b_k := k^2$. ∎

(c) If $(a_k) \sim (b_k)$ and $b_k \to c$, it follows that $a_k \to c$.

Proof This follows immediately from $a_k = (a_k/b_k) \cdot b_k$. ∎

(d) Suppose (a_k) and (b_k) are series in $\mathbb{C}^\times$ and $(a_k - b_k)$ is bounded. If $|b_k| \to \infty$ it then follows that $(a_k) \sim (b_k)$.

Proof We have $|(a_k/b_k) - 1| = |(a_k - b_k)/b_k| \to 0$ as $k \to \infty$. ∎

With the help of the Euler–Maclaurin sum formula, we can construct important examples of asymptotically equivalent series. For that, we require the following result.

6.12 Lemma *Suppose $z \in \mathbb{C}$ is such that $\operatorname{Re} z > 1$. Then the limit*

$$\int_1^\infty \frac{\widetilde{B}_k(x)}{x^z} \, dx := \lim_n \int_1^n \frac{\widetilde{B}_k(x)}{x^z} \, dx$$

exists in $\mathbb{C}$ and

$$\left| \int_1^\infty \frac{\widetilde{B}_k(x)}{x^z} \, dx \right| \le \frac{\|\widetilde{B}_k\|_\infty}{\operatorname{Re} z - 1} \quad \text{for } k \in \mathbb{N} \ .$$

Proof Suppose $1 \le m \le n$. Then we have

$$\left| \int_1^n \frac{\widetilde{B}_k(x)}{x^z} \, dx - \int_1^m \frac{\widetilde{B}_k(x)}{x^z} \, dx \right| = \left| \int_m^n \frac{\widetilde{B}_k(x)}{x^z} \, dx \right| \le \|\widetilde{B}_k\|_\infty \int_m^n x^{-\operatorname{Re} z} \, dx$$

$$= \frac{\|\widetilde{B}_k\|_\infty}{\operatorname{Re} z - 1} \left(\frac{1}{m^{\operatorname{Re} z - 1}} - \frac{1}{n^{\operatorname{Re} z - 1}} \right) \ .$$

This estimate shows that $\left(\int_1^n \big(\widetilde{B}_k(x)/x^z\big)\, dx\right)_{n\in\mathbb{N}^\times}$ is a Cauchy series in $\mathbb{C}$ and that the stated limit exists. In the above estimate, we set $m = 1$ and take the limit $n \to \infty$, thus proving the lemma's second part. ∎

6.13 Examples (a) (the formula of de Moivre and Stirling) For $n \to \infty$, we have

$$n! \sim \sqrt{2\pi n}\, n^n e^{-n}\ .$$

Proof (i) We set $a := 1,\ b := n,\ m := 1$ and $f(x) := \log x$. Then, from the Euler–Maclaurin sum formula, we have

$$\sum_{k=1}^n \log k - \int_1^n \log x\, dx = \frac{1}{2}\log n + \frac{B_2}{2}\frac{1}{x}\Big|_1^n + \frac{1}{3}\int_1^n \frac{\widetilde{B}_3(x)}{x^3}\, dx\ .$$

We note $\log x = \big[x(\log x - 1)\big]'$ and $B_2 = 1/6$, which give

$$\log n! - n\log n - \frac{1}{2}\log n + n = \log\big(n!\, n^{-n-1/2}e^n\big)$$

$$= 1 + \frac{1}{12}\Big(\frac{1}{n} - 1\Big) + \frac{1}{3}\int_1^n \frac{\widetilde{B}_3(x)}{x^3}\, dx\ .$$

Now Lemma 6.12 implies that the sequence $\big(\log(n!\, n^{-n-1/2}e^n)\big)_{n\in\mathbb{N}}$ converges. Thus there is an $A > 0$ such that $(n!\, n^{-n-1/2}e^n) \to A$ as $n \to \infty$. Hence

$$n! \sim A n^{n+1/2} e^{-n}\ (n \to \infty)\ . \tag{6.7}$$

(ii) To determine the value of A, apply the Wallis product representation of $\pi/2$:

$$\prod_{k=1}^{\infty} \frac{(2k)^2}{(2k-1)(2k+1)} = \frac{\pi}{2}$$

(see Example 5.5(d)). Therefore we have also

$$\lim_n \left(\prod_{k=1}^n \frac{(2k)^2}{(2k-1)(2k+1)} \cdot \frac{\prod_{k=1}^n (2k)^2}{\prod_{k=1}^n (2k)^2}\right) = \lim_n \frac{2^{4n}(n!)^4}{(2n)!\,(2n+1)!} = \pi/2\ . \tag{6.8}$$

On the other hand, from (6.7), we have

$$\frac{(n!)^2}{(2n)!} \sim \frac{A^2 n^{2n+1} e^{-2n}}{A(2n)^{2n+1/2} e^{-2n}} = 2^{-2n-1} A\sqrt{2n}\ ,$$

whence the asymptotic equivalence

$$\frac{2^{4n}(n!)^4}{2n\big[(2n)!\big]^2} \sim \frac{A^2}{4} \tag{6.9}$$

follows. We note finally that $2n\big[(2n)!\big]^2 \sim (2n)!\,(2n+1)!$ and $A > 0$, and it then follows from (6.8), (6.9), and Remark 6.11(c) that $A = \sqrt{2\pi}$. Inserting this into (6.7), we are done. ∎

(b) (Euler's constant) The limit

$$C := \lim_{n \to \infty} \left[\sum_{k=1}^{n} \frac{1}{k} - \log n \right]$$

exists in $\mathbb{R}$ and is called **Euler's constant** or the **Euler–Mascheroni constant**.[6] In addition,

$$\sum_{k=1}^{n} \frac{1}{k} \sim \log n \quad (n \to \infty) .$$

Proof   For $a := 1$, $b := n$, $m := 0$, and $f(x) := 1/x$ it follows from the Euler–Maclaurin sum formula that

$$\sum_{k=1}^{n} \frac{1}{k} = \int_{1}^{n} \frac{dx}{x} + \frac{1}{2} \left(1 + \frac{1}{n} \right) - \int_{1}^{n} \frac{\tilde{B}_1(x)}{x^2} \, dx .$$

From Lemma 6.12, C is in $\mathbb{R}$. The desired asymptotic equivalence follows from Remark 6.11(d). ∎

The Riemann ζ function

We now apply the Euler–Maclaurin sum formula to examples which important in the theory of the distribution of prime numbers.

Suppose $s \in \mathbb{C}$ satisfies $\operatorname{Re} s > 1$, and define $f(x) := 1/x^s$ for $x \in [1, \infty)$. Then

$$f^{(k)}(x) = (-1)^k s(s+1) \cdot \dots \cdot (s+k-1) x^{-s-k} = (-1)^k \binom{s+k-1}{k} k! \, x^{-s-k}$$

for $k \in \mathbb{N}^{\times}$. From the Euler–Maclaurin sum formula (with $a = 1$ and $b = n$) it follows for $m \in \mathbb{N}$ that

$$\sum_{k=1}^{n} \frac{1}{k^s} = \int_{1}^{n} \frac{dx}{x^s} + \frac{1}{2} \left(1 + \frac{1}{n^s} \right) - \sum_{j=1}^{m} \frac{B_{2j}}{2j} \binom{s+2j-2}{2j-1} x^{-s-2j+1} \Big|_{1}^{n}$$
$$- \binom{s+2m}{2m+1} \int_{1}^{n} \tilde{B}_{2m+1}(x) \, x^{-(s+2m+1)} \, dx . \tag{6.10}$$

Because $|n^{-s}| = n^{-\operatorname{Re} s} \to 0$ as $n \to \infty$ and $\operatorname{Re} s > 0$ we get

$$\int_{1}^{n} \frac{dx}{x^s} = \frac{1}{s-1} \left(1 - \frac{1}{n^{s-1}} \right) \to \frac{1}{s-1}$$

as well as

$$|n^{-s-2j+1}| \to 0 \quad \text{for } j = 1, \dots, m ,$$

---

[6]The initial decimal digits of C are 0.577 215 664 901 ...

as $n \to \infty$. We note in addition Lemma 6.12, so that, in the limit $n \to \infty$, (6.10) becomes

$$\sum_{n=1}^{\infty} \frac{1}{n^s} = \frac{1}{2} + \frac{1}{s-1} + \sum_{j=1}^{m} \frac{B_{2j}}{2j} \binom{s+2j-2}{2j-1}$$
$$- \binom{s+2m}{2m+1} \int_1^{\infty} \widetilde{B}_{2m+1}(x)\, x^{-(s+2m+1)}\, dx \ . \tag{6.11}$$

From this, it follows in particular that the series $\sum 1/n^s$ converges for every $s \in \mathbb{C}$ with $\operatorname{Re} s > 1$.[7]

The formula (6.11) permits another conclusion. For that, we set

$$F_m(s) := \int_1^{\infty} \widetilde{B}_{2m+1}(x)\, x^{-(s+2m+1)}\, dx \quad \text{for } m \in \mathbb{N} \ ,$$

and $H_m := \{ z \in \mathbb{C} \ ; \ \operatorname{Re} z > -2m \}$. According to Lemma 6.12, $F_m : H_m \to \mathbb{C}$ is well defined, and it is not hard to see that F_m is also analytic (see Exercise 4).

We now consider the map

$$G_m : H_m \to \mathbb{C} \ , \quad s \mapsto \sum_{j=1}^{m} \frac{B_{2j}}{2j} \binom{s+2j-2}{2j-1} - \binom{s+2m}{2m+1} F_m(s) \quad \text{for } m \in \mathbb{N} \ ,$$

and get the following property:

6.14 Lemma   *For $n > m$, G_n is an extension of G_m.*

Proof   With $H := \{ z \in \mathbb{C} \ ; \ \operatorname{Re} z > 1 \}$ it follows from (6.11) that

$$G_m(s) = G_n(s) = \sum_{k=1}^{\infty} \frac{1}{k^s} - \frac{1}{2} - \frac{1}{s-1} \quad \text{for } s \in H \ .$$

Both F_k and G_k are analytic on H_k for every $k \in \mathbb{N}$. The claim therefore follows from the identity theorem for analytic functions (Theorem V.3.13). ∎

6.15 Theorem   *The function*

$$\{ z \in \mathbb{C} \ ; \ \operatorname{Re} z > 1 \} \to \mathbb{C} \ , \quad s \mapsto \sum_{n=1}^{\infty} \frac{1}{n^s}$$

*has a uniquely analytic continuation ζ on $\mathbb{C} \backslash \{1\}$, called the **Riemann ζ function**. For $m \in \mathbb{N}$ and $s \in \mathbb{C}$ with $\operatorname{Re} s > -2m$ and $s \neq 1$, we have*

$$\zeta(s) = \frac{1}{2} + \frac{1}{s-1} + \sum_{j=1}^{m} \frac{B_{2j}}{2j} \binom{s+2j-2}{2j-1} - \binom{s+2m}{2m+1} F_m(s) \ .$$

Proof   This follows immediately from (6.11) and Lemma 6.14. ∎

---

[7]Note also Remark 6.16(a).

6.16 Remarks   (a) The series $\sum 1/n^s$ converges absolutely for every $s \in \mathbb{C}$ with $\operatorname{Re} s > 1$.

Proof   For $s \in \mathbb{C}$ with $\operatorname{Re} s > 1$, we have

$$|\zeta(s)| = \left| \sum_{n=1}^{\infty} \frac{1}{n^s} \right| \le \sum_{n=1}^{\infty} \left| \frac{1}{n^s} \right| = \sum_{n=1}^{\infty} \frac{1}{n^{\operatorname{Re} s}} = \zeta(\operatorname{Re} s) ,$$

and the claim follows from the majorant criterion. ∎

(b) (product representation of the ζ function)   *We denote the sequence of prime numbers by (p_k) with $p_1 < p_2 < p_3 < \cdots$. Then*

$$\zeta(s) = \prod_{k=1}^{\infty} \frac{1}{1 - p_k^{-s}} \quad \text{for } \operatorname{Re} s > 1 .$$

Proof   Because $|1/p_k^s| = 1/p_k^{\operatorname{Re} s} < 1$ we have (the geometric series)

$$\frac{1}{1 - p_k^{-s}} = \sum_{j=0}^{\infty} \frac{1}{p_k^{js}} \quad \text{for } k \in \mathbb{N}^\times .$$

It therefore follows for every $m \in \mathbb{N}$ that

$$\prod_{k=1}^{m} \frac{1}{1 - p_k^{-s}} = \prod_{k=1}^{m} \sum_{j=0}^{\infty} \frac{1}{p_k^{js}} = {\sum}' \frac{1}{n^s} ,$$

where, after "multiplying out", the series $\sum'$ contains all numbers of the form $1/n^s$, whose prime factor decomposition $n = q_1^{\nu_1} \cdot \ldots \cdot q_\ell^{\nu_\ell}$ has no other prime numbers from $p_1, \ldots, p_m$. Therefore $\sum'(1/n^s)$ indeed contains all numbers $n \in \mathbb{N}$ with $n \le p_m$. The absolute convergence of the series $\sum_n (1/n^s)$ then implies

$$\left| \zeta(s) - \prod_{k=1}^{m} \frac{1}{1 - p_k^{-s}} \right| = \left| \sum_{n=1}^{\infty} \frac{1}{n^s} - {\sum}' \frac{1}{n^s} \right| \le \sum_{n > p_m} \frac{1}{n^{\operatorname{Re} s}} .$$

From Exercise I.5.8(b) (a theorem of Euclid), it follows that $p_m \to \infty$ for $m \to \infty$. Therefore, from (a), the remaining series $\left(\sum_{n > p_m} (1/n^{\operatorname{Re} s}) \right)$ is a null sequence. ∎

(c) The Riemann ζ function has *no* zeros in $\{ z \in \mathbb{C} \, ; \, \operatorname{Re} z > 1 \}$.

Proof   Suppose $s \in \mathbb{C}$ satisfies $\operatorname{Re} s > 1$ and $\zeta(s) = 0$. Because $\operatorname{Re} s > 1$, we then have that $\lim_k (1 - p_k^{-s}) = 1$. Hence there is an $m_0 \in \mathbb{N}$ that $\log(1 - p_k^{-s})$ is well defined for $k \ge m_0$, and we find

$$\log \left(\prod_{k=m_0}^{N} \frac{1}{1 - p_k^{-s}} \right) = - \sum_{k=m_0}^{N} \log(1 - p_k^{-s}) \quad \text{for } N \ge m_0 . \tag{6.12}$$

The convergence of $\left(\prod_{k=m_0}^{N} (1 - p_k^{-s})^{-1} \right)_{N \in \mathbb{N}}$, the continuity of the logarithm, and (6.12) imply that the series $\sum_{k \ge m_0} \log(1 - p_k^{-s})$ converges. In particular, the remaining terms

$\left(\sum_{k=m}^{\infty} \log(1 - p_k^{-s})\right)_{m \geq m_0}$ form a null sequence. Then it follows from (6.12) that

$$\lim_{m \to \infty} \prod_{k=m}^{\infty} (1 - p_k^{-s})^{-1} = 1 .$$

Consequently there exists an $m_1 \geq m_0$ such that $\prod_{k=m_1+1}^{\infty}(1-p_k^{-s})^{-1} \neq 0$. Now because $\prod_{k=1}^{m_1}(1 - p_k^{-s})^{-1}$ is not zero, it follows that

$$0 = \zeta(s) = \prod_{k=1}^{m_1} \frac{1}{1 - p_k^{-s}} \prod_{k=m_1+1}^{\infty} \frac{1}{1 - p_k^{-s}} \neq 0 ,$$

which is impossible. ∎

(d) The series $\sum 1/p_k$ diverges.

Proof   From the considerations in the proof of (b), we have the estimate

$$\prod_{p \leq m} \left(1 - \frac{1}{p}\right)^{-1} \geq \sum_{n=1}^{m} \frac{1}{n} \quad \text{for } n \in \mathbb{N}^{\times} ,$$

where the product has all prime numbers $\leq m$. Now we consider the sum a Riemann sum and find that function $x \mapsto 1/x$ is decreasing:

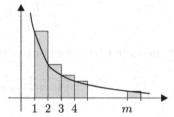

$$\sum_{n=1}^{m} \frac{1}{n} > \int_{1}^{m} \frac{dx}{x} = \log m .$$

Because the logarithm is monotone, it follows that

$$\sum_{p \leq m} \log\left(1 - \frac{1}{p}\right)^{-1} > \log \log m \quad \text{for } m \geq 2 . \tag{6.13}$$

When $|z| < 1$, we can use the series expansion of the logarithm (Theorem V.3.9)

$$\log(1 - z)^{-1} = -\log(1 - z) = \sum_{k=1}^{\infty} \frac{z^k}{k}$$

to get

$$\sum_{p \leq m} \log\left(1 - \frac{1}{p}\right)^{-1} = \sum_{p \leq m} \sum_{k=1}^{\infty} \frac{1}{k p^k} \leq \sum_{p \leq m} \frac{1}{p} + R , \tag{6.14}$$

where

$$R := \sum_{p \leq m} \sum_{k=2}^{\infty} \frac{1}{k p^k} \leq \frac{1}{2} \sum_{p \leq m} \sum_{k=2}^{\infty} \frac{1}{p^k} = \frac{1}{2} \sum_{p \leq m} \frac{1}{p(p-1)} < \frac{1}{2} \sum_{j=2}^{\infty} \frac{1}{j(j-1)} = \frac{1}{2} .$$

It now follows from (6.13) and (6.14) that

$$\sum_{p \leq m} \frac{1}{p} \geq \log \log m - \frac{1}{2} \quad \text{for } m \geq 2 , \tag{6.15}$$

which, since the sum is over all prime numbers $\leq m$, proves the claim. ∎

(e) The preceding statements contain information about the density of prime numbers. Because $\sum 1/p_k$ diverges (according to (6.15), at least as fast as $\log\log m$), the sequence (p_k) cannot go too quickly to ∞. In studies of the prime number distribution, $\pi(x)$ for $x \in \mathbb{R}^\times$ denotes the number of prime numbers $\leq x$. The famous **prime number theorem** says

$$\pi(n) \sim n/\log n \quad (n \to \infty) \ .$$

To get more information, one can study the asymptotic behavior of the relative error

$$r(n) := \frac{\pi(n) - n/\log n}{n/\log n} = \frac{\pi(n)}{n/\log n} - 1$$

as $n \to \infty$. It is possible to show that

$$r(n) = O\Big(\frac{1}{\log n}\Big) \quad (n \to \infty) \ .$$

It however only *conjectured* that for every $\varepsilon > 0$ there is a distinctly better error estimate:

$$r(n) = O\Big(\frac{1}{n^{1/2-\varepsilon}}\Big) \quad (n \to \infty) \ .$$

This conjecture is equivalent to the celebrated **Riemann hypothesis**:

The ζ function has no zeros s such that $\mathrm{Re}\,s > 1/2$.

We know from (c) that ζ has no zeros in $\{\, z \in \mathbb{C} \ ; \ \mathrm{Re}\,z > 1 \,\}$. It is also known that the set of zeros of ζ with real part ≤ 0 correspond to $-2\mathbb{N}^\times$. Calling these zeros trivial, one can conclude from the properties of the ζ function that the Riemann hypothesis is equivalent to the following statement:

All nontrivial zeros of the Riemann ζ function

lie on the line $\mathrm{Re}\,z = 1/2$.

It is known that ζ has infinitely many zeros on this line and, for a certain large value of K, there are no nontrivial zeros s with $|\,\mathrm{Im}\,s| < K$ for which $\mathrm{Re}\,s \neq 1/2$. Proving the Riemann hypothesis stands today as one of the greatest efforts in mathematics.

For a proof of the prime number theorem, for the stated asymptotic error estimates, and for further investigations, consult the literature of analytic number theory, for example [Brü95], [Pra78], [Sch69]. ∎

The trapezoid rule

In most practical applications, a given integral must be approximated numerically, because no antiderivative is known. From the definition of integrals $\int_\alpha^\beta f\,dx$, every jump continuous function f can be integrated approximately as a Riemann sum over rectangles whose (say) top-left corners meet at the graph of f. Such an approximation tends to leave gaps between the rectangles and the function, and this slows its convergence as the mesh decreases. If the function is sufficiently smooth, one can anticipate that (in the real-valued case) the area under the graph of f can be better estimated by trapezoids.

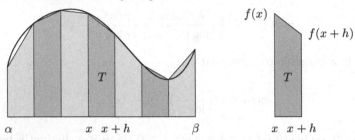

Here $h\big[f(x+h)+f(x)\big]/2$ is the (oriented) area of the trapezoid T.

The next theorem shows that if f is smooth, the "trapezoid rule" for small steps h gives a good approximation to $\int f$. As usual, we take $-\infty < \alpha < \beta < \infty$ and E to be a Banach space.

6.17 Theorem   *Suppose* $f \in C^2\big([\alpha,\beta], E\big)$, $n \in \mathbb{N}^\times$, *and* $h := (\beta - \alpha)/n$. *Then*

$$\int_\alpha^\beta f(x)\,dx = \Big[\frac{1}{2}f(\alpha) + \sum_{k=1}^{n-1} f(\alpha + kh) + \frac{1}{2}f(\beta)\Big]h + R(f,h) \ ,$$

where

$$|R(f,h)| \le \frac{\beta - \alpha}{12}h^2 \,\|f''\|_\infty \ .$$

Proof   From the additivity of integrals and the substitution rule, it follows after defining $x(t) := \alpha + kh + th$ and $g_k(t) := f(\alpha + kh + th)$ for $t \in [0,1]$ that

$$\int_\alpha^\beta f(x)\,dx = \sum_{k=0}^{n-1} \int_{\alpha+kh}^{\alpha+(k+1)h} f(x)\,dx = h\sum_{k=0}^{n-1} \int_0^1 g_k(t)\,dt \ .$$

Formula (6.6) implies[8]

$$\int_0^1 g_k(t)\,dt = \frac{1}{2}\big[g_k(0) + g_k(1)\big] - \int_0^1 B_1(t)g_k'(t)\,dt \ .$$

---

[8]Note that the formula (6.6) is entirely elementary and the theory of the Bernoulli polynomials is not needed for this derivation.

We set $u := g'_k$ and $v' := B_1(X)$, and thus $u' = g''_k$ and $v = -X(1-X)/2$, as well as $v(0) = v(1) = 0$. Integrating by parts, we therefore get

$$-\int_0^1 B_1(t)g'_k(t)\,dt = \int_0^1 v(t)g''_k(t)\,dt \ .$$

Defining

$$R(f,h) := h\sum_{k=0}^{n-1}\int_0^1 v(t)g''_k(t)\,dt \ ,$$

we arrive at the representation

$$\int_\alpha^\beta f(x)\,dx = h\left[\frac{1}{2}f(\alpha) + \sum_{k=1}^{n-1}f(\alpha+kh) + \frac{1}{2}f(\beta)\right] + R(f,h) \ .$$

To estimate the remainder $R(f,h)$, we note first that

$$\left|\int_0^1 v(t)g''_k(t)\,dt\right| \le \|g''_k\|_\infty \int_0^1 \frac{t(1-t)}{2}\,dt = \frac{1}{12}\|g''_k\|_\infty \ .$$

The chain rule implies

$$\|g''_k\|_\infty = \max_{t\in[0,1]}|g''_k(t)| = h^2 \max_{\alpha+kh\le x\le\alpha+(k+1)h}|f''(x)| \le h^2\|f''\|_\infty \ .$$

Consequently,

$$|R(f,h)| \le h\sum_{k=0}^{n-1}\left|\int_0^1 v(t)g''_k(t)\,dt\right| \le \frac{h^3}{12}\|f''\|_\infty\sum_{k=0}^{n-1}1 = \frac{\beta-\alpha}{12}h^2\|f''\|_\infty \ ,$$

which shows the dependence on h. ∎

The trapezoid rule constitutes the simplest "quadrature formula" for numerically approximating integrals. For further exploration and more efficient methods, see lectures and books about numerical methods (see also Exercise 8).

Exercises

1 Prove the statements (iv) and (v) in Proposition 6.6.

2 Compute B_8, B_{10}, and B_{12}, as well as $B_4(X)$, $B_5(X)$, $B_6(X)$ and $B_7(X)$.

3 Show that for $n \in \mathbb{N}^\times$

  (i) $B_{2n+1}(X)$ has in $[0,1]$ only the zeros 0, $1/2$, and 1;

  (ii) $B_{2n}(X)$ has in $[0,1]$ only two zeros x_{2m} and x'_{2m}, with $x_{2m} + x'_{2m} = 1$.

4 Denote by $\widetilde{B}_n$ the 1-periodic continuation of $B_n(X)|[0,1)$ onto $\mathbb{R}$. Then show these:

(i) $\widetilde{B}_n \in C^{n-2}(\mathbb{R})$ for $n \geq 2$.

(ii) $\int_k^{k+1} \widetilde{B}_n(x)\, dx = 0$ for $k \in \mathbb{Z}$ and $n \in \mathbb{N}$.

(iii) For every $n \in \mathbb{N}$ map

$$F_n : \{z \in \mathbb{C} \; ; \; \mathrm{Re}\, z > -2n\} \to \mathbb{C}, \quad s \mapsto \int_1^\infty \widetilde{B}_{2n+1}(x) x^{-(s+2n+1)}\, dx$$

is analytic.

5 Prove Remark 6.11(a).

6 Show for the ζ function that $\lim_{n \to \infty} \zeta(n) = 1$.

7 Suppose $h > 0$ and $f \in C^4([-h, h], \mathbb{R})$. Show

$$\left| \int_{-h}^h f(x)\, dx - \frac{h}{3}\big(f(-h) + 4f(0) + f(h)\big) \right| \leq \frac{h^5}{90} \|f^{(4)}\|_\infty .$$

(Hints: Integrate by parts and use the mean value theorem for integrals.)

8 Suppose $f \in C^4([\alpha, \beta], \mathbb{R})$ and let $\alpha_j := \alpha + jh$ for $0 \leq j \leq 2n$ with $h := (\beta - \alpha)/2n$ and $n \in \mathbb{N}^\times$. Further let[9]

$$S(f, [\alpha, \beta], h) := \frac{h}{3}\left(f(\alpha) + f(\beta) + 2\sum_{j=1}^{n-1} f(\alpha_{2j}) + 4\sum_{j=1}^n f(\alpha_{2j-1}) \right).$$

This is **Simpson's rule**, another rule for approximating integrals. Show it satisfies the error estimate

$$\left| \int_\alpha^\beta f(x)\, dx - S(f, [\alpha, \beta], h) \right| \leq (\beta - \alpha)\frac{h^4}{180} \|f^{(4)}\|_\infty .$$

(Hint: Exercise 7.)

9 Calculate the error in

$$\int_0^1 \frac{dx}{1 + x^2} = \frac{\pi}{4} \quad \text{and} \quad \int_1^2 \frac{dx}{x} = \log 2$$

when using the trapezoid rule for $n = 2, 3, 4$ and Simpson's rule for $n = 1, 2$.

10 Show that Simpson's rule with a single inner grid point gives the correct value of $\int_\alpha^\beta p$ for any polynomial $p \in \mathbb{R}[X]$ of degree at most three.

---

[9] As usual, we assign the "empty sum" $\sum_{j=1}^0 \cdots$ the value 0.

7 Fourier series

At the end of Section V.4, we asked about the connection between trigonometric series and continuous periodic functions. With the help of the available integration theory, we can now study this connection more precisely. In doing so, we will get a first look at the vast *theory of Fourier series*, wherein we must concentrate on only the most important theorems and techniques. For simplicity, we treat only the case of piecewise continuous 2π-periodic functions, because studying more general classes of functions will call for the Lebesgue integration theory, which we will not see until Volume III.

The theory of Fourier series is closely related to the theory of inner product spaces, in particular, with the L_2 space. For these reasons we first consider the L_2 structure of spaces of piecewise continuous functions as well as orthonormal bases of inner product spaces. The results will be of fundamental importance for determining the how Fourier series converge.

The L_2 scalar product

Suppose $I := [\alpha, \beta]$ is a compact perfect interval. From Exercise 4.3 we know that $\int_\alpha^\beta f \, dx$ for $f \in \mathcal{SC}(I)$ is independent of the value of f on its discontinuities. Therefore we can fix this value arbitrarily for the purposes of studying integrals. On these grounds, we will consider only "normalized" functions, as follows. We call $f : I \to \mathbb{C}$ **normalized** when

$$f(x) = \big(f(x+0) + f(x-0)\big)/2 \quad \text{for } x \in \overset{\circ}{I} \,, \tag{7.1}$$

and

$$f(\alpha) = f(\beta) = \big(f(\alpha+0) + f(\beta-0)\big)/2 \,. \tag{7.2}$$

Denote by $SC(I)$ the set of normalized, piecewise continuous functions $f : I \to \mathbb{C}$.[1]

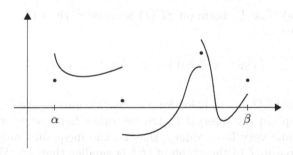

The meaning of the assignment (7.2) will be clarified in the context of periodic functions.

---

[1] We naturally write $SC[\alpha, \beta]$ instead of $SC([\alpha, \beta])$, and $C[\alpha, \beta]$ instead of $C([\alpha, \beta])$ etc.

7.1 Proposition $SC(I)$ *is a vector subspace of* $S(I)$, *and*

$$(f\,|\,g)_2 := \int_\alpha^\beta f\bar{g}\,dx \quad \text{for } f,g \in SC(I)\,,$$

defines a scalar product $(\cdot\,|\,\cdot)_2$ *on* $SC(I)$.

Proof   The first claim is clear. The linearity of integrals implies that $(\cdot\,|\,\cdot)_2$ is a sesquilinear form on $SC(I)$. Corollary 4.10 implies $\overline{(f\,|\,g)}_2 = (g\,|\,f)_2$ for $f,g \in SC(I)$. From the positivity of integrals, we have

$$(f\,|\,f)_2 = \int_\alpha^\beta |f|^2\,dx \geq 0 \quad \text{for } f \in SC(I)\,.$$

If $f \in SC(I)$ is not the zero function, then there is, by the normalizability, a point a at which f, and therefore also $|f|^2$, is continuous and for which $|f(a)|^2 > 0$. It then follows from Proposition 4.8 that $(f\,|\,f) > 0$. Thus $(\cdot\,|\,\cdot)_2$ is an inner product on $SC(I)$. ∎

The norm

$$f \mapsto \|f\|_2 := \sqrt{(f\,|\,f)_2} = \left(\int_\alpha^\beta |f|^2\,dx\right)^{1/2}$$

induced by $(\cdot\,|\,\cdot)_2$ is called the **L_2 norm**, and we call $(\cdot\,|\,\cdot)_2$ itself the **L_2 scalar product** on $SC(I)$. In the following, we always equip $SC(I)$ with this scalar product and understand

$$SC(I) := \big(SC(I),(\cdot\,|\,\cdot)_2\big)$$

unless explicitly stated otherwise. Thus $SC(I)$ is an inner product space.

7.2 Remarks   (a) The L_2 norm on $SC(I)$ is weaker[2] than the supremum norm. Precisely, we have

$$\|f\|_2 \leq \sqrt{\beta - \alpha}\,\|f\|_\infty \quad \text{for } f \in SC(I)\,.$$

(b) The inequality $\|f\|_\infty < \varepsilon$ holds for $f \in SC(I)$, and so the entire graph of f lies in an "ε-strip" in the interval I. On the other hand, even when $\|f\|_2 < \varepsilon$, f itself can assume very large values, that is, the inequality only requires that the (oriented) area in I of the graph of $|f|^2$ is smaller than ε^2. We say that the function f is smaller than ε in its **quadratic mean**. Convergence in the L_2 norm is also called **convergence** in the **quadratic mean**.

---

[2]See Exercise 4.5.

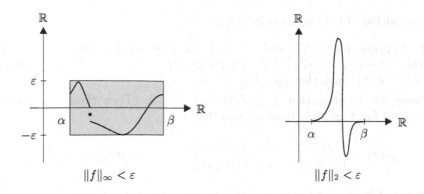

$$\|f\|_\infty < \varepsilon \qquad\qquad\qquad \|f\|_2 < \varepsilon$$

(c) Although the sequence (f_n) may converge in $SC(I)$ — and therefore also in its quadratic mean — to zero, the values $\big(f_n(x)\big)$ for $x \in I$ need not converge to zero.

Proof   For $n \in \mathbb{N}^\times$ suppose $j, k \in \mathbb{N}$ are the unique numbers with $n = 2^k + j$ and $j < 2^k$. Then we define $f_n \in SC[0,1]$ by $f_0 := \mathbf{1}$ and

$$f_n(x) := \begin{cases} 1 , & x \in \big(j2^{-k}, (j+1)2^{-k}\big) , \\ 0 & \text{otherwise} . \end{cases}$$

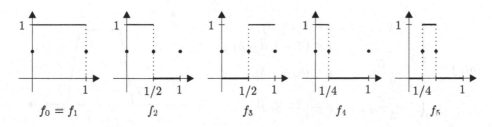

$$f_0 = f_1 \qquad\quad f_2 \qquad\qquad f_3 \qquad\qquad f_4 \qquad\qquad f_5$$

Because $\|f_n\|_2^2 = 2^{-k}$, (f_n) converges in $SC[0,1]$ to 0; however, because the "bump" in f continues to sweep the interval from left to right each time it narrows, $\big(f_n(x)\big)$ does not converge for any $x \in [0,1]$. ∎

Approximating in the quadratic mean

It is obvious that $C(I)$ is not dense in $SC(I)$ in the supremum norm. However in L_2 norm, the situation is different, as the following theorem about *approximating in the quadratic mean* shows. Here we set

$$C_0(I) := \big\{ u \in C(I) \,;\, u(\alpha) = u(\beta) = 0 \big\} .$$

$C_0(I)$ is obviously a closed vector subspace of $C(I)$ (in the maximum norm), and it is therefore a Banach space.

7.3 Proposition $C_0(I)$ *is dense in* $SC(I)$.

Proof Suppose $f \in SC(I)$ and $\varepsilon > 0$. From Theorem 1.2, there is a $g \in \mathcal{T}(I)$ such that $\|f - g\|_\infty < \varepsilon/2\sqrt{\beta - \alpha}$. Therefore[3] $\|f - g\|_2 < \varepsilon/2$. It then suffices to find an $h \in C_0(I)$ such that $\|g - h\|_2 < \varepsilon/2$.

Suppose therefore that $f \in \mathcal{T}(I)$ and $\varepsilon > 0$. Then there is a partition $(\alpha_0, \ldots, \alpha_n)$ for f and functions g_j such that

$$g_j(x) = \begin{cases} f(x) , & x \in (\alpha_j, \alpha_{j+1}) , \\ 0 , & x \in I \backslash (\alpha_j, \alpha_{j+1}) , \end{cases} \qquad \text{for } 0 \leq j \leq n - 1 ,$$

and $f(x) = \sum_{j=0}^{n-1} g_j(x)$ for $x \neq \alpha_j$. On the other hand, by the triangle inequality, it suffices to find $h_0, \ldots, h_{n-1} \in C_0(I)$ such that $\|g_j - h_j\|_2 < \varepsilon/n$.

Thus we can assume that f is a nontrivial staircase function "with only one step". In other words, there are numbers $\bar{\alpha}, \bar{\beta} \in I$ and $y \in \mathbb{C}^\times$ with $\bar{\alpha} < \bar{\beta}$ which can be chosen so that

$$f(x) = \begin{cases} y , & x \in (\bar{\alpha}, \bar{\beta}) , \\ 0 , & x \in I \backslash [\bar{\alpha}, \bar{\beta}] , \end{cases}$$

Let $\varepsilon > 0$. Then we choose $\delta > 0$ such that $\delta < (\bar{\beta} - \bar{\alpha})/2 \wedge \varepsilon^2/|y|^2$ and define

$$g(x) := \begin{cases} 0 , & x \in I \backslash (\bar{\alpha}, \bar{\beta}) , \\ y , & x \in (\bar{\alpha} + \delta, \bar{\beta} - \delta) , \\ \dfrac{x - \bar{\alpha}}{\delta} y , & x \in [\bar{\alpha}, \bar{\alpha} + \delta] , \\ \dfrac{\bar{\beta} - x}{\delta} y , & x \in [\bar{\beta} - \delta, \bar{\beta}] . \end{cases}$$

Then g belongs to $C_0(I)$, and we have

$$\|f - g\|_2^2 = \int_\alpha^\beta |f - g|^2 \, dx \leq \delta |y|^2 < \varepsilon^2 . \blacksquare$$

7.4 Corollary

(i) $(C(I), (\cdot | \cdot)_2)$ *is not a Hilbert space.*

(ii) *The maximum norm on* $C(I)$ *is strictly stronger than the* L_2 *norm.*

Proof (i) Suppose $f \in SC(I) \backslash C(I)$. Then, by Proposition 7.3, there is a sequence (f_j) in $C(I)$ that converges in the L_2 norm to f. Therefore, according to Proposition II.6.1, (f_j) is a Cauchy sequence in $SC(I)$ and therefore also in

---

[3]Of course, $\|u\|_2$ is defined for every $u \in \mathcal{S}(I)$.

$E := \big(C(I), \|\cdot\|_2\big)$. Were E a Hilbert space, and therefore complete, then there would be a $g \in E$ with $f_j \to g$ in E. Therefore (f_j) would converge in $SC(I)$ to both f and g, and we would have $f \neq g$, which is impossible. Therefore E is not complete.

(ii) This follows from (i) and the Remarks 7.2(a) and II.6.7(a). ∎

Orthonormal systems

We recall now some ideas from the theory of inner product spaces (see Exercise II.3.10). Let $E := \big(E, (\cdot|\cdot)\big)$ be an inner product space. Then $u, v \in E$ are **orthogonal** if $(u|v) = 0$, and we also write $u \perp v$. A subset M of E is said to be an **orthogonal system** if every pair of distinct elements of M are orthogonal. If in addition $\|u\| = 1$ for $u \in M$, we say M is an **orthonormal system** (ONS).

7.5 Examples (a) For $k \in \mathbb{Z}$, let

$$\mathsf{e}_k(t) := \frac{1}{\sqrt{2\pi}} e^{ikt} \quad \text{for } t \in \mathbb{R} .$$

Then $\{\, \mathsf{e}_k \;;\; k \in \mathbb{Z} \,\}$ is an ONS in the inner product space $SC[0, 2\pi]$.

Proof Because the exponential function is $2\pi i$-periodic, we find

$$(\mathsf{e}_j|\mathsf{e}_k)_2 = \int_0^{2\pi} \mathsf{e}_j \bar{\mathsf{e}}_k \, dt = \frac{1}{2\pi} \int_0^{2\pi} e^{i(j-k)t} \, dt = \begin{cases} 1 , & j = k , \\[2mm] \dfrac{-i}{2\pi(j-k)} e^{i(j-k)t} \Big|_0^{2\pi} = 0 , & j \neq k , \end{cases}$$

as desired. ∎

(b) Suppose

$$\mathsf{c}_0(t) := \frac{1}{\sqrt{2\pi}} , \quad \mathsf{c}_k(t) :- \frac{1}{\sqrt{\pi}} \cos(kt)$$

and

$$\mathsf{s}_k(t) := \frac{1}{\sqrt{\pi}} \sin(kt)$$

for $t \in \mathbb{R}$ and $k \in \mathbb{N}^\times$. Then $\{\, \mathsf{c}_0, \mathsf{c}_k, \mathsf{s}_k \;;\; k \in \mathbb{N}^\times \,\}$ is an orthonormal system in the inner product space $SC\big([0, 2\pi], \mathbb{R}\big)$.

Proof Euler's formula (III.6.1) implies $\mathsf{e}_k = (\mathsf{c}_k + i\mathsf{s}_k)/\sqrt{2}$ for $k \in \mathbb{N}^\times$. Then

$$2(\mathsf{e}_j|\mathsf{e}_k)_2 = (\mathsf{c}_j|\mathsf{c}_k)_2 + (\mathsf{s}_j|\mathsf{s}_k)_2 + i\big[(\mathsf{s}_j|\mathsf{c}_k)_2 - (\mathsf{c}_j|\mathsf{s}_k)_2\big] \qquad (7.3)$$

for $j, k \in \mathbb{N}^\times$. Because $(\mathsf{e}_j|\mathsf{e}_k)_2$ is real, we find

$$(\mathsf{s}_j|\mathsf{c}_k)_2 = (\mathsf{c}_j|\mathsf{s}_k)_2 \quad \text{for } j, k \in \mathbb{N}^\times . \qquad (7.4)$$

Integration by parts yields

$$(\mathsf{s}_j|\mathsf{c}_k)_2 = \frac{1}{\pi} \int_0^{2\pi} \sin(jt)\cos(kt) \, dt = -\frac{j}{k}(\mathsf{c}_j|\mathsf{s}_k)_2$$

for $j, k \in \mathbb{N}^\times$. Therefore we get with (7.4)

$$(1 + j/k)(s_j \,|\, c_k)_2 = 0 \quad \text{for } j, k \in \mathbb{N}^\times .$$

We can then state

$$(s_j \,|\, c_k)_2 = 0 \quad \text{for } j, k \in \mathbb{N} ,$$

because this relation is trivially true for the remaining cases $j = 0$ or $k = 0$.

Using the Kronecker symbol, we get from (7.3) and (a) that

$$(c_j \,|\, c_k)_2 + (s_j \,|\, s_k)_2 = 2\delta_{jk} \quad \text{for } j, k \in \mathbb{N}^\times . \tag{7.5}$$

Integration by parts yields

$$(s_j \,|\, s_k)_2 = \frac{j}{k}(c_j \,|\, c_k)_2 \quad \text{for } j, k \in \mathbb{N}^\times . \tag{7.6}$$

Therefore

$$(1 + j/k)(c_j \,|\, c_k)_2 = 2\delta_{jk} \quad \text{for } j, k \in \mathbb{N}^\times .$$

From this and (7.6), it follows that

$$(c_j \,|\, c_k)_2 = (s_j \,|\, s_k)_2 = 0 \quad \text{for } j \neq k ,$$

and

$$\|c_k\|_2^2 = \|s_k\|_2^2 = 1 \quad \text{for } k \in \mathbb{N}^\times .$$

Finally, $(c_0 \,|\, c_j)_2 = (c_0 \,|\, s_j)_2 = 0$ is obvious for $j \in \mathbb{N}^\times$ and $\|c_0\|_2^2 = 1$. ∎

Integrating periodic functions

We now gather elementary but useful properties of periodic functions.

7.6 Remarks Suppose $f : \mathbb{R} \to \mathbb{C}$ is p-periodic for some $p > 0$.

(a) If $f \,|\, [0, p]$ is jump continuous, then f is jump continuous on every compact subinterval I of $\mathbb{R}$. Similarly, if $f \,|\, [0, p]$ belongs to $SC[0, p]$, $f \,|\, I$ belongs to $SC(I)$.

(b) If $f \,|\, [0, p]$ is jump continuous, we have

$$\int_0^p f \, dx = \int_\alpha^{\alpha+p} f \, dx \quad \text{for } \alpha \in \mathbb{R} .$$

Proof From the additivity of integrals, we have

$$\int_\alpha^{\alpha+p} f \, dx = \int_0^p f \, dx + \int_p^{\alpha+p} f \, dx - \int_0^\alpha f \, dx .$$

Using the substitution $y(x) := x - p$ in the second integral on the right hand side, we get, using the p-periodicity of f, that

$$\int_p^{\alpha+p} f \, dx = \int_0^\alpha f(y + p) \, dy = \int_0^\alpha f \, dy . \blacksquare$$

From Remark V.4.12(b), we may confine our study of periodic functions to the case $p = 2\pi$. So we now take $I := [0, 2\pi]$ and define the L_2 scalar product on this interval.

Fourier coefficients

Consider

$$T_n : \mathbb{R} \to \mathbb{C}, \quad t \mapsto \frac{a_0}{2} + \sum_{k=1}^{n} [a_k \cos(kt) + b_k \sin(kt)], \tag{7.7}$$

a trigonometric polynomial. Using

$$c_0 := a_0/2, \quad c_k := (a_k - ib_k)/2, \quad c_{-k} := (a_k + ib_k)/2 \tag{7.8}$$

for $1 \le k \le n$, we can write T_n in the form

$$T_n(t) = \sum_{k=-n}^{n} c_k e^{ikt} \quad \text{for } t \in \mathbb{R}, \tag{7.9}$$

(see (V.4.5) and (V.4.6)).

7.7 Lemma *The coefficients c_k can be found as*

$$c_k = \frac{1}{2\pi} \int_0^{2\pi} T_n(t) e^{-ikt} \, dt = \frac{1}{\sqrt{2\pi}} (T_n \,|\, e_k)_2 \quad \text{for } -n \le k \le n.$$

Proof   We form the inner product in $SC(I)$ between $T_n \in C(I)$ and e_j. Then it follows from (7.9) and Example 7.5(a) that

$$(T_n \,|\, e_j)_2 = \sqrt{2\pi} \sum_{k=-n}^{n} c_k (e_k \,|\, e_j)_2 = \sqrt{2\pi} \, c_j \quad \text{for } -n \le j \le n. \blacksquare$$

Suppose now (a_k) and (b_k) are series in $\mathbb{C}$, and c_k is defined for $k \in \mathbb{Z}$ through (7.8). Then we can write the trigonometric series

$$\frac{a_0}{2} + \sum_{k \ge 1} [a_k \cos(k\cdot) + b_k \sin(k\cdot)] = \frac{a_0}{2} + \sqrt{\pi} \sum_{k \ge 1} (a_k c_k + b_k s_k)$$

in the form

$$\sum_{k \in \mathbb{Z}} c_k e^{ik\cdot} = \sqrt{2\pi} \sum_{k \in \mathbb{Z}} c_k e_k. \tag{7.10}$$

Convention In the rest of this section, we understand $\sum_{k \in \mathbb{Z}} g_k$ to be the sequence of partial sums $\left(\sum_{k=-n}^{n} g_k \right)_{n \in \mathbb{N}}$, rather than the sequence of the sum of the two single series $\sum_{k \geq 0} g_k$ and $\sum_{k > 1} g_{-k}$.

The next theorem shows when the coefficients of the trigonometric series (7.1) can be extracted from the function it converges to.

7.8 Proposition *If the trigonometric series (7.10) converges uniformly on $\mathbb{R}$, it converges to a continuous 2π-periodic function f, and*

$$c_k = \frac{1}{2\pi} \int_0^{2\pi} f(t) e^{-ikt}\, dt = \frac{1}{\sqrt{2\pi}} (f \,|\, e_k)_2 \quad \text{for } k \in \mathbb{Z} .$$

Proof Because the partial sum T_n is continuous and 2π-periodic, the first claim follows at once from Theorem V.2.1. Therefore define

$$f(t) := \sum_{n=-\infty}^{\infty} c_n e^{int} := \lim_{n \to \infty} T_n(t) \quad \text{for } t \in \mathbb{R} .$$

The series $(T_n \bar{e}_k)_{n \in \mathbb{N}}$ converges uniformly for every $k \in \mathbb{Z}$ and therefore converges in $C(I)$ to $f \bar{e}_k$. Thus we get from Proposition 4.1 and Lemma 7.7

$$(f \,|\, e_k)_2 = \int_0^{2\pi} f \bar{e}_k\, dt = \lim_{n \to \infty} \int_0^{2\pi} T_n \bar{e}_k\, dt = \sqrt{2\pi}\, c_k \quad \text{for } k \in \mathbb{Z} . \blacksquare$$

Classical Fourier series

We denote by $SC_{2\pi}$ the vector subspace of $B(\mathbb{R})$ that contains all 2π-periodic functions $f : \mathbb{R} \to \mathbb{C}$ with $f \,|\, I \in SC(I)$ and that is endowed with scalar product

$$(f \,|\, g)_2 := \int_0^{2\pi} f \bar{g}\, dx \quad \text{for } f, g \in SC_{2\pi} .$$

Then

$$\widehat{f}_k := \frac{1}{2\pi} \int_0^{2\pi} f(t) e^{-ikt}\, dt = \frac{1}{\sqrt{2\pi}} (f \,|\, e_k)_2 \tag{7.11}$$

is well defined for $f \in SC_{2\pi}$ and $k \in \mathbb{Z}$ and is called the k-th (**classical**) **Fourier coefficient** of f. The trigonometric series

$$\mathsf{S}f := \sum_{k \in \mathbb{Z}} \widehat{f}_k e^{ik\cdot} = \sum_{k \in \mathbb{Z}} (f \,|\, e_k) e_k \tag{7.12}$$

is the (**classical**) **Fourier series** of f. Its n-th partial sum

$$S_n f := \sum_{k=-n}^{n} \widehat{f}_k e^{ik\cdot}$$

is its n-th **Fourier polynomial**.

7.9 Remarks (a) $SC_{2\pi}$ is an inner product space, and $C_{2\pi}$, the space of the continuous 2π-periodic functions[4] $f : \mathbb{R} \to \mathbb{C}$, is dense in $SC_{2\pi}$.

Proof   That a periodic function is determined by its values on its periodic interval follows from the proof of Theorems 7.1 and 7.3. Here we note that the 2π-periodic extension of $g \in C_0(I)$ belongs to $C_{2\pi}$. ∎

(b) The Fourier series Sf can be expressed in the form

$$Sf = \frac{a_0}{2} + \sum_{k\geq 1}\left(a_k \cos(k\cdot) + b_k \sin(k\cdot)\right) = \frac{a_0}{2} + \sqrt{\pi}\sum_{k\geq 1}(a_k c_k + b_k s_k) \ ,$$

where

$$a_k := a_k(f) := \frac{1}{\pi}\int_0^{2\pi} f(t)\cos(kt)\,dt = \frac{1}{\sqrt{\pi}}(f \,|\, c_k)$$

and

$$b_k := b_k(f) := \frac{1}{\pi}\int_0^{2\pi} f(t)\sin(kt)\,dt = \frac{1}{\sqrt{\pi}}(f \,|\, s_k)$$

for $k \in \mathbb{N}^\times$. Also

$$a_0 := a_0(f) := \frac{1}{\pi}\int_0^{2\pi} f(t)\,dt = \sqrt{\frac{2}{\pi}}\,(f \,|\, c_0) \ .$$

Proof   This follows from (7.8) and Euler's formula $\sqrt{2}\,e_k = c_k + i s_k$. ∎

(c) If f is even, then Sf is a purely a **cosine series**:

$$Sf = \frac{a_0}{2} + \sum_{k\geq 1} a_k \cos(k\cdot) = \frac{a_0}{2} + \sqrt{\pi}\sum_{k\geq 1} a_k c_k$$

with

$$a_k = a_k(f) = \frac{2}{\pi}\int_0^{\pi} f(t)\cos(kt)\,dt \quad \text{for } k \in \mathbb{N} \ .$$

If f is odd, then Sf is purely a **sine series**:

$$Sf = \sum_{k\geq 1} b_k \sin(k\cdot) = \sqrt{\pi}\sum_{k\geq 1} b_k s_k$$

---

[4]See Section V.4.

with

$$b_k = b_k(f) = \frac{2}{\pi} \int_0^\pi f(t)\sin(kt)\,dt \quad \text{for } k \in \mathbb{N} . \tag{7.13}$$

Proof The cosine is even and the sine is odd. Then if f is even, fc_k is even and fs_k is odd. If f is odd, fc_k is odd and fs_k is even. Consequently, from Remark 7.6(b) and Exercise 4.11,

$$a_k = \frac{1}{\pi} \int_{-\pi}^\pi f(t)\cos(kt)\,dt = \frac{2}{\pi} \int_0^\pi f(t)\cos(kt)\,dt$$

and

$$b_k = \frac{1}{\pi} \int_{-\pi}^\pi f(t)\sin(kt)\,dt = 0$$

if f is even. If f is odd, then $a_k = 0$ and (7.13) gives the b_k. ∎

7.10 Examples (a) Define $f \in SC_{2\pi}$ by $f(t) = \text{sign}(t)$ for $t \in (-\pi, \pi)$.

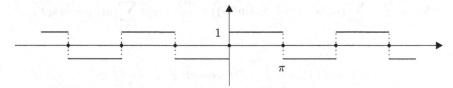

Then

$$Sf = \frac{4}{\pi} \sum_{k\geq 0} \frac{\sin\big((2k+1)\cdot\big)}{2k+1} .$$

The figure graphs a sequence of Fourier polynomials in $[-\pi, \pi]$:

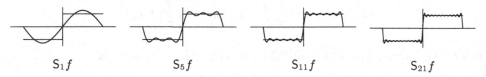

$S_1 f$ $S_5 f$ $S_{11} f$ $S_{21} f$

(On the basis of these graphs, we might think that Sf converges pointwise to f.)
Proof Because f is odd, and because

$$\frac{2}{\pi} \int_0^\pi f(t)\sin(kt)\,dt = \frac{2}{\pi} \int_0^\pi \sin(kt)\,dt = -\frac{2}{k\pi}\cos(kt)\Big|_0^\pi = \begin{cases} \dfrac{4}{k\pi}, & k \in 2\mathbb{N}+1, \\ 0, & k \in 2\mathbb{N}, \end{cases}$$

the claim follows from Remark 7.9(c). ∎

(b) Suppose $f \in C_{2\pi}$ such $f(t) = |t|$ for $|t| \leq \pi$. In other words,

$$f(t) = 2\pi\,\text{zigzag}(t/2\pi) \quad \text{for } t \in \mathbb{R} ,$$

(see Exercise III.1.1).

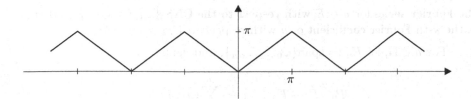

Then

$$Sf = \frac{\pi}{2} - \frac{4}{\pi} \sum_{k \geq 0} \frac{\cos\big((2k+1)\cdot\big)}{(2k+1)^2} \ .$$

This series converges normally on $\mathbb{R}$ (that is, it is norm convergent).

Proof Because f is even, and because integration by parts integration gives

$$a_k = \frac{2}{\pi} \int_0^\pi t\cos(kt)\,dt = -\frac{2}{\pi k} \int_0^\pi \sin(kt)\,dt = \frac{2}{\pi k^2}\big((-1)^k - 1\big) \ ,$$

for $k \geq 1$, the first claim follows from Remark 7.9(c). Now, from Example II.7.1(b), $\sum (2k+1)^{-2}$ is a convergent majorant for $\sum c_{2k+1}/(2k+1)^2$. The second claim then follows from the Weierstrass majorant criterion (Theorem V.1.6). ■

(c) Suppose $z \in \mathbb{C}\backslash\mathbb{Z}$, and $f_z \in C_{2\pi}$ is defined by $f_z(t) := \cos(zt)$ for $|t| \leq \pi$. Then

$$Sf_z = \frac{\sin(\pi z)}{\pi}\Big(\frac{1}{z} + \sum_{n \geq 1}(-1)^n\Big(\frac{1}{z+n} + \frac{1}{z-n}\Big)\cos(n\cdot)\Big) \ .$$

This series converges normally on $\mathbb{R}$.

Proof Since f is even, Sf_z is a cosine series. Using the first addition theorem from Proposition III.6.3(i), we find

$$a_n = \frac{2}{\pi} \int_0^\pi \cos(zt)\cos(nt)\,dt = \frac{1}{\pi}\int_0^\pi \Big(\cos\big((z+n)t\big) + \cos\big((z-n)t\big)\Big)\,dt$$

$$= (-1)^n\frac{\sin(\pi z)}{\pi}\Big(\frac{1}{z+n} + \frac{1}{z-n}\Big)$$

for $n \in \mathbb{N}$. Therefore Sf_z has the stated form.

For $n > 2\,|z|$, we have $|a_n| < |\sin(\pi z)|\,|z|/|z^2 - n^2| < 2\,|\sin(\pi z)|\,|z|/n^2$. The normal convergence again follows from the Weierstrass majorant criterion. ■

Bessel's inequality

Let $\big(E, (\cdot|\cdot)\big)$ be an inner product space, and let $\{\varphi_k \ ; \ k \in \mathbb{N}\}$ be an ONS in[5] E. In a generalization of the classical Fourier series, we call the series in E

$$\sum_k (u|\varphi_k)\varphi_k$$

---

[5]Note that this implies that E is infinite-dimensional (see Exercise II.3.10(a)).

the **Fourier series** for $u \in E$ **with respect to the ONS** $\{ \varphi_k \; ; \; k \in \mathbb{N} \}$, and $(u \,|\, \varphi_k)$ is the k-th **Fourier coefficient** of u with respect to $\{ \varphi_k \; ; \; k \in \mathbb{N} \}$.

For $n \in \mathbb{N}$, let $E_n := \operatorname{span}\{\varphi_0, \dots, \varphi_n\}$, and define

$$P_n : E \to E_n \; , \quad u \mapsto \sum_{k=0}^{n} (u \,|\, \varphi_k)\varphi_k \; .$$

The next theorem shows that $P_n u$ for each $u \in E$ yields the unique element of E_n that is closest in distance to u, and that $u - P_n u$ is perpendicular to E_n.[6]

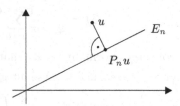

7.11 Proposition   For $u \in E$ and $n \in \mathbb{N}$, we have

 (i) $u - P_n u \in E_n^\perp$;

 (ii) $\|u - P_n u\| = \min_{v \in E_n} \|u - v\| = \operatorname{dist}(u, E_n)$ and

$$\|u - P_n u\| < \|u - v\| \quad \text{for } v \in E_n \text{ and } v \neq P_n u \; ;$$

(iii) $\|u - P_n u\|^2 = \|u\|^2 - \sum_{k=0}^{n} \big|(u \,|\, \varphi_k)\big|^2$.

Proof   (i) For $0 \le j \le n$, we find

$$(u - P_n u \,|\, \varphi_j) = (u \,|\, \varphi_j) - \sum_{k=0}^{n} (u \,|\, \varphi_k)(\varphi_k \,|\, \varphi_j) = (u \,|\, \varphi_j) - (u \,|\, \varphi_j) = 0 \; ,$$

because $(\varphi_k \,|\, \varphi_j) = \delta_{kj}$.

   (ii) Suppose $v \in E_n$. Because $P_n u - v \in E_n$, it follows then from (i) (see (II.3.7)) that

$$\|u - v\|^2 = \|(u - P_n u) + (P_n u - v)\|^2 = \|(u - P_n u)\|^2 + \|(P_n u - v)\|^2 \; .$$

Therefore $\|u - v\| > \|u - P_n u\|$ for every $v \in E_n$ with $v \neq P_n u$.

   (iii) Because

$$\|u - P_n u\|^2 = \|u\|^2 - 2\operatorname{Re}(u \,|\, P_n u) + \|P_n u\|^2 \; ,$$

the claim follows from $(\varphi_j \,|\, \varphi_k) = \delta_{jk}$. ∎

---

[6]See Exercise II.3.12.

7.12 Corollary (Bessel's inequality) *For every $u \in E$, the series of squares of Fourier coefficients converges, and*

$$\sum_{k=0}^{\infty} \left| (u \,|\, \varphi_k) \right|^2 \leq \|u\|^2 .$$

Proof From Proposition 7.11(iii), we get

$$\sum_{k=0}^{n} \left| (u \,|\, \varphi_k) \right|^2 \leq \|u\|^2 \quad \text{for } n \in \mathbb{N} .$$

Thus the claim is implied by Theorem II.7.7. ∎

7.13 Remarks (a) According to (7.11), the relation $(f \,|\, e_k)_2 = \sqrt{2\pi} \,\widehat{f}_k$ holds for $f \in SC_{2\pi}$ between the k-th Fourier coefficient of f with respect to the ONS $\{ e_k \;;\; k \in \mathbb{Z} \}$ and the classical k-th Fourier coefficient $\widehat{f}_k$. This difference in normalization is historical. (Of course, it is irrelevant that, in the classical case, the ONS is indexed with $k \in \mathbb{Z}$ rather than $k \in \mathbb{N}$.)

(b) For $n \in \mathbb{N}$, we have

$$P_n \in \mathcal{L}(E) , \quad P_n^2 = P_n , \quad \operatorname{im}(P_n) = E_n . \tag{7.14}$$

A linear vector space map A satisfying $A^2 = A$ is called a **projection**. Therefore P_n is a continuous linear projection from E onto E_n, and because $u - P_n u$ is orthogonal to E_n for every $u \in E$, P_n is an **orthogonal projection**. Proposition 7.11 then says every u has a uniquely determined closest element from E_n; that element is obtained by orthogonally projecting u onto E_n.

Proof We leave the simple proof of (7.14) to you. ∎

Complete orthonormal systems

The ONS $\{ \varphi_k \;;\; k \in \mathbb{N} \}$ in E is said to be **complete** or an **orthonormal basis** (ONB) of E if, for every $u \in E$, Bessel's inequality becomes the equality

$$\|u\|^2 = \sum_{k=0}^{\infty} \left| (u \,|\, \varphi_k) \right|^2 \quad \text{for } u \in E . \tag{7.15}$$

This statement is called a **completeness relation** or **Parseval's theorem**.

The following theorem clarifies the meaning of a complete orthonormal system.

7.14 Theorem   *The ONS $\{\varphi_k \ ; \ k \in \mathbb{N}\}$ in E is complete if and only if for every $u \in E$ the Fourier series $\sum(u_k \,|\, \varphi_k)\varphi_k$ converges to u, that is, if*

$$u = \sum_{k=0}^{\infty}(u \,|\, \varphi_k)\varphi_k \quad \text{for } u \in E \ .$$

Proof   According to Proposition 7.11(iii), $P_n u \to u$, and therefore

$$u = \lim_{n \to \infty} P_n u = \sum_{k=0}^{\infty}(u \,|\, \varphi_k)\varphi_k$$

if and only if Parseval's theorem holds. ∎

After these general considerations, which, besides giving the subject a geometrical interpretation, are also of theoretical and practical interest, we can return to the classical theory of Fourier series.

7.15 Theorem   *The functions $\{e_k \ ; \ k \in \mathbb{Z}\}$ form an ONB on $SC_{2\pi}$.*

Proof   Suppose $f \in SC_{2\pi}$ and $\varepsilon > 0$. Then Remark 7.9(a) supplies a $g \in C_{2\pi}$ such that $\|f - g\|_2 < \varepsilon/2$. Using the Weierstrass approximation theorem (Corollary V.4.17), we find an $n := n(\varepsilon)$ and a trigonometric polynomial T_n such that

$$\|g - T_n\|_\infty < \varepsilon/2\sqrt{2\pi} \ .$$

From Remark 7.2(a) and Proposition 7.11(ii) we then have

$$\|g - S_n g\|_2 \leq \|g - T_n\|_2 < \varepsilon/2 \ ,$$

and therefore

$$\|f - S_n f\|_2 \leq \|f - S_n g\|_2 \leq \|f - g\|_2 + \|g - S_n g\|_2 < \varepsilon \ ,$$

where, in the first step of the second row, we have used the minimum property, that is, Proposition 7.11(ii). Finally, Proposition 7.11(iii) gives

$$0 \leq \|f\|_2^2 - \sum_{k=0}^{m}\left|(f \,|\, e_k)_2\right|^2 \leq \|f\|_2^2 - \sum_{k=0}^{n}\left|(f \,|\, e_k)_2\right|^2 = \|f - S_n f\|_2^2 < \varepsilon$$

for $m \geq n$. This implies the completeness relation holds for every $f \in SC_{2\pi}$. ∎

7.16 Corollary   *For every $f \in SC_{2\pi}$, the Fourier series Sf converges in the quadratic mean to f, and Parseval's theorem reads*

$$\frac{1}{2\pi}\int_0^{2\pi}|f|^2\,dt = \sum_{k=-\infty}^{\infty}\left|\widehat{f}_k\right|^2 \ .$$

7.17 Remarks   **(a)** The real ONS $\{ c_0, c_k, s_k \ ; \ k \in \mathbb{N}^\times \}$ is an ONB in the space $SC_{2\pi}$. Defining Fourier coefficients by $a_k := a_k(f)$ and $b_k := b_k(f)$, we have

$$\frac{1}{\pi} \int_0^{2\pi} |f|^2 \, dt = \frac{a_0^2}{2} + \sum_{k=1}^\infty (a_k^2 + b_k^2)$$

for real-valued $f \in SC_{2\pi}$.

Proof   Example 7.5(b), Remark 7.9(b), and Euler's formula

$$\sqrt{2} \, (f \, | \, e_k) = (f \, | \, c_k) - i \, (f \, | \, s_k) \, ,$$

imply

$$2 \left| (f \, | \, e_k) \right|^2 = (f \, | \, c_k)^2 + (f \, | \, s_k)^2 = \pi (a_k^2 + b_k^2) \quad \text{for } k \in \mathbb{Z}^\times \, ,$$

having additionally used that $a_{-k} = a_k$ and $b_{-k} = -b_k$. ∎

(b) (Riemann's lemma)   For $f \in SC[0, 2\pi]$, we have

$$\int_0^{2\pi} f(t) \sin(kt) \, dt \to 0 \ \text{and} \ \int_0^{2\pi} f(t) \cos(kt) \, dt \to 0 \quad (k \to \infty) \, .$$

Proof   This follows immediately from the convergence of the series in (a). ∎

(c) Suppose $\ell_2(\mathbb{Z})$ is the set of all series $\boldsymbol{x} := (x_k)_{k \in \mathbb{Z}} \in \mathbb{C}^{\mathbb{Z}}$ such that

$$|\boldsymbol{x}|_2^2 := \sum_{k=-\infty}^\infty |x_k|^2 < \infty \, .$$

Then $\ell_2(\mathbb{Z})$ is a vector subspace of $\mathbb{C}^{\mathbb{Z}}$ and a Hilbert space with the scalar product $(\boldsymbol{x}, \boldsymbol{y}) \mapsto (\boldsymbol{x} \, | \, \boldsymbol{y})_2 := \sum_{k=-\infty}^\infty x_k \bar{y}_k$. Parseval's theorem then implies that the map

$$SC_{2\pi} \to \ell_2(\mathbb{Z}) \, , \quad f \mapsto \left(\sqrt{2\pi} \, \hat{f}_k \right)_{k \in \mathbb{Z}}$$

is a linear isometry. This isometry is however not surjective and therefore not an isometric isomorphism, as we shall see in Volume III in the context of the Lebesgue integration theory. Hence there is an orthogonal series $\sum_{k \in \mathbb{Z}} c_k e_k$ such that $\sum_{k=-\infty}^\infty |c_k|^2 < \infty$, which nevertheless does not converge to any function in $f \in SC_{2\pi}$. From this it also follows that $SC_{2\pi}$ — and therefore $SC[0, 2\pi]$ — is not complete and is therefore not a Hilbert space. In Volume III, we will remedy this defect by learning how to extend $SC[0, 2\pi]$ to a complete Hilbert space, called $L_2(0, 2\pi)$.

Proof   We leave the proof of some of these statements as Exercise 10. ∎

Parseval's theorem can be used to calculate the limit of various series, as we demonstrate in the following examples.[7]

---

[7]See also Application 7.23(a).

7.18 Examples **(a)**

$$\frac{\pi^2}{8} = \sum_{k=0}^{\infty} \frac{1}{(2k+1)^2} = 1 + \frac{1}{3^2} + \frac{1}{5^2} + \frac{1}{7^2} + \cdots$$

Proof This follows from Example 7.10(a) and Remark 7.17(a). ∎

(b) The series $\sum_{k \geq 0} 1/(2k+1)^4$ has the value $\pi^4/96$.

Proof This is a consequence of Example 7.10(b) and Remark 7.17(a). ∎

Piecewise continuously differentiable functions

We now turn to the question of the uniform convergence of Fourier series. To get a simple sufficient criterion, we must require more regularity of the functions we consider.

Suppose $J := [\alpha, \beta]$ is a compact perfect interval. We say $f \in \mathcal{SC}(J)$ is **piecewise continuously differentiable** (or has piecewise continuous derivatives) if there is a partition $(\alpha_0, \alpha_1, \ldots, \alpha_n)$ of J such that $f_j := f|(\alpha_j, \alpha_{j+1})$ for $0 \leq j \leq n-1$ has a uniformly continuous derivative.

7.19 Lemma *The function $f \in \mathcal{SC}(J)$ is piecewise continuously differentiable if and only if there is a partition $(\alpha_0, \ldots, \alpha_n)$ of J with these properties:*

(i) *$f|(\alpha_j, \alpha_{j+1}) \in C^1(\alpha_j, \alpha_{j+1})$ for $0 \leq j \leq n-1$.*

(ii) *For $0 \leq j \leq n-1$ and $1 \leq k \leq n$, the limits $f'(\alpha_j + 0)$ and $f'(\alpha_k - 0)$ exist.*

Proof "$\Rightarrow$" Because f is piecewise continuously differentiable, (i) follows by definition, and (ii) is a consequence of Theorem 2.1.

"$\Leftarrow$" According to Proposition III.2.24, if the partition satifies properties (i) and (ii), then $f_j' \in C(\alpha_j, \alpha_{j+1})$ has a continuous extension on $[\alpha_j, \alpha_{j+1}]$. Therefore Theorem III.3.13 implies f_j' is uniformly continuous. ∎

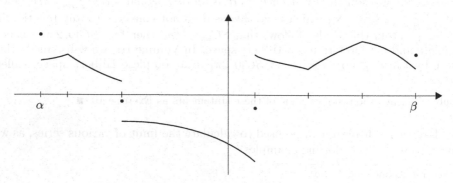

If $f \in \mathcal{SC}(J)$ is piecewise continuously differentiable, Lemma 7.19 guarantees the existence of a partition $(\alpha_0, \ldots, \alpha_n)$ of J and a unique normalized piecewise continuous function f', which we call the **normalized derivative**, such that

$$f' | (\alpha_{j-1}, \alpha_j) = [f | (\alpha_{j-1}, \alpha_j)]' \quad \text{for } 0 \le j \le n-1 .$$

Finally, we call $f \in SC_{2\pi}$ **piecewise continuously differentiable** if $f | [0, 2\pi]$ has these properties.

7.20 Remarks   (a) If $f \in SC_{2\pi}$ is piecewise continuously differentiable, then f' belongs to $SC_{2\pi}$.

Proof   This is a direct consequence of the definition of the normalization on the interval boundaries. ∎

(b) (derivative rule) If $f \in C_{2\pi}$ is piecewise continuously differentiable, then

$$\widehat{f'_k} = ik\widehat{f_k} \quad \text{for } k \in \mathbb{Z} ,$$

where $\widehat{f'_k} := \widehat{(f')_k}$.

Proof   Suppose $0 < \alpha_1 < \cdots < \alpha_{n-1} < 2\pi$ are all the discontinuities of f' in $(0, 2\pi)$. Then it follows from the additivity of integrals and from integration by parts (with $\alpha_0 := 0$ and $\alpha_n := 2\pi$) that

$$2\pi\widehat{f'_k} = \int_0^{2\pi} f'(t)e^{-ikt}\, dt = \sum_{j=0}^{n-1} \int_{\alpha_j}^{\alpha_{j+1}} f'(t)e^{-ikt}\, dt$$

$$= \sum_{j=0}^{n-1} f(t)e^{-ikt} \Big|_{\alpha_j}^{\alpha_{j+1}} + ik \sum_{j=0}^{n-1} \int_{\alpha_j}^{\alpha_{j+1}} f(t)e^{-ikt}\, dt$$

$$= ik \int_0^{2\pi} f(t)e^{-ikt}\, dt = 2\pi ik\widehat{f_k}$$

for $k \in \mathbb{Z}$, where the first sum in the third equality vanishes by the continuity of f and its 2π-periodicity. ∎

Uniform convergence

We can now prove the following simple criterion for the uniform and absolute convergence of Fourier series.

7.21 Theorem   *Suppose $f \colon \mathbb{R} \to \mathbb{C}$ is 2π-periodic, continuous, and piecewise continuously differentiable. The Fourier series $\mathsf{S}f$ in $\mathbb{R}$ converges normally to f.*

Proof   The Remarks 7.20(a) and (b), the Cauchy–Schwarz inequality for series[8]

---

[8]That is, the Hölder inequality for $p = 2$.

from Exercise IV.2.18, and the completeness relation implies

$$\sum_{\substack{k=-\infty \\ k\neq 0}}^{\infty} |\widehat{f_k}| = \sum_{\substack{k=-\infty \\ k\neq 0}}^{\infty} \frac{1}{k} |\widehat{f_k'}| \leq \left(2\sum_{k=1}^{\infty} \frac{1}{k^2}\right)^{1/2} \left(\sum_{k=-\infty}^{\infty} |\widehat{f_k'}|^2\right)^{1/2} \leq \sqrt{\frac{1}{\pi}\sum_{k=1}^{\infty}\frac{1}{k^2}} \, \|f'\|_2 \ .$$

Therefore, because $\left\| \widehat{f_k} e^{ik\cdot} \right\|_\infty = |\widehat{f_k}|$, Sf has the convergent majorant $\sum_{k\in\mathbb{Z}} |\widehat{f_k}|$. The Weierstrass majorant criterion of Theorem V.1.6 then implies the normal convergence of the Fourier series of f. We denote the value of Sf by g. Then g is continuous and 2π-periodic, and we have $\|S_n f - g\|_\infty \to 0$ for $n \to \infty$. Because the maximum norm is stronger than the L_2 norm, we have also $\lim \|S_n f - g\|_2 = 0$, that is, Sf converges in $SC_{2\pi}$ to g. According to Corollary 7.16, Sf converges in $SC_{2\pi}$ to f. Then it follows from the uniqueness of its limit that Sf converges normally to f. ∎

7.22 Examples   (a) For $|t| \leq \pi$, we have

$$|t| = \frac{\pi}{2} - \frac{4}{\pi}\sum_{k=0}^{\infty} \frac{\cos\big((2k+1)t\big)}{(2k+1)^2} \ ,$$

and the series converges normally on $\mathbb{R}$.

Proof   It follows from the previous theorem and Example 7.10(b) that the (normalized) derivative f' of the function $f \in C_{2\pi}$ with $f(t) = |t|$ for $|t| \leq \pi$ is given by map of Example 7.10(a). ∎

(b) (partial fraction decomposition of the cotangent)   For $z \in \mathbb{C}\backslash\mathbb{Z}$, we have

$$\pi\cot(\pi z) = \frac{1}{z} + \sum_{n=1}^{\infty}\left(\frac{1}{z+n} + \frac{1}{z-n}\right) . \tag{7.16}$$

Proof   From Theorem 7.21 and Example 7.10(c) we get

$$\cos(zt) = \frac{\sin(\pi z)}{\pi}\left(\frac{1}{z} + \sum_{n=1}^{\infty}(-1)^n\left(\frac{1}{z+n} + \frac{1}{z-n}\right)\cos(nt)\right)$$

for $|t| \leq \pi$. Finally, put $t = \pi$. ∎

The partial fraction decomposition of the cotangent has the following two interesting and beautiful consequences which, like the partial fraction decomposition itself, go back to Euler.

7.23 Applications   (a) (the Euler formula for $\zeta(2k)$)   For $k \in \mathbb{N}^\times$, we have

$$\zeta(2k) = \sum_{n=1}^{\infty} \frac{1}{n^{2k}} = \frac{(-1)^{k+1}(2\pi)^{2k}}{2(2k)!} B_{2k} \ .$$

In particular, we find

$$\zeta(2) = \pi^2/6 \ , \quad \zeta(4) = \pi^4/90 \ , \quad \zeta(6) = \pi^6/945 \ .$$

Proof   From (7.16), we get

$$\pi z \cot(\pi z) = 1 + 2z^2 \sum_{n=1}^{\infty} \frac{1}{z^2 - n^2} \quad \text{for } z \in \mathbb{C} \backslash \mathbb{Z} \ . \tag{7.17}$$

If $|z| \leq r < 1$, then $|n^2 - z^2| \geq n^2 - r^2 > 0$ for $n \in \mathbb{N}^\times$. This implies that the series in (7.17) converges normally on $r\mathbb{B}$. The geometric series gives

$$\frac{1}{z^2 - n^2} = -\frac{1}{n^2} \sum_{k=0}^{\infty} \left(\frac{z^2}{n^2}\right)^k \quad \text{for } z \in \mathbb{B} \text{ and } n \in \mathbb{N}^\times \ .$$

Thus we get from (7.17) that

$$\pi z \cot(\pi z) = 1 - 2z^2 \sum_{n=1}^{\infty} \frac{1}{n^2} \sum_{k=0}^{\infty} \left(\frac{z^2}{n^2}\right)^k = 1 - 2 \sum_{k=1}^{\infty} \left(\sum_{n=1}^{\infty} \frac{1}{n^{2k}}\right) z^{2k} \quad \text{for } z \in \mathbb{B} \ , \tag{7.18}$$

where, in the last step, we have exchanged the order of summation, which we can do because the double series theorem allows it (as you may verify).

Now we get the stated relation from Application 6.5 using the identity theorem for power series.[9] ∎

(b) (the product representation of the sine)   For $z \in \mathbb{C}$, we have

$$\sin(\pi z) = \pi z \prod_{n=1}^{\infty} \left(1 - \frac{z^2}{n^2}\right) \ .$$

Putting $z = 1/2$ here gives the Wallis product (see Example 5.5(d)).

Proof   We set

$$f(z) := \frac{\sin(\pi z)}{\pi z} \quad \text{for } z \in \mathbb{C}^\times \ ,$$

and $f(0) := 0$. Then it follows from the power series expansion of the sine that

$$f(z) = \sum_{k=0}^{\infty} (-1)^k \frac{\pi^{2k} z^{2k}}{(2k+1)!} \ . \tag{7.19}$$

The convergence radius of this power series is ∞. Then, according to Proposition V.3.5, f is analytic on $\mathbb{C}$.

As for the next, we set

$$g(z) := \prod_{n=1}^{\infty} \left(1 - \frac{z^2}{n^2}\right) \quad \text{for } z \in \mathbb{C} \ .$$

---

[9]This verifies the power series expansion in Application 6.5 for $|z| < 1$.

We want to show that g is also analytic on $\mathbb{C}$. So we fix $N \in \mathbb{N}^\times$ and $z \in N\mathbb{B}$ and then consider

$$\sum_{n=2N}^{m} \log\left(1 - \frac{z^2}{n^2}\right) = \log \prod_{n=2N}^{m} \left(1 - \frac{z^2}{n^2}\right) . \tag{7.20}$$

From the power series expansion of the logarithm (see Theorem V.3.9), we get

$$\left|\log\left(1 - \frac{z^2}{n^2}\right)\right| \le \sum_{k=1}^{\infty} \frac{|z|^{2k}}{kn^{2k}} \le \frac{|z|^2}{n^2} \frac{1}{1 - (z/n)^2} \le \frac{4}{3} \frac{N^2}{n^2}$$

for $z \in N\mathbb{B}$ and $n \ge 2N$. Therefore, by the Weierstrass majorant criterion, the series

$$\sum_{n \ge 2N} \log\left(1 - \frac{z^2}{n^2}\right)$$

converges absolutely and uniformly on $N\mathbb{B}$.

From (7.20) and the continuity of the exponential function, we can conclude that

$$\prod_{n=2N}^{\infty} \left(1 - \frac{z^2}{n^2}\right) = \lim_{m\to\infty} \prod_{n=2N}^{m} \left(1 - \frac{z^2}{n^2}\right)$$
$$= \lim_{m\to\infty} \exp \sum_{n=2N}^{m} \log\left(1 - \frac{z^2}{n^2}\right) = \exp \sum_{n=2N}^{\infty} \log\left(1 - \frac{z^2}{n^2}\right) , \tag{7.21}$$

and that the convergence is uniform on $N\mathbb{B}$. Consequently, the product

$$\prod_{n=1}^{\infty} \left(1 - \frac{z^2}{n^2}\right) = \prod_{n=1}^{2N-1} \left(1 - \frac{z^2}{n^2}\right) \prod_{n=2N}^{\infty} \left(1 - \frac{z^2}{n^2}\right) \tag{7.22}$$

also converges uniformly on $N\mathbb{B}$. Observing this holds for every $N \in \mathbb{N}^\times$, we see g is well defined on $\mathbb{C}$, and, using

$$g_m(z) := \prod_{n=1}^{m} \left(1 - \frac{z^2}{n^2}\right) \quad \text{for } z \in \mathbb{C} \text{ and } m \in \mathbb{N}^\times ,$$

we find that $g_m \to g$ locally uniformly on $\mathbb{C}$. Because the "partial product" g_m is analytic on $\mathbb{C}$, it then follows from the Weierstrass convergence theorem for analytic functions[10] (Theorem VIII.5.27) that g is also analytic on $\mathbb{C}$.

Because of the identity theorem for analytic functions, the proof will be complete if we can show that $f(x) = g(x)$ for $x \in J := (-1/\pi, 1/\pi)$.

From (7.19) and Corollary II.7.9 to the Leibniz criterion for alternating series, we get

$$|f(x) - 1| \le \pi^2 x^2/6 < 1 \quad \text{for } x \in J .$$

Therefore, by Theorem V.3.9, $\log f$ is analytic on J, and we find

$$(\log f)'(x) = \pi \cot(\pi x) - 1/x \quad \text{for } x \in J\setminus\{0\} .$$

---

[10]Naturally the proof of Theorem VIII.5.27 is independent of the product representation of the sine, so that we may anticipate this step.

From (7.17) we can read off

$$(\log f)'(x) = \sum_{n=1}^{\infty} \frac{2x}{x^2 - n^2} \qquad \text{for } x \in J \backslash \{0\} \ , \tag{7.23}$$

where the right hand series converges normally for every $r \in (0,1)$ in $[-r, r]$. In particular, it follows that (7.23) is valid for all $x \in J$.

From formulas (7.21) and (7.22), we learn the relation

$$\log g(x) = \sum_{n=1}^{\infty} \log\left(1 - \frac{x^2}{n^2}\right) \quad \text{for } x \in J \ .$$

Now it follows from Corollary V.2.9 (the differentiability of the limit of a series of functions) that $\log g$ is differentiable on J and that

$$(\log g)'(x) = \sum_{n=1}^{\infty} \frac{2x}{x^2 - n^2} \quad \text{for } x \in J \ .$$

Consequently, $(\log f)' = (\log g)'$ on J. Finally, because $\log f(0) = 0 = \log g(0)$ and from the uniqueness theorem for antiderivatives (Remark V.3.7(a)), we get $\log f = \log g$ on the entire J. ∎

We have studied criteria governing whether Fourier series converge uniformly or in the quadratic mean. It is also natural to inquire about their pointwise convergence. On that subject, there are a multitude of results and simple, sufficient criteria coming from classical analysis. Some of the simplest of these criteria are presented in common textbooks (see for example [BF87], [Bla91], [Kön92], [Wal92]). For example, it is easy to verify that the Fourier series of Example 7.10(a) converges to the given "square wave" function.

We will not elaborate here, having confined our attention to the essential results of classical Fourier series in one variable. Of perhaps greater significance is the L_2 theory of Fourier series, which, as we have attempted to show above, is very general. It finds uses in many problems in mathematics and physics — for example, in the theory of differential equations in one and more variables — and plays an extremely important role in modern analysis and its applications. You will likely meet the Hilbert space theory of orthogonality in further studies of analysis and its applications, and we also will give an occasional further look at this theory.

Exercises

1 Suppose $f \in SC_{2\pi}$ has $\int_0^{2\pi} f = 0$. Show that there is a $\xi \in [0, 2\pi]$ such that $f(\xi) = 0$.

2 Verify that for $f \in C_{2\pi}$ defined by $f(t) := |\sin(t)|$ on $t \in [0, 2\pi]$, we have

$$Sf = \frac{2}{\pi} - \frac{4}{\pi} \sum_{k=1}^{\infty} \frac{\cos(k\cdot)}{4k^2 - 1} \ .$$

3   For $n \in \mathbb{N}$,

$$D_n := \sqrt{2\pi} \sum_{k=-n}^{n} e_k = 1 + 2\cos + 2\cos(2\cdot) + \cdots + 2\cos(n\cdot)$$

is called the **Dirichlet kernel** of degree n. Show that

$$D_n(t) = \frac{\sin\big((n+1/2)t\big)}{\sin(t/2)} \quad \text{for } t \in \mathbb{R} \qquad \text{and} \qquad \int_0^{2\pi} D_n(t)\,dt = 1 \text{ for } n \in \mathbb{N}.$$

4   Define $f \in SC_{2\pi}$ by

$$f(t) := \begin{cases} \dfrac{\pi - t}{2}, & 0 < t < 2\pi, \\ 0, & t = 0. \end{cases}$$

Prove that

$$Sf = \sum_{k=1}^{\infty} \frac{\sin(k\cdot)}{k}.$$

(Hints: $I_n(t) := \int_\pi^t D_n(s)\,ds = (t-\pi) + 2(\sin t + \cdots + \sin(nt)/n)$ for $t \in (0, 2\pi)$. Exercise 3 and integration by parts result in $|I_n(t)| \le 2\big[(n+1/2)\sin(t/2)\big]^{-1}$ for $t \in (0, 2\pi)$.)

5   Show that the 1-periodic extension of $\widetilde{B}_n$ from $B_n(X)\,|\,[0,1)$ to $\mathbb{R}$ has the following Fourier expansion:

$$\widetilde{B}_{2n} = 2\frac{(-1)^{n-1}(2n)!}{(2\pi)^{2n}} \sum_{k=1}^{\infty} \frac{\cos(2\pi k\cdot)}{k^{2n}} \qquad \text{for } n \in \mathbb{N}^{\times}, \tag{7.24}$$

$$\widetilde{B}_{2n+1} = 2\frac{(-1)^{n-1}(2n+1)!}{(2\pi)^{2n+1}} \sum_{k=1}^{\infty} \frac{\sin(2\pi k\cdot)}{k^{2n+1}} \qquad \text{for } n \in \mathbb{N}. \tag{7.25}$$

(Hints: It suffices to prove (7.24) and (7.25) on $[0,1]$. For the restrictions U_{2n} and U_{2n+1} of the series in (7.24) and (7.25), respectively, to $[0,1]$, show that $U'_{m+1} = (m+1)U_m$ for $m \in \mathbb{N}^{\times}$. From Exercise 4, it follows that $U_1(t) = B_1(t)$. Thus Proposition 6.6(iii) shows that for $m \ge 2$, there is a constant c_m such that $U_m(t) = B_m(t) + c_m$ for $t \in [0,1]$. Finally verify $\int_0^1 U_m(t)\,dt = 0$ (Proposition 4.1(ii)) and $\int_0^1 B_m(t)\,dt = 0$ (Exercise 6.4(iii)). Then it follows that $c_m = 0$ for $m \ge 2$.)

6   Verify the asymptotic equivalences

$$|B_{2n}| \sim 2\frac{(2n)!}{(2\pi)^{2n}} \sim 4\sqrt{\pi n}\Big(\frac{n}{\pi e}\Big)^{2n}.$$

(Hints: Exercise 5 shows

$$B_{2n} = \widetilde{B}_{2n}(0) = (-1)^{n-1}2\frac{(2n)!}{(2\pi)^{2n}} \sum_{k=1}^{\infty} \frac{1}{k^{2n}}.$$

Further note Exercise 6.6, Application 7.23(a) and Exercise 6.3(a).)

7 Prove **Wirtinger's inequality**: For any $-\infty < a < b < \infty$ and $f \in C^1([a,b])$ such that $f(a) = f(b) = 0$, we have

$$\int_a^b |f|^2 \le [(b-a)^2/\pi^2] \int_a^b |f'|^2 .$$

The constant $(b-a)^2/\pi^2$ cannot be made smaller.
(Hint: Using the substitution $x \mapsto \pi(x-a)/(b-a)$, it suffices to consider $a = 0$ and $b = \pi$. For $g : [-\pi, \pi] \to \mathbb{K}$ such that $g(x) := f(x)$ for $x \in [0, \pi]$ and $g(x) := -f(-x)$ for $x \in [-\pi, 0]$, show that $g \in C^1([-\pi, \pi])$ and that g is odd. Thus it follows from Parseval's theorem and Remark 7.20(b) that

$$\frac{1}{2\pi}\int_{-\pi}^{\pi}|g'|^2 = \sum_{k\in\mathbb{Z}}|\widehat{g'_k}|^2 \ge \sum_{k\neq 0}|\widehat{g'_k}|^2 = \sum_{k\neq 0}|ik\widehat{g_k}|^2 \ge \sum_{k\neq 0}|\widehat{g_k}|^2 = \frac{1}{2\pi}\int_{-\pi}^{\pi}|g|^2$$

because $\widehat{g}(0) = 0$. The claim then follows from the construction of g.)

8 For $f, g \in SC_{2\pi}$,

$$f * g : \mathbb{R} \to \mathbb{K}, \quad x \mapsto \frac{1}{2\pi}\int_0^{2\pi} f(x-y)g(y)\, dy$$

is called the **convolution** of f with g. Show that
  (i) $f * g = g * f$;
  (ii) if $f, g \in C_{2\pi}$, then $f * g \in C_{2\pi}$;
  (iii) $e_n * e_m = \delta_{nm}e_n$ for $m, n \in \mathbb{Z}$.

9 Denote by D_n the n-th Dirichlet kernel. Show that if $f \in SC_{2\pi}$, then $S_n f = D_n * f$.

10 Verify the statements of Remark 7.17(c):
  (i) $\ell_2(\mathbb{Z})$ is a Hilbert space;
  (ii) $SC_{2\pi} \to \ell_2(\mathbb{Z})$, $f \mapsto \left(\sqrt{2\pi}\,\widehat{f_k}\right)_{k\in\mathbb{Z}}$ is a linear isometry.

11 Prove the **generalized Parseval's theorem**

$$\frac{1}{2\pi}\int_0^{2\pi} f\bar{g}\, dt = \sum_{k=-\infty}^{\infty} \widehat{f_k}\overline{\widehat{g_k}} \quad \text{for } f, g \in SC_{2\pi} .$$

(Hint: Note that

$$z\bar{w} = \frac{1}{4}\left(|z+w|^2 - |z-w|^2 + i\,|z+iw|^2 - i\,|z-iw|^2\right) \quad \text{for } z, w \in \mathbb{C},$$

and apply Corollary 7.16.)

8 Improper integrals

Until now, we have only been able to integrate jump continuous functions on *compact* intervals. However, we have seen already in Lemma 6.12 that it is meaningful to extend the concept of integration to the case of continuous functions defined on noncompact intervals. Considering the area of a set F that lies between the graph of a continuous function $f : \mathbb{R}^+ \to \mathbb{R}^+$ and the positive half axis, we will show it is finite if f goes to zero sufficiently quickly as $x \to \infty$.

To calculate F, we first calculate the area of F in the interval $[0, n]$, that is, we truncate the area at $x = n$, and we then let n go to ∞. The following development will refine this idea and extend the scope of integration to functions defined on an arbitrary interval.[1]

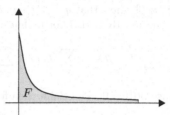

In this section, we assume

- J is an interval with $\inf J = a$ and $\sup J = b$;
  $E := (E, |\cdot|)$ is a Banach space.

Admissible functions

The map $f : J \to E$ is said to be **admissible** if its restriction to every compact interval J is jump continuous.

8.1 Examples (a) Every $f \in C(J, E)$ is admissible.

(b) If a and b are finite, then every $f \in \mathcal{S}([a, b], E)$ is admissible.

(c) If $f : J \to E$ is admissible, then so is $|f| : J \to \mathbb{R}$. ∎

Improper integrals

An admissible function $f : J \to E$ is said to be (**improperly**) **integrable**, if there is $c \in (a, b)$ for which the limits

$$\lim_{\alpha \to a+0} \int_\alpha^c f \quad \text{and} \quad \lim_{\beta \to b-0} \int_c^\beta f$$

exist in E.

---

[1]We confine our presentation in this section to a relatively elementary generalization of integration, saving the complete integration theory for Volume III.

8.2 Lemma *If $f \colon J \to E$ improperly integrable, then the limits*

$$\lim_{\alpha \to a+0} \int_\alpha^c f \quad \text{and} \quad \lim_{\beta \to b-0} \int_c^\beta f$$

exist for every $c \in (a, b)$. In addition,

$$\lim_{\alpha \to a+0} \int_\alpha^c f + \lim_{\beta \to b-0} \int_c^\beta f = \lim_{\alpha \to a+0} \int_\alpha^{c'} f + \lim_{\beta \to b-0} \int_{c'}^\beta f$$

for every choice of $c, c' \in (a, b)$.

Proof By assumption, there is a $c \in (a, b)$ such that

$$e_{a,c} := \lim_{\alpha \to a+0} \int_\alpha^c f \quad \text{and} \quad e_{c,b} := \lim_{\beta \to b-0} \int_c^\beta f$$

exist in E. Suppose now $c' \in (a, b)$. Because of Theorem 4.4, $\int_\alpha^{c'} f = \int_\alpha^c f + \int_c^{c'} f$ for every $\alpha \in (a, b)$. Thus, the limit $e_{a,c'} := \lim_{\alpha \to a+0} \int_\alpha^{c'} f$ exists in E and is given by $e_{a,c'} = e_{a,c} + \int_c^{c'} f$. It follows likewise that

$$e_{c',b} := \lim_{\beta \to b-0} \int_{c'}^\beta f$$

exists in E with $e_{c',b} = e_{c,b} + \int_{c'}^c f$. Therefore we find

$$e_{a,c'} + e_{c',b} = e_{a,c} + \int_c^{c'} f + e_{c,b} + \int_{c'}^c f = e_{a,c} + e_{c,b} \;. \; \blacksquare$$

Suppose $f \colon J \to E$ is improperly integrable and $c \in (a, b)$. We then call

$$\int_a^b f\, dx := \int_a^b f(x)\, dx := \lim_{\alpha \to a+0} \int_\alpha^c f + \lim_{\beta \to b-0} \int_c^\beta f$$

the (**improper**) **integral** of f over J. We may also use the notations

$$\int_a^b f \quad \text{and} \quad \lim_{\substack{\alpha \to a+0 \\ \beta \to b-0}} \int_\alpha^\beta f\, dx \;.$$

Instead of saying "f is improperly integrable" we may also say, "The integral $\int_a^b f\, dx$ **exists** or **converges**." Lemma 8.2 shows that the improper integral is well defined, that is, it is independence of the choice of $c \in (a, b)$.

8.3 Proposition *For $-\infty < a < b < \infty$ and $f \in \mathcal{S}([a,b], E)$, the improper integral coincides with the Cauchy–Riemann integral.*

Proof This follows from Remark 8.1(b), the continuity of integrals as functions of their limits, and Proposition 4.4. ∎

8.4 Examples **(a)** Suppose $a > 0$ and $s \in \mathbb{C}$. Then

$$\int_a^\infty \frac{dx}{x^s} \text{ exists} \iff \operatorname{Re} s > 1 \ ,$$

and

$$\int_a^\infty \frac{dx}{x^s} = \frac{a^{1-s}}{s-1} \quad \text{for } \operatorname{Re} s > 1 \ .$$

Proof According to Remark 8.1(a), the function $(a,\infty) \to \mathbb{C}$ $x \mapsto 1/x^s$ is admissible for every $s \in \mathbb{C}$.

(i) We first consider the case $s \neq 1$. Then

$$\int_a^\beta \frac{dx}{x^s} = \frac{1}{1-s} x^{1-s} \Big|_a^\beta = \frac{1}{1-s} \left(\frac{1}{\beta^{s-1}} - \frac{1}{a^{s-1}} \right) \ .$$

From $|\beta^{1-s}| = \beta^{1-\operatorname{Re} s}$, it follows that

$$\frac{1}{\beta^{s-1}} \to 0 \ (\beta \to \infty) \iff \operatorname{Re} s > 1$$

and that $\lim_{\beta \to \infty} 1/\beta^{s-1}$ does not exist for $\operatorname{Re} s < 1$. Finally, the limit does not exist if $\operatorname{Re} s = 1$ because $s = 1 + i\tau$ for $\tau \in \mathbb{R}^\times$ implies $\beta^{s-1} = \beta^{i\tau} = e^{i\tau \log \beta}$ and because the exponential function is $2\pi i$-periodic.

(ii) If $s = 1$ then

$$\lim_{\beta \to \infty} \int_a^\beta \frac{dx}{x} = \lim_{\beta \to \infty} (\log \beta - \log a) = \infty \ .$$

Therefore the function $x \mapsto 1/x$ is not integrable on (a, ∞). ∎

(b) Suppose $b > 0$ and $s \in \mathbb{C}$. Then

$$\int_0^b \frac{dx}{x^s} \text{ exists} \iff \operatorname{Re} s < 1 \ ,$$

and $\int_0^b x^{-s}\,dx = b^{1-s}/(1-s)$ for $\operatorname{Re} s < 1$.

Proof The proof is just like the proof of (a). ∎

(c) The integral $\int_0^\infty x^s\,dx$ does not exist for any $s \in \mathbb{C}$.

(d) The integral

$$\int_{-\infty}^{\infty} x \, dx = \lim_{\substack{\alpha \to -\infty \\ \beta \to \infty}} \int_{\alpha}^{\beta} x \, dx = \lim_{\substack{\alpha \to -\infty \\ \beta \to \infty}} \frac{x^2}{2} \Big|_{\alpha}^{\beta}$$

does not exist.[2]

(e) We have

$$\int_0^{\infty} \frac{dx}{1 + x^2} = \frac{\pi}{2} \quad \text{and} \quad \int_{-1}^{1} \frac{dx}{\sqrt{1 - x^2}} = \pi .$$

Proof   This follows because

$$\lim_{\beta \to \infty} \arctan \Big|_0^{\beta} = \frac{\pi}{2} \quad \text{and} \quad \lim_{\substack{\alpha \to -1+0 \\ \beta \to 1-0}} \arcsin \Big|_{\alpha}^{\beta} = \pi ,$$

where we have used Example 4.15(a). ∎

The integral comparison test for series

In light of Section 6, we anticipate a connection between infinite series and improper integrals.[3] The following theorem produces such a relationship in an important case.

8.5 Proposition   *Suppose $f \colon [1, \infty) \to \mathbb{R}^+$ is admissible and decreasing. Then*

$$\sum_{n \geq 1} f(n) < \infty \iff \int_1^{\infty} f(x) \, dx \text{ exists } .$$

Proof   (i) The assumption implies

$$f(n) \leq f(x) \leq f(n - 1)$$

for $x \in [n - 1, n]$ and $n \in \mathbb{N}$ with $n \geq 2$. Therefore we have

$$f(n) \leq \int_{n-1}^{n} f(x) \, dx \leq f(n - 1) ,$$

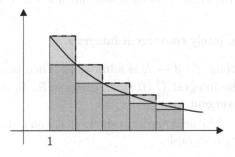

and we find after summing over n that

$$\sum_{n=2}^{N} f(n) \leq \int_1^{N} f(x) \, dx \leq \sum_{n=1}^{N-1} f(n) \quad \text{for } N \geq 2 . \tag{8.1}$$

---

[2]Note that $\int_{-\infty}^{\infty} x \, dx$ does not exist, even though $\lim_{\gamma \to \infty} \int_{-\gamma}^{\gamma} x \, dx = 0$! It important when verifying the convergence of an integral to check that the upper and lower limits can be passed independently.

[3]See also the proof of Remark 6.16(d).

(ii) The series $\sum f(n)$ converges. Then the bound

$$\int_1^N f\,dx \leq \sum_{n=1}^{\infty} f(n) \quad \text{for } N \geq 2 \;.$$

follows from (8.1) and $f \geq 0$. Therefore the function $\beta \mapsto \int_1^\beta f(x)\,dx$ is increasing and bounded. Then $\int_1^\infty f\,dx$ by Proposition III.5.3.

(iii) The integral $\int_1^\infty f\,dx$ converges. Then it follows from (8.1) that

$$\sum_{n=2}^N f(n) \leq \int_1^\infty f(x)\,dx \quad \text{for } N \geq 2 \;.$$

Therefore Theorem II.7.7 implies $\sum_{n\geq 1} f(n)$ exists also. ∎

8.6 Example   For $s \in \mathbb{R}$, the series

$$\sum_{n\geq 2} \frac{1}{n(\log n)^s}$$

converges if and only if $s > 1$.

Proof   If $s \leq 0$, then $\sum 1/n$ is a divergent minorant. It therefore suffices to consider the case $s > 0$. Because

$$\int_2^\beta \frac{dx}{x(\log x)^s} = \begin{cases} \left[(\log \beta)^{1-s} - (\log 2)^{1-s}\right]/(1-s)\,, & s \neq 1\,, \\ \log(\log \beta) - \log(\log 2)\,, & s = 1\,, \end{cases}$$

$\int_2^\infty dx/x(\log x)^s$ exists if and only if $s > 1$. Then Proposition 8.5 finishes the proof. ∎

Absolutely convergent integrals

Suppose $f : J \to E$ is admissible. Then we say f is **absolutely integrable** (over J) if the integral $\int_a^b |f(x)|\,dx$ exists in $\mathbb{R}$. In this case, we also say $\int_a^b f$ is **absolutely convergent**.

The next theorem shows that every absolutely integrable function is (improperly) integrable.

8.7 Proposition   *If f is absolutely integrable, then the integral $\int_a^b f\,dx$ exists in E.*

Proof   Suppose $c \in (a,b)$. Because $\int_a^b |f|\,dx$ exists in $\mathbb{R}$, there is for every $\varepsilon > 0$ a $\delta > 0$ such that

$$\left| \int_{\alpha_1}^{\alpha_2} |f| \right| = \left| \int_{\alpha_1}^{c} |f| - \int_{\alpha_2}^{c} |f| \right| < \varepsilon \quad \text{for } \alpha_1, \alpha_2 \in (a, a+\delta) \;.$$

Proposition 4.3 then implies

$$\left|\int_{\alpha_1}^{c} f - \int_{\alpha_2}^{c} f\right| = \left|\int_{\alpha_1}^{\alpha_2} f\right| \le \left|\int_{\alpha_1}^{\alpha_2} |f|\right| < \varepsilon \quad \text{for } \alpha_1, \alpha_2 \in (a, a + \delta) \ . \tag{8.2}$$

Suppose now (α'_j) is a sequence in (a, c) such that $\lim \alpha'_j = a$. Due to (8.2), $(\int_{\alpha'_j}^{c} f \, dx)$ is then a Cauchy sequence in E. Hence there is an $e' \in E$ such that $\int_{\alpha'_j}^{c} f \, dx \to e'$ as $j \to \infty$. If (α''_j) is another sequence in (a, c) such then $\lim \alpha''_j = a$, then, analogously, there is an $e'' \in E$ such that $\lim_j \int_{\alpha''_j}^{c} f \, dx = e''$ in E.

Now we choose $N \in \mathbb{N}$ with $\alpha'_j, \alpha''_j \in (a, a + \delta)$ for $j \ge N$. From (8.2), we then get $\left|\int_{\alpha'_j}^{c} f - \int_{\alpha''_j}^{c} f\right| < \varepsilon$ for $j \ge N$, after which taking the limit $j \to \infty$ gives the inequality $|e' - e''| \le \varepsilon$. Because this holds for every $\varepsilon > 0$, it follows that $e' = e''$, and we have proved that $\lim_{\alpha \to a+0} \int_{\alpha}^{c} f \, dx$ exists in E. Analogously, one shows that $\lim_{\beta \to b-0} \int_{c}^{b} f \, dx$ exists in E. Therefore the improper integral $\int_{a}^{b} f \, dx$ exists in E. ∎

The majorant criterion

Similarly to the situation with series, the existence of an integrable majorant of f secures the absolute convergence of $\int_{a}^{b} f \, dx$.

8.8 Theorem *Suppose $f : J \to E$ and $g : J \to \mathbb{R}^+$ is admissible with*

$$|f(x)| \le g(x) \quad \text{for } x \in (a, b) \ .$$

If g is integrable, then f is absolutely integrable.

Proof Suppose $c \in (a, b)$ and $\alpha_1, \alpha_2 \in (a, c)$. Then using Corollary 4.6, we have

$$\left|\int_{\alpha_1}^{c} |f| - \int_{\alpha_2}^{c} |f|\right| = \left|\int_{\alpha_1}^{\alpha_2} |f|\right| \le \left|\int_{\alpha_1}^{\alpha_2} g\right| = \left|\int_{\alpha_1}^{c} g - \int_{\alpha_2}^{c} g\right| \ .$$

If $\int_{a}^{b} g \, dx$ exists in $\mathbb{R}$, there is for every $\varepsilon > 0$ a $\delta > 0$ such that $\left|\int_{\alpha_1}^{c} g - \int_{\alpha_2}^{c} g\right| < \varepsilon$ for $\alpha_1, \alpha_2 \in (a, a + \delta)$, and we get

$$\left|\int_{\alpha_1}^{c} |f| - \int_{\alpha_2}^{c} |f|\right| < \varepsilon \quad \text{for } \alpha_1, \alpha_2 \in (a, a + \delta) \ .$$

This statement makes possible a corresponding reiteration of the proof Theorem 8.7, from which the absolute convergence of $\int_{a}^{b} f \, dx$ follows. ∎

8.9 Examples Suppose $f : (a, b) \to E$ is admissible.

(a) If f is real-valued with $f \geq 0$, then f is absolutely integrable if and only if f is integrable.

(b) (i) Suppose $a > 0$ and $b = \infty$. If there are numbers $\varepsilon > 0$, $M > 0$, and $c > a$ such that

$$|f(x)| \leq \frac{M}{x^{1+\varepsilon}} \quad \text{for } x \geq c ,$$

then $\int_c^\infty f$ is absolutely convergent.

  (ii) Suppose $a = 0$ and $b > 0$. If there are numbers $\varepsilon > 0$, $M > 0$ and $c \in (0, b)$ such that

$$|f(x)| \leq \frac{M}{x^{1-\varepsilon}} \quad \text{for } x \in (0, c) ,$$

then $\int_0^c f$ is absolutely convergent.

Proof   This follows from Theorem 8.8 and Examples 8.4(a) and (b). ∎

(c) The integral $\int_0^\infty \left(\sin(x)/(1 + x^2) \right) dx$ converges absolutely.

Proof   Obviously, $x \mapsto 1/(1 + x^2)$ is a majorant of $x \mapsto \sin(x)/(1 + x^2)$. In addition, Example 8.4(e) implies the integral $\int_0^\infty dx/(1 + x^2)$ exists. ∎

8.10 Remarks   **(a)** Suppose $f_n, f \in \mathcal{S}([a, b], E)$ and (f_n) converges uniformly to f. Then, we have proved in Theorem 4.1 that $\left(\int_a^b f_n \right)_{n \in \mathbb{N}}$ converges in E to $\int_a^b f$. The analogous statement does not hold for improper integrals.

Proof   We consider the sequence of functions

$$f_n(x) := \frac{1}{n} e^{-x/n} \quad \text{for } x \in \mathbb{R}^+ \text{ and } n \in \mathbb{N}^\times .$$

Then every f_n belongs to $C(\mathbb{R}^+, \mathbb{R})$. In addition, the sequence (f_n) converges uniformly to 0, since $\|f_n\|_\infty = 1/n$.

  On the other hand, we have

$$\int_0^\infty f_n(x)\, dx = \int_0^\infty \frac{1}{n} e^{-x/n}\, dx = \lim_{\substack{\alpha \to 0+ \\ \beta \to \infty}} \left(-e^{-x/n} \right) \big|_\alpha^\beta = 1 .$$

Altogether, it follows that $\lim_n \int_0^\infty f_n\, dx = 1$, but $\int_0^\infty \lim_n f_n\, dx = 0$. ∎

(b) When a sequence of functions is also the integrand in an improper integral, one can inquire whether it is valid to exchange the order of the corresponding limits. We refrain from this now and instead develop the results in Volume III as part of the more general *Lebesgue integration theory*, to which the necessary analysis is better suited than the (simpler) Cauchy–Riemann integration theory. In the framework of Lebesgue theory, we will find a very general and flexible criterion for when taking limits can be exchanged with integration. ∎

Exercises

1 Prove the convergence of these improper integrals:

(i) $\displaystyle\int_1^\infty \frac{\sin x}{x^2}\,dx$,   (ii) $\displaystyle\int_0^\infty \frac{\sin x}{x}\,dx$,   (iii) $\displaystyle\int_1^\infty \frac{\sin(1/x)}{x}\,dx$.

In which cases do the integrals converge absolutely?

2 Which of these improper integrals have a value?

(i) $\displaystyle\int_0^1 \frac{\arcsin x}{\sqrt{1-x^2}}\,dx$,   (ii) $\displaystyle\int_1^\infty \frac{\log x}{x^2}\,dx$,   (iii) $\displaystyle\int_0^\infty \frac{\log x}{1+x^2}\,dx$,   (iv) $\displaystyle\int_0^\infty x^n e^{-x}\,dx$.

(Hint for (iii): Consider $\int_1^\infty$ and $\int_0^1$.)

3 Show that $\int_0^\pi \log \sin x\,dx = -\pi \log 2$. (Hint: $\sin 2x = 2\sin x \cos x$.)

4 Suppose $-\infty < a < b \leq \infty$ and $f : [a,b) \to E$ is admissible. Show then that $\int_a^b f$ exists if and only if for every $\varepsilon > 0$ there is a $c \in [a,b)$ such that $\left|\int_\alpha^\beta f\right| < \varepsilon$ for $\alpha, \beta \in [c,b)$.

5 Show that the integral $\int_0^\infty \sqrt{t}\cos(t^2)\,dt$ converges.[4]

6 Suppose $-\infty < a < b \leq \infty$, and $f : [a,b) \to \mathbb{R}$ is admissible.

(i) If $f \geq 0$, show $\int_a^b f(x)\,dx$ if and only if $K := \sup_{c \in [a,b)} \left|\int_a^c f\right| < \infty$.

(ii) Suppose $f \in C([a,b),\mathbb{R})$ and $g \in C^1([a,b),\mathbb{R})$. Show that if $K < \infty$ and $g(x)$ tends monotonically to 0 as $x \to b-0$, then $\int_a^b fg$ converges.

7 Suppose that $f \in C^1([a,\infty),\mathbb{R})$ with $a \in \mathbb{R}$, and suppose that f' is increasing with $\lim_{x\to\infty} f'(x) = \infty$. Prove that $\int_a^\infty \sin(f(x))\,dx$ converges.
(Hint: Substitute $y = f(x)$ and recall Theorem III.5.7.)

8 The function $f \in C([0,\infty),\mathbb{R})$ satisfies $\sup_{0<a<b<\infty}\left|\int_a^b f\right| < \infty$. Show then that

$$\int_0^\infty \frac{f(ax) - f(bx)}{x}\,dx = f(0)\log\left(\frac{a}{b}\right) \quad \text{for } 0 < a < b.$$

(Hint: Let $c > 0$. The existence of $\xi(c) \in [ac,bc]$ such that

$$\int_c^\infty \frac{f(ax) - f(bx)}{x}\,dx = \int_{ac}^{bc} \frac{f(x)}{x}\,dx = f(\xi(c))\log\left(\frac{a}{b}\right)$$

follows from Theorem 4.16.)

9 Suppose $-\infty < a < 0 < b < \infty$, and $f : [a,0) \cup (0,b] \to \mathbb{R}$ is admissible. If

$$\lim_{\varepsilon \to 0}\left(\int_a^{-\varepsilon} f(x)\,dx + \int_\varepsilon^b f(x)\,dx\right)$$

exists in $\mathbb{R}$, then this limit is called the **Cauchy principal value** of $\int_a^b f$, and we write[5] $PV \int_a^b f$. Compute

$$PV \int_{-1}^1 \frac{dx}{x} \ , \quad PV \int_{-\pi/2}^{\pi/2} \frac{dx}{\sin x} \ , \quad PV \int_{-1}^1 \frac{dx}{x(6 + x - x^2)} \ .$$

---

[4] This exercise shows in particular that the convergence of $\int_0^\infty f$ does not allow one to conclude that $f(x) \to 0$ for $x \to \infty$. Compare this result to Theorem II.7.2.

[5] PV stands for "principal value".

9   The gamma function

In this section, which closes for the time being the theory of integration, we will
study one of the most important nonelementary functions of mathematics, namely,
the gamma function. We will derive its essential properties and, in particular, show
that it interpolates the factorial $n \mapsto n!$, that is, it allows this function to take
arguments that are not whole numbers. In addition, we will prove a refinement
of the de Moivre–Stirling asymptotic form of $n!$ and — as a by-product of the
general theory — calculate the value of an important improper integral, namely,
the Gaussian error integral.

Euler's integral representation

We introduce the gamma function for $z \in \mathbb{C}$ with $\operatorname{Re} z > 0$ by considering a
parameter-dependent improper integral.

9.1 Lemma   *For $z \in \mathbb{C}$ such that $\operatorname{Re} z > 0$, the integral*

$$\int_0^\infty t^{z-1} e^{-t}\, dt$$

converges absolutely.

Proof   First, we remark that the function $t \mapsto t^{z-1} e^{-t}$ on $(0, \infty)$ is admissible.
(i) We consider the integral on its lower limit. For $t \in (0, 1]$, we have

$$\left| t^{z-1} e^{-t} \right| = t^{\operatorname{Re} z - 1} e^{-t} \le t^{\operatorname{Re} z - 1} .$$

From Example 8.9(b), it then follows that $\int_0^1 t^{z-1} e^{-t}\, dt$ is absolutely convergent.
(ii) For $m \in \mathbb{N}^\times$, we have the estimate

$$\frac{t^m}{m!} \le \sum_{k=0}^\infty \frac{t^k}{k!} = e^t \quad \text{for } t \ge 0 .$$

After multiplying by $t^{\operatorname{Re} z - 1}$, we find

$$\left| t^{z-1} e^{-t} \right| = t^{\operatorname{Re} z - 1} e^{-t} \le \frac{m!}{t^{m - \operatorname{Re} z + 1}} .$$

We choose $m \in \mathbb{N}^\times$ such that $m > \operatorname{Re} z$ and get from Example 8.9(b) the absolute
convergence of $\int_1^\infty t^{z-1} e^{-t}\, dt$. ∎

For $z \in \mathbb{C}$ with $\operatorname{Re} z > 0$, the integral

$$\Gamma(z) := \int_0^\infty t^{z-1} e^{-t}\, dt \tag{9.1}$$

is called the **second Euler integral** or **Euler's gamma integral**, and (9.1) defines the
gamma function for $[\operatorname{Re} z > 0] := \{ z \in \mathbb{C} \; ; \; \operatorname{Re} z > 0 \}$.

9.2 Theorem   *The gamma function satisfies the **functional equation***

$$\Gamma(z+1) = z\Gamma(z) \quad \text{for } \operatorname{Re} z > 0 \,.$$

In particular,

$$\Gamma(n+1) = n! \quad \text{for } n \in \mathbb{N} \,.$$

Proof   Integrating by parts with $u(t) := t^z$ and $v'(t) := e^{-t}$ gives

$$\int_\alpha^\beta t^z e^{-t}\, dt = -t^z e^{-t}\Big|_\alpha^\beta + z \int_\alpha^\beta t^{z-1} e^{-t}\, dt \quad \text{for } 0 < \alpha < \beta < \infty \,.$$

From Proposition III.6.5(iii) and the continuity of the power function on $[0, \infty)$, we get $t^z e^{-t} \to 0$ as $t \to 0$ and as $t \to \infty$. Consequently, we have

$$\Gamma(z+1) = \lim_{\substack{\alpha \to 0+ \\ \beta \to \infty}} \int_\alpha^\beta t^z e^{-t}\, dt = z \int_0^\infty t^{z-1} e^{-t}\, dt = z\Gamma(z)$$

and therefore the stated functional equation.

   Finally, completing the induction, we have

$$\Gamma(n+1) = n\Gamma(n) = n(n-1) \cdot \cdots \cdot 1 \cdot \Gamma(1) = n!\,\Gamma(1) \quad \text{for } n \in \mathbb{N} \,.$$

Because $\Gamma(1) = \int_0^\infty e^{-t}\, dt = -e^{-t}\big|_0^\infty = 1$, we are done.[1] ∎

The gamma function on $\mathbb{C} \backslash (-\mathbb{N})$

For $\operatorname{Re} z > 0$ and $n \in \mathbb{N}^\times$, it follows from Theorem 9.2 that

$$\Gamma(z+n) = (z+n-1)(z+n-2) \cdot \cdots \cdot (z+1) z \Gamma(z) \,,$$

or

$$\Gamma(z) = \frac{\Gamma(z+n)}{z(z+1) \cdot \cdots \cdot (z+n-1)} \,. \tag{9.2}$$

The right side of (9.2) is defined for every $z \in \mathbb{C} \backslash (-\mathbb{N})$ with $\operatorname{Re} z > -n$. This formula suggests an extension of the function $z \mapsto \Gamma(z)$ onto $\mathbb{C} \backslash (-\mathbb{N})$.

9.3 Theorem   (and Definition)   *For $z \in \mathbb{C} \backslash (-\mathbb{N})$*

$$\Gamma_n(z) := \frac{\Gamma(z+n)}{z(z+1) \cdot \cdots \cdot (z+n-1)}$$

is independent of $n \in \mathbb{N}^\times$ if $n > -\operatorname{Re} z$. Therefore we can define through

$$\Gamma(z) := \Gamma_n(z) \quad \text{for } z \in \mathbb{C} \backslash (-\mathbb{N}) \text{ and } n > -\operatorname{Re} z$$

---

[1]Naturally $f\big|_a^\infty$ is an abbreviation for $\lim_{b \to \infty} f\big|_a^b$

an extension to $\mathbb{C}\backslash(-\mathbb{N})$ *of the function given in* (9.1). *This extension is called the* **gamma function**. *It satisfies the functional equation*

$$\Gamma(z+1) = z\Gamma(z) \quad \text{for } z \in \mathbb{C}\backslash(-\mathbb{N}) \ .$$

Proof Due to (9.2), Γ_n agrees on $[\operatorname{Re} z > 0]$ with the function defined in (9.1). Also, for $m, n \in \mathbb{N}$ such that $n > m > -\operatorname{Re} z$, we have

$$\Gamma(z+n) = (z+n-1) \cdot \cdots \cdot (z+m)\Gamma(z+m) \ .$$

Therefore

$$
\begin{aligned}
\Gamma_n(z) &= \frac{\Gamma(z+n)}{z(z+1)\cdot\cdots\cdot(z+n-1)} \\
&= \frac{(z+n-1)\cdot\cdots\cdot(z+m)\Gamma(z+m)}{z(z+1)\cdot\cdots\cdot(z+m-1)(z+m)\cdot\cdots\cdot(z+n-1)} \\
&= \frac{\Gamma(z+m)}{z(z+1)\cdot\cdots\cdot(z+m-1)} = \Gamma_m(z) \ .
\end{aligned}
$$

Thus the functions Γ_n and Γ_m agree where they are both defined. This shows that the gamma function is well defined and agrees with the Euler gamma integral when $[\operatorname{Re} z > 0]$.

For $z \in \mathbb{C}\backslash(-\mathbb{N})$ and $n \in \mathbb{N}^\times$ such $\operatorname{Re} z > -n$, we have

$$\Gamma(z+1) = \frac{\Gamma(z+n+1)}{(z+1)\cdot\cdots\cdot(z+n)} = \frac{z(z+n)\Gamma(z+n)}{z(z+1)\cdot\cdots\cdot(z+n)} = z\Gamma(z)$$

since $(z+n)\Gamma(z+n) = \Gamma(z+n+1)$. ∎

Gauss's representation formula

The gamma function has another representation, with important applications, due to Gauss.

9.4 Theorem *For* $z \in \mathbb{C}\backslash(-\mathbb{N})$, *we have*

$$\Gamma(z) = \lim_{n\to\infty} \frac{n^z n!}{z(z+1)\cdot\cdots\cdot(z+n)} \ .$$

Proof (i) We consider first the case $\operatorname{Re} z > 0$. Because

$$\Gamma(z) = \lim_{n\to\infty} \int_0^n t^{z-1}e^{-t}\,dt$$

and, according to Theorem III.6.23,

$$e^{-t} = \lim_{n \to \infty} \left(1 - \frac{t}{n}\right)^n \quad \text{for } t \geq 0 \,, \tag{9.3}$$

we can guess the formula

$$\Gamma(z) = \lim_{n \to \infty} \int_0^n t^{z-1}\left(1 - \frac{t}{n}\right)^n dt \quad \text{for Re } z > 0 \,. \tag{9.4}$$

We prove it below.

(ii) Suppose Re $z > 0$. We integrate (9.4) by parts using $u(t) := (1 - t/n)^n$ and $v'(t) := t^{z-1}$:

$$\int_0^n t^{z-1}\left(1 - \frac{t}{n}\right)^n dt = \frac{1}{z} \int_0^n t^z \left(1 - \frac{t}{n}\right)^{n-1} dt \,.$$

Another integration by parts gives

$$\int_0^n t^{z-1}\left(1 - \frac{t}{n}\right)^n dt = \frac{1}{z} \cdot \frac{n-1}{n(z+1)} \int_0^n t^{z+1}\left(1 - \frac{t}{n}\right)^{n-2} dt \,,$$

and, completing the induction, we find

$$\int_0^n t^{z-1}\left(1 - \frac{t}{n}\right)^n dt = \frac{1}{z} \cdot \frac{n-1}{n(z+1)} \cdot \frac{n-2}{n(z+2)} \cdot \cdots \cdot \frac{1}{n(z+n-1)} \int_0^n t^{z+n-1} dt$$

$$= \frac{1}{z} \cdot \frac{n-1}{n(z+1)} \cdot \frac{n-2}{n(z+2)} \cdot \cdots \cdot \frac{1}{n(z+n-1)} \cdot \frac{n^{z+n}}{z+n}$$

$$= \frac{n!\, n^z}{z(z+1) \cdot \cdots \cdot (z+n)} \,.$$

The claim then follows in this case from (9.4).

(iii) We set

$$\gamma_n(z) := \frac{n!\, n^z}{z(z+1) \cdot \cdots \cdot (z+n)} \quad \text{for } z \in \mathbb{C}\backslash(-\mathbb{N}) \,. \tag{9.5}$$

Then we have

$$\gamma_n(z) = \frac{1}{z(z+1) \cdot \cdots \cdot (z+k-1)}\left(1 + \frac{z+1}{n}\right)\left(1 + \frac{z+2}{n}\right) \cdot \cdots$$

$$\cdot \left(1 + \frac{z+k}{n}\right)\gamma_n(z+k)$$

for every $k \in \mathbb{N}^\times$. We now choose $k > -\operatorname{Re} z$, and so, from (ii) and Theorem 9.3, we get

$$\lim_n \gamma_n(z) = \frac{\Gamma(z+k)}{z(z+1) \cdot \cdots \cdot (z+k-1)} = \Gamma(z) \,.$$

Then it only remains to show (9.4).

(iv) To prove (9.4) for constant z, we can set $f(t) := t^{z-1}e^{-t}$ and

$$f_n(t) := \begin{cases} t^{z-1}(1 - t/n)^n , & 0 < t \leq n , \\ 0 , & t > n . \end{cases}$$

Then it follows from (9.3) that

$$\lim_{n \to \infty} f_n(t) = f(t) \quad \text{for } t > 0 . \tag{9.6}$$

Because the sequence $\big((1 - t/n)^n\big)_{n \in \mathbb{N}}$ converges monotonically to e^{-t} for $t > 0$ (see (v)), we also have

$$|f_n(t)| \leq g(t) \quad \text{for } t > 0 \text{ and } n \in \mathbb{N} , \tag{9.7}$$

where we define the integrable function $t \mapsto g(t) := t^{\operatorname{Re} z-1}e^{-t}$ on $(0, \infty)$. Now it follows from (9.6), (9.7), and the convergence theorem of Lebesgue (proved in Volume III) that

$$\lim_{n \to \infty} \int_0^n t^{z-1}(1 - t/n)^n \, dt = \lim_{n \to \infty} \int_0^\infty f_n(t) \, dt = \int_0^\infty f(t) \, dt = \Gamma(z) ,$$

and therefore (9.4).

To avoid looking ahead to the theorem of Lebesgue, we will directly verify (9.4). But first, we must prove that the convergence in (9.3) is monotonic and locally uniform.

(v) From the Taylor expansion of the logarithm (Application IV.3.9(d) or Theorem V.3.9), we have

$$\frac{\log(1 - s)}{s} = -\frac{1}{s}\sum_{k=1}^\infty \frac{s^k}{k} = -1 - \sum_{k=1}^\infty \frac{s^k}{k+1} \quad \text{for } s \in (0, 1) ,$$

and we find

$$\frac{\log(1 - s)}{s} \uparrow -1 \quad (s \to 0+) . \tag{9.8}$$

We now set in (9.8) an $s := t/n$ such that $t > 0$, and therefore

$$n \log\Big(1 - \frac{t}{n}\Big) = t\Big[\frac{n}{t} \log\Big(1 - \frac{t}{n}\Big)\Big] \uparrow -t \quad (n \to \infty) .$$

The monotonicity and the continuity of the exponential function then shows that the sequence $\big([1 - t/n]^n\big)_{n \in \mathbb{N}}$ converges to e^{-t} for every $t \geq 0$.

The uniform convergence of this sequence on compact subintervals of $\mathbb{R}^+$ follows from the uniform continuity of the map $s \mapsto \big[\log(1-s)\big]/s$ on such intervals (see Exercise III.3.11). Namely, suppose $T > 0$ and $\varepsilon > 0$. Then there is a $\delta > 0$ such that

$$\left| \frac{\log(1-s)}{s} + 1 \right| < \varepsilon \quad \text{for } 0 < s < \delta . \tag{9.9}$$

Now, we indeed have $t/n < \delta$ for $n > T/\delta$ and $t \in [0,T]$. Defining $s := t/n$, it therefore follows from (9.9) that

$$\left| n \log\left(1 - \frac{t}{n}\right) + t \right| \leq T \left| \frac{n}{t} \log\left(1 - \frac{t}{n}\right) + 1 \right| < T\varepsilon \quad \text{for } 0 \leq t \leq T .$$

Thus the series $\left(n \log(1 - t/n)\right)_{n \in \mathbb{N}}$ converges uniformly to $-t$ on $[0,T]$. Because the exponential function is uniformly continuous on $[0,T]$, the series $\left((1 - t/n)^n\right)_{n \in \mathbb{N}}$ converges uniformly in $t \in [0,T]$ to e^{-t}.

(vi) Finally, we prove (9.4). Suppose then $\varepsilon > 0$. The proof of Lemma 9.1 shows that there is an $N_0 \in \mathbb{N}$ such that

$$\int_{N_0}^{\infty} t^{\operatorname{Re} z - 1} e^{-t} \, dt < \varepsilon/3 .$$

Therefore, from (v), we have

$$\int_{N_0}^{n} t^{\operatorname{Re} z - 1} \left(1 - \frac{t}{n}\right)^n dt \leq \int_{N_0}^{n} t^{\operatorname{Re} z - 1} e^{-t} \, dt \leq \int_{N_0}^{\infty} t^{\operatorname{Re} z - 1} e^{-t} \, dt < \frac{\varepsilon}{3}$$

for $n \geq N_0$. Finally, there is, for the same reasons as in (v), an $N \in \mathbb{N}$ such that

$$\int_{0}^{N_0} t^{\operatorname{Re} z - 1} \left(e^{-t} - \left(1 - \frac{t}{n}\right)^n\right) dt \leq \frac{\varepsilon}{3} \quad \text{for } n \geq N .$$

For $n \geq N_0 \vee N$, we now get

$$\left| \Gamma(z) - \int_0^n t^{z-1} \left(1 - \frac{t}{n}\right)^n dt \right|$$

$$\leq \left| \Gamma(z) - \int_0^{N_0} t^{z-1} e^{-t} \, dt \right| + \left| \int_0^{N_0} t^{z-1} e^{-t} \, dt - \int_0^n t^{z-1}\left(1 - \frac{t}{n}\right)^n dt \right|$$

$$\leq \int_{N_0}^{\infty} t^{\operatorname{Re} z - 1} e^{-t} \, dt + \int_0^{N_0} t^{\operatorname{Re} z - 1}\left(e^{-t} - \left(1 - \frac{t}{n}\right)^n\right) dt$$

$$+ \int_{N_0}^{n} t^{\operatorname{Re} z - 1}\left(1 - \frac{t}{n}\right)^n dt \leq \varepsilon ,$$

which shows (9.4). ∎

The reflection formula

As an application of the Gauss product representation, we will deduce an important relationship between the gamma function and the sine. To that end, we first prove a product form of $1/\Gamma$, which is due to Weierstrass.

9.5 Proposition   *For $z \in \mathbb{C}\backslash(-\mathbb{N})$, we have*

$$\frac{1}{\Gamma(z)} = ze^{Cz} \prod_{k=1}^{\infty} \left(1 + \frac{z}{k}\right) e^{-z/k} \ , \tag{9.10}$$

where C is the Euler–Mascheroni constant. The infinite product converges absolutely and is locally uniform. In particular, the gamma function has no zeros.

Proof   Obviously

$$\frac{1}{\gamma_n(z)} = z\exp\left[z\Big(\sum_{k=1}^{n}\frac{1}{k} - \log n\Big)\right] \prod_{k=1}^{n} \left(\frac{z+k}{k}\, e^{-z/k}\right)$$

for $z \in \mathbb{C}\backslash(-\mathbb{N})$. Defining $a_k(z) := (1 + z/k)e^{-z/k}$, we have for $|z| \le R$

$$|a_k(z) - 1| = \left|(1 + z/k)\Big[1 - z/k + \sum_{j=2}^{\infty}\frac{(-1)^j}{j!}\Big(\frac{z}{k}\Big)^j\Big] - 1\right| \le c/k^2 \ , \tag{9.11}$$

with an appropriate constant $c := c(R)$. Hence there is a constant $K \in \mathbb{N}^{\times}$ such that $|a_k(z) - 1| \le 1/2$ for $k \ge K$ and $|z| \le R$. From this, it follows that

$$\prod_{k=1}^{n}\frac{z+k}{k}\, e^{-z/k} = \Big(\prod_{k=1}^{K} a_k(z)\Big)\exp\Big(\sum_{k=K+1}^{n} \log a_k(z)\Big) \tag{9.12}$$

for $n > K$ and $|z| \le R$. From the power series representation of the logarithm, we get the existence of a constant $M \ge 1$ such that $|\log(1 + \zeta)| \le M\,|\zeta|$ for $|\zeta| \le 1/2$. Thus we find using (9.11) (which changes everything) the estimate

$$\sum_{k>K} |\log a_k(z)| \le M \sum_{k>K} |a_k(z) - 1| \le cM \sum_{k} k^{-2}$$

for $|z| \le R$. The series $\sum_{k>K} \log a_k(z)$ then converges absolutely and uniformly for $|z| \le R$ due to the Weierstrass majorant criterion. From (9.12) and the properties of the exponential function, this holds also for the infinite product appearing in (9.10). The claim now follows from Theorem 9.4 and Example 6.13(b). $\blacksquare$

We can use this product representation to get the previously mentioned relationship between the sine and the gamma function.

9.6 Theorem (reflection formula for the Γ function[2]) *For $z \in \mathbb{C}\backslash\mathbb{Z}$, we have*

$$\Gamma(z)\Gamma(1-z) = \frac{\pi}{\sin(\pi z)} .$$

Proof   The representation (9.10) implies (see Proposition II.2.4(ii))

$$\frac{1}{\Gamma(z)}\frac{1}{\Gamma(1-z)} = \frac{1}{-z\Gamma(z)\Gamma(-z)} = z\prod_{k=1}^{\infty}\left(1 - \frac{z^2}{k^2}\right)$$

for $z \in \mathbb{C}\backslash\mathbb{Z}$. The claim now follows from Application 7.23(b). ∎

Using the reflection formula, we can compute the value of some important improper integrals.

9.7 Application   (Gaussian error integral[3])   The identity says

$$\int_{-\infty}^{\infty} e^{-x^2}\, dx = \Gamma(1/2) = \sqrt{\pi} .$$

Proof   Using the substitution $x = \sqrt{t}$, we find[4]

$$\Gamma(1/2) = \int_{0}^{\infty} e^{-t}\frac{dt}{\sqrt{t}} = 2\int_{0}^{\infty} e^{-x^2}\, dx = \int_{-\infty}^{\infty} e^{-x^2}\, dx ,$$

because $x \mapsto e^{-x^2}$ is even. The claim then follows from the reflection formula. ∎

In Volume III, we will learn an easier method for calculating the Gaussian error integral, which uses a multivariable integration.

The logarithmic convexity of the gamma function

Using

$$\varphi_n(z) := z\log n - \sum_{k=0}^{n}\log(z+k) + \log(n!) \quad\text{for } n \in \mathbb{N}^{\times},\quad z \in \mathbb{C}\backslash(-\mathbb{N}),$$

we get from (9.5) the representation $\gamma_n = e^{\varphi_n}$. Then, by taking the logarithmic derivative γ_n'/γ_n of γ_n, we find the simple form

$$\psi_n(z) := \frac{\gamma_n'(z)}{\gamma_n(z)} = \log n - \sum_{k=0}^{n}\frac{1}{z+k}, \tag{9.13}$$

---

[2]So called because its left side is even about $z = -1/2$.

[3]The name "error integral" comes from the theory of probability, in which the function $x \mapsto e^{-x^2}$ plays a fundamental role.

[4]This substitution is carried out in the proper integral $\int_{\alpha}^{\beta}$, and then the limits $\alpha \to 0+$ and $\beta \to \infty$ are taken.

and, further,

$$\psi_n'(z) = \sum_{k=0}^{n} \frac{1}{(z+k)^2} \tag{9.14}$$

for $n \in \mathbb{N}^{\times}$ and $z \in \mathbb{C}\backslash(-\mathbb{N})$.

We can now prove the following forms of the first two logarithmic derivatives of the gamma function.

9.8 Proposition $\Gamma \in C^2\big(\mathbb{C}\backslash(-\mathbb{N})\big)$. *Then we have*

$$\frac{\Gamma'(z)}{\Gamma(z)} = -C - \frac{1}{z} - \sum_{k=1}^{\infty}\Big(\frac{1}{z+k} - \frac{1}{k}\Big) \tag{9.15}$$

and

$$\Big(\frac{\Gamma'}{\Gamma}\Big)'(z) = \sum_{k=0}^{\infty} \frac{1}{(z+k)^2} \tag{9.16}$$

for $z \in \mathbb{C}\backslash(-\mathbb{N})$, where C is the Euler–Mascheroni constant.

Proof We first must show that the sequence (ψ_n) and (ψ_n') on $\mathbb{C}\backslash(-\mathbb{N})$ is locally uniformly convergent. So we consider

$$\psi_n(z) = \Big(\log n - \sum_{k=1}^{n}\frac{1}{k}\Big) - \frac{1}{z} - \sum_{k=1}^{n}\Big(\frac{1}{z+k} - \frac{1}{k}\Big) \ .$$

Thus, we must show the local uniform convergence of the series

$$\sum_{k}\Big(\frac{1}{z+k} - \frac{1}{k}\Big) = \sum_{k} \frac{-z}{k(z+k)} \quad \text{and} \quad \sum_{k}(z+k)^{-2} \tag{9.17}$$

on $\mathbb{C}\backslash(-\mathbb{N})$.

Suppose $z_0 \in \mathbb{C}\backslash(-\mathbb{N})$ and $0 < r < \text{dist}(z_0, -\mathbb{N})$. Then, for $z \in \mathbb{B}(z_0, r)$, we have the estimate

$$|z + k| \geq |z_0 + k| - |z - z_0| \geq k - |z_0| - r \geq k/2$$

for $k \in \mathbb{N}^{\times}$ such that $k \geq k_0 := 2(|z_0| + r)$. Consequently, we find

$$\Big|\frac{z}{k(z+k)}\Big| \leq \frac{k_0}{k^2} \ , \quad \frac{1}{|z+k|^2} \leq \frac{4}{k^2}$$

for $z \in \mathbb{B}(z_0, r)$ and $k \geq k_0$. Thus follows the uniformly convergence on $\mathbb{B}(z_0, r)$ — and accordingly the local uniform convergence on $\mathbb{C}\backslash(-\mathbb{N})$ — of the series (9.17) from the Weierstrass majorant criterion (Theorem V.1.6). Therefore we have

$$\psi(z) := \lim_{n\to\infty} \psi_n(z) = -C - \frac{1}{z} - \sum_{k=1}^{\infty}\Big(\frac{1}{z+k} - \frac{1}{z}\Big) \quad \text{for } z \in \mathbb{C}\backslash(-\mathbb{N}) \ . \tag{9.18}$$

From the theorem of the differentiability of the limits of series of functions (Theorem V.2.8), it further follows that ψ is continuously differentiable on $\mathbb{C}\backslash(-\mathbb{N})$, with

$$\psi'(z) = \lim_{n\to\infty} \psi'_n(z) = \sum_{k=0}^{\infty} \frac{1}{(z+k)^2} \quad \text{for } z \in \mathbb{C}\backslash(-\mathbb{N}) . \tag{9.19}$$

For $z \in \mathbb{C}\backslash(-\mathbb{N})$, we can also write $\varphi_n(z)$ in the form

$$\varphi_n(z) = -\log z + z\Big(\log n - \sum_{k=1}^{n}\frac{1}{k}\Big) + \sum_{k=1}^{n}\Big(\frac{z}{k} - \log\Big(1+\frac{z}{k}\Big)\Big) .$$

Then

$$\log(1+z/k) = z/k + O\big((z/k)^2\big) \quad (k \to \infty)$$

and Example 6.13(b) imply that $\varphi_n(z)$ converges to

$$\varphi(z) := -\log z - Cz + \sum_{k=1}^{\infty}\Big(\frac{z}{k} - \log\Big(1+\frac{z}{k}\Big)\Big)$$

as $n \to \infty$. Thus we get from $\gamma_n = e^{\varphi_n}$ and Theorem 9.4 that

$$\Gamma(z) = e^{\varphi(z)} \quad \text{for } z \in \mathbb{C}\backslash(-\mathbb{N}) . \tag{9.20}$$

Because $\varphi'_n = \psi_n$ and the sequence (ψ_n) is locally uniformly convergent, Theorem V.2.8 then guarantees that φ is continuously differentiable with $\varphi' = \psi$. Now it follows from (9.20) that Γ is also continuously differentiable with

$$\Gamma' = \varphi'e^{\varphi} = \psi\Gamma . \tag{9.21}$$

Then (9.15) follows from (9.18), and (9.16) is a consequence of (9.19). Finally, that $\Gamma \in C^2\big(\mathbb{C}\backslash(-\mathbb{N})\big)$ follows from (9.21) and the continuous differentiability of ψ. ∎

9.9 Remarks (a) The above theorem and Theorem VIII.5.11 imply that Γ is analytic on $\mathbb{C}\backslash(-\mathbb{N})$.

(b) Because $(\Gamma'/\Gamma)' = \big(\Gamma''\Gamma - (\Gamma')^2\big)/\Gamma^2$, we get from (9.16) that

$$\Gamma''(x)\Gamma(x) > \big(\Gamma'(x)\big)^2 \geq 0 \quad \text{for } x \in \mathbb{R}\backslash(-\mathbb{N}) .$$

Therefore $\operatorname{sign}\big(\Gamma''(x)\big) = \operatorname{sign}\big(\Gamma(x)\big)$ for $x \in \mathbb{R}\backslash(-\mathbb{N})$. From Gauss's formula of Theorem 9.4, we read off that

$$\operatorname{sign}\big(\Gamma(x)\big) = \begin{cases} 1 , & x > 0 , \\ (-1)^k , & -k < x < -k+1 , \quad k \in \mathbb{N} . \end{cases}$$

Therefore Γ is convex on $(0,\infty)$ and the intervals $(-2k, -2k+1)$ but concave on the intervals $(-2k-1, -2k)$, where $k \in \mathbb{N}$.

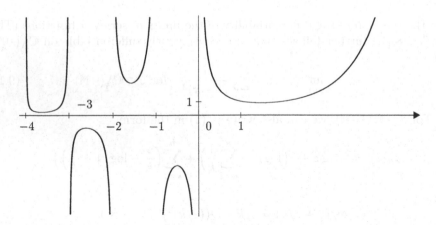

(c) A function f that is everywhere strictly positive on a perfect interval is said to be **log convex** if $\log f$ is convex. If f is twice differentiable, then, according to Corollary IV.2.13, f is log convex if and only if $f''f - (f')^2 \geq 0$. Therefore the gamma function is log convex on $(0, \infty)$. One can show that $\Gamma \,|\, (0, \infty)$ is the unique function $f : (0, \infty) \rightarrow (0, \infty)$ that satisfies the conditions

  (i) f is log convex,

 (ii) $f(x+1) = xf(x)$ and for $x > 0$, and

(iii) $f(1) = 1$.

For a proof and a construction of the theory of the gamma function on the basis of these properties, we refer to [Art31] (see also [Kön92] and Exercise 6). ∎

Stirling's formula

The de Moivre–Stirling formula describes the asymptotic behavior of the factorial $n \mapsto n!$ as $n \rightarrow \infty$. Example 6.13(a) says

$$\Gamma(n) = (n-1)! \sim \sqrt{2\pi}\, n^{n-\frac{1}{2}} e^{-n} \ . \tag{9.22}$$

The following theorem refines and amplifies this statement.

9.10 Theorem (Stirling's formula) *For every $x > 0$, there is a $\theta(x) \in (0, 1)$ such that*

$$\Gamma(x) = \sqrt{2\pi}\, x^{x-1/2} e^{-x} e^{\theta(x)/12x} \ .$$

Proof For γ_n, we get from (9.5) that

$$\log \gamma_n(x) = \log n! + x \log n - \sum_{k=0}^{n} \log(x+k) \quad \text{for } x > 0 \ .$$

To the sum, we apply the Euler–Maclaurin formula (Theorem 6.9) with $a := 0$, $b := n$ and $m := 0$ and find

$$\sum_{k=0}^{n} \log(x+k) = \int_0^n \log(x+t)\,dt + \frac{1}{2}\big[\log x + \log(x+n)\big] + R_n(x) \; ,$$

where

$$R_n(x) := \int_0^n \frac{\widetilde{B}_1(t)}{x+t}\,dt \; .$$

Integrating by parts with $u(t) := \log(x+t)$ and $v' = 1$ then gives

$$\int_0^n \log(x+t)\,dt = (x+t)\big[\log(x+t) - 1\big]\Big|_0^n = (x+n)\log(x+n) - n - x\log x \; ,$$

and thus

$$\log \gamma_n(x) = \left(x - \frac{1}{2}\right)\log x + \log n! + n + x\log n - \left(x+n+\frac{1}{2}\right)\log(x+n) - R_n(x) \; .$$

Because $\log(x+n) = \log n + \log(1 + x/n)$, it follows that

$$\left(x+n+\frac{1}{2}\right)\log(x+n) = x\log n + \log\big(n^{(n+\frac{1}{2})}\big) + \left(x+\frac{1}{2}\right)\log\left(1+\frac{x}{n}\right)$$
$$+ \log\left(\left(1+\frac{x}{n}\right)^n\right) \; ,$$

and we get

$$\log \gamma_n(x) = \left(x - \frac{1}{2}\right)\log x - \log\left(\left(1+\frac{x}{n}\right)^n\right) - \left(x+\frac{1}{2}\right)\log\left(1+\frac{x}{n}\right)$$
$$+ \log\left[\frac{n!}{n^{n+1/2}e^{-n}}\right] - R_n(x) \; . \tag{9.23}$$

To estimate $R_n(x)$, we consider

$$-R_n(x) = -\sum_{k=0}^{n-1}\int_k^{k+1}\frac{\widetilde{B}_1(t)}{x+t}\,dt = -\sum_{k=0}^{n-1}\int_0^1\frac{B_1(t)}{x+t+k}\,dt = \sum_{k=0}^{n-1}g(x+k)$$

with

$$g(x) := -\int_0^1 \frac{t-1/2}{x+t}\,dt = \left(x+\frac{1}{2}\right)\log\left(\frac{x+1}{x}\right) - 1 = \frac{1}{2y}\log\frac{1+y}{1-y} - 1 \; ,$$

where we have set $y := 1/(2x+1)$. For $x > 0$ we have $0 < y < 1$. Therefore, from

$$\log\big((1+y)/(1-y)\big) = \log(1+y) - \log(1-y)$$

and the power series expansion of the logarithm

$$\frac{1}{2}\log\frac{1+y}{1-y} = \sum_{k=0}^{\infty}\frac{y^{2k+1}}{2k+1} \; ,$$

we have

$$0 < g(x) = \sum_{k=1}^{\infty} \frac{y^{2k}}{2k+1} < \frac{1}{3} \sum_{k=1}^{\infty} y^{2k} = \frac{1}{3} \frac{y^2}{1-y^2}$$

$$= \frac{1}{12x(x+1)} = \frac{1}{12x} - \frac{1}{12(x+1)} =: h(x) \quad \text{for } x > 0 .$$

Thus $\sum_k h(x+k)$ represents a convergent majorant of the series $\sum_k g(x+k)$. Therefore

$$R(x) := \lim_{n \to \infty} R_n(x) = -\sum_{k=0}^{\infty} g(x+k) ,$$

exists, and we have

$$0 < -R(x) < \frac{1}{12x} \quad \text{for } x > 0 ,$$

since $1/12x$ is the value of the series $\sum_k h(x+k)$ and $g(x) < h(x)$ for $x > 0$.

Now we pass the limit $n \to \infty$ in (9.23) and find from Theorem 9.4 that

$$\log \Gamma(x) = \left(x - \frac{1}{2}\right) \log x - x + \log \sqrt{2\pi} - R(x) ,$$

where we have reused the de Moivre–Stirling formula (9.22). The theorem now follows with $\theta(x) := -12xR(x)$. ∎

Stirling's formula allows the approximate calculation of the gamma function, and, in particular, it approximates large arguments of the factorial $n!$. If one approximates $\Gamma(x)$ as $\sqrt{2\pi}\, x^{x-1/2} e^{-x}$, the relative error is smaller than $e^{1/12x} - 1$. This error diminishes rapidly as x grows large.

The Euler beta integral

The integral

$$\int_0^1 t^{p-1}(1-t)^{q-1}\, dt , \tag{9.24}$$

called the **first Euler integral**, plays an important role in the context of the gamma function.

9.11 Proposition   *The integral (9.24) converges absolutely for $p, q \in [\operatorname{Re} z > 0]$.*

Proof   Because

$$\int_0^1 |t^{p-1}(1-t)^{q-1}|\, dt = \int_0^1 t^{\operatorname{Re} p-1}(1-t)^{\operatorname{Re} q-1}\, dt ,$$

it suffices to consider the case $p, q \in \mathbb{R}$.

(i) Suppose $q \in \mathbb{R}$. Then there are numbers $0 < m < M$ such that

$$m \leq (1-t)^{q-1} \leq M \quad \text{for } 0 \leq t \leq 1/2 .$$

Consequently, it follows from Example 8.9(b) that

$$\int_0^{1/2} t^{p-1}(1-t)^{q-1} \, dt \text{ exists} \Longleftrightarrow p > 0 .$$

(ii) Suppose now $p > 0$. Then there are numbers $0 < m' < M'$ such that

$$m' \leq t^{p-1} \leq M' \quad \text{for } 1/2 \leq t \leq 1 .$$

Therefore $\int_{1/2}^1 t^{p-1}(1-t)^{q-1} \, dt$ exists if and only if $\int_{1/2}^1 (1-t)^{q-1} \, dt$ exists. Using the substitution $s := 1 - t$, it follows[5] from Example 8.9(b) that

$$\int_{1/2}^1 (1-t)^{q-1} \, dt = \int_0^{1/2} s^{q-1} \, ds \text{ exists} \Longleftrightarrow q > 0 ,$$

as claimed. ∎

Using these theorems, the **Euler beta function** B is well defined through

$$\mathsf{B} : [\operatorname{Re} z > 0] \times [\operatorname{Re} z > 0] \to \mathbb{C} , \quad (p,q) \mapsto \int_0^1 t^{p-1}(1-t)^{q-1} \, dt .$$

9.12 Remarks (a) The gamma function and the beta function are connected by the functional equation

$$\frac{\Gamma(p)\Gamma(q)}{\Gamma(p+q)} = \mathsf{B}(p,q) \quad \text{for } p, q \in [\operatorname{Re} z > 0] . \tag{9.25}$$

A proof of this statement for $p, q \in (0, \infty)$ is outlined in Exercises 12 and 13. We treat the general case in Volume III.

(b) For $p, q \in [\operatorname{Re} z > 0]$, we have

$$\mathsf{B}(p,q) = \mathsf{B}(q,p) \quad \text{and} \quad \mathsf{B}(p+1,q) = \frac{p}{q}\mathsf{B}(p,q+1) .$$

Proof The first statement follows immediately from (9.25). Likewise (9.25) and Theorem 9.2 imply

$$\mathsf{B}(p+1,q) = \frac{p\Gamma(p)\Gamma(q)}{(p+q)\Gamma(p+q)} = \frac{p}{p+q}\mathsf{B}(p,q) ,$$

and then also, by symmetry, $\mathsf{B}(p, q+1) = (q/(p+q))\mathsf{B}(p,q)$. ∎

---

[5]As already noted for improper integrals, we make these substitutions before passing any limits to infinity.

(c) Using the functional equation (9.25) and the properties of the gamma function, the beta function can be defined for all $p, q \in \mathbb{C}$ such that $p, q, (p+q) \notin -\mathbb{N}$. ∎

Exercises

1 Prove that

$$\Gamma\left(n + \frac{1}{2}\right) = \sqrt{\pi} \prod_{k=1}^{n} \frac{2k-1}{2} = 1 \cdot 3 \cdot \cdots \cdot (2n-1) \frac{\sqrt{\pi}}{2^n} \quad \text{for } n \in \mathbb{N}^{\times} .$$

2 Show that Γ belongs to $C^{\infty}(\mathbb{C} \backslash (-\mathbb{N}), \mathbb{C})$ with

$$\Gamma^{(n)}(z) = \int_0^{\infty} t^{z-1} (\log t)^n e^{-t} \, dt \quad \text{for } n \in \mathbb{N} \text{ and } \operatorname{Re} z > 0 .$$

(Hint: Consider (9.4).)

3 Let C be the Euler–Mascheroni constant. Verify these statements:
  (i) $\int_0^{\infty} e^{-t} \log t \, dt = -C$.
  (ii) $\int_0^{\infty} e^{-t} t \log t \, dt = 1 - C$.
  (iii) $\int_0^{\infty} e^{-t} t^{n-1} \log t \, dt = (n-1)! \sum_{k=1}^{n-1} (1/k - C)$ for $n \in \mathbb{N}$ and $n \geq 2$.
  (iv) $\int_0^{\infty} (\log t)^2 e^{-t} \, dt = C^2 + \pi^2/6$.
(Hint: To compute the integrals in (iv), consider Application 7.23(a).)

4 Show that the gamma function has the representation

$$\Gamma(z) = \int_0^1 (-\log t)^{z-1} \, dt$$

for $z \in \mathbb{C}$ such that $\operatorname{Re} z > 0$.

5 Suppose f and g are log convex. Show that fg is also log convex.

6 Suppose the function $f : (0, \infty) \to (0, \infty)$ is differentiable[6] and satisfies
  (i) f is log convex;
  (ii) $f(x+1) = xf(x)$ for $x > 0$;
  (iii) $f(1) = 1$.
Show then that $f = \Gamma | (0, \infty)$.
(Hints: (α) Define $h := \log(\Gamma/f)$. Using (iii), it suffices to verify $h' = 0$.
(β) From (ii) and Theorem 9.2, it follows that h is 1-periodic.
(γ) From (i) it follows that $(\log f)'$ is increasing (see Theorem IV.2.12). This implies

$$0 \leq (\log f)'(x+y) - (\log f)'(x) \leq (\log f)'(x+1) - (\log f)'(x) = 1/x$$

for $0 < y \leq 1$, where the last step follows from reusing (ii). An analogous estimate holds also for $(\log \Gamma)'$. Therefore

$$-1/x \leq h'(x+y) - h'(x) \leq 1/x \quad \text{for } y \in (0, 1] \text{ and } x \in (0, \infty) .$$

---

[6] One can do without the differentiability of f (see [Art31], [Kön92]).

Then that h, h', and $h'(\cdot + y)$ are 1-periodic implies that

$$-1/(x+n) \le h'(x+y) - h'(x) \le 1/(x+n) \quad \text{for } x, y \in (0, 1] \text{ and } n \in \mathbb{N}^\times .$$

For $n \to \infty$, it now follows that $h' = 0$.)

7 Prove the **Legendre duplication formula**

$$\Gamma\Big(\frac{x}{2}\Big)\Gamma\Big(\frac{x+1}{2}\Big) = \frac{\sqrt{\pi}}{2^{x-1}}\Gamma(x) \quad \text{for } x \in (0, \infty) .$$

(Hint: Consider $h(x) := 2^x \Gamma(x/2)\,\Gamma\big((x+1)/2\big)$ and apply Exercise 6.)

8 For $x \in (-1, 1)$, verify the power series expansion

$$(\log \Gamma)(1+x) = -Cx + \sum_{k=2}^{\infty}(-1)^k \frac{\zeta(k)}{k} x^k .$$

(Hint: Expand $\log(1 + x/n)$ as a power series for $n \in \mathbb{N}^\times$, and look to Proposition 9.5.)

9 Show $\int_0^1 \log \Gamma(x)\,dx = \log \sqrt{2\pi}$. (Hint: Exercise 8.4 and Theorem 9.6).

10 Verify that for $0 < a < b$, we have

$$\int_a^b \frac{dx}{\sqrt{(x-a)(b-x)}} = \pi .$$

11 For fixed $q \in (0, \infty)$, show the function $(0, \infty) \to (0, \infty)$, $p \mapsto \mathsf{B}(p, q)$ is differentiable and log convex. (Hint: Show $\partial_p^2\big(\log\big(t^{p-1}(1-t)^{q-1}\big)\big) = 0$ for $p, q \in (0, \infty)$ and $t \in (0, 1)$.)

12 Prove (without using (9.25)) that for $p, q \in (0, \infty)$, we have

$$\mathsf{B}(p+1, q) = p\mathsf{B}(p, q)/(p+q) .$$

13 For $p, q \in (0, \infty)$, show

$$\frac{\Gamma(p)\,\Gamma(q)}{\Gamma(p+q)} = \mathsf{B}(p, q) .$$

(Hints: Fix $q \in (0, \infty)$ and let

$$f: (0, \infty) \to (0, \infty) , \quad p \mapsto \mathsf{B}(p, q)\,\Gamma(p+q)/\Gamma(q) .$$

According to Exercise 11 and Proposition 9.8, f is differentiable, and Exercises 5 and 11 show that f is log convex. Further, it follows from Exercise 12 that f satisfies the functional equation $f(p+1) = pf(p)$. Then $f(1) = 1$ and Exercise 6 implies the claim.)

Chapter VII

Multivariable differential calculus

In Volume I, we used the differential calculus to extract deep insight about the "fine structure" of functions. In that process, the idea of linear approximations proved to be extremely effective. However, we have until now concentrated on functions of one variable.

This chapter is mainly devoted to extending the differential calculus to functions of multiple variables, but we will also further explore the simple idea of linear approximations. Indeed — in contrast to the one-dimensional case — the issues here are intrinsically more complicated, because linear maps in the multidimensional case show a considerably richer structure than those in one dimension.

As before, we prefer to work in a coordinate-free representation. In other words, we develop the differential calculus for maps between Banach spaces. This representation is conceptually simple and actually makes many expressions look much simpler. The classical formulas for the derivatives in the usual coordinate representation follow easily from the general results using the product structure of finite-dimensional Euclidean spaces.

Because linear maps between Banach spaces underlie the linear approximation and therefore also differentiability, the first section is devoted to studying spaces of linear operators. Special interest falls naturally on the finite-dimensional case, which we develop using basic rules for matrices from linear algebra.

As an application of these results, we will study the exponential function in the algebra of endomorphisms of Banach spaces and derive from it the basic facts about systems of ordinary differential equations and then second order differential equations with constant coefficients.

Section 2 establishes another central concept, the Fréchet derivative. Beyond this, we consider directional derivatives, which arise naturally from partial derivatives and the representation of the Fréchet derivative by means of the Jacobi matrix. Finally, we examine the connection between the differentiability of functions of a complex variable and the total differentiability of the corresponding real representation. We characterize complex differentiability by the Cauchy–Riemann equations.

In Section 3, we put together the rules for differentiation and derive the important mean value theorem through a simple generalization of the one variable case.

Before we can turn to higher derivatives, we must clarify the connection between multilinear maps and linear-map-valued linear operators. The simple results developed in Section 4 will build the foundation for a clear representation of higher derivatives in multiple variables.

In Section 5, we explain the concept of higher order derivatives. In particular, we prove the fundamental Taylor's formula — both for maps between Banach spaces and for functions of finitely many variables. Generalizing the criterion in the one-dimensional case, we will specify sufficient conditions for the presence of local extrema of functions of multiple variables.

Section 6 plays a special role. Here, we shall see the first reward for developing the differential calculus on Banach spaces. Using it, we will explain the calculus of variations and derive the important Euler–Lagrange differential equations. In the process, we hope you will appreciate the power of abstract approach, in which functions are regarded as points in an infinite-dimensional Banach space. This simple geometrical idea of "functions of functions" proves to have wide-ranging consequences. Here, in the calculus of variations, we seek to minimize a certain function of a function, whose form is usually determined by a physical principle, such as "least action". The minima, if present, occur at critical points, and this criterion leads to the Euler–Lagrange differential equation(s), whose importance for mathematics and physics cannot be overstated.

After this excursion into "higher analysis", we prove in Section 7 perhaps the most important theorem of differential calculus, namely, the inverse function theorem. Equivalent to this result is the implicit function theorem, which we derive in the subsequent sections. Without extra trouble, we can also prove this theorem for maps between Banach spaces. And again, the coordinate-free representation yields significant gains in clarity and elegance.

In Section 8, we give a glimpse of the theory of nonlinear ordinary differential equations. Using the implicit function theorem, we discuss first order scalar differential equations. We also prove the Picard–Lindelöf theorem, which is the fundamental existence and uniqueness theorem for ordinary differential equations.

In the balance of this chapter, we will illustrate the importance of the implicit function theorem in the finite-dimensional case. With its help, we will characterize submanifolds of $\mathbb{R}^n$; these are subsets of $\mathbb{R}^n$ that locally "look like" $\mathbb{R}^m$ for $m < n$. Through numerous examples of curves, surfaces, and higher-dimensional submanifolds, we will see that this resemblance can be better visualized and more precisely described through their tangential spaces. Here, we will concentrate on submanifolds of Euclidean vector spaces, as they are conceptually simpler than abstract manifolds (which are not defined as subsets of $\mathbb{R}^n$). However, we will lay the groundwork for analysis on general manifolds, which we will then tackle in Volume III. As a practical application of these geometric ideas, we will study constrained minimization problems and their solution by the method of Lagrange multipliers. After deriving the relevant rules, we will treat two nontrivial examples.

1 Continuous linear maps

As we just said, differentiation in the multivariable case depends on the idea of local approximations by affine functions. Of course, from Proposition I.12.8, the nontrivial part of an affine function between vectors spaces is just its linear map. Thus we will concentrate next on linear maps between normed vector spaces. After deducing a few of their fundamental properties, we will use them to study the exponential map and the theory of linear differential equations with constant coefficients.

In the following, suppose

- $E = (E, \|\cdot\|)$ and $F = (F, \|\cdot\|)$ are normed vector spaces[1] over the field $\mathbb{K}$.

The completeness of $\mathcal{L}(E, F)$

From Section VI.2, we know already that $\mathcal{L}(E, F)$, the space of all bounded linear maps from E to F, is a normed vector space. Now we explore the completeness of this space.

1.1 Theorem   *If F is a Banach space, then so is $\mathcal{L}(E, F)$.*

Proof   (i) Suppose (A_n) is a Cauchy sequence in $\mathcal{L}(E, F)$. Then for every $\varepsilon > 0$, there is an $N(\varepsilon) \in \mathbb{N}$ such that $\|A_n - A_m\| < \varepsilon$ for $m, n \geq N(\varepsilon)$. In particular, for every $x \in E$, we have

$$\|A_n x - A_m x\| \leq \|A_n - A_m\| \, \|x\| \leq \varepsilon \, \|x\| \text{ for } m, n \geq N(\varepsilon) . \qquad (1.1)$$

Therefore $(A_n x)$ is a Cauchy sequence in F. Because F is complete, there is a unique $y \in F$ such that $\lim A_n x = y$. We can then define a map $A : E \to F$ by $x \mapsto \lim A_n x$. From the linearity of limits it follows at once that A is linear.

(ii) Because it is a Cauchy sequence, (A_n) is bounded in $\mathcal{L}(E, F)$ (see Proposition II.6.3). Thus there is an $\alpha \geq 0$ such that $\|A_n\| \leq \alpha$ for all $n \in \mathbb{N}$. From this, it follows that

$$\|A_n x\| \leq \|A_n\| \, \|x\| \leq \alpha \, \|x\| \text{ for } n \in \mathbb{N} \text{ and } x \in E .$$

Leaving out the middle term, the limit $n \to \infty$ implies the estimate $\|Ax\| \leq \alpha \, \|x\|$ for every $x \in E$. Therefore A belongs to $\mathcal{L}(E, F)$.

(iii) Finally we prove that the sequence (A_n) in $\mathcal{L}(E, F)$ converges to A. From (1.1), it follows that

$$\|A_n x - A_m x\| \leq \varepsilon   \text{ for } n, m \geq N(\varepsilon) \text{ and } x \in \mathbb{B}_E .$$

---

[1]When necessary, we distinguish the norms in E and F by corresponding indices.

For $m \to \infty$ this implies

$$\|A_n x - Ax\| \leq \varepsilon \quad \text{for } n \geq N(\varepsilon) \text{ and } x \in \mathbb{B}_E \ ,$$

and, as claimed, we get the inequality

$$\|A_n - A\| = \sup_{\|x\| \leq 1} \|A_n x - Ax\| \leq \varepsilon \quad \text{for } n \geq N(\varepsilon)$$

by forming the supremum over $\mathbb{B}_E$. ∎

1.2 Corollary

  (i) $\mathcal{L}(E, \mathbb{K})$ is a Banach space.
  (ii) If E is Banach space, then $\mathcal{L}(E)$ is a Banach algebra with unity.

Proof   (i) is clear, and (ii) follows from Remark VI.2.4(g). ∎

Finite-dimensional Banach spaces

The normed vector spaces E and F are called (**topologically**) **isomorphic** if there is a continuous linear isomorphism A from E to F such that A^{-1} is also continuous, that is, if A belongs to $\mathcal{L}(F, E)$. Then A is a **topological isomorphism** from E to F. We denote by

$$\mathcal{L}\mathrm{is}(E, F)$$

the set of all topological isomorphisms from E to F, and we write $E \cong F$ if $\mathcal{L}\mathrm{is}(E, F)$ is not empty.[2] Also, we set

$$\mathcal{L}\mathrm{aut}(E) := \mathcal{L}\mathrm{is}(E, E) \ .$$

Thus $\mathcal{L}\mathrm{aut}(E)$ is the set of all **topological automorphisms** of E.

1.3 Remarks   (a) The spaces $\mathcal{L}(\mathbb{K}, F)$ and F are isometrically[3] isomorphic. More precisely, using the isometric isomorphism

$$\mathcal{L}(\mathbb{K}, F) \to F \ , \quad A \mapsto A1 \ , \tag{1.2}$$

we **canonically** identify $\mathcal{L}(\mathbb{K}, F)$ and F as $\mathcal{L}(\mathbb{K}, F) = F$.

Proof   It is clear that the map (1.2) is linear and injective. For $v \in F$, consider $A_v \in \mathcal{L}(\mathbb{K}, F)$ with $A_v x := xv$. Then $A_v 1 = v$. Therefore $A \mapsto A1$ is a vector space isomorphism. Furthermore,

$$\|Ax\|_F = |x| \, \|A1\|_F \leq \|A1\|_F \quad \text{for } x \in \mathbb{B}_{\mathbb{K}} \text{ and } A \in \mathcal{L}(\mathbb{K}, F) \ .$$

---

[2] Note that in normed vector spaces $\cong$ always means "topologically isomorphic" and not just that there exists a vector space isomorphism.
[3] See Example III.1.3(o).

This implies $\|A\|_{\mathcal{L}(\mathbb{K},F)} = \|A1\|_F$. Therefore (1.2) is an isometry. ∎

(b) $\mathcal{L}\mathrm{aut}(E)$ is a group, the **group of topological automorphisms** of E, where the multiplication is defined through the composition of the two linear maps.

Proof This follows from Remark VI.2.4(g). ∎

(c) If E and F are isomorphic, then E is a Banach space if and only if F is.

Proof It is easy to see that a topological isomorphism maps Cauchy sequences to Cauchy sequence and maps convergent sequences to convergent sequences. ∎

(d) Suppose E and F are Banach spaces, and $A \in \mathcal{L}(E,F)$ is bijective. Then A is a topological isomorphism, that is, $A \in \mathcal{L}\mathrm{is}(E,F)$.

Proof This theorem is a standard part of functional analysis (see the Banach homomorphism theorem, also known as the open mapping theorem). ∎

1.4 Theorem *Suppose $\{b_1, \ldots, b_n\}$ is a basis of E. Then*

$$T : E \to \mathbb{K}^n , \quad e = \sum_{j=1}^n x_j b_j \to (x_1, \ldots, x_n) \tag{1.3}$$

is a topological isomorphism, that is, every finite-dimensional normed vector space is topologically isomorphic to a Euclidean vector space.

Proof Obviously, T is well defined, linear, and bijective, and

$$T^{-1}x = \sum_{j=1}^n x^j b_j \quad \text{for } x = (x^1, \ldots, x^n) \in \mathbb{K}^n ,$$

(see Remark I.12.5). From the Cauchy–Schwarz inequality of Corollary II.3.9, we have

$$\|T^{-1}x\| \le \sum_{j=1}^n |x^j|\,\|b_j\| \le \left(\sum_{j=1}^n |x^j|^2 \right)^{1/2} \left(\sum_{j=1}^n \|b_j\|^2 \right)^{1/2} = \beta\,|x|$$

with $\beta := \left(\sum_{j=1}^n \|b_j\|^2 \right)^{1/2}$. Therefore T^{-1} belongs to $\mathcal{L}(\mathbb{K}^n, E)$.

We set $|x|_{\bullet} := \|T^{-1}x\|$ for $x \in \mathbb{K}^n$. It is not difficult to see that $|\cdot|_{\bullet}$ is a norm on $\mathbb{K}^n$ (see Exercise II.3.1). In Example III.3.9(a), we have seen that all norms are equivalent on $\mathbb{K}^n$. Hence there is an $\alpha > 0$ such that $|x| \le \alpha\,|x|_{\bullet}$ for $x \in \mathbb{K}^n$. Thus we have

$$|Te| \le \alpha\,|Te|_{\bullet} = \alpha\,\|e\| \quad \text{for } e \in E ,$$

that is, $T \in \mathcal{L}(E, \mathbb{K}^n)$, and we are done. ∎

1.5 Corollary   *If E is finite-dimensional, these statements hold:*

(i) *All norms on E are equivalent.*

(ii) *E is complete and therefore a Banach space.*

Proof   Let $A := T^{-1}$ with the T of (1.3).

(i) For $j = 1, 2$, let $\|\cdot\|_j$ be norms on E. Then $x \mapsto |x|_{(j)} := \|Ax\|_j$ are norms on $\mathbb{K}^n$. Hence, there is from Example III.3.9(a) an $\alpha \geq 1$ such that

$$\alpha^{-1} |x|_{(1)} \leq |x|_{(2)} \leq \alpha |x|_{(1)} \quad \text{for } x \in \mathbb{K}^n \ .$$

Because $\|e\|_j = |A^{-1}e|_{(j)}$ it follows from this that

$$\alpha^{-1} \|e\|_1 \leq \|e\|_2 \leq \alpha \|e\|_1 \quad \text{for } e \in E \ .$$

Hence $\|\cdot\|_1$ and $\|\cdot\|_2$ are equivalent norms on E.

(ii) This follows from Remark 1.3(c) and Theorems 1.4 and II.6.5. ∎

1.6 Theorem   *Let E be finite-dimensional. Then $\mathrm{Hom}(E, F) = \mathcal{L}(E, F)$. In other words, every linear operator on a finite-dimensional normed vector space is continuous.*

Proof   We define $T \in \mathcal{L}\mathrm{is}(E, \mathbb{K}^n)$ through (1.3). Then there exists a $\tau > 0$ such that $|Te| \leq \tau \|e\|$ for $e \in E$.

Now suppose $A \in \mathrm{Hom}(E, F)$. Then $Ae = \sum_{j=1}^n x^j Ab_j$. Consequently, the Cauchy–Schwarz inequality (with $x_e := Te$) says

$$\|Ae\| \leq \left(\sum_{j=1}^n \|Ab_j\|^2 \right)^{1/2} |x_e| = \alpha |x_e| \quad \text{for } e \in E \ ,$$

where we have set $\alpha := \left(\sum_j \|Ab_j\|^2 \right)^{1/2}$. Thus we get

$$\|Ae\| \leq \alpha |x_e| = \alpha |Te| \leq \alpha \tau \|e\| \quad \text{for } e \in E \ .$$

Finally, according to Theorem VI.2.5, A is continuous. ∎

1.7 Remarks   **(a)** The statements of Corollary 1.5 and Theorem 1.6 do not hold for infinite-dimensional normed vector spaces.

Proof   (i) We set $E := BC^1((-1, 1), \mathbb{R})$. In the normed vector space $(E, \|\cdot\|_\infty)$, we consider the sequence (u_n) with $u_n(t) := \sqrt{t^2 + 1/n}$ for $t \in [-1, 1]$ and $n \in \mathbb{N}^\times$.

It is easy to see that (u_n) is a Cauchy sequence in $(E, \|\cdot\|_\infty)$. We now assume $(E, \|\cdot\|_\infty)$ is complete. Then there is a $u \in E$ such that $\lim \|u_n - u\|_\infty = 0$. In particular, (u_n) converges pointwise to u.

Obviously, $u_n(t) = \sqrt{t^2 + 1/n} \to |t|$ as $n \to \infty$ for $t \in [-1, 1]$. Therefore, it follows from the uniqueness of limits (with pointwise convergence) that

$$(t \mapsto |t|) = u \in BC^1\big((-1, 1), \mathbb{R}\big) \ ,$$

which is false. The contradiction shows $(E, \|\cdot\|_\infty)$ is not complete.

Finally, we know from Exercise V.2.10 that $\big(BC^1((-1, 1), \mathbb{R}), \|\cdot\|_\bullet\big)$ with the norm $\|u\|_\bullet := \|u\|_\infty + \|u'\|_\infty$ is a Banach space. Thus, due to Remark II.6.7(a), $\|\cdot\|$ and $\|\cdot\|_\bullet$ cannot be equivalent.

(ii) We set

$$E := \big(C^1([0, 1], \mathbb{R}), \|\cdot\|_\infty\big) \quad \text{and} \quad F := \big(C([0, 1], \mathbb{R}), \|\cdot\|_\infty\big)$$

and consider $A : E \to F$, $u \mapsto u'$. Then E and F are normed vector spaces, and A is linear. We assume that A is bounded and therefore continuous. Because (u_n) with

$$u_n(t) := (1/n) \sin(n^2 t) \quad \text{for } n \in \mathbb{N}^\times \text{ and } t \in [0, 1]$$

converges to zero in E, we have $Au_n \to 0$ in F.

From $(Au_n)(t) = n \cos(n^2 t)$, we have $(Au_n)(0) = n$. But, since (Au_n) converges to 0 in F and since uniformity implies pointwise convergence, this is not possible. Thus A does not belong to $\mathcal{L}(E, F)$, that is, $\operatorname{Hom}(E, F) \backslash \mathcal{L}(E, F) \neq \emptyset$. ∎

(b) Every finite-dimensional inner product space is a Hilbert space.

(c) Suppose E is a finite-dimensional Banach space. Then there is an equivalent **Hilbert norm** on E, that is, a norm induced by a scalar product. In other words, every finite-dimensional Banach space can be renormed into a Hilbert space.[4]

Proof Suppose $n := \dim E$ and $T \in \mathcal{L}\mathrm{is}(E, \mathbb{K}^n)$. Then

$$(x \mid y)_E := (Tx \mid Ty) \quad \text{for } x, y \in E$$

defines a scalar product on E. The claim then follows from Corollary 1.5(i). ∎

Matrix representations

Let $m, n \in \mathbb{N}^\times$. We denote by $\mathbb{K}^{m \times n}$ the set of all $(m \times n)$ matrices

$$[a_k^j] = \begin{bmatrix} a_1^1 & \cdots & a_n^1 \\ \vdots & & \vdots \\ a_1^m & \cdots & a_n^m \end{bmatrix}$$

with entries a_k^j in $\mathbb{K}$. Here the upper index is the row number, and the lower index is the column number.[5] When required for clarity, we will write $[a_k^j]$ as

---

[4] Assuming a finite dimension is important here.

[5] Instead of a_k^j, we will occasionally write a_{jk} or a^{jk}, where the first index is always the row number and the second is always the column number.

$$[a_k^j]_{\substack{1 \le j \le m \\ 1 \le k \le n}} \; .$$

Finally, we set

$$|M| := \sqrt{\sum_{j=1}^{m} \sum_{k=1}^{n} |a_k^j|^2} \quad \text{and} \quad M := [a_k^j] \in \mathbb{K}^{m \times n} \; . \qquad (1.4)$$

We will assume you are familiar with the basic ideas of matrices in linear algebra and, in particular, you know that $\mathbb{K}^{m \times n}$ is a $\mathbb{K}$-vector space of dimension mn with respect to the usual matrix addition and multiplication by scalars.

1.8 Proposition
  (i) *By* (1.4), *we define a norm* $|\cdot|$ *on* $\mathbb{K}^{m \times n}$, *called the* **Hilbert-Schmidt norm**. *Therefore*
$$\mathbb{K}^{m \times n} := (\mathbb{K}^{m \times n}, |\cdot|)$$

   is a Banach space.
  (ii) *The map*

$$\mathbb{K}^{m \times n} \to \mathbb{K}^{mn} \; , \quad [a_k^j] \mapsto (a_1^1, \ldots, a_n^1, a_1^2, \ldots, a_n^2, \ldots, a_1^m, \ldots, a_n^m)$$

   is an isometric isomorphism.

Proof   We leave the simple proof to you.[6] ∎

In the following, we will always implicitly endow $\mathbb{K}^{m \times n}$ with Hilbert–Schmidt norm.

Suppose E and F are finite-dimensional, and suppose

$$\mathcal{E} = \{e_1, \ldots, e_n\} \quad \text{and} \quad \mathcal{F} = \{f_1, \ldots, f_m\}$$

are (ordered) bases of E and F. According to Theorem 1.6, $\mathrm{Hom}(E, F) = \mathcal{L}(E, F)$. We will represent the action of $A \in \mathcal{L}(E, F)$ on vectors in E using the bases $\mathcal{E}$ and $\mathcal{F}$. First, for $k = 1, \ldots, n$, there are unique numbers $a_k^1, \ldots, a_k^m$ such that $Ae_k = \sum_{j=1}^{m} a_k^j f_j$. Thus from the linearity of A, we have for every $x = \sum_{k=1}^{n} x^k e_k \in E$ that

$$Ax = \sum_{k=1}^{n} x^k Ae_k = \sum_{j=1}^{m} \left(\sum_{k=1}^{n} a_k^j x^k \right) f_j = \sum_{j=1}^{m} (Ax)^j f_j$$

with

$$(Ax)^j := \sum_{k=1}^{n} a_k^j x^k \quad \text{for } j = 1, \ldots, m \; .$$

---

[6]See also Exercise II.3.14.

We set

$$[A]_{\mathcal{E},\mathcal{F}} := [a_k^j] \in \mathbb{K}^{m \times n}$$

and call $[A]_{\mathcal{E},\mathcal{F}}$ the **representation matrix** of A in the bases $\mathcal{E}$ of E and $\mathcal{F}$ of F. (If $E = F$, we simply write $[A]_{\mathcal{E}}$ for $[A]_{\mathcal{E},\mathcal{E}}$.)

Now let $[a_k^j] \in \mathbb{K}^{m \times n}$. For $x = \sum_{k=1}^n x^k e_k \in E$ we set

$$Ax := \sum_{j=1}^m \Big(\sum_{k=1}^n a_k^j x^k \Big) f_j \ .$$

Then $A := (x \mapsto Ax)$ is a linear map from E to F whose representation matrix is $[A]_{\mathcal{E},\mathcal{F}} = [a_k^j]$.

The next theorem summarizes the most important properties of the representation matrices of linear maps between finite-dimensional vector spaces.

1.9 Theorem

(i) *We set $n := \dim E$ and $m := \dim F$, and let $\mathcal{E}$ and $\mathcal{F}$ be respective bases of E and F. Then the* **matrix representation**

$$\mathcal{L}(E, F) \to \mathbb{K}^{m \times n} \ , \quad A \mapsto [A]_{\mathcal{E},\mathcal{F}} \tag{1.5}$$

is a topological isomorphism.

(ii) *Suppose G is a finite-dimensional normed vector space and $\mathcal{G}$ is a basis of G. Then,[7]*

$$[AB]_{\mathcal{E},\mathcal{G}} = [A]_{\mathcal{F},\mathcal{G}}[B]_{\mathcal{E},\mathcal{F}} \quad \text{for } A \in \mathcal{L}(F, G) \text{ and } B \in \mathcal{L}(E, F) \ .$$

Proof (i) The map (1.5) is clearly linear. The above considerations show that it is bijective also (see also [Gab96, Chapter D.5.5]). In particular, $\dim \mathcal{L}(E, F) = \dim \mathbb{K}^{m \times n}$. Because $\mathbb{K}^{m \times n} \cong \mathbb{K}^{mn}$, the space $\mathcal{L}(E, F)$ has dimension mn. Hence we can apply Theorem 1.6, and we find

$$\big(A \mapsto [A]_{\mathcal{E},\mathcal{F}} \big) \in \mathcal{L}\big(\mathcal{L}(E, F), \mathbb{K}^{m \times n} \big) \text{ and } \big([A]_{\mathcal{E},\mathcal{F}} \mapsto A \big) \in \mathcal{L}\big(\mathbb{K}^{m \times n}, \mathcal{L}(E, F) \big) \ .$$

Therefore $A \mapsto [A]_{\mathcal{E},\mathcal{F}}$ is a topological isomorphism.

(ii) This fact is well known in linear algebra (for example, [Gab96, Chapter D.5.5]). ∎

In analysis, we will consider all maps of metric spaces in the space of continuous linear maps between two Banach spaces E and F. If these are finite-dimensional, then, according to Theorem 1.9, we can consider the maps to be matrices. In this case, it is easy to prove from the above results that the maps are continuous. As the next important corollary shows, it suffices to verify that the entries in the representation matrix are continuous.

---

[7] For two composed linear maps, we usually write AB instead of $A \circ B$.

1.10 Corollary *Suppose X is a metric space and E and F are finite-dimensional Banach spaces with respective dimensions n and m and ordered bases $\mathcal{E}$ and $\mathcal{F}$. Let $A(\cdot)\colon X \to \mathcal{L}(E,F)$, and let $\left[a_k^j(x)\right]$ be the representation matrix $[A(x)]_{\mathcal{E},\mathcal{F}}$ of $A(x)$ for $x \in X$. Then*

$$A(\cdot) \in C\big(X, \mathcal{L}(E,F)\big) \iff \Big(a_k^j(\cdot) \in C(X,\mathbb{K}),\ 1 \le j \le m,\ 1 \le k \le n\Big).$$

Proof This follows immediately from Theorems 1.9, Proposition 1.8(ii), and Proposition III.1.10. ∎

Convention In $\mathbb{K}^n$, we use the standard basis $\{e_1,\ldots,e_n\}$ defined in Example I.12.4(a) unless we say otherwise. Also, we denote by $[A]$ the representation matrix of $A \in \mathcal{L}(\mathbb{K}^n,\mathbb{K}^m)$.

The exponential map

Suppose E is a Banach space and $A \in \mathcal{L}(E)$. From Corollary 1.2(ii), we know that $\mathcal{L}(E)$ is a Banach algebra. Thus A^k belongs to $\mathcal{L}(E)$ for $k \in \mathbb{N}$, and

$$\|A^k\| \le \|A\|^k \quad \text{for } k \in \mathbb{N}.$$

Therefore the exponential series $\sum_k \alpha^k/k!$ is for every $\alpha \ge \|A\|$ a majorant of the **exponential series**

$$\sum_k A^k/k! \tag{1.6}$$

in $\mathcal{L}(E)$. Hence from the majorant criterion (Theorem II.8.3), the series (1.6) converges absolutely in $\mathcal{L}(E)$. We call its value

$$e^A := \sum_{k=0}^{\infty} \frac{A^k}{k!}$$

the **exponential** of A, and

$$\mathcal{L}(E) \to \mathcal{L}(E), \quad A \mapsto e^A$$

is the **exponential map** in $\mathcal{L}(E)$.

If A and $t \in \mathbb{K}$, then tA also belongs to $\mathcal{L}(E)$. Therefore the map

$$U := U_A \colon \mathbb{K} \to \mathcal{L}(E), \quad t \mapsto e^{tA} \tag{1.7}$$

is well defined. In the next theorem, we gather the most important properties of the function U.

1.11 Theorem *Suppose E is a Banach space and $A \in \mathcal{L}(E)$. Then*

(i) *$U \in C^\infty(\mathbb{K}, \mathcal{L}(E))$ and $\dot{U} = AU$;*

(ii) *U is a group homomorphism from the additive group $(\mathbb{K}, +)$ to the multi-plicative group $\mathcal{L}\mathrm{aut}(E)$.*

Proof (i) It is enough to show that for every $r \geq 1$, we have

$$U \in C^1(r\mathbb{B}_{\mathbb{C}}, \mathcal{L}(E)) \quad \text{and} \quad \dot{U}(t) = AU(t) \text{ for } |t| < r .$$

The conclusion then follows by induction.

For $n \in \mathbb{N}$ and $|t| < r$, define $f_n(t) := (tA)^n/n!$. Then $f_n \in C^1(r\mathbb{B}_{\mathbb{C}}, \mathcal{L}(E))$, and $\sum f_n$ converges pointwise to $U \,|\, r\mathbb{B}_{\mathbb{C}}$. Also $\dot{f}_n = Af_{n-1}$, and thus

$$\|\dot{f}_n(t)\| \leq \|A\| \, \|f_{n-1}(t)\| \leq (r\,\|A\|)^n/(n-1) \quad \text{for } t \in r\mathbb{B}_{\mathbb{C}} \text{ and } n \in \mathbb{N}^\times .$$

Therefore a scalar exponential series is a majorant series for $\sum \dot{f}_n$. Hence $\sum \dot{f}_n$ converges normally to $AU \,|\, r\mathbb{B}_{\mathbb{C}}$ on $r\mathbb{B}_{\mathbb{C}}$, and the theorem then follows from Corollary V.2.9.

(ii) For $s, t \in \mathbb{K}$, sA and tA commute in $\mathcal{L}(E)$. The binomial theorem implies (Theorem I.8.4)

$$(sA + tA)^n = \sum_{k=0}^{n} \binom{n}{k} (sA)^k (tA)^{n-k} .$$

Using the absolute convergence of the exponential series, we get, as in the proof of Example II.8.12(a),

$$U(s + t) = e^{(s+t)A} = e^{sA+tA} = e^{sA}e^{tA} = U(s)U(t) \quad \text{for } s, t \in \mathbb{K} . \ \blacksquare$$

1.12 Remarks Suppose $A \in \mathcal{L}(E)$ and $s, t \in \mathbb{K}$.

(a) The exponentials e^{sA} and e^{tA} commute: $e^{sA}e^{tA} = e^{tA}e^{sA}$.

(b) $\|e^{tA}\| \leq e^{|t|\,\|A\|}$.

Proof This follows from Remark II.8.2(c). $\blacksquare$

(c) $(e^{tA})^{-1} = e^{-tA}$.

(d) $\partial_t^n e^{tA} = A^n e^{tA} = e^{tA} A^n$ for $n \in \mathbb{N}$.

Proof By induction, this follows immediately from Theorem 1.11(i). $\blacksquare$

1.13 Examples **(a)** If $E = \mathbb{K}$, the exponential map agrees with the exponential function.

(b) Suppose $N \in \mathcal{L}(E)$ is **nilpotent**, that is, there is an $m \in \mathbb{N}$ such that $N^{m+1} = 0$. Then

$$e^N = \sum_{k=0}^{m} \frac{N^k}{k!} = 1 + N + \frac{N^2}{2!} + \cdots + \frac{N^m}{m!} ,$$

where 1 denotes the unit element of $\mathcal{L}(E)$.

(c) Suppose E and F are Banach spaces, $A \in \mathcal{L}(E)$, and $B \in \mathcal{L}(F)$. Also suppose

$$A \otimes B : E \times F \to E \times F , \quad (x, y) \mapsto (Ax, By) .$$

Then

$$e^{A \otimes B} = e^A \otimes e^B .$$

Proof This follows easily from $(A \otimes B)^n = A^n \otimes B^n$. ∎

(d) For $A \in \mathcal{L}(\mathbb{K}^m)$ and $B \in \mathcal{L}(\mathbb{K}^n)$, we have the relation

$$[e^{A \otimes B}] = \begin{bmatrix} [e^A] & 0 \\ 0 & [e^B] \end{bmatrix} .$$

Proof This follows from (c). ∎

(e) Suppose $\lambda_1, \ldots, \lambda_m \in \mathbb{K}$, and define $\mathrm{diag}(\lambda_1, \ldots, \lambda_m) \in \mathcal{L}(\mathbb{K}^m)$ by

$$[\mathrm{diag}(\lambda_1, \ldots, \lambda_m)] := \begin{bmatrix} \lambda_1 & & 0 \\ & \ddots & \\ 0 & & \lambda_m \end{bmatrix} =: \mathrm{diag}[\lambda_1, \ldots, \lambda_m] .$$

Then

$$e^{\mathrm{diag}(\lambda_1, \ldots, \lambda_m)} = \mathrm{diag}(e^{\lambda_1}, \ldots, e^{\lambda_m}) .$$

Proof This follows by induction using (a) and (d). ∎

(f) For $A \in \mathcal{L}(\mathbb{R}^2)$ such that

$$[A] = \begin{bmatrix} 0 & -1 \\ 1 & 0 \end{bmatrix} ,$$

we have

$$[e^{tA}] = \begin{bmatrix} \cos t & -\sin t \\ \sin t & \cos t \end{bmatrix} \quad \text{for } t \in \mathbb{R} .$$

Proof Exercise I.11.10 implies easily that

$$(tA)^{2n+1} = (-1)^n t^{2n+1} A , \quad (tA)^{2n} = (-1)^n t^{2n} 1_{\mathbb{R}^2} \quad \text{for } t \in \mathbb{R} \text{ and } n \in \mathbb{N} .$$

Because an absolutely convergent series can be rearranged arbitrarily, we get

$$\sum_{k=0}^{\infty} \frac{(tA)^k}{k!} = \sum_{n=0}^{\infty} \frac{(tA)^{2n}}{(2n)!} + \sum_{n=0}^{\infty} \frac{(tA)^{2n+1}}{(2n+1)!}$$

$$= \Big(\sum_{n=0}^{\infty} (-1)^n \frac{t^{2n}}{(2n)!}\Big) 1_{\mathbb{R}^2} + \Big(\sum_{n=0}^{\infty} (-1)^n \frac{t^{2n+1}}{(2n+1)!}\Big) A ,$$

which proves the claim. ∎

Linear differential equations

For $a \in E$, the map $u(\cdot, a) \colon \mathbb{R} \to E$, $t \mapsto e^{tA}a$ is smooth and satisfies the "differential equation" $\dot{x} = Ax$, which follows at once from Theorem 1.11(i). We will now study such differential equations more thoroughly.

Suppose $A \in \mathcal{L}(E)$ and $f \in C(\mathbb{R}, E)$. Then

$$\dot{x} = Ax + f(t) \quad \text{for } t \in \mathbb{R} , \tag{1.8}$$

is called a **first order linear differential equation** in E. If $f = 0$, (1.8) is **homogeneous**; otherwise, it is **inhomogeneous**. Should A be independent of the "time" t, we say that $\dot{x} = Ax$ is a linear homogeneous differential equation with **constant coefficients**. By a **solution** to (1.8), we mean a function $x \in C^1(\mathbb{R}, E)$ that satisfies (1.8) pointwise, that is, $\dot{x}(t) = A\big(x(t)\big) + f(t)$ for $t \in \mathbb{R}$.

Let $a \in E$. Then

$$\dot{x} = Ax + f(t) \text{ for } t \in \mathbb{R} , \quad \text{with} \quad x(0) = a \tag{1.9$_a$}$$

is an **initial value problem** for the differential equation (1.8), and a is an **initial value**. A **solution** to this initial value problem is a function $x \in C^1(\mathbb{R}, E)$ that satisfies (1.9)$_a$ pointwise.

1.14 Remarks (a) By specifying the argument t of f in (1.8) and (1.9)$_a$ (but not of x), we are adopting a less precise, symbolic notation, which we do because it emphasizes that f generally depends on t. For a full description of (1.8) or (1.9)$_a$, one must always specify what is meant by a solution to such a differential equation.

(b) For simplicity, we consider here only the case that f is defined on all of $\mathbb{R}$. We leave to you the obvious modifications needed when f is defined and continuous only on a perfect subinterval. ∎

(c) Because $C^1(\mathbb{R}, E) \subset C(\mathbb{R}, E)$ and the derivative operator ∂ is linear (see Theorem IV.1.12), the map

$$\partial - A \colon C^1(\mathbb{R}, E) \to C(\mathbb{R}, E) , \quad u \mapsto \dot{u} - Au$$

is linear. Defining

$$\ker(\partial - A) = \big\{ x \in C^1(\mathbb{R}, E) ; \ x \text{ is a solution of } \dot{x} = Ax \big\} =: V$$

shows the **solution space** V of homogeneous differential equations $\dot{x} = Ax$ to be a vector subspace of $C^1(\mathbb{R}, E)$.

(d) The collective solutions to (1.8) form an affine subspace $u + V$ of $C^1(\mathbb{R}, E)$, where u is (any) solution to (1.8).

Proof For $w := u + v \in u + V$, we have $(\partial - A)w = (\partial - A)u + (\partial - A)v = f$. Therefore w is a solution to (1.8). If, conversely, w is a solution to (1.8), then $v := u - w$ belongs to V because $(\partial - A)v = (\partial - A)u - (\partial - A)w = f - f = 0$. ∎

(e) For $a \in E$ and $x \in C(\mathbb{R}, E)$, let

$$\Phi_a(x)(t) := a + \int_0^t \big(Ax(s) + f(s)\big)\, ds \ .$$

Obviously, Φ_a is an affine map from $C(\mathbb{R}, E)$ to $C^1(\mathbb{R}, E)$, and x is a solution of $(1.9)_a$ if and only if x is a fixed point of Φ_a in[8] $C(\mathbb{R}, E)$, that is, when x solves the linear **integral equation**

$$x(t) = a + \int_0^t \big(Ax(s) + f(s)\big)\, ds \quad \text{for } t \in \mathbb{R} \ . \ \blacksquare$$

Gronwall's lemma

An important aid to exploring differential equations is the following estimate, known as **Gronwall's lemma**.

1.15 Lemma (Gronwall) *Suppose J is an interval, $t_0 \in J$, and $\alpha, \beta \in [0, \infty)$. Further, suppose $y \colon J \to [0, \infty)$ is continuous and satisfies*

$$y(t) \le \alpha + \beta \left| \int_{t_0}^t y(\tau)\, d\tau \right| \quad \text{for } t \in J \ . \tag{1.10}$$

Then

$$y(t) \le \alpha e^{\beta\,|t - t_0|} \quad \text{for } t \in J \ .$$

Proof (i) We consider first the case $t \ge t_0$ and set

$$h \colon [t_0, t] \to \mathbb{R}^+ \ , \quad s \mapsto \beta e^{\beta(t_0 - s)} \int_{t_0}^s y(\tau)\, d\tau \quad \text{for } t \in J \ .$$

From Theorem VI.4.12 and (1.10), it follows that

$$h'(s) = -\beta h(s) + \beta e^{\beta(t_0 - s)} y(s) \le \alpha \beta e^{\beta(t_0 - s)} = \frac{d}{ds}\big(-\alpha e^{\beta(t_0 - s)}\big) \quad \text{for } s \in [t_0, t] \ .$$

Integrating this inequality from to t_0 to t gives

$$h(t) = \beta e^{\beta(t_0 - t)} \int_{t_0}^t y(\tau)\, d\tau \le \alpha - \alpha e^{\beta(t_0 - t)} \ ,$$

where we have used $h(t_0) = 0$. Thus we have

$$\beta \int_{t_0}^t y(\tau)\, d\tau \le \alpha e^{\beta(t - t_0)} - \alpha \ ,$$

---

[8]Of course, we identify Φ_a with $i \circ \Phi_a$, where i denotes the inclusion of $C^1(\mathbb{R}, E)$ in $C(\mathbb{R}, E)$.

and (1.10) finishes the case.

(ii) When $t \leq t_0$, we set

$$h(s) := -\beta e^{\beta(s-t_0)} \int_s^{t_0} y(\tau)\, d\tau \quad \text{for } s \in [t, t_0] \ ,$$

and proceed as in (i). ∎

Using Gronwall's lemma, we can easily prove the next statement about the solution space of equation (1.8).

1.16 Proposition

(i) *Suppose $x \in C^1(\mathbb{R}, E)$ solves $\dot{x} = Ax$. Then*

$$x = 0 \Longleftrightarrow x(t_0) = 0 \text{ for some } t_0 \in \mathbb{R} \ .$$

(ii) *The initial value problem $(1.9)_a$ has at most one solution.*

(iii) *The initial value problem $\dot{x} = Ax$, $x(0) = a$ has the unique solution $t \mapsto e^{tA}a$ for every $a \in E$.*

(iv) *The solution space $V \subset C^1(\mathbb{R}, E)$ of $\dot{x} = Ax$ is a vector space isomorphic to E. An isomorphism is given by*

$$E \to V \ , \quad a \mapsto (t \mapsto e^{tA}a) \ .$$

(v) *Suppose $x_1, \ldots, x_m$ solves $\dot{x} = Ax$. Then these statements are equivalent:*

(α) *$x_1, \ldots, x_m$ are linearly independent in $C^1(\mathbb{R}, E)$.*

(β) *The $x_1(t), \ldots, x_m(t)$ are linearly independent in E for every $t \in \mathbb{R}$.*

(γ) *There is a $t_0 \in \mathbb{R}$ such that $x_1(t_0), \ldots, x_m(t_0)$ are linearly independent in E.*

Proof (i) It suffices to check the implication "$\Longleftarrow$". So let $t_0 \in \mathbb{R}$ with $x(t_0) = 0$. Then

$$x(t) = A \int_{t_0}^t x(\tau)\, d\tau \quad \text{for } t \in \mathbb{R}$$

(see Remark 1.14(e) and Exercise VI.3.3). Then $y := \|x\|$ satisfies the inequality

$$y(t) \leq \|A\| \left\| \int_{t_0}^t x(\tau)\, d\tau \right\| \leq \|A\| \left| \int_{t_0}^t \|x(\tau)\|\, d\tau \right| = \|A\| \left| \int_{t_0}^t y(\tau)\, d\tau \right| \quad \text{for } t \in \mathbb{R} \ ,$$

and the claim follows from Gronwall's lemma.

(ii) If u and v solve $(1.9)_a$, then $w = u - v$ also solves $\dot{x} = Ax$ with $w(0) = 0$. From (i), it follows that $w = 0$.

(iii) This follows from Theorem 1.11 and (ii).

(iv) This is a consequence of (iii).

(v) Due to (iv) and Theorem 1.11(ii), it suffices to prove the implication "$(\alpha)\Rightarrow(\beta)$". Hence suppose for $t_0 \in \mathbb{R}$ that $\lambda_j \in \mathbb{K}$ satisfy $\sum_{j=1}^{m} \lambda_j x_j(t_0) = 0$ for $j = 1,\ldots,m$. Because $\sum_{j=1}^{m} \lambda_j x_j$ solves $\dot{x} = Ax$, it follows from (i) that $\sum_{j=1}^{m} \lambda_j x_j = 0$. By assumption, $x_1,\ldots,x_m$ in $C^1(\mathbb{R}, E)$ are linearly independent. Therefore, by the linear independence of $x_1(t_0),\ldots,x_m(t_0)$ in E, we have $\lambda_1 = \cdots = \lambda_m = 0$. ∎

The variation of constants formula

From Proposition 1.16(iii), the initial value problem for the homogeneous equation $\dot{x} = Ax$ has a unique solution for every initial value. To solve the inhomogeneous equation (1.8), we refer to Remark 1.14(d): If the solution space V of the homogeneous equation $\dot{x} = Ax$ is known, any solution to (1.8) in fact gives rise to all the solutions.

To construct such a **particular solution**, suppose $x \in C^1(\mathbb{R}, E)$ solves (1.8) and

$$T \in C^1\big(\mathbb{R}, \mathcal{L}(E)\big) \quad \text{such that } T(t) \in \mathcal{L}\mathrm{aut}(E) \qquad (1.11)$$

for $t \in \mathbb{R}$. Further define

$$y(t) := T^{-1}(t)x(t), \qquad \text{where} \quad T^{-1}(t) := \big[T(t)\big]^{-1} \text{ and } t \in \mathbb{R}.$$

The freedom to choose the "transformation" T can be used to find the "simplest possible" differential equation for y. First, the product rule[9] and that x solves (1.8) give

$$\begin{aligned}
\dot{y}(t) &= (T^{-1})^{\cdot}(t)x(t) + T^{-1}(t)\dot{x}(t) \\
&= (T^{-1})^{\cdot}(t)x(t) + T^{-1}(t)Ax(t) + T^{-1}(t)f(t) \qquad (1.12) \\
&= (T^{-1})^{\cdot}(t)T(t)y(t) + T^{-1}(t)AT(t)y(t) + T^{-1}(t)f(t).
\end{aligned}$$

Next, by differentiating the identity

$$T^{-1}(t)T(t) = \mathrm{id}_E \quad \text{for } t \in \mathbb{R},$$

(using the product rule), we have

$$(T^{-1})^{\cdot}(t)T(t) + T^{-1}(t)\dot{T}(t) = 0 \quad \text{for } t \in \mathbb{R},$$

and thus

$$(T^{-1})^{\cdot}(t) = -T^{-1}(t)\dot{T}(t)T^{-1}(t) \quad \text{for } t \in \mathbb{R}.$$

---

[9]It is easy to verify that the proof of the product rule in Theorem IV.1.6(ii) is also valid in the current, more general, situation. See also Example 4.8(b).

Together with (1.12), this becomes

$$\dot{y}(t) = T^{-1}(t)\left[AT(t) - \dot{T}(t)\right]y(t) + T^{-1}(t)f(t) \quad \text{for } t \in \mathbb{R} \ . \tag{1.13}$$

According to Theorem 1.11, $t \mapsto T(t) := e^{tA}$ satisfies $AT - \dot{T} = 0$ and (1.11). With this choice of T, it follows from (1.13) that

$$\dot{y}(t) = T^{-1}(t)f(t) \quad \text{for } t \in \mathbb{R} \ .$$

Therefore

$$y(t) = y(0) + \int_0^t e^{-\tau A}f(\tau)\, d\tau \quad \text{for } t \in \mathbb{R} \ .$$

Because $y(0) = x(0)$, we have

$$x(t) = e^{tA}x(0) + \int_0^t e^{(t-\tau)A}f(\tau)\, d\tau \quad \text{for } t \in \mathbb{R} \ . \tag{1.14}$$

Now we can easily prove an existence and uniqueness result for $(1.9)_a$.

1.17 Theorem   *For every $a \in E$ and $f \in C(\mathbb{R}, E)$, there is a unique solution $u(\,\cdot\,; a)$ to*

$$\dot{x} = Ax + f(t) \text{ and } x(0) = a$$

which is given through the **variation of constants formula**

$$u(t; a) = e^{tA}a + \int_0^t e^{(t-\tau)A}f(\tau)\, d\tau \quad \text{for } t \in \mathbb{R} \ . \tag{1.15}$$

Proof   Due to Proposition 1.16(ii), we need only show that (1.15) solves $(1.9)_a$. This, however, follows immediately from Theorems 1.11 and VI.4.12. ∎

1.18 Remark   To clarify the name "variation of constants formula", we consider the simplest case, namely, the $\mathbb{R}$-valued equation

$$\dot{x} = ax + f(t) \quad \text{for } t \in \mathbb{R} \ . \tag{1.16}$$

If $a \in \mathbb{R}$, the homogeneous equation $\dot{x} = ax$ has a solution $v(t) := e^{ta}$. Thus the **general solution** of the homogeneous equation is given by cv, where $c \in \mathbb{R}$ is a constant. That is, $\{\,cv \ ; \ c \in \mathbb{R}\,\}$ is the solution space of the homogeneous equation.

To specify a particular solution to (1.16), we "vary the constant", that is, we make the ansatz $x(t) := c(t)v(t)$ with a to-be-determined function c. Because x must satisfy the equation (1.16), we have

$$\dot{c}v + c\dot{v} = acv + f \ ,$$

and thus, from $c\dot{v} = c(av)$, we have $\dot{c} = f/v$. Then (1.14) follows by integration. ∎

Determinants and eigenvalues

We next assemble some important results from linear algebra.

1.19 Remarks (a) The **determinant** $\det[a_k^j]$ of $[a_k^j] \in \mathbb{K}^{m \times m}$ is defined through the signature formula

$$\det[a_k^j] := \sum_{\sigma \in S_m} \text{sign}(\sigma) a_{\sigma(1)}^1 \cdot \cdots \cdot a_{\sigma(m)}^m$$

(see [Gab96, Chapter A3]), where sign is the sign function from Exercise I.9.6.[10]

(b) Let E be a finite-dimensional Banach space. The **determinant function**

$$\det : \mathcal{L}(E) \to \mathbb{K}$$

can be expressed for $A \in \mathcal{L}(E)$ as $\det(A) := \det([A]_{\mathcal{E}})$, where $\mathcal{E}$ is a basis of E. Then det is well defined, that is, independent of the particular basis $\mathcal{E}$, and we have

$$\det(1_E) = 1 \text{ and } \det(AB) = \det(A) \det(B) \quad \text{for } A, B \in \mathcal{L}(E) , \tag{1.17}$$

and

$$\det(A) \neq 0 \iff A \in \mathcal{L}\text{aut}(E) . \tag{1.18}$$

Letting $\lambda - A := \lambda 1_E - A$ and $m := \dim(E)$, we have

$$\det(\lambda - A) = \sum_{k=0}^{m} (-1)^k \alpha_{m-k} \lambda^{m-k} \quad \text{for } \lambda \in \mathbb{C} , \tag{1.19}$$

with $\alpha_k \in \mathbb{C}$ and

$$\alpha_m = 1 , \quad \alpha_{m-1} = \text{tr}(A) , \quad \alpha_0 = \det(A) . \tag{1.20}$$

The zeros of the **characteristic polynomial** (1.19) are the **eigenvalues** of A, and the set of all eigenvalues is the **spectrum** $\sigma(A)$ of A.

The number $\lambda \in \mathbb{K}$ is an eigenvalue of A if and only if the **eigenvalue equation**

$$Ax = \lambda x \tag{1.21}$$

has a nontrivial solution $x \in E \setminus \{0\}$, that is, if $\ker(\lambda - A) \neq \{0\}$. For[11] $\lambda \in \sigma(A) \cap \mathbb{K}$, the **geometric multiplicity** of λ is given by $\dim(\ker(\lambda - A))$, and every

---

[10]sign(σ) is also called the **signature** of the permutation σ.

[11]Note that the spectrum of A is also a subset of $\mathbb{C}$ if E is a real Banach space. If $\mathbb{K} = \mathbb{R}$ and λ belongs to $\sigma(A) \setminus \mathbb{R}$, then the equation (1.21) is not meaningful. In this case, we must allow for the **complexification** of E and A, so that λ can be allowed as a solution to the eigenvalue equation (see, for example, [Ama95, § 12]). Then the geometric multiplicity of $\lambda \in \sigma(A)$ is defined in every case.

$x \in \ker(\lambda - A)\backslash\{0\}$ is an **eigenvector of A with eigenvalue λ**. The number of times λ appears as a zero of the characteristic polynomial is the **algebraic multiplicity** of λ. The geometric multiplicity is no greater than the algebraic multiplicity. Finally, λ is a **simple eigenvalue** of A if its algebraic multiplicity is 1 and, therefore, if λ is a simple zero of the characteristic polynomial. If the geometric and the algebraic multiplicity are equal, then λ is **semisimple**.

Every $A \in \mathcal{L}(E)$ has exactly m eigenvalues $\lambda_1, \ldots, \lambda_m$ if they are counted according to their (algebraic) multiplicity. The coefficients α_k of the characteristic polynomial of A are the elementary symmetric functions of $\lambda_1, \ldots, \lambda_m$. In particular,

$$\operatorname{tr}(A) = \lambda_1 + \cdots + \lambda_m \text{ and } \det(A) = \lambda_1 \cdot \cdots \cdot \lambda_m . \tag{1.22}$$

If $\mathcal{E}$ is a basis of E such that $[A]_{\mathcal{E}}$ is an (upper) triangular matrix,

$$[A]_{\mathcal{E}} = \begin{bmatrix} a_1^1 & a_2^1 & \cdots & a_m^1 \\ & a_2^2 & \cdots & a_m^2 \\ & & \ddots & \vdots \\ 0 & & & a_m^m \end{bmatrix}, \tag{1.23}$$

that is, $a_k^j = 0$ for $k < j$, then the eigenvalues of A appear with their algebraic multiplicity along the diagonal.

Proof   The properties (1.17)–(1.20) of the determinants of square matrices are standard in linear algebra, and we assume you know them.

If $\mathcal{F}$ is another basis of E, there is an invertible matrix $T \in \mathbb{K}^{m \times m}$ such that

$$[A]_{\mathcal{F}} = T[A]_{\mathcal{E}} T^{-1}$$

(for example, [Koe83, Chapter 9.3.2]). Therefore $\det([A]_{\mathcal{F}}) = \det([A]_{\mathcal{E}})$ follows from the matrix rules (1.17). This shows that $\det(A)$ is well defined for $A \in \mathcal{L}(E)$. Now it is obvious that (1.17)–(1.20) hold for $A \in \mathcal{L}(E)$.

The existence of eigenvalues of A and the statement that A has exactly m eigenvalues if they are counted by their algebraic multiplicity both follow from the fundamental theorem of algebra (see Example III.3.9(b)). ∎

(c) Suppose $p \in K[X]$ is a polynomial of degree $n \geq 1$ over the field K. Then p **splits** in K if there are $k, \nu_1, \ldots, \nu_k \in \mathbb{N}^\times$ and $a, \lambda_1, \ldots, \lambda_k \in K$ such that

$$p = a \prod_{j=1}^{k} (X - \lambda_j)^{\nu_j} .$$

The fundamental theorem of algebra implies that every nonconstant polynomial $p \in \mathbb{C}[X]$ splits in $\mathbb{C}$ (see Example III.3.9(b)). In general, a polynomial need not split. For example, if K is an ordered field, then $X^2 + 1$ does not split in K.

(d) Suppose $\lambda \in \mathbb{K}$ and $p \in \mathbb{N}^{\times}$. Then the $J(\lambda, p) \in \mathbb{K}^{p \times p}$ defined by

$$
J(\lambda, 1) := [\lambda] \quad \text{and} \quad J(\lambda, p) :=
\begin{bmatrix}
\lambda & 1 & & & \\
 & \ddots & \ddots & & 0 \\
 & & \ddots & \ddots & \\
 & 0 & & \ddots & 1 \\
 & & & & \lambda
\end{bmatrix}
$$

for $p \geq 2$ is called the **elementary Jordan matrix** of size p for λ.

For $\mu = \alpha + i\omega \in \mathbb{C}$ such that $\alpha \in \mathbb{R}$ and $\omega > 0$, let

$$
A_2(\mu) :=
\begin{bmatrix}
\alpha & -\omega \\
\omega & \alpha
\end{bmatrix}
\in \mathbb{R}^{2 \times 2},
$$

and let 1_2 denote the unity element in $\mathbb{R}^{2 \times 2}$. Then we call the $\widetilde{J}(\mu, p) \in \mathbb{R}^{2p \times 2p}$ defined by

$$
\widetilde{J}(\mu, 1) := A_2(\mu) \quad \text{and} \quad \widetilde{J}(\mu, p) :=
\begin{bmatrix}
A_2(\mu) & 1_2 & & & \\
 & \ddots & \ddots & & 0 \\
 & & \ddots & \ddots & \\
 & 0 & & \ddots & 1_2 \\
 & & & & A_2(\mu)
\end{bmatrix}
$$

for $p \geq 2$ the **extended Jordan matrix** of size $2p$ for μ.

(e) Suppose E is a finite-dimensional Banach space over $\mathbb{K}$ and $A \in \mathcal{L}(E)$. Also suppose the characteristic polynomial of A splits in $\mathbb{K}$. In linear algebra, one shows that there is then a basis $\mathcal{E}$ of E and $p_1, \ldots, p_r \in \mathbb{N}^{\times}$ such that

$$
[A]_{\mathcal{E}} =
\begin{bmatrix}
J(\lambda_1, p_1) & & \\
 & \ddots & 0 \\
0 & & J(\lambda_r, p_r)
\end{bmatrix}
\in \mathbb{K}^{n \times n},
\tag{1.24}
$$

where $\{\lambda_1, \ldots, \lambda_r\} = \sigma(A)$. One calls (1.24) the **Jordan normal form** of A. It is unique up to ordering of the blocks $J(\lambda_j, p_j)$. The basis $\mathcal{E}$ is called the **Jordan basis** of E for A. If A has only semisimple eigenvalues, then $r = \dim(E)$ and $p_j = 1$ for $j = 1, \ldots, r$. Such an A is said to be **diagonalizable**.

(f) The characteristic polynomial of

$$
A =
\begin{bmatrix}
0 & 1 & 0 \\
-1 & 0 & 0 \\
0 & 0 & 1
\end{bmatrix}
\in \mathbb{R}^{3 \times 3}
$$

reads

$$
p = X^3 - X^2 + X - 1 = (X - 1)(X^2 + 1).
$$

Therefore p does not split in $\mathbb{R}$, and $\sigma(A) = \{1, i, -i\}$.

In this case, A cannot be put into the form (1.24). However, it can be put into an extended Jordan normal form.

(g) Suppose E is a finite-dimensional *real* Banach space and the characteristic polynomial p of $A \in \mathcal{L}(E)$ does not split in $\mathbb{R}$. Then there is a nonreal eigenvalue μ of A. Because the characteristic polynomial $p = \sum_{k=0}^{n} a_k X^k$ has only real coefficients, it follows from

$$0 = \overline{0} = \overline{p(\mu)} = \overline{\sum_{k=0}^{n} a_k \mu^k} = \sum_{k=0}^{n} a_k \overline{\mu}^k = p(\overline{\mu})$$

that $\overline{\mu}$ is also an eigenvalue of A.

Suppose now

$$\sigma := \{\mu_1, \overline{\mu}_1, \ldots, \mu_\ell, \overline{\mu}_\ell\} \subset \sigma(A) \quad \text{for } \ell \geq 1 ,$$

is the set of all nonreal eigenvalues of A, where, without loss of generality, we may assume that $\mu_j = \alpha_j + i\omega_j$ with $\alpha_j \in \mathbb{R}$ and $\omega_j > 0$. If A also has real eigenvalues, we denote them by $\lambda_1, \ldots, \lambda_k$.

Then there is a basis $\mathcal{E}$ of E and $p_1, \ldots, p_r, q_1, \ldots, q_s \in \mathbb{N}^\times$ such that

$$[A]_{\mathcal{E}} = \begin{bmatrix} J(\lambda_1, p_1) & & & & & \\ & \ddots & & & 0 & \\ & & J(\lambda_r, p_r) & & & \\ & & & \widetilde{J}(\mu_1, q_1) & & \\ & 0 & & & \ddots & \\ & & & & & \widetilde{J}(\mu_s, q_s) \end{bmatrix} \in \mathbb{R}^{n \times n} . \qquad (1.25)$$

Here, $\lambda_j \in \sigma(A) \setminus \sigma$ for $j = 1, \ldots, r$ and $\mu_k \in \sigma$ for $k = 1, \ldots, s$. We call (1.25) the **extended Jordan normal form** of A. It is unique up to the ordering of the constituent matrices. (see [Gab96, Chapter A5]). ∎

Fundamental matrices

Suppose E is a vector space of dimension n over $\mathbb{K}$ and $A \in \mathcal{L}(E)$. Also denote by $V \subset C^1(\mathbb{R}, E)$ the solution space of homogeneous linear differential equations $\dot{x} = Ax$. By Proposition 1.16(iv), V and E are isomorphic and, in particular, have the same dimension. We call every basis of V a **fundamental system** for $\dot{x} = Ax$.

1.20 Examples **(a)** Let

$$A := \begin{bmatrix} a & b \\ 0 & c \end{bmatrix} \in \mathbb{K}^{2 \times 2} .$$

To get a fundamental system for $\dot{x} = Ax$, we solve first the initial value problem

$$\dot{x}^1 = ax^1 + bx^2 , \qquad x^1(0) = 1 ,$$
$$\dot{x}^2 = cx^2 , \qquad\qquad x^2(0) = 0 . \tag{1.26}$$

From Application IV.3.9(b), it follows that $x^2 = 0$ and $x^1(t) = e^{at}$ for $t \in \mathbb{R}$. Therefore $x_1 := \big(t \mapsto (e^{at}, 0)\big)$ is the solution of (1.26). We get a second solution x_2 linearly independent of x_1 by solving

$$\dot{x}^1 = ax^1 + bx^2 , \qquad x^1(0) = 0 ,$$
$$\dot{x}^2 = cx^2 , \qquad\qquad x^2(0) = 1 .$$

In this case, we have $x_2^2(t) = e^{ct}$ for $t \in \mathbb{R}$, and the variation of constants formula gives

$$x_2^1(t) = \int_0^t e^{a(t-\tau)} b e^{c\tau} \, d\tau = b e^{at} \int_0^t e^{\tau(c-a)} \, d\tau$$
$$= \begin{cases} b(e^{ct} - e^{at})/(c - a) , & a \neq c , \\ bt e^{ct} , & a = c . \end{cases}$$

Thus

$$x_1(t) = (e^{at}, 0) , \qquad x_2(t) = \begin{cases} \big(b(e^{ct} - e^{at})/(c - a), e^{ct}\big) , & a \neq c , \\ \big(bt e^{at}, e^{at}\big) , & a = c , \end{cases}$$

where $t \in \mathbb{R}$, is a fundamental system for $\dot{x} = Ax$.

(b) Suppose $\omega > 0$ and

$$A := \begin{bmatrix} 0 & -\omega \\ \omega & 0 \end{bmatrix} \in \mathbb{K}^{2 \times 2} .$$

From

$$\dot{x}^1 = -\omega x^2 \text{ and } \dot{x}^2 = \omega x^1$$

it follows that $\ddot{x}^j + \omega^2 x^j = 0$ for $j = 0, 1$. Exercise IV.3.2(a) shows that

$$x_1(t) = \big(\cos(\omega t), \sin(\omega t)\big) \text{ and } x_2(t) = \big(-\sin(\omega t), \cos(\omega t)\big) \quad \text{for } t \in \mathbb{R} ,$$

is fundamental system for $\dot{x} = Ax$. ∎

If $A \in \mathbb{K}^{n \times n}$ and $\{x_1, \dots, x_n\}$ is a fundamental system for $\dot{x} = Ax$, we call the map

$$X : \mathbb{R} \to \mathbb{K}^{n \times n} , \qquad t \mapsto [x_1(t), \dots, x_n(t)] = \begin{bmatrix} x_1^1(t) & \cdots & x_n^1(t) \\ \vdots & & \vdots \\ x_1^n(t) & \cdots & x_n^n(t) \end{bmatrix}$$

the **fundamental matrix** for $\dot{x} = Ax$. If $X(0) = 1_n$, we call X the **principal fundamental matrix**.

1.21 Remarks   Let $A \in \mathbb{K}^{n \times n}$.

(a) The map

$$\mathbb{K}^{n \times n} \to \mathbb{K}^{n \times n} \ , \quad X \mapsto AX$$

is linear. Thus, in addition to the linear differential equation $\dot{x} = Ax$ in $\mathbb{K}^n$, we can consider the linear differential equation $\dot{X} = AX$ in $\mathbb{K}^{n \times n}$, which is then solved by any fundamental matrix for $\dot{x} = Ax$.

(b) Suppose X is a fundamental matrix for $\dot{x} = Ax$ and $t \in \mathbb{R}$. Then

   (i) $X(t) \in \mathcal{L}\mathrm{aut}(\mathbb{K}^n)$;

   (ii) $X(t) = e^{tA} X(0)$.

Proof   (i) is a consequence of Proposition 1.16(v).
(ii) follows from (a), Theorem 1.11, and Proposition 1.16(iii). $\blacksquare$

(c) The map $t \mapsto e^{tA}$ is the unique principal fundamental matrix for $\dot{x} = Ax$.

Proof   This follows from (b). $\blacksquare$

(d) For $s, t \in \mathbb{R}$, we have

$$e^{(t-s)A} = X(t) X^{-1}(s) \ .$$

Proof   This also follows from (b). $\blacksquare$

(e) Using the Jordan normal form and Examples 1.13(b)–(e), we can calculate e^{tA}. We demonstrate this for $n = 2$.

    The characteristic polynomial of

$$A := \begin{bmatrix} a & b \\ c & d \end{bmatrix} \in \mathbb{R}^{2 \times 2}$$

gives $\det(\lambda - A) = \lambda^2 - \lambda(a+d) + ad - bc$ and has zeros

$$\lambda_{1,2} = \left(a + d \pm \sqrt{D}\right)/2 \ , \quad \text{where} \quad D := (a-d)^2 + 4bc \ .$$

    <u>1. Case: $\mathbb{K} = \mathbb{C}$, $D < 0$</u>   Here A has two simple, complex conjugate eigenvalues $\lambda := \lambda_1$ and $\overline{\lambda} = \lambda_2$. According to Remark 1.19(e), there is a basis $\mathcal{B}$ of $\mathbb{C}^2$ such that $[A]_{\mathcal{B}} = \mathrm{diag}[\lambda, \overline{\lambda}]$, and Example 1.13(e) implies

$$e^{t[A]_{\mathcal{B}}} = \mathrm{diag}[e^{\lambda t}, e^{\overline{\lambda} t}] \quad \text{for } t \in \mathbb{R} \ .$$

Denote by $T \in \mathcal{L}\mathrm{aut}(\mathbb{C}^2)$ the basis change from the standard basis to $\mathcal{B}$. Then, using Theorem 1.9, we have

$$[A]_{\mathcal{B}} = [TAT^{-1}] = [T]A[T]^{-1} \ .$$

Therefore, from Exercise 11, we get

$$e^{t[A]_B} = [T]e^{tA}[T]^{-1} ,$$

and we find

$$e^{tA} = [T]^{-1}\operatorname{diag}[e^{\lambda t}, e^{\bar{\lambda} t}][T] \quad \text{for } t \in \mathbb{R} .$$

2. Case: $\mathbb{K} = \mathbb{C}$ or $\mathbb{K} = \mathbb{R}$, $D > 0$ Then A has two simple, real eigenvalues λ_1 and λ_2, and it follows as in the first case that

$$e^{tA} = [T]^{-1}\operatorname{diag}[e^{\lambda_1 t}, e^{\lambda_2 t}][T] \quad \text{for } t \in \mathbb{R} .$$

3. Case: $\mathbb{K} = \mathbb{C}$ or $\mathbb{K} = \mathbb{R}$, $D = 0$ Now A has the eigenvalue $\lambda := (a+d)/2$ with algebraic multiplicity 2. Thus, according to Remark 1.19(e), there is a basis $\mathcal{B}$ of $\mathbb{K}^2$ such that

$$[A]_B = \begin{bmatrix} \lambda & 1 \\ 0 & \lambda \end{bmatrix} .$$

Because

$$\begin{bmatrix} \lambda & 0 \\ 0 & \lambda \end{bmatrix} \quad \text{and} \quad \begin{bmatrix} 0 & 1 \\ 0 & 0 \end{bmatrix}$$

commute, it follows from Exercise 11 and Example 1.13(e) that

$$e^{t[A]_B} = e^{\lambda t}\exp\left(t\begin{bmatrix} 0 & 1 \\ 0 & 0 \end{bmatrix}\right) = e^{\lambda t}\left(1_2 + t\begin{bmatrix} 0 & 1 \\ 0 & 0 \end{bmatrix}\right) = e^{\lambda t}\begin{bmatrix} 1 & t \\ 0 & 1 \end{bmatrix} .$$

Thus we get

$$e^{tA} = e^{\lambda t}[T]^{-1}\begin{bmatrix} 1 & t \\ 0 & 1 \end{bmatrix}[T] \quad \text{for } t \in \mathbb{R} .$$

4. Case: $\mathbb{K} = \mathbb{R}$, $D < 0$ A has two complex conjugate eigenvalues $\lambda := \alpha + i\omega$ and $\bar{\lambda}$, where $\alpha := (a+d)/2$ and $\omega := \sqrt{-D}/2$. From Remark 1.19(g), there is a basis $\mathcal{B}$ of $\mathbb{R}^2$ such that

$$[A]_B = \begin{bmatrix} \alpha & -\omega \\ \omega & \alpha \end{bmatrix} .$$

The matrices

$$\begin{bmatrix} \alpha & 0 \\ 0 & \alpha \end{bmatrix} \quad \text{and} \quad \begin{bmatrix} 0 & -\omega \\ \omega & 0 \end{bmatrix}$$

commute. Thus from Example 1.13(f) we have

$$\exp\left(t\begin{bmatrix} \alpha & -\omega \\ \omega & \alpha \end{bmatrix}\right) = e^{t\alpha}\exp\left(t\omega\begin{bmatrix} 0 & -1 \\ 1 & 0 \end{bmatrix}\right) = e^{t\alpha}\begin{bmatrix} \cos(\omega t) & -\sin(\omega t) \\ \sin(\omega t) & \cos(\omega t) \end{bmatrix} .$$

From this we derive

$$e^{tA} = e^{\alpha t}[T]^{-1}\begin{bmatrix} \cos(\omega t) & -\sin(\omega t) \\ \sin(\omega t) & \cos(\omega t) \end{bmatrix}[T] \quad \text{for } t \in \mathbb{R} ,$$

where T is the basis change from the standard basis to $\mathcal{B}$ (see Example 1.13(f)). ∎

Second order linear differential equations

Until now, we have considered only first order linear differential equations, in which there was a linear relationship between the sought-for function and its derivative. Many applications, however, require differential equations of higher order. Second order occurs frequently, and we shall now concentrate on that case.[12]

In the following, let

- $b, c \in \mathbb{R}$ and $g \in C(\mathbb{R}, \mathbb{K})$.

For a **second order linear differential equation with constant coefficients** given by

$$\ddot{u} + b\dot{u} + cu = g(t) \quad \text{for } t \in \mathbb{R} , \tag{1.27}$$

we say $u \in C^2(\mathbb{R}, \mathbb{K})$ is a **solution** if it satisfies (1.27) pointwise. The results below show that (1.27) is equivalent to a first order differential equation of the form (1.8) in the $(u, \dot{u})$ plane, called the **phase plane**.

1.22 Lemma

(i) If $u \in C^2(\mathbb{R}, \mathbb{K})$ solves (1.27), then $(u, \dot{u}) \in C^1(\mathbb{R}, \mathbb{K}^2)$ solves the equation

$$\dot{x} = Ax + f(t) \tag{1.28}$$

in $\mathbb{K}^2$, where

$$A := \begin{bmatrix} 0 & 1 \\ -c & -b \end{bmatrix} \text{ and } f := (0, g) . \tag{1.29}$$

(ii) If $x \in C^1(\mathbb{R}, \mathbb{K}^2)$ solves (1.28), then $u := \mathrm{pr}_1 x$ solves (1.27).

Proof   (i) We set $x := (u, \dot{u})$ and get from (1.27) that

$$\dot{x} = \begin{bmatrix} \dot{u} \\ \ddot{u} \end{bmatrix} = \begin{bmatrix} \dot{u} \\ -cu - b\dot{u} + g(t) \end{bmatrix} = Ax + f(t) .$$

(ii) Let $x = (u, v) \in C^1(\mathbb{R}, \mathbb{K}^2)$ solve (1.28). Then

$$\dot{u} = v , \quad \dot{v} = -cu - bv + g(t) . \tag{1.30}$$

The first equation says u belongs to $C^2(\mathbb{R}, \mathbb{K})$, and together the equations imply that u solves the differential equation (1.27). ∎

Lemma 1.22 allows us to apply our knowledge of first order differential equations to (1.27). We will need the eigenvalues of A. From $\det(\lambda - A) = \lambda(\lambda + b) + c$, we find that the eigenvalues λ_1 and λ_2 of A are the zeros of the polynomial

---

[12]For differential equations of even higher order, consult the literature of ordinary differential equations, for example [Ama95].

$X^2 + bX + c$, called the **characteristic polynomial** of $\ddot{u} + b\dot{u} + cu = 0$, and we get

$$\lambda_{1,2} = \left(-b \pm \sqrt{D}\right)/2 \ , \quad \text{where} \quad D = b^2 - 4c \ .$$

Now we consider the initial value problem

$$\ddot{u} + b\dot{u} + cu = g(t) \text{ for } t \in \mathbb{R} \ , \quad \text{with} \quad u(0) = a_1 \ , \quad \dot{u}(0) = a_2 \ , \qquad (1.31)$$

where $(a_1, a_2) \in \mathbb{K}^2$. From the above, we can easily prove a basic theorem.

1.23 Theorem

(i) *For every $(a_1, a_2) \in \mathbb{K}^2$, there is a unique solution $u \in C^2(\mathbb{R}, \mathbb{K})$ to the initial value problem (1.31).*

(ii) *The set of solutions to $\ddot{u} + b\dot{u} + cu = 0$ span a two-dimensional vector subspace V of $C^2(\mathbb{R}, \mathbb{K})$ through*

$$\{e^{\lambda_1 t}, e^{\lambda_2 t}\} \quad \text{if } D > 0 \text{ or } (D < 0 \text{ and } \mathbb{K} = \mathbb{C}) \ ,$$
$$\{e^{\alpha t}, te^{\alpha t}\} \quad \text{if } D = 0 \ ,$$
$$\{e^{\alpha t}\cos(\omega t), e^{\alpha t}\sin(\omega t)\} \quad \text{if } D < 0 \text{ and } \mathbb{K} = \mathbb{R} \ ,$$

where $\alpha := -b/2$ and $\omega := \sqrt{-D}/2$.

(iii) *The set of solutions to (1.27) forms a two-dimensional affine subspace $v + V$ of $C^2(\mathbb{R}, \mathbb{K})$, where v is any solution of (1.27).*

(iv) *Suppose the equations $\ddot{u} + b\dot{u} + cu = 0$ and $w := u_1\dot{u}_2 - \dot{u}_1 u_2$ have linearly independent solutions $u_1, u_2 \in V$. Then*

$$v(t) := \int_0^t \frac{u_1(\tau)g(\tau)}{w(\tau)} \, d\tau \, u_2(t) - \int_0^t \frac{u_2(\tau)g(\tau)}{w(\tau)} \, d\tau \, u_1(t) \quad \text{for } t \in \mathbb{R}$$

solves (1.27).

Proof (i) We find this immediately from Lemma 1.22 and Theorem 1.17.

(ii) and (iii) These are implied by Lemma 1.22, Proposition 1.16(iv) and Remarks 1.14(d) and 1.21(e).

(iv) Let A and f be as in (1.29). Then by Lemma 1.22 and Remark 1.21(a),

$$X := \begin{bmatrix} u_1 & u_2 \\ \dot{u}_1 & \dot{u}_2 \end{bmatrix}$$

is a fundamental matrix for $\dot{x} = Ax$. Because

$$e^{(t-\tau)A} = X(t)X^{-1}(\tau) \quad \text{for } t, \tau \in \mathbb{R}$$

(see Remark 1.21(d)), it follows from the variation of constants formula of Theorem 1.17 that

$$y(t) = X(t) \int_0^t X^{-1}(\tau)f(\tau)\,d\tau \quad \text{for } t \in \mathbb{R} , \tag{1.32}$$

solves $\dot{x} = Ax + f(t)$. Because $\det(X) = w$, we have

$$X^{-1} = \frac{1}{w} \begin{bmatrix} \dot{u}_2 & -u_2 \\ -\dot{u}_1 & u_1 \end{bmatrix} ,$$

where the function w has no zeros, due to Proposition 1.16(v). Therefore we conclude that $X^{-1}f = (-u_2 g/w, u_1 g/w)$. The theorem then follows from Lemma 1.22 and (1.32). ∎

1.24 Examples (a) The characteristic polynomial $X^2 - 2X + 10$ of

$$\ddot{u} - 2\dot{u} + 10u = 0$$

has zeros $1 \pm 3i$. Thus $\{e^t \cos(3t), e^t \sin(3t)\}$ is a fundamental system.[13] The general solution therefore reads

$$u(t) = e^t\big(a_1 \cos(3t) + a_2 \sin(3t)\big) \quad \text{for } t \in \mathbb{R} ,$$

where $a_1, a_2 \in \mathbb{R}$.

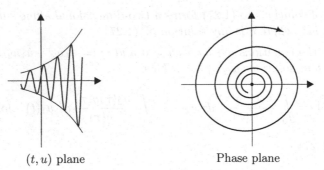

(t, u) plane   Phase plane

(b) To solve the initial value problem

$$\ddot{u} - 2\dot{u} + u = e^t , \quad u(0) = 0 , \quad \dot{u}(0) = 1 , \tag{1.33}$$

we first find a fundamental system for the homogeneous equation $\ddot{u} - 2\dot{u} + u = 0$. The associated characteristic polynomial is

$$X^2 - 2X + 1 = (X - 1)^2 .$$

---

[13] As with first order linear differential equations, we call any basis of the solution space a **fundamental system**.

Then, according to Theorem 1.23(iii), $u_1(t) := e^t$ and $u_1(t) := te^t$ for $t \in \mathbb{R}$ form a fundamental system. Because

$$w(\tau) = u_1(\tau)\dot{u}_2(\tau) - \dot{u}_1(\tau)u_2(\tau) = e^{2\tau} \quad \text{for } \tau \in \mathbb{R} ,$$

the equation

$$v(t) = \int_0^t \frac{u_1 g}{w} \, d\tau \, u_2(t) - \int_0^t \frac{u_2 g}{w} \, d\tau \, u_1(t) = \frac{t^2}{2} e^t \quad \text{for } t \in \mathbb{R} ,$$

is particular solution of (1.33). Thus the general solution reads

$$x(t) = a_1 e^t + a_2 t e^t + \frac{t^2}{2} e^t \quad \text{for } t \in \mathbb{R} \text{ and } a_1, a_2 \in \mathbb{R} .$$

Because $x(0) = a_1$ and $\dot{x}(0) = a_1 + a_2$, we find finally that the unique solution of (1.33) is

$$u(t) = (t + t^2/2)e^t \quad \text{for } t \in \mathbb{R} .$$

(c) The differential equation of the **harmonic oscillator** (or the **undamped oscillator**) reads

$$\ddot{u} + \omega_0^2 u = 0 ,$$

where $\omega_0 > 0$ is the (angular) **frequency**. From Theorem 1.23, the general solution has the form

$$u(t) = a_1 \cos(\omega_0 t) + a_2 \sin(\omega_0 t) \quad \text{for } t \in \mathbb{R} ,$$

where $a_1, a_2 \in \mathbb{R}$. Obviously, all the solutions are periodic with period $2\pi/\omega_0$, and they rotate about the center of the phase plane.

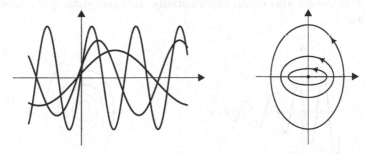

(d) The differential equation of the **damped oscillator** is

$$\ddot{u} + 2\alpha \dot{u} + \omega_0^2 u = 0 .$$

Here, $\alpha > 0$ is the **damping constant**, and $\omega_0 > 0$ is the frequency of the undamped oscillator. The zeros of the characteristic polynomial are

$$\lambda_{1,2} = -\alpha \pm \sqrt{\alpha^2 - \omega_0^2} .$$

(i) (Strong damping: $\alpha > \omega_0$) In this case, there are two negative eigenvalues $\lambda_1 < \lambda_2 < 0$, and the general solution reads

$$u(t) = a_1 e^{\lambda_1 t} + a_2 e^{\lambda_2 t} \quad \text{for } t \in \mathbb{R} \text{ and } a_1, a_2 \in \mathbb{R} \, .$$

Thus all solutions fade away exponentially. In the phase plane, there is a "stable node" at the origin.

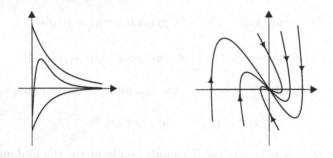

(ii) (Weak damping: $\alpha < \omega_0$) In this case, the characteristic polynomial has two complex conjugate eigenvalues $\lambda_{1,2} = -\alpha \pm i\omega$, where $\omega := \sqrt{\omega_0^2 - \alpha^2}$. Thus the general solution, according to Theorem 1.22, is

$$u(t) = e^{-\alpha t}\big(a_1 \cos(\omega t) + a_2 \sin(\omega t)\big) \quad \text{for } t \in \mathbb{R} \text{ and } a_1, a_2 \in \mathbb{R} \, .$$

In this case, solutions also damp exponentially, and the phase plane's origin is a "stable vortex".

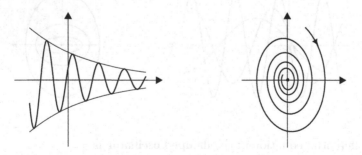

(iii) (Critical damping: $\alpha = \omega_0$) Here, $\lambda = -\alpha$ is an eigenvalue of the characteristic polynomial with algebraic multiplicity 2. Therefore the general solution reads

$$u(t) = e^{-\alpha t}(a_1 + a_2 t) \quad \text{for } t \in \mathbb{R} \text{ and } a_1, a_2 \in \mathbb{R} \, .$$

The phase plane's origin is now called a "stable virtual node". ∎

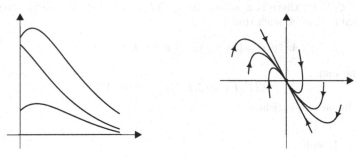

Exercises

1 Suppose E and F_j are Banach spaces and $A_j \in \mathrm{Hom}(E, F_j)$ for $j = 1, \ldots, m$. Also let $F := \prod_{j=1}^{m} F_j$. For $A := \left(x \mapsto (A_1 x, \ldots, A_m x)\right) \in \mathrm{Hom}(E, F)$, show that

$$A \in \mathcal{L}(E, F) \iff \left(A_j \in \mathcal{L}(E, F_j),\ j = 1, \ldots, m\right) .$$

2 For a square matrix $A := [a_k^j] \in \mathbb{K}^{m \times m}$, the **trace** is defined by

$$\mathrm{tr}(A) := \sum_{j=1}^{m} a_j^j .$$

Show

   (i) the map $\mathrm{tr} \colon \mathbb{K}^{m \times m} \to \mathbb{K}$ is linear;

   (ii) $\mathrm{tr}(AB) = \mathrm{tr}(BA)$ for $A, B \in \mathbb{K}^{m \times m}$.

3 Two matrices $A, B \in \mathbb{K}^{m \times m}$ are **similar** if there is an invertible matrix $S \in \mathbb{K}^{m \times m}$ (that is, if S belongs to $\mathcal{L}\mathrm{aut}(\mathbb{K}^m)$) such that $A = SBS^{-1}$. Show that $\mathrm{tr}(A) = \mathrm{tr}(B)$ if A and B are similar.

4 Suppose E is a finite-dimensional normed vector space. Prove that for $A \in \mathcal{L}(E)$, we have

$$\mathrm{tr}\left([A]_{\mathcal{E}}\right) = \mathrm{tr}\left([A]_{\mathcal{F}}\right) ,$$

where $\mathcal{E}$ and $\mathcal{F}$ are bases of E.

    From this it follows that the **trace** of $A \in \mathcal{L}(E)$ is well defined through

$$\mathrm{tr}(A) := \mathrm{tr}\left([A]_{\mathcal{E}}\right) ,$$

where $\mathcal{E}$ is any basis of E.

Show also that

   (i) $\mathrm{tr} \in \mathcal{L}\left(\mathcal{L}(E), \mathbb{K}\right)$;

   (ii) $\mathrm{tr}(AB) = \mathrm{tr}(BA)$ for $A, B \in \mathcal{L}(E)$.

5 Suppose $(E, (\cdot|\cdot)_E)$ and $(F, (\cdot|\cdot)_F)$ are finite-dimensional Hilbert spaces. Verify that to each $A \in \mathcal{L}(E, F)$ there is a unique $A^* \in \mathcal{L}(F, E)$, called the **adjoint operator** (or, simply, adjoint) A^* of A, such that

$$(Ax\,|\,y)_F = (x\,|\,A^*y)_E \quad \text{for } x \in E \text{ and } y \in F .$$

Show that the map

$$\mathcal{L}(E, F) \to \mathcal{L}(F, E) , \quad A \mapsto A^*$$

is conjugate linear and satisfies

(i) $(AB)^* = B^* A^*$,

(ii) $(A^*)^* = A$, and

(iii) $(C^{-1})^* = (C^*)^{-1}$

for $A, B \in \mathcal{L}(E, F)$ and $C \in \mathcal{L}\text{is}(E, F)$. For $A = [a_k^j] \in \mathbb{K}^{m \times n}$, show $A^* = [\overline{a_j^k}] \in \mathbb{K}^{n \times m}$. This matrix is the **Hermitian conjugate** or **Hermitian adjoint** of the matrix A.

  If $A = A^*$, (and therefore $E = F$), we say A is **self adjoint** or **symmetric**. Show that $A^* A$ and $A A^*$ are self adjoint for $A \in \mathcal{L}(E, F)$.

6 Let $E := (E, (\cdot|\cdot))$ be a finite-dimensional Hilbert space. Then show for $A \in \mathcal{L}(E)$ that

(i) $\text{tr}(A^*) = \overline{\text{tr}(A)}$;

(ii) $\text{tr}(A) = \sum_{j=1}^n (A\varphi_j\,|\,\varphi_j)$, where $\{\varphi_1, \dots, \varphi_n\}$ is an orthonormal basis of E.

7 Suppose E and F are finite-dimensional Hilbert spaces. Prove that

(i) $\mathcal{L}(E, F)$ is a Hilbert space with the inner product

$$(A, B) \mapsto A : B := \text{tr}(B^* A) ;$$

(ii) if $E := \mathbb{K}^n$ and $F := \mathbb{K}^m$, then

$$|A|_{\mathcal{L}(E,F)} \leq \sqrt{A : A} = \|A\| \quad \text{for } A \in \mathcal{L}(E, F) .$$

8 Suppose $[g_{jk}] \in \mathbb{R}^{n \times n}$ is symmetric and **positive definite**, that is, $g_{jk} = g_{kj}$ and there is a $\gamma > 0$ such that

$$\sum_{j,k=1}^n g_{jk} \xi^j \xi^k \geq \gamma |\xi|^2 \quad \text{for } \xi \in \mathbb{R}^n .$$

Show that

$$(x\,|\,y)^g := \sum_{j,k=1}^n g_{jk} x^j y^k \quad \text{for } x, y \in \mathbb{R}^n$$

defines a scalar product on $\mathbb{R}^n$.

9 Suppose E is a Banach space and $A, B \in \mathcal{L}(E)$ commute, that is, $AB = BA$. Prove $Ae^B = e^B A$ and $e^{A+B} = e^A e^B$.

10 Find $A, B \in \mathbb{K}^{2 \times 2}$ such that

(i) $AB \neq BA$ and $e^{A+B} \neq e^A e^B$;

(ii) $AB \neq BA$ and $e^{A+B} = e^A e^B$.

(Hint for (ii): $e^{2k\pi i} = 1$ for $k \in \mathbb{Z}$.)

11   Suppose E is a Banach space, $A \in \mathcal{L}(E)$ and $B \in \mathcal{L}\mathrm{aut}(E)$. Show

$$e^{BAB^{-1}} = Be^A B^{-1} .$$

12   Suppose E is a finite-dimensional Hilbert space and $A \in \mathcal{L}(E)$. Show
  (i) $(e^A)^* = e^{A^*}$;
  (ii) if A is self adjoint, then so is e^A;
  (iii) if A is **anti self adjoint** or **skew-symmetric**, that is $A^* = -A$, then $[e^A]^* e^A = 1$.

13   Calculate e^A for $A \in \mathcal{L}(\mathbb{R}^4)$ having the representation matrices

$$\begin{bmatrix} 1 & 1 & 1 & 1 \\ 1 & 1 & 1 & 1 \\ 1 & 1 & 1 & 1 \\ 1 & 1 & 1 & 1 \end{bmatrix}, \quad \begin{bmatrix} 2 & 1 & 0 & 0 \\ & 2 & 1 & 0 \\ 0 & & 2 & 1 \\ & & & 2 \end{bmatrix}, \quad \begin{bmatrix} 2 & 1 & 0 & 0 \\ & 2 & 1 & 0 \\ 0 & & 2 & 0 \\ & & & 2 \end{bmatrix}, \quad \begin{bmatrix} 2 & 1 & 0 & 0 \\ & 2 & 0 & 0 \\ 0 & & 2 & 0 \\ & & & 2 \end{bmatrix}.$$

14   Determine the $A \in \mathbb{R}^{3\times3}$ that gives

$$e^{tA} = \begin{bmatrix} \cos t & -\sin t & 0 \\ \sin t & \cos t & 0 \\ 0 & 0 & 1 \end{bmatrix} \quad \text{for } t \in \mathbb{R} .$$

15   Find the general solution to

$$\dot{x} = 3x - 2y + t ,$$
$$\dot{y} = 4x - y + t^2 .$$

16   Solve the initial value problem

$$\dot{x} = z - y , \quad x(0) = 0 ,$$
$$\dot{y} = z + e^t , \quad y(0) = 3/2 ,$$
$$\dot{z} = z - x , \quad z(0) = 1 .$$

17   Suppose $A = [a_k^j] \in \mathbb{R}^{n\times n}$ is a **Markov matrix**, that is, it has $a_k^j \geq 0$ for all $j \neq k$ and $\sum_{k=1}^{n} a_k^j = 1$ for $j = 1, \ldots, n$. Also suppose

$$H_c := \{ (x^1, \ldots, x^n) \in \mathbb{R}^n ; \sum_{j=1}^{n} x^j = c \} \quad \text{for } c \in \mathbb{R} .$$

Show that every solution of $\dot{x} = Ax$ with initial values in H_c remains in H_c for all time, that is, $e^{tA} H_c \subset H_c$ for $t \in \mathbb{R}$.

18   Suppose $b, c \in \mathbb{R}$, and $z \in C^2(\mathbb{R}, \mathbb{C})$ solves $\ddot{u} + b\dot{u} + cu = 0$. Show that $\mathrm{Re}\, z$ and $\mathrm{Im}\, z$ are real solutions of $\ddot{u} + b\dot{u} + cu = 0$.

19   Let $b, c \in \mathbb{R}$, and suppose u solves $\ddot{u} + b\dot{u} + cu = g(t)$. Show that
  (i) $k \in \mathbb{N} \cup \{\infty\}$ and $g \in C^k(\mathbb{R}, \mathbb{K})$ implies $u \in C^{k+2}(\mathbb{R}, \mathbb{K})$;

(ii) if $g \in C^\omega(\mathbb{R})$, then $u \in C^\omega(\mathbb{R})$.

20 The differential equation for the **damped driven oscillator** is

$$\ddot{u} + 2\alpha\dot{u} + \omega_0^2 u = c\sin(\omega t) \quad \text{for } t \in \mathbb{R} ,$$

where $\alpha, \omega_0, \omega > 0$ and $c \in \mathbb{R}$. Find its general solution.

21 Find the general solution on $(-\pi/2, \pi/2)$ for

(i) $\ddot{u} + u = 1/\cos t$;

(ii) $\ddot{u} + 4u = 2\tan t$.

22 Suppose E is a finite-dimensional Banach space and $A \in \mathcal{L}(E)$. Show these statements are equivalent:

(i) Every solution u of $\dot{x} = Ax$ in E satisfies $\lim_{t \to \infty} u(t) = 0$.

(ii) $\operatorname{Re}\lambda < 0$ for $\lambda \in \sigma(A)$.

In the remaining exercises, E is a Banach space, and $A \in \mathcal{L}(E)$.

23 If $\lambda \in \mathbb{K}$ has $\operatorname{Re}\lambda > \|A\|$, then

$$(\lambda - A)^{-1} = \int_0^\infty e^{-t(\lambda - A)} \, dt .$$

(Hint: Set $R(\lambda) := \int_0^\infty e^{-t(\lambda - A)} \, dt$ and calculate $(\lambda - A)R(\lambda)$.)

24 Show the exponential map has the representation

$$e^{tA} = \lim_{n \to \infty} \left(1 - \frac{t}{n}A\right)^{-n} .$$

(Hint: The proof of Theorem III.6.23.)

25 Let C be a closed, convex subset of E. Show these are equivalent:

(i) $e^{tA}C \subset C$ for all $t \in \mathbb{R}$;

(ii) $(\lambda - A)^{-1}C \subset C$ for all $\lambda \in \mathbb{K}$ such that $\operatorname{Re}\lambda > \|A\|$.

(Hint: Exercises 23 and 24.)

2 Differentiability

In this section, we explain the central concepts of "differentiability", "derivative", and "directional derivative" of a function[1] $f : X \subset E \to F$. Here E and F are Banach spaces, and X is a subset of E. After doing so, we will illustrate their power and flexibility by making specific assumptions about the spaces E and F.

If $E = \mathbb{R}^n$, we explain the notion of partial derivatives and investigate the connection between differentiability and the existence of partial derivatives.

In applications, the case $E = \mathbb{R}^n$ and $F = \mathbb{R}$ is particularly important. Here we introduce the concept of the gradient and clarify with the help of the Riesz representation theorem the relationship between the derivative and the gradient.

In the case $E = F = \mathbb{C}$, we derive the important Cauchy–Riemann equations. These connect the complex differentiability of $f = u + iv : \mathbb{C} \to \mathbb{C}$ and the differentiability of $(u, v) : \mathbb{R}^2 \to \mathbb{R}^2$.

Finally, we show that these new ideas agree with the well-established $E = \mathbb{K}$ from Section IV.1.

In the following, suppose

- $E = (E, \|\cdot\|)$ and $F = (F, \|\cdot\|)$ are Banach spaces over the field $\mathbb{K}$; X is an open subset of E.

The definition

A function $f : X \to F$ is **differentiable** at $x_0 \in X$ if there is an $A_{x_0} \in \mathcal{L}(E, F)$ such that

$$\lim_{x \to x_0} \frac{f(x) - f(x_0) - A_{x_0}(x - x_0)}{\|x - x_0\|} = 0 . \tag{2.1}$$

The next theorem reformulates this definition and gives the first properties of differentiable functions.

2.1 Proposition *Let $f : X \to F$ and $x_0 \in X$.*
 (i) *These statements are equivalent:*
   (α) *f is differentiable at x_0.*
   (β) *There exist $A_{x_0} \in \mathcal{L}(E, F)$ and $r_{x_0} : X \to F$, where r_{x_0} is continuous at x_0 and satisfies $r_{x_0}(x_0) = 0$, such that*

$$f(x) = f(x_0) + A_{x_0}(x - x_0) + r_{x_0}(x)\|x - x_0\| \quad \text{for } x \in X .$$

   (γ) *There exists $A_{x_0} \in \mathcal{L}(E, F)$ such that*

$$f(x) = f(x_0) + A_{x_0}(x - x_0) + o(\|x - x_0\|) \quad (x \to x_0) . \tag{2.2}$$

---
[1]The notation $f : X \subset E \to F$ is a shorthand for $X \subset E$ and $f : X \to F$.

(ii) *If f is differentiable at x_0, then f is continuous at x_0.*

(iii) *Suppose f is differentiable at x_0. Then the linear operator $A_{x_0} \in \mathcal{L}(E, F)$ from (2.1) is uniquely defined.*

Proof (i) The proof is just like the proof of Theorem IV.1.1 and is left to you.

(ii) If f is differentiable at x_0, then its continuity at x_0 follows directly from (iβ).

(iii) Suppose $B \in \mathcal{L}(E, F)$, and

$$f(x) = f(x_0) + B(x - x_0) + o(\|x - x_0\|) \quad (x \to x_0) . \tag{2.3}$$

Subtracting (2.3) from (2.2) and dividing the result by $\|x - x_0\|$, we find

$$\lim_{x \to x_0} (A_{x_0} - B)\left(\frac{x - x_0}{\|x - x_0\|} \right) = 0 .$$

Suppose $y \in E$ has $\|y\| = 1$ and $x_n := x_0 + y/n$ for $n \in \mathbb{N}^\times$. Then because $\lim x_n = x_0$ and $(x_n - x_0)/\|x_n - x_0\| = y$, we find

$$(A_{x_0} - B)y = \lim_n (A_{x_0} - B)\left(\frac{x_n - x_0}{\|x_n - x_0\|} \right) = 0 .$$

Since this holds for every $y \in \partial \mathbb{B}_E$, it follows that $A_{x_0} = B$. $\blacksquare$

The derivative

Suppose $f : X \to F$ is differentiable at $x_0 \in X$. Then we denote by $\partial f(x_0)$ the linear operator $A_{x_0} \in \mathcal{L}(E, F)$ uniquely determined by Proposition 2.1. This is called the **derivative** of f at x_0 and will also be written

$$Df(x_0) \quad \text{or} \quad f'(x_0) .$$

Therefore $\partial f(x_0) \in \mathcal{L}(E, F)$, and

$$\lim_{x \to x_0} \frac{f(x) - f(x_0) - \partial f(x_0)(x - x_0)}{\|x - x_0\|} = 0 .$$

If $f : X \to F$ is differentiable at every point $x \in X$, we say f is **differentiable** and call the map[2]

$$\partial f : X \to \mathcal{L}(E, F) , \quad x \mapsto \partial f(x)$$

the **derivative** of f.

When $\mathcal{L}(E, F)$ is a Banach space, we can meaningfully speak of the continuity of the derivative. If ∂f is continuous, that is, $\partial f \in C(X, \mathcal{L}(E, F))$, we call f **continuously differentiable**. We set

$$C^1(X, F) := \{ f : X \to F \ ; \ f \text{ is continuously differentiable} \} .$$

---

[2]To clearly distinguish the derivative $\partial f(x_0) \in \mathcal{L}(E, F)$ at a point x_0 from the derivative $\partial f : X \to \mathcal{L}(E, F)$, we may also speak of the **derivative function** $\partial f : X \to \mathcal{L}(E, F)$.

2.2 Remarks (a) These statements are equivalent:

(i) $f : X \to F$ is differentiable at $x_0 \in X$.

(ii) There is a $\partial f(x_0) \in \mathcal{L}(E, F)$ such that

$$f(x) = f(x_0) + \partial f(x_0)(x - x_0) + o(\|x - x_0\|) \quad (x \to x_0) .$$

(iii) There is a $\partial f(x_0) \in \mathcal{L}(E, F)$ and an $r_{x_0} : X \to F$ that is continuous at x_0 and satisfies $r_{x_0}(x_0) = 0$, such that

$$f(x) = f(x_0) + \partial f(x_0)(x - x_0) + r_{x_0}(x) \|x - x_0\| \quad \text{for } x \in X .$$

(b) From (a), it follows that f is differentiable at x_0 if and only if f is approximated at x_0 by the affine map

$$g : E \to F , \quad x \mapsto f(x_0) + \partial f(x_0)(x - x_0)$$

such that the error goes to zero faster than x goes to x_0, that is,

$$\lim_{x \to x_0} \frac{\|f(x) - g(x)\|}{\|x - x_0\|} = 0 .$$

Therefore, *the map f is differentiable at x_0 if and only if it is approximately linear at x_0* (see Corollary IV.1.3).

(c) The concepts of "differentiability" and "derivative" are independent of the choice of equivalent norms in E and F.

Proof This follows, for example, from Remark II.3.13(d). ∎

(d) Instead of saying "differentiable" we sometimes say "totally differentiable" or "Fréchet differentiable".

(e) (the case $E = \mathbb{K}$) In Remark 1.3(a), we saw that the map $\mathcal{L}(\mathbb{K}, F) \to F$, $A \mapsto A1$ is an isometric isomorphism, with which we canonically identify $\mathcal{L}(\mathbb{K}, F)$ and F. This identification shows that if $E = \mathbb{K}$, the new definitions of differentiability and derivative agree with those of Section IV.1.

(f) $C^1(X, F) \subset C(X, F)$, that is, *every continuously differentiable function is also continuous.* ∎

2.3 Examples (a) For $A \in \mathcal{L}(E, F)$, the function $A = (x \mapsto Ax)$ is continuously differentiable, and $\partial A(x) = A$ for $x \in E$.

Proof Because $Ax = Ax_0 + A(x - x_0)$, the claim follows from Remark 2.2(a). ∎

(b) Suppose $y_0 \in F$. Then the constant map $k_{y_0} : E \to F$, $x \mapsto y_0$ is continuously differentiable, and $\partial k_{y_0} = 0$.

Proof This is obvious because $k_{y_0}(x) = k_{y_0}(x_0)$ for $x, x_0 \in E$. ∎

(c) Suppose H is a Hilbert space, and define $b\colon H \to \mathbb{K}$, $x \mapsto \|x\|^2$. Then b is continuously differentiable, and

$$\partial b(x) = 2\operatorname{Re}(x\,|\,\cdot)\quad\text{for }x \in H\ .$$

Proof For every choice of $x, x_0 \in H$ such that $x \neq x_0$, we have

$$\|x\|^2 - \|x_0\|^2 = \|x - x_0 + x_0\|^2 - \|x_0\|^2 = \|x - x_0\|^2 + 2\operatorname{Re}(x_0\,|\,x - x_0)$$

and therefore
$$\frac{b(x) - b(x_0) - 2\operatorname{Re}(x_0\,|\,x - x_0)}{\|x - x_0\|} = \|x - x_0\|\ .$$

This implies $\partial b(x_0)h = 2\operatorname{Re}(x_0\,|\,h)$ for $h \in H$. ∎

Directional derivatives

Suppose $f\colon X \to F$, $x_0 \in X$ and $v \in E\backslash\{0\}$. Because X is open, there is an $\varepsilon > 0$ such that $x_0 + tv \in X$ for $|t| < \varepsilon$. Therefore the function

$$(-\varepsilon, \varepsilon) \to F\ ,\quad t \mapsto f(x_0 + tv)$$

is well defined. When this function is differentiable at the point 0, we call its derivative the **directional derivative** of f at the point x_0 in the direction v and denote it by $D_v f(x_0)$. Thus

$$D_v f(x_0) = \lim_{t \to 0}\frac{f(x_0 + tv) - f(x_0)}{t}\ .$$

2.4 Remark The function

$$f_{v,x_0}\colon (-\varepsilon, \varepsilon) \to F\ ,\quad t \mapsto f(x_0 + tv)$$

can be viewed as a "curve" in $E \times F$, which lies "on the graph of f". Then $D_v f(x_0)$ represents for $\|v\| = 1$ the slope of the tangent to this curve at the point $\big(x_0, f(x_0)\big)$ (see Remark IV.1.4(a)). ∎

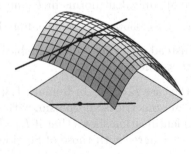

The next result says that f has a directional derivative at x_0 in every direction if f is differentiable there.

2.5 Proposition *Suppose $f\colon X \to F$ is differentiable at $x_0 \in X$. Then $D_v f(x_0)$ exists for every $v \in E\backslash\{0\}$, and $D_v f(x_0) = \partial f(x_0)v$.*

Proof For $v \in E\backslash\{0\}$, it follows from Remark 2.2(a) that

$$f(x_0 + tv) = f(x_0) + \partial f(x_0)(tv) + o(\|tv\|) = f(x_0) + t\partial f(x_0)v + o(|t|)$$

as $t \to 0$. The theorem now follows from Remark IV.3.1(c). ∎

2.6 Remark The converse of Proposition 2.5 is false, that is, a function having a directional derivative in every direction need not be differentiable.

Proof We consider a function $f : \mathbb{R}^2 \to \mathbb{R}$ defined by

$$f(x,y) := \begin{cases} \dfrac{x^2 y}{x^2 + y^2} \,, & (x,y) \neq (0,0) \,, \\[2mm] 0 \,, & (x,y) = (0,0) \,. \end{cases}$$

For every $v = (\xi, \eta) \in \mathbb{R}^2 \backslash \{(0,0)\}$, we have

$$f(tv) = \frac{t^3 \xi^2 \eta}{t^2(\xi^2 + \eta^2)} = t f(v) \,.$$

Thus

$$D_v f(0) = \lim_{t \to 0} f(tv)/t = f(v) \,.$$

If f were differentiable at 0, Proposition 2.5 would imply that $\partial f(0)v = D_v f(0) = f(v)$ for every $v \in \mathbb{R}^2 \backslash \{(0,0)\}$. Because $v \mapsto \partial f(0)v$ is linear, but f is not, this is not possible. Therefore f is not differentiable at 0. ∎

Partial derivatives

If $E = \mathbb{R}^n$, the derivatives in the direction of the coordinate axes are particularly important, for practical and historical reasons. It is convenient to introduce a specific notation for them. We thus write ∂_k or $\partial/\partial x^k$ for the derivatives in the direction of the standard basis vectors[3] e_k for $k = 1, \ldots, n$. Thus

$$\partial_k f(x_0) := \frac{\partial f}{\partial x^k}(x_0) := D_{e_k} f(x_0) = \lim_{t \to 0} \frac{f(x_0 + t e_k) - f(x_0)}{t} \quad \text{for } 1 \leq k \leq n \,,$$

and $\partial_k f(x_0)$ is called the **partial derivative** with respect to x^k of f at x_0. The function f is said to be **partially differentiable at x_0** if all $\partial_1 f(x_0), \ldots, \partial_n f(x_0)$ exist, and it is called [**continuously**] **partially differentiable** if it is partially differentiable at every point of X [if $\partial_k f : X \to F$ is continuous for $1 \leq k \leq n$].

2.7 Remarks (a) When the k-th partial derivative exists at $x_0 \in X$, we have

$$\partial_k f(x_0) = \lim_{h \to 0} \frac{f(x_0^1, \ldots, x_0^{k-1}, x_0^k + h, x_0^{k+1}, \ldots, x_0^n) - f(x_0)}{h} \quad \text{for } 1 \leq k \leq n \,.$$

Therefore f is partially differentiable at x_0 in the k-th coordinate if and only if the function $t \mapsto f(x_0^1, \ldots, x_0^{k-1}, t, x_0^{k+1}, \ldots, x_0^n)$ of one real variable is differentiable at x_0^k.

---

[3]Occasionally, we also write ∂_{x^k} for $\partial/\partial x^k$.

(b) The partial derivative $\partial_k f(x_0)$ can be defined if x_0^k is a cluster point of the set

$$\left\{ \xi \in \mathbb{R} \; ; \; (x_0^1, \ldots, x_0^{k-1}, \xi, x_0^{k+1}, \ldots, x_0^n) \in X \right\} .$$

In particular, X must not be open.

(c) If f is differentiable at x_0, then f is partially differentiable at x_0.

Proof   This is a special case of Proposition 2.5. ∎

(d) If f is partially differentiable at x_0, it does not follow that f is differentiable at x_0.

Proof   Remark 2.6. ∎

(e) That f is partially differentiable at x_0 does not imply it is continuous there.

Proof   We consider $f \colon \mathbb{R}^2 \to \mathbb{R}$ with

$$f(x,y) := \begin{cases} \dfrac{xy}{(x^2 + y^2)^2} \,, & (x,y) \neq (0,0) \,, \\[2mm] 0 \,, & (x,y) = (0,0) \,. \end{cases}$$

Then $f(h,0) = f(0,h) = 0$ for all $h \in \mathbb{R}$. Therefore $\partial_1 f(0,0) = \partial_2 f(0,0) = 0$. Then f is partially differentiable at $(0,0)$.

Because $f(0,0) = 0$ and $f(1/n, 1/n) = n^2/4$ for $n \in \mathbb{N}^\times$, we see that f is not continuous at $(0,0)$. ∎

(f) To clearly distinguish "differentiability" from "partial differentiability", we sometimes also say f is **totally differentiable** [at x_0] if f is differentiable [at x_0]. ∎

Next, we seek concrete representations for $\partial f(x_0)$ when $E = \mathbb{R}^n$ or $F = \mathbb{R}^m$. These will allow us explicitly calculate partial derivatives.

2.8 Theorem

(i) *Suppose $E = \mathbb{R}^n$ and $f \colon X \to F$ is differentiable at x_0. Then*

$$\partial f(x_0)h = \sum_{k=1}^n \partial_k f(x_0)h^k \quad \text{for } h = (h^1, \ldots, h^n) \in \mathbb{R}^n \,.$$

(ii) *Suppose E is a Banach space and $f = (f^1, \ldots, f^m) \colon X \to \mathbb{K}^m$. Then f is differentiable at x_0 if and only if all of the coordinate functions f^j for $1 \le j \le m$ are differentiable at x_0. Then*

$$\partial f(x_0) = \big(\partial f^1(x_0), \ldots, \partial f^m(x_0)\big) \,,$$

that is, vectors are differentiated componentwise.

Proof (i) Because $h = \sum_k h^k e_k$ for $h = (h^1, \ldots, h^n) \in \mathbb{R}^n$, it follows from the linearity of $\partial f(x_0)$ and Proposition 2.5 that

$$\partial f(x_0) h = \sum_{k=1}^n h^k \partial f(x_0) e_k = \sum_{k=1}^n \partial_k f(x_0) h^k .$$

(ii) For $A = (A^1, \ldots, A^m) \in \operatorname{Hom}(E, \mathbb{K}^m)$, we have

$$A \in \mathcal{L}(E, \mathbb{K}^m) \Longleftrightarrow A^j \in \mathcal{L}(E, \mathbb{K}) \text{ for } 1 \le j \le m ,$$

(see Exercise 1.1). Now it follows from Proposition II.3.14 and Proposition 2.1(iii) that

$$\lim_{x \to x_0} \frac{f(x) - f(x_0) - \partial f(x_0)(x - x_0)}{\|x - x_0\|} = 0$$

is equivalent to

$$\lim_{x \to x_0} \frac{f^j(x) - f^j(x_0) - \partial f^j(x_0)(x - x_0)}{\|x - x_0\|} = 0 \text{ for } 1 \le j \le m ,$$

as the theorem requires. ∎

The Jacobi matrix

Suppose X is open in $\mathbb{R}^n$ and $f = (f^1, \ldots, f^m) \colon X \to \mathbb{R}^m$ is partially differentiable at x_0. We then call

$$\left[\partial_k f^j(x_0)\right] = \begin{bmatrix} \partial_1 f^1(x_0) & \cdots & \partial_n f^1(x_0) \\ \vdots & & \vdots \\ \partial_1 f^m(x_0) & \cdots & \partial_n f^m(x_0) \end{bmatrix}$$

the **Jacobi matrix** of f at x_0.

2.9 Corollary *If f is differentiable at x_0, every coordinate function f^j is partially differentiable at x_0 and*

$$\left[\partial f(x_0)\right] = \left[\partial_k f^j(x_0)\right] = \begin{bmatrix} \partial_1 f^1(x_0) & \cdots & \partial_n f^1(x_0) \\ \vdots & & \vdots \\ \partial_1 f^m(x_0) & \cdots & \partial_n f^m(x_0) \end{bmatrix} ,$$

that is, the representation matrix (in the standard basis) of the derivative of f is the Jacobi matrix of f.

Proof   For $k = 1, \ldots, n$, we have $\partial f(x_0)e_k = \sum_{j=1}^{m} a_k^j e_j$ for unique $a_k^j \in \mathbb{R}$. From the linearity of $\partial f(x_0)$ and Proposition 2.8, it follows that

$$\partial f(x_0)e_k = \big(\partial f^1(x_0)e_k, \ldots, \partial f^m(x_0)e_k\big) = \big(\partial_k f^1(x_0), \ldots, \partial_k f^m(x_0)\big)$$
$$= \sum_{j=1}^{m} \partial_k f^j(x_0)e_j .$$

Therefore $a_k^j = \partial_k f^j$. ∎

A differentiability criterion

Remark 2.6 shows that the existence of all partial derivatives of a function does not imply its total differentiability. To guarantee differentiability, extra assumptions are required. One such assumption is that the partial derivatives are continuous. As the next theorem — the fundamental differentiability criterion — shows, this means the map is actually continuously differentiable.

2.10 Theorem   *Suppose X is open in $\mathbb{R}^n$ and F is a Banach space. Then $f : X \to F$ is continuously differentiable if and only if f has continuous partial derivatives.*

Proof   "$\Rightarrow$" follows easily from Proposition 2.5.
    "$\Leftarrow$" Let $x \in X$. We define a linear map $A(x) \colon \mathbb{R}^n \to F$ through

$$h := (h^1, \ldots, h^n) \mapsto A(x)h := \sum_{k=1}^{n} \partial_k f(x)h^k .$$

Theorem 1.6 says $A(x)$ belongs to $\mathcal{L}(\mathbb{R}^n, F)$. Our goal is to show $\partial f(x) = A(x)$, that is,

$$\lim_{h \to 0} \frac{f(x+h) - f(x) - A(x)h}{|h|} = 0 .$$

We choose $\varepsilon > 0$ with $\mathbb{B}(x, \varepsilon) \subset X$ and set $x_0 := x$ and $x_k := x_0 + \sum_{j=1}^{k} h^j e_j$ for $1 \leq k \leq n$ and $h = (h^1, \ldots, h^n) \in \mathbb{B}(x, \varepsilon)$. Then we get

$$f(x+h) - f(x) = \sum_{k=1}^{n} \big(f(x_k) - f(x_{k-1})\big) ,$$

and the fundamental theorem of calculus implies

$$f(x+h) - f(x) = \sum_{k=1}^{n} h^k \int_0^1 \partial_k f(x_{k-1} + th^k e_k)\, dt .$$

With this, we find the representation

$$f(x+h) - f(x) - A(x)h = \sum_{k=1}^{n} h^k \int_0^1 \big(\partial_k f(x_{k-1} + th^k e_k) - \partial_k f(x)\big)\, dt ,$$

which gives the bound

$$\|f(x+h) - f(x) - A(x)h\|_F \leq |h|_\infty \sum_{k=1}^{n} \sup_{|y-x|_\infty \leq |h|_\infty} \|\partial_k f(y) - \partial_k f(x)\|_F \ .$$

The continuity of $\partial_k f$ implies

$$\lim_{h\to 0} \left(\sum_{k=1}^{n} \sup_{|y-x|_\infty \leq |h|_\infty} \|\partial_k f(y) - \partial_k f(x)\|_F \right) = 0 \ .$$

Therefore

$$f(x+h) - f(x) - A(x)h = o(|h|_\infty) \quad (h \to 0) \ .$$

Thus, from Remarks 2.2(a) and 2.2(c), f is differentiable at x, and $\partial f(x) = A(x)$. Using

$$\|(\partial f(x) - \partial f(y))h\|_F \leq \sum_{k=1}^{n} \|\partial_k f(x) - \partial_k f(y)\|_F \, |h^k|$$

$$\leq \left(\sum_{k=1}^{n} \|\partial_k f(x) - \partial_k f(y)\|_F \right) |h|_\infty$$

and the equivalence of the norms of $\mathbb{R}^n$, the continuity of ∂f follows from that of $\partial_k f$ for $1 \leq k \leq n$. ∎

2.11 Corollary *Let X be open in $\mathbb{R}^n$. Then $f : X \to \mathbb{R}^m$ is continuously differentiable if and only if every coordinate function $f^j : X \to \mathbb{R}$ has continuous partial derivatives, and then*

$$[\partial f(x)] = [\partial_k f^j(x)] \in \mathbb{R}^{m \times n} \ .$$

2.12 Example The function

$$f : \mathbb{R}^3 \to \mathbb{R}^2 \ , \quad (x, y, z) \mapsto \left(e^x \cos y, \sin(xz) \right)$$

is continuously differentiable, and

$$[\partial f(x, y, z)] = \begin{bmatrix} e^x \cos y & -e^x \sin y & 0 \\ z \cos(xz) & 0 & x \cos(xz) \end{bmatrix}$$

for $(x, y, z) \in \mathbb{R}^3$. ∎

The derivative of a real-valued function of several real variables has an important geometric interpretation, which we prepare for next.

The Riesz representation theorem

Let E be a normed vector space on $\mathbb{K}$. We call $E' := \mathcal{L}(E, \mathbb{K})$ the (continuous) **dual space** of E. The elements of E' are the **continuous linear forms** on E.

2.13 Remarks **(a)** By convention, we give E' the operator norm on $\mathcal{L}(E, \mathbb{K})$, that is,

$$\|f\| := \sup_{\|x\| \leq 1} |f(x)| \quad \text{for } f \in E' .$$

Then Corollary 1.2 guarantees that $E' := (E', \|\cdot\|)$ is a Banach space. We also say that the norm for E' is the **norm dual** to the norm for E (or simply the dual norm).

(b) Let $(E, (\cdot|\cdot))$ be an inner product space. For $y \in E$, we set

$$f_y(x) := (x|y) \quad \text{for } x \in E .$$

Then f_y is a continuous linear form on E, and $\|f_y\|_{E'} = \|y\|_E$.

Proof Because the inner product is linear in its first argument, we have $f_y \in \text{Hom}(E, \mathbb{K})$. Because $f_0 = 0$, it suffices to consider the case $y \in E \backslash \{0\}$.

From the Cauchy–Schwarz inequality, we get

$$|f_y(x)| = |(x|y)| \leq \|x\|\,\|y\| \quad \text{for } x \in E .$$

Therefore $f_y \in \mathcal{L}(E, \mathbb{K}) = E'$ with $\|f_y\| \leq \|y\|$. From $f_y(y/\|y\|) = (y/\|y\| \,|\, y) = \|y\|$, it follows that

$$\|f_y\| = \sup_{\|x\| \leq 1} |f_y(x)| \geq |f_y(y/\|y\|)| = \|y\| .$$

Altogether, we find $\|f_y\| = \|y\|$. ∎

(c) Let $(E, (\cdot|\cdot))$ be an inner product space. Then the map

$$T : E \to E' , \quad y \mapsto f_y := (\cdot|y)$$

is conjugate linear and an isometry.

Proof This follows from $T(y + \lambda z) = (\cdot|y + \lambda z) = (\cdot|y) + \overline{\lambda}(\cdot|z) = Ty + \overline{\lambda}Tz$ and from (b). ∎

According to Remark 2.13(b), every $y \in E$ has a corresponding continuous linear form f_y on E. The next theorem shows for finite-dimensional Hilbert spaces that there are no other continuous linear forms.[4] In other words, every $f \in E'$ determines a unique $y \in E$ such that $f = f_y$.

---

[4]From Theorem 1.6 we know that every linear form on a finite-dimensional Hilbert space is continuous. Moreover, Corollary 1.5 guarantees that every finite-dimensional inner product space is a Hilbert space.

2.14 Theorem (Riesz representation theorem) *Suppose* $(E,(\cdot|\cdot))$ *is a finite-dimensional Hilbert space. Then, for every* $f \in E'$, *there is a unique* $y \in E$ *such that* $f(x) = (x|y)$ *for all* $x \in E$. *Therefore*

$$T : E \to E' , \quad y \mapsto (\cdot|y)$$

is a bijective conjugate linear isometry.

Proof From Remark 2.13(c), we know that $T : E \to E'$ is a conjugate linear isometry. In particular, T is injective. Letting $n := \dim(E)$, it follows from Theorem 1.9 and Proposition 1.8(ii) that

$$E' = \mathcal{L}(E,\mathbb{K}) \cong \mathbb{K}^{1 \times n} \cong \mathbb{K}^n ,$$

and thus $\dim(E') = \dim(E)$. The **rank formula** of linear algebra, namely,

$$\dim\big(\ker(T)\big) + \dim\big(\operatorname{im}(T)\big) = \dim E \tag{2.4}$$

(see, for example, [Gab96, ????? D.5.4]), now implies the surjectivity of T. ∎

2.15 Remarks (a) With the notation of Theorem 2.14, we set

$$[\cdot|\cdot] : E' \times E' \to \mathbb{K} , \quad [f|g] := (T^{-1}g|T^{-1}f) .$$

Then $(E',[\cdot|\cdot])$ is a Hilbert space.

Proof We leave the simple check to you. ∎

(b) Suppose $(E,(\cdot|\cdot))$ is a *real* finite-dimensional Hilbert space. Then the map $E \to E'$, $y \mapsto (\cdot|y)$ is an isometric isomorphism. In particular, $(\mathbb{R}^m)'$ is isometrically isomorphic to $\mathbb{R}^m$ by the **canonical isomorphism**, where $(\cdot|\cdot)$ denotes the Euclidean inner product on $\mathbb{R}^m$. Very often, one identifies the spaces $\mathbb{R}^m$ and $(\mathbb{R}^m)'$ using this isomorphism.

(c) The Riesz representation theorem also holds for infinite-dimensional Hilbert spaces. For a proof in full generality, we direct you to the literature on functional analysis or encourage you to take a course in the subject. ∎

The gradient

Suppose X is open in $\mathbb{R}^n$ and $f : X \to \mathbb{R}$ is differentiable at $x_0 \in X$. Then we also call the derivative $\partial f(x_0)$ the **differential** of f at x_0, and write it as[5] $df(x_0)$. The differential of f at x_0 is therefore a (continuous) linear form on $\mathbb{R}^n$. Using the Riesz representation theorem, there is a unique $y \in \mathbb{R}^n$ such that

$$df(x_0)h = (h|y) = (y|h) \quad \text{for } h \in \mathbb{R}^n .$$

---

[5]In Section VIII.3, we will clarify how this notation relates to the same notation in Remark VI.5.2 for the induced differential.

This $y \in \mathbb{R}^n$ uniquely determined by f and x_0 is called the **gradient** of f at x_0. We denote it by $\nabla f(x_0)$ or $\operatorname{grad} f(x_0)$. The differential and the gradient of f at x_0 are therefore linked by the fundamental relationship

$$df(x_0)h = \big(\nabla f(x_0) \,|\, h\big) \quad \text{for } h \in \mathbb{R}^n .$$

Note that the differential $df(x_0)$ is a linear form on $\mathbb{R}^n$, while the gradient $\nabla f(x_0)$ is a vector in $\mathbb{R}^n$.

2.16 Proposition   *We have*

$$\nabla f(x_0) = \big(\partial_1 f(x_0), \ldots, \partial_n f(x_0)\big) \in \mathbb{R}^n .$$

Proof   From Proposition 2.5, we have $df(x_0)e_k = \partial_k f(x_0)$ for $k = 1, \ldots, n$. Now it follows that

$$\big(\nabla f(x_0) \,|\, h\big) = df(x_0)h = df(x_0) \sum_{k=1}^{n} h^k e_k = \sum_{k=1}^{n} \partial_k f(x_0) h^k = (y \,|\, h)$$

for $h = (h^1, \ldots, h^n) \in \mathbb{R}^n$ with $y := \big(\partial_1 f(x_0), \ldots, \partial_n f(x_0)\big) \in \mathbb{R}^n$. The theorem follows because this is true for every $h \in \mathbb{R}^n$. $\blacksquare$

2.17 Remarks   **(a)** We call the point x_0 a **critical point** of f if $\nabla f(x_0) = 0$.

(b) Suppose x_0 is not a critical point of f. We set $h_0 := \nabla f(x_0)/|\nabla f(x_0)|$. The proof of Remark 2.13(b) then shows that

$$df(x_0)h_0 = |\nabla f(x_0)| = \max_{\substack{h \in \mathbb{R}^n \\ |h| \le 1}} df(x_0)h .$$

Because $df(x_0)h$ is the directional derivative of f at x_0 in the direction h, we have

$$\begin{aligned} f(x_0 + th_0) &= f(x_0) + t\,df(x_0)h_0 + o(t) \\ &= f(x_0) + t \max_{|h|=1} df(x_0)h + o(t) \end{aligned}$$

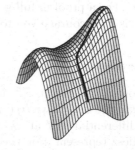

as $t \to 0$, that is, the vector $\nabla f(x_0)$ points in the direction along which f has the largest directional derivative f, that is, in the direction of **steepest ascent** of f.

If $n = 2$, the graph of f,

$$M := \big\{ \, (x, f(x)) \; ; \; x \in X \, \big\} \subset \mathbb{R}^2 \times \mathbb{R} \cong \mathbb{R}^3 ,$$

can be interpreted as a surface in $\mathbb{R}^3$.[6] We now imagine taking a point m on a path in M as follows: We start at a point $(x_0, f(x_0)) \in M$ such that x_0 is not a critical point of f. We then move m so that the projection of its "velocity vector" is always along the gradient of f. The route m takes when moved in this way is called the **curve of steepest ascent**. If we move m against the gradient, its route is called the curve of steepest descent.

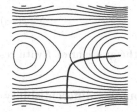

(c) The remark above shows that $\nabla f(x_0)$ has acquired a geometric interpretation independent of the special choice of coordinates or scalar product. However, the *representation* of $\nabla f(x_0)$ does depend on the choice of scalar product. In particular, the representation from Proposition 2.16 holds only when $\mathbb{R}^n$ is given the Euclidean scalar product.

Proof Suppose $[g_{jk}] \in \mathbb{R}^{n \times n}$ is a symmetric, positive definite matrix, that is, $g_{jk} = g_{kj}$ for $1 \le j, k \le n$ and there is a $\gamma > 0$ such that

$$\sum_{j,k=1}^{n} g_{jk} \xi^j \xi^k \ge \gamma |\xi|^2 \quad \text{for } \xi \in \mathbb{R}^n \ .$$

Then

$$(x \,|\, y)^g := \sum_{j,k=1}^{n} g_{jk} x^j y^k \quad \text{for } x, y \in \mathbb{R}^n$$

defines a scalar product on $\mathbb{R}^n$ (see Exercise 1.8).[7] Thus Theorem 2.14 says there exists a unique $y \in \mathbb{R}^n$ such that $df(x_0)h = (y \,|\, h)^g$ for $h \in \mathbb{R}^n$. We call $\nabla^g f(x_0) := y$ the **gradient of f at x_0 with respect to the scalar product** $(\cdot \,|\, \cdot)^g$. To determine its components, we consider

$$\sum_{j=1}^{n} \partial_j f(x_0) h^j = df(x_0)h = (y \,|\, h)^g = \sum_{j,k=1}^{n} y_{jk} y^k h^j \quad \text{for } h \in \mathbb{R}^n \ .$$

Therefore,

$$\sum_{k=1}^{n} g_{jk} y^k = \partial_j f(x_0) \quad \text{for } j = 1, \ldots, n \ . \tag{2.5}$$

By assumption, the matrix g is invertible, and its inverse is also symmetric and positive definite.[8] We denote the entries of the inverse $[g_{jk}]$ by g^{jk}, that is $[g^{jk}] = [g_{jk}]^{-1}$. From (2.5), it then follows that

$$y^k = \sum_{j=1}^{n} g^{kj} \partial_j f(x_0) \quad \text{for } k = 1, \ldots, n \ .$$

---

[6] See Corollary 8.9.

[7] The converse also holds: Every scalar product on $\mathbb{R}^n$ is of the form $(\cdot \,|\, \cdot)^g$ (see, for example, [Art93, Section VII.1]).

[8] See Exercise 5.18.

This means that

$$\nabla^g f(x_0) = \Big(\sum_{k=1}^n g^{1k} \partial_k f(x_0), \dots, \sum_{k=1}^n g^{nk} \partial_k f(x_0) \Big) \qquad (2.6)$$

is the gradient of f with respect to the scalar product induced by $g = [g_{jk}]$. ∎

Complex differentiability

With the identification $\mathcal{L}(\mathbb{C}) = \mathbb{C}$ and $\mathbb{C}^{1\times 1} = \mathbb{C}$, we can identify $A \in \mathcal{L}(\mathbb{C})$ with its matrix representation. Therefore, $Az = a \cdot z$ for $z \in \mathbb{C}$ and some $a \in \mathbb{C}$. As usual, we identify $\mathbb{C} = \mathbb{R} + i\mathbb{R}$ with $\mathbb{R}^2$:

$$\mathbb{C} = \mathbb{R} + i\mathbb{R} \ni z = x + iy \longleftrightarrow (x, y) \in \mathbb{R}^2 , \quad \text{where} \quad x = \operatorname{Re} z \text{ and } y = \operatorname{Im} z .$$

For the action of A with respect to this identification, we find by the identification $a = \alpha + i\beta \longleftrightarrow (\alpha, \beta)$ that

$$Az = a \cdot z = (\alpha + i\beta)(x + iy) = (\alpha x - \beta y) + i(\beta x + \alpha y) ,$$

and therefore

$$Az \longleftrightarrow \begin{bmatrix} \alpha & -\beta \\ \beta & \alpha \end{bmatrix} \begin{bmatrix} x \\ y \end{bmatrix} . \qquad (2.7)$$

Suppose X is open at $\mathbb{C}$. For $f\colon X \to \mathbb{C}$, we set $u := \operatorname{Re} f$ and $v := \operatorname{Im} f$. Then

$$\mathbb{C}^X \ni f = u + iv \longleftrightarrow (u, v) =: F \in \big(\mathbb{R}^2 \big)^X ,$$

where, in the last term, X is understood to be a subset of $\mathbb{R}^2$. In this notation, the following fundamental theorem gives the connection between complex[9] and total differentiability.

2.18 Theorem *The function f is complex differentiable at $z_0 = x_0 + iy_0$ if and only if $F := (u, v)$ is totally differentiable at (x_0, y_0) and satisfies the* **Cauchy–Riemann equations**[10]

$$u_x = v_y , \quad u_y = -v_x ,$$

at (x_0, y_0). In that case,

$$f'(z_0) = u_x(x_0, y_0) + iv_x(x_0, y_0) .$$

Proof (i) Suppose f is complex differentiable at z_0. We set

$$A := \begin{bmatrix} \alpha & -\beta \\ \beta & \alpha \end{bmatrix} ,$$

---

[9]Recall that f is complex differentiable at a point z_0 if and only if $\big(f(z_0 + h) - f(z_0) \big)/h$ has a limit in $\mathbb{C}$ as $h \to 0$ in $\mathbb{C}$.

[10]For functions f in two or three real variables, common notations are $f_x := \partial_1 f$, $f_y := \partial_2 f$ and $f_z := \partial_3 f$.

where $\alpha := \operatorname{Re} f'(z_0)$ and $\beta := \operatorname{Im} f'(z_0)$. Then for $h = \xi + i\eta \longleftrightarrow (\xi, \eta)$, we have

$$\lim_{(\xi,\eta)\to(0,0)} \frac{|F(x_0 + \xi, y_0 + \eta) - F(x_0, y_0) - A(\xi, \eta)|}{|(\xi, \eta)|}$$

$$= \lim_{h\to 0} \left| \frac{f(z_0 + h) - f(z_0) - f'(z_0)h}{h} \right| = 0 .$$

Therefore F is totally differentiable at (x_0, y_0) (see Remark II.2.1(a)) with

$$[\partial F(x_0, y_0)] = \begin{bmatrix} \partial_1 u(x_0, y_0) & \partial_2 u(x_0, y_0) \\ \partial_1 v(x_0, y_0) & \partial_2 v(x_0, y_0) \end{bmatrix} = \begin{bmatrix} \alpha & -\beta \\ \beta & \alpha \end{bmatrix} .$$

Thus we have the Cauchy–Riemann equations

$$\partial_1 u(x_0, y_0) = \partial_2 v(x_0, y_0) , \quad \partial_2 u(x_0, y_0) = -\partial_1 v(x_0, y_0) . \tag{2.8}$$

(ii) If F is totally differentiable at (x_0, y_0) and (2.8) holds, we may set

$$a := \partial_1 u(x_0, y_0) + i\,\partial_1 v(x_0, y_0) .$$

Then because of (2.7), we have

$$\lim_{h\to 0} \left| \frac{f(z_0 + h) - f(z_0) - ah}{h} \right|$$

$$= \lim_{(\xi,\eta)\to(0,0)} \frac{|F(x_0 + \xi, y_0 + \eta) - F(x_0, y_0) - \partial F(x_0, y_0)(\xi, \eta)|}{|(\xi, \eta)|} = 0 .$$

Consequently, f is complex differentiable at z_0 with $f'(z_0) = a$. ∎

2.19 Examples   (a) According to Example IV.1.13(b), $f : \mathbb{C} \to \mathbb{C}$, $z \mapsto z^2$ is everywhere complex differentiable with $f'(z) = 2z$. Because

$$f(x + iy) = (x + iy)^2 = x^2 - y^2 + i(2xy) \longleftrightarrow \big(u(x,y), v(x,y)\big) ,$$

we have

$$u_x = v_y = 2x \text{ and } u_y = -v_x = -2y .$$

Therefore the Cauchy–Riemann equations are satisfied (as they must be), and $f'(z) = f'(x + iy) = 2x + i2y = 2(x + iy) = 2z$.

(b) The map $f : \mathbb{C} \to \mathbb{C}$ for $z \mapsto \bar{z}$ is complex differentiable nowhere,[11] because, with

$$f(x + iy) = x - iy \longleftrightarrow \big(u(x,y), v(x,y)\big) ,$$

the Cauchy–Riemann equations are never satisfied:

$$u_x = 1 , \quad u_y = 0 , \quad v_x = 0 , \quad v_y = -1 .$$

---

[11] See Exercise IV.1.4.

The above map gives a very simple example of a continuous but nowhere differentiable complex-valued function of one complex variable. Note, though, that the "same" function

$$F = (u, v) \colon \mathbb{R}^2 \to \mathbb{R}^2 , \quad (x, y) \mapsto (x, -y)$$

is totally differentiable everywhere with the constant derivative

$$[\partial F(x_0, y_0)] = \begin{bmatrix} 1 & 0 \\ 0 & -1 \end{bmatrix}$$

for $(x_0, y_0) \in \mathbb{R}^2$. Therefore F is actually continuously differentiable. ■

Exercises

1 Calculate $\partial_2 f(1, y)$ for $f \colon (0, \infty)^2 \to \mathbb{R}$ with

$$f(x, y) := \left(((x^x)^x)^x \right)^y + \log(x) \arctan\Big(\arctan\big(\arctan(\sin(\cos(xy) - \log(x + y))) \big) \Big) .$$

2 Suppose $f \colon \mathbb{R}^2 \to \mathbb{R}$ is defined by

$$f(x, y) := \begin{cases} \sqrt{x^2 + y^2} , & y > 0 , \\ x , & y = 0 , \\ -\sqrt{x^2 + y^2} , & y < 0 . \end{cases}$$

Show

(a) f is not differentiable at $(0, 0)$;

(b) every directional derivative of f exists at $(0, 0)$.

3 At which points is

$$\mathbb{R}^2 \to \mathbb{R} , \quad (x, y) \mapsto \begin{cases} x^3 / \sqrt{x^2 + y^2} , & (x, y) \neq (0, 0) , \\ 0 , & (x, y) = (0, 0) , \end{cases}$$

differentiable?

4 Let $f \colon \mathbb{R}^2 \to \mathbb{R}$ be defined by

$$f(x, y) := \begin{cases} xy / \sqrt{x^2 + y^2} , & (x, y) \neq (0, 0) , \\ 0 , & (x, y) = (0, 0) . \end{cases}$$

Show:

(a) $\partial_1 f(0, 0)$ and $\partial_2 f(0, 0)$ exist.

(b) $D_v f(0, 0)$ does not exist for $v \in \mathbb{R}^2 \setminus \{e_1, e_2\}$.

(c) f is not differentiable at $(0, 0)$.

5 Calculate the Jacobi matrix of

$$\mathbb{R}^3 \to \mathbb{R}, \qquad (x,y,z) \mapsto 3x^2 y + e^{xz^2} + 4z^3 ;$$
$$\mathbb{R}^2 \to \mathbb{R}^3, \qquad (x,y) \mapsto \bigl(xy, \cosh(xy), \log(1+x^2)\bigr) ;$$
$$\mathbb{R}^3 \to \mathbb{R}^3, \qquad (x,y,z) \mapsto \bigl(\log(1+x^2+z^2), z^2+y^2-x^2, \sin(xz)\bigr) ;$$
$$\mathbb{R}^3 \to \mathbb{R}^3, \qquad (x,y,z) \mapsto (x \sin y \cos z, x \sin y \sin z, x \cos y) .$$

6 Suppose X is open at $\mathbb{R}^n$, F is a Banach space, and $f : X \to F$. Also suppose $x_0 \in X$, and choose $\varepsilon > 0$ such that $\mathbb{B}(x_0, \varepsilon) \subset X$. Finally, let

$$x_k(h) := x_0 + h^1 e_1 + \cdots + h^k e_k \quad \text{for } k = 1, \dots, n \text{ and } h \in \mathbb{B}(x_0, \varepsilon) .$$

Prove that f is differentiable at x_0 if and only if, for every $h \in \mathbb{B}(x_0, \varepsilon)$ such that $h^k \neq 0$ for $1 \leq k \leq n$, the limit

$$\lim_{\substack{h^k \to 0 \\ h^k \neq 0}} \frac{f\bigl(x_k(h)\bigr) - f\bigl(x_{k-1}(h)\bigr)}{h^k} \quad \text{for } 1 \leq k \leq n$$

exists in F.

7 Prove Remark 2.15(a). (Hint: Theorem 2.14.)

8 Determine the gradients of these functions:

$$\mathbb{R}^m \to \mathbb{R}, \qquad x \mapsto (x_0 \,|\, x) ;$$
$$\mathbb{R}^m \setminus \{0\} \to \mathbb{R}, \qquad x \mapsto |x| ;$$
$$\mathbb{R}^m \to \mathbb{R}, \qquad x \mapsto |(x_0 \,|\, x)|^2 ;$$
$$\mathbb{R}^m \setminus \{0\} \to \mathbb{R}, \qquad x \mapsto 1/|x| .$$

9 At which points is $\mathbb{C} \to \mathbb{C}$, $z \mapsto z\,|z|$ differentiable? If possible, find the derivative.

10 At which points is $\mathbb{R}^m \to \mathbb{R}^m$, $x \mapsto x\,|x|^k$ with $k \in \mathbb{N}$ differentiable? Calculate the derivative wherever it exists.

11 Find all differentiable functions $f : \mathbb{C} \to \mathbb{C}$ such that $f(\mathbb{C}) \subset \mathbb{R}$.

12 For $p \in [1, \infty]$, let $f_p : \mathbb{R}^m \to \mathbb{R}$, $x \mapsto |x|_p$. Where is f_p differentiable? If possible, find $\nabla f_p(x)$.

13 Suppose X is open in $\mathbb{C}$ and $f : X \to \mathbb{C}$ is differentiable. Also define $f^*(z) := \overline{f(\bar{z})}$ for $z \in X^* := \{ \bar{z} \in \mathbb{C} \,;\, z \in X \}$. Show that $f^* : X^* \to \mathbb{C}$ is differentiable.

3 Multivariable differentiation rules

We collect here the most important rules for multidimensional differentiation. As usual, we assume

- E and F are Banach spaces over the field $\mathbb{K}$;
  X is an open subset of E.

Linearity

The next theorem shows that — as in the one-variable case — differentiation is linear.

3.1 Proposition   *Suppose $f, g : X \to F$ are differentiable at x_0 and $\alpha \in \mathbb{K}$. Then $f + \alpha g$ is also differentiable at x_0, and*

$$\partial(f + \alpha g)(x_0) = \partial f(x_0) + \alpha \partial g(x_0) \ .$$

Proof   By assumption, we can write

$$f(x) = f(x_0) + \partial f(x_0)(x - x_0) + r_{x_0}(x) \, \|x - x_0\| \ ,$$
$$g(x) = g(x_0) + \partial g(x_0)(x - x_0) + s_{x_0}(x) \, \|x - x_0\| \ ,$$

where the functions $r_{x_0}, s_{x_0} : X \to F$ are continuous and vanish at x_0. Thus,

$$(f + \alpha g)(x) = (f + \alpha g)(x_0) + \big[\partial f(x_0) + \alpha \partial g(x_0)\big](x - x_0) + t_{x_0}(x) \, \|x - x_0\|$$

where $t_{x_0} := r_{x_0} + \alpha s_{x_0}$. Proposition 2.1 then implies the theorem. ∎

3.2 Corollary   *$C^1(X, F)$ is a vector subspace of $C(X, F)$, and*

$$\partial : C^1(X, F) \to C\big(X, \mathcal{L}(E, F)\big) \ , \quad f \mapsto \partial f$$

is linear.

The chain rule

In Chapter IV, we saw the great importance of the chain rule for functions of one variable, and, as we shall see here, the same is true in the multivariable case.

3.3 Theorem   (chain rule)   *Suppose Y is open in F and G is a Banach space. Also suppose $f : X \to F$ is differentiable at x_0 and $g : Y \to G$ is differentiable at $y_0 := f(x_0)$ and that $f(X) \subset Y$. Then $g \circ f : X \to G$ is differentiable at x_0, and the derivative is given by*

$$\partial(g \circ f)(x_0) = \partial g\big(f(x_0)\big)\partial f(x_0) \ .$$

Proof[1]   If $A := \partial g\big(f(x_0)\big)\partial f(x_0)$, then $A \in \mathcal{L}(E, G)$. Proposition 2.1 implies

$$
\begin{aligned}
f(x) &= f(x_0) + \partial f(x_0)(x - x_0) + r(x)\,\|x - x_0\| && \text{for } x \in X \ , \\
g(y) &= g(y_0) + \partial g(y_0)(y - y_0) + s(y)\,\|y - y_0\| && \text{for } y \in Y \ ,
\end{aligned}
\tag{3.1}
$$

where $r : X \to F$ and $s : Y \to G$ are continuous at x_0 and at y_0, respectively, and also $r(x_0) = 0$ and $s(y_0) = 0$. We define $t : X \to G$ through $t(x_0) := 0$ and

$$
t(x) := \partial g\big(f(x_0)\big)r(x) + s\big(f(x)\big)\left\|\partial f(x_0)\frac{x - x_0}{\|x - x_0\|} + r(x)\right\| \quad \text{for } x \neq x_0 \ .
$$

Then t is continuous at x_0. From (3.1) and with $y := f(x)$, we derive the relation

$$
\begin{aligned}
(g \circ f)(x) &= g\big(f(x_0)\big) + A(x - x_0) + \partial g\big(f(x_0)\big)r(x)\,\|x - x_0\| \\
&\quad + s\big(f(x)\big)\left\|\partial f(x_0)(x - x_0) + r(x)\,\|x - x_0\|\right\| \\
&= (g \circ f)(x_0) + A(x - x_0) + t(x)\,\|x - x_0\| \ .
\end{aligned}
$$

The theorem then follows from Proposition 2.1. ∎

3.4 Corollary   (chain rule in coordinates)   *Suppose X is open in $\mathbb{R}^n$ and Y is open in $\mathbb{R}^m$. Also suppose $f : X \to \mathbb{R}^m$ is differentiable at x_0 and $g : Y \to \mathbb{R}^\ell$ is differentiable at $y_0 := f(x_0)$ and also $f(X) \subset Y$. Then $h := g \circ f : X \to \mathbb{R}^\ell$ is differentiable at x_0 and*

$$
\big[\partial h(x_0)\big] = \big[\partial g\big(f(x_0)\big)\big]\big[\partial f(x_0)\big] \ ,
\tag{3.2}
$$

that is, the Jacobi matrix of the composition $h = g \circ f$ is the product of the Jacobi matrices of g and f.

Proof   This follows from Theorem 3.3, Corollary 2.9, and Theorem 1.9(ii). ∎

3.5 Remark   With the notation of Corollary 3.4, the entries of the Jacobi matrix of h are

$$
\frac{\partial h^j(x_0)}{\partial x^k} = \sum_{i=1}^{m} \frac{\partial g^j\big(f(x_0)\big)}{\partial y^i}\,\frac{\partial f^i(x_0)}{\partial x^k} \quad \text{for } 1 \leq j \leq \ell \text{ and } 1 \leq k \leq n \ .
$$

With this fundamental formula, we can calculate partial derivatives of composed functions in many concrete cases.

Proof   Using the rule for multiplying matrices, this follows immediately from (3.2). ∎

---

[1]See the proof of Theorem IV.1.7.

3.6 Examples   **(a)** We consider the map

$$f : \mathbb{R}^2 \to \mathbb{R}^3 , \qquad (x, y) \mapsto (x^2, xy, xy^2) ,$$
$$g : \mathbb{R}^3 \to \mathbb{R}^2 , \quad (\xi, \eta, \zeta) \mapsto \big(\sin \xi, \cos(\xi\eta\zeta)\big) .$$

For $h := g \circ f : \mathbb{R}^2 \to \mathbb{R}^2$, we then have $h(x, y) = \big(\sin x^2, \cos(x^4 y^3)\big)$. Thus h is continuously differentiable, and

$$[\partial h(x, y)] = \begin{bmatrix} 2x \cos x^2 & 0 \\ -4x^3 y^3 \sin(x^4 y^3) & -3x^4 y^2 \sin(x^4 y^3) \end{bmatrix} . \tag{3.3}$$

As for the Jacobi matrices of g and f, we respectively find

$$\begin{bmatrix} \cos \xi & 0 & 0 \\ -\eta\zeta \sin(\xi\eta\zeta) & -\xi\zeta \sin(\xi\eta\zeta) & -\xi\eta \sin(\xi\eta\zeta) \end{bmatrix} \quad \text{and} \quad \begin{bmatrix} 2x & 0 \\ y & x \\ y^2 & 2xy \end{bmatrix} .$$

Now we can easily verify that the product of these two matrices agrees with (3.3) at the position $(\xi, \eta, \zeta) = f(x, y)$.

(b) Suppose X is open in $\mathbb{R}^n$ and $f \in C^1(X, \mathbb{R})$. Also let I be an open interval in $\mathbb{R}$, and suppose $\varphi \in C^1(I, \mathbb{R}^n)$ with $\varphi(I) \subset X$. Then $f \circ \varphi$ belongs to $C^1(I, \mathbb{R})$, and

$$(f \circ \varphi)^{\cdot}(t) = \big(\nabla f(\varphi(t)) \,\big|\, \dot\varphi(t)\big) \quad \text{for } t \in I .$$

Proof   From the chain rule, we get

$$(f \circ \varphi)^{\cdot}(t) = df(\varphi(t))\dot\varphi(t) = \big(\nabla f(\varphi(t)) \,\big|\, \dot\varphi(t)\big) \quad \text{for } t \in I . \ \blacksquare$$

(c) We use the above notation and consider a path $\varphi : I \to X$, which stays within a "level set" of f, that is, there is a $y \in \mathrm{im}(f)$ such that $f(\varphi(t)) = y$ for each $t \in I$. From (b), it then follows that

$$\big(\nabla f(\varphi(t)) \,\big|\, \dot\varphi(t)\big) = 0$$

for $t \in I$. This shows the gradient $\nabla f(x)$ at the point $x = \varphi(t)$ is orthogonal to the path φ and, therefore, also to the tangent of φ through $(t, \varphi(t))$ (see Remark IV.1.4(a)). Somewhat imprecisely, we say, "The gradient is orthogonal to level sets". $\blacksquare$

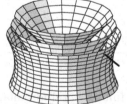

Level sets of the function
$(x, y, z) \mapsto x^2 + y^2 - z^2$

The product rule

The product rule for real-valued functions is another application of the chain rule.[2]

3.7 Proposition *Suppose $f, g \in C^1(X, \mathbb{R})$. Then fg also belongs to $C^1(X, \mathbb{R})$, and the product rule says*

$$\partial(fg) = g\partial f + f\partial g . \tag{3.4}$$

Proof For

$$m : \mathbb{R}^2 \to \mathbb{R} , \quad (\alpha, \beta) \mapsto \alpha\beta$$

we have $m \in C^1(\mathbb{R}^2, \mathbb{R})$ and $\nabla m(\alpha, \beta) = (\beta, \alpha)$. Setting $F := m \circ (f, g)$, we get $F(x) = f(x)g(x)$ for $x \in X$. From the chain rule, it follows that $F \in C^1(X, \mathbb{R})$ with

$$\partial F(x) = \partial m\big(f(x), g(x)\big) \circ \big(\partial f(x), \partial g(x)\big) = g(x)\partial f(x) + f(x)\partial g(x) . \ \blacksquare$$

3.8 Corollary *In the case $E = \mathbb{R}^n$, we have*

$$d(fg) = gdf + fdg \quad \text{and} \quad \nabla(fg) = g\nabla f + f\nabla g .$$

Proof The first formula is another way of writing (3.4). Because

$$\big(\nabla(fg)(x)\,\big|\,h\big) = d(fg)(x)h = f(x)dg(x)h + g(x)df(x)h$$
$$= \big(f(x)\nabla g(x) + g(x)\nabla f(x)\,\big|\,h\big)$$

for $x \in X$ and $h \in \mathbb{R}^n$, the second formula also holds. $\blacksquare$

The mean value theorem

As in the one-variable case, the multivariable case has a mean value theorem, which means we can use the derivative to estimate the difference of function values.

In what follows, we use the notation $[\![x, y]\!]$ introduced in Section III.4 for the straight path $\{ x + t(y - x) \ ; \ t \in [0, 1] \}$ between the points $x, y \in E$.

3.9 Theorem (mean value theorem) *Suppose $f : X \to F$ is differentiable. Then*

$$\|f(x) - f(y)\| \leq \sup_{0 \leq t \leq 1} \big\|\partial f\big(x + t(y - x)\big)\big\| \, \|y - x\| \tag{3.5}$$

for all $x, y \in X$ such that $[\![x, y]\!] \subset X$.

---

[2]See also Example 4.8(b).

Proof   We set $\varphi(t) := f(x + t(y - x))$ for $t \in [0, 1]$. Because $[\![x, y]\!] \subset X$, we know φ is defined. The chain rule shows that φ is differentiable with

$$\dot\varphi(t) = \partial f(x + t(y - x))(y - x) \ .$$

From Theorem IV.2.18, the mean value theorem for vector-valued functions of one variable, it follows that

$$\|f(y) - f(x)\| = \|\varphi(1) - \varphi(0)\| \leq \sup_{0 \leq t \leq 1} \|\dot\varphi(t)\| \ .$$

We finish the theorem by also considering

$$\|\dot\varphi(t)\| \leq \|\partial f(x + t(y - x))\| \, \|y - x\| \quad \text{for } t \in [0, 1] \ . \ \blacksquare$$

Under the somewhat stronger assumption of continuous differentiability, we can prove a very useful variant of the mean value theorem:

3.10 Theorem   (mean value theorem in integral form)   *Let $f \in C^1(X, F)$. Then we have*

$$f(y) - f(x) = \int_0^1 \partial f(x + t(y - x))(y - x) \, dt \tag{3.6}$$

for $x, y \in X$ such that $[\![x, y]\!] \subset X$.

Proof   The auxiliary function φ from the previous proof is now continuously differentiable. Thus we can apply the fundamental theorem of calculus (Corollary VI.4.14), and it gives

$$f(y) - f(x) = \varphi(1) - \varphi(0) = \int_0^1 \dot\varphi(t) \, dt = \int_0^1 \partial f(x + t(y - x))(y - x) \, dt \ . \ \blacksquare$$

3.11 Remarks   Suppose $f : X \to F$ is differentiable.

(a) If ∂f is continuous, the representation (3.6) gives the estimate (3.5).

Proof   Apply Proposition VI.4.3 and the definition of the operator norm. $\blacksquare$

(b) If X is convex and $\partial f : X \to \mathcal{L}(E, F)$ is bounded, then f is Lipschitz continuous.

Proof   Let $\alpha := \sup_{x \in X} \|\partial f(x)\|$. Then from Theorem 3.9, we get

$$\|f(y) - f(x)\| \leq \alpha \, \|y - x\| \quad \text{for } x, y \in X \ . \ \blacksquare \tag{3.7}$$

(c) If X is connected and $\partial f = 0$, then f is constant.

Proof   Suppose $x_0 \in X$ and $r > 0$ such that $\mathbb{B}(x_0, r) \subset X$. Letting $y_0 := f(x_0)$, it then follows from (3.7) that $f(x) = y_0$ for all $x \in \mathbb{B}(x_0, r)$. Because x_0 was arbitrary, f is locally constant. Therefore $f^{-1}(y_0)$ is nonempty and open in X. Also, from Proposition 2.1(ii) and Example III.2.22(a), $f^{-1}(y_0)$ is closed in X. The claim then follows from Remark III.4.3. ∎

The differentiability of limits of sequences of functions

The theorem of the differentiability of limits of sequences of functions (Theorem V.2.8) can now be easily extended to the general case.

3.12 Theorem   *Let $f_k \in C^1(X, F)$ for $k \in \mathbb{N}$. Also assume there are functions $f \in F^X$ and $g \colon X \to \mathcal{L}(E, F)$ such that*

(i) *(f_k) converges pointwise to f;*

(ii) *(∂f_k) locally converges uniformly to g.*

Then f belongs to $C^1(X, F)$, and we have $\partial f = g$.

Proof   Under the application of Theorem 3.9, the proof of Theorem V.2.8 remains valid word-for-word. ∎

Necessary condition for local extrema

In the important Theorem IV.2.1, we gave a necessary criterion for the existence of a local extremum of a function of one real variable. With the methods developed here, we can now treat many variables.

3.13 Theorem   *Suppose the map $f \colon X \to \mathbb{R}$ has a local extremal point at x_0 and all its directional derivatives exist there. Then*

$$D_v f(x_0) = 0 \quad \text{for } v \in E \backslash \{0\} \ .$$

Proof   Let $v \in E \backslash \{0\}$. We choose $r > 0$ such that $x_0 + tv \in X$ for $t \in (-r, r)$ and consider a function of one real variable:

$$\varphi \colon (-r, r) \to \mathbb{R} \ , \quad t \mapsto f(x_0 + tv) \ .$$

Then φ is differentiable at 0 and has a local extremum at 0. From Theorem IV.2.1, we have $\dot{\varphi}(0) = D_v f(x_0) = 0$. ∎

3.14 Remarks   (a) Suppose $f \colon X \to \mathbb{R}$ is differentiable at x_0. Then x_0 is called a **critical point** of f if $df(x_0) = 0$. If f has a local extremum at x_0, then x_0 is a critical point. When $E = \mathbb{R}^n$, this definition agrees with that of Remark 2.17(a).

Proof   This follows from Proposition 2.5 and Theorem 3.13. ∎

(b) A critical point x_0 is *not* necessarily a local extremal point.[3]

Proof Consider

$$f : \mathbb{R}^2 \to \mathbb{R} , \quad (x,y) \mapsto x^2 - y^2 .$$

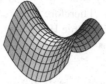

Then $\nabla f(0,0) = 0$, but $(0,0)$ is not
an extremal point of f. Instead, it is
a "saddle point". ∎

Exercises

1 We say a function $f : E \to F$ is **positively homogeneous** of degree $\alpha \in \mathbb{R}$ if

$$f(tx) = t^\alpha f(x) \quad \text{for } t > 0 \text{ and } x \in E \backslash \{0\} .$$

Show that if $f \in C^1(E,F)$ is positively homogeneous of degree 1, then $f \in \mathcal{L}(E,F)$.

2 Suppose $f : \mathbb{R}^m \to \mathbb{R}$ is differentiable at $\mathbb{R}^m \backslash \{0\}$. Prove that f is positively homogeneous of degree α if it satisfies the **Euler homogeneity relation**

$$\big(\nabla f(x) \big| x \big) = \alpha f(x) \quad \text{for } x \in \mathbb{R}^m \backslash \{0\} .$$

3 Suppose X is open in $\mathbb{R}^m$ and $f \in C^1(X,\mathbb{R}^n)$. Show that

$$g : X \to \mathbb{R} , \quad x \mapsto \sin\big(|f(x)|^2 \big)$$

is continuously differentiable, and determine ∇g.

4 For $f \in C^1(\mathbb{R}^n, \mathbb{R})$ and $A \in \mathcal{L}(\mathbb{R}^n)$, show

$$\nabla(f \circ A)(x) = A^* \nabla f(Ax) \quad \text{for } x \in \mathbb{R}^n .$$

5 Suppose X is open in $\mathbb{R}^k$ and Y is open and bounded in $\mathbb{R}^n$. Further suppose $f \in C\big(\overline{X \times Y}, \mathbb{R} \big)$ is differentiable in $X \times Y$ and there is a $\xi \in C^1(X,\mathbb{R}^n)$ such that $\operatorname{im}(\xi) \subset Y$ and

$$m(x) := \min_{y \in \overline{Y}} f(x,y) = f\big(x, \xi(x) \big) \text{ for } x \in X .$$

Calculate the gradient of $m : X \to \mathbb{R}$.

6 For $g \in C^1(\mathbb{R}^m, \mathbb{R})$ and $f_j \in C^1(\mathbb{R}, \mathbb{R})$ with $j = 1, \dots, m$, calculate the gradient of

$$\mathbb{R}^m \to \mathbb{R} , \quad (x_1, \dots, x_m) \mapsto g\big(f_1(x_1), \dots, f_m(x_m) \big) .$$

7 The function $f \in C^1(\mathbb{R}^2, \mathbb{R})$ satisfies $\partial_1 f = \partial_2 f$ and $f(0,0) = 0$. Show that there is a $g \in C(\mathbb{R}^2, \mathbb{R})$ such that $f(x,y) = g(x,y)(x+y)$ for $(x,y) \in \mathbb{R}^2$.

8 Verify for the exponential map that

$$\exp \in C^1 \big(\mathcal{L}(E), \mathcal{L}(E) \big) \quad \text{and} \quad \partial \exp(0) = \operatorname{id}_{\mathcal{L}(E)} .$$

---

[3]Compare this also with Remark IV.2.2(c).

4 Multilinear maps

Let E and F be Banach spaces, and let X be an open subset of E. For $f \in C^1(X, F)$, we have $\partial f \in C(X, \mathcal{L}(E, F))$. We set $g := \partial f$ and $\mathbb{F} := \mathcal{L}(E, F)$. According to Theorem 1.1, $\mathbb{F}$ is a Banach space, and we can therefore study the differentiability of the map $g \in C(X, \mathbb{F})$. If g is [continuously] differentiable, we say f is twice [continuously] differentiable, and $\partial^2 f := \partial g$ is the second derivative of f. Thus

$$\partial^2 f(x) \in \mathcal{L}(E, \mathcal{L}(E, F)) \quad \text{for } x \in X .$$

Theorem 1.1 says that $\mathcal{L}(E, \mathcal{L}(E, F))$ is also a Banach space. Thus we can study the third derivative $\partial^3 f := \partial(\partial^2 f)$ (if it exists) and likewise find

$$\partial^3 f(x) \in \mathcal{L}(E, \mathcal{L}(E, \mathcal{L}(E, F))) \quad \text{for } x \in X .$$

In this notation, the spaces occupied by $\partial^n f$ evidently become increasingly complex with increasing n. However, things are not as bad as they seem. We will show that the complicated-looking space $\mathcal{L}(E, \mathcal{L}(E, \ldots, \mathcal{L}(E, F) \cdots))$ (with E occurring n-times) is isometrically isomorphic to the space of all continuous multilinear maps from $E \times \cdots \times E$ (again with n E's) to F. Multilinear maps are therefore the proper setting for understanding higher derivatives.

Continuous multilinear maps

In the following, $E_1, \ldots, E_m$ for $m \geq 2$, E, and F are Banach spaces over the field $\mathbb{K}$. A map $\varphi \colon E_1 \times \cdots \times E_m \to F$ is **multilinear** or, equivalently, **m-linear**[1] if for every $k \in \{1, \ldots, m\}$ and every choice of $x_j \in E_j$ for $j = 1, \ldots, m$ with $j \neq k$, the map

$$\varphi(x_1, \ldots, x_{k-1}, \cdot, x_{k+1}, \ldots, x_m) \colon E_k \to F$$

is linear, that is, φ is multilinear if it is linear in every variable.

First we show that a multilinear map is continuous if and only if it is bounded.[2]

4.1 Proposition *For the m-linear map $\varphi \colon E_1 \times \cdots \times E_m \to F$, these statements are equivalent:*

(i) *φ is continuous.*

(ii) *φ is continuous at 0.*

(iii) *φ is bounded on bounded sets.*

(iv) *There is an $\alpha \geq 0$ such that*

$$\|\varphi(x_1, \ldots, x_m)\| \leq \alpha \|x_1\| \cdot \cdots \cdot \|x_m\| \quad \text{for } x_j \in E_j , \quad 1 \leq j \leq m .$$

---

[1] When $m = 2$ [or $m = 3$] one speaks of **bilinear** [or **trilinear**] maps.

[2] Compare with Theorem VI.2.5.

Proof The implication "(i)⇒(ii)" is clear.

"(ii)⇒(iii)" Supposing $B \subset E_1 \times \cdots \times E_m$ is bounded, there exists, according to Example II.3.3(c) and Remark II.3.2(a), a $\beta > 0$ such that $\|x_j\| \leq \beta$ for $(x_1, \ldots, x_m) \in B$ and $1 \leq j \leq m$. Because φ is continuous at 0, there is a $\delta > 0$ such that

$$\|\varphi(y_1, \ldots, y_m)\| \leq 1 \quad \text{for } y_j \in E_j, \quad \|y_j\| \leq \delta, \quad 1 \leq j \leq m .$$

For $(x_1, \ldots, x_m) \in B$ and $1 \leq j \leq m$, we set $y_j := \delta x_j / \beta$. Then $\|y_j\| \leq \delta$, and

$$(\delta/\beta)^m \|\varphi(x_1, \ldots, x_m)\| = \|\varphi(y_1, \ldots, y_m)\| \leq 1 .$$

Therefore $\varphi(B)$ is bounded.

"(iii)⇒(iv)" By assumption, there is an $\alpha > 0$ such that

$$\|\varphi(x_1, \ldots, x_m)\| \leq \alpha \quad \text{for } (x_1, \ldots, x_m) \in \overline{\mathbb{B}} .$$

For $y_j \in E_j \setminus \{0\}$, we set $x_j := y_j / \|y_j\|$. Then $(x_1, \ldots, x_m)$ belongs to $\overline{\mathbb{B}}$, and we find

$$\frac{1}{\|y_1\| \cdot \cdots \cdot \|y_m\|} \|\varphi(y_1, \ldots, y_m)\| = \|\varphi(x_1, \ldots, x_m)\| \leq \alpha ,$$

which proves the claim.

"(iv)⇒(i)" Suppose $y = (y_1, \ldots, y_m)$ is a point of $E_1 \times \cdots \times E_m$ and (x^n) is a sequence in $E_1 \times \cdots \times E_m$ with $\lim_n x^n = y$. Letting $(x_1^n, \ldots, x_m^n) := x^n$, we have by assumption that

$$\|\varphi(y_1, \ldots, y_m) - \varphi(x_1^n, \ldots, x_m^n)\|$$

$$\leq \|\varphi(y_1 - x_1^n, y_2, \ldots, y_m)\| + \|\varphi(x_1^n, y_2 - x_2^n, y_3, \ldots, y_m)\|$$

$$+ \cdots + \|\varphi(x_1^n, x_2^n, \ldots, y_m - x_m^n)\|$$

$$\leq \alpha \Big(\|y_1 - x_1^n\| \|y_2\| \cdot \cdots \cdot \|y_m\| + \|x_1^n\| \|y_2 - x_2^n\| \cdot \cdots \cdot \|y_m\|$$

$$+ \cdots + \|x_1^n\| \|x_2^n\| \cdot \cdots \cdot \|y_m - x_m^n\| \Big) .$$

Because the sequence (x^n) is bounded in $E_1 \times \cdots \times E_m$, this estimate shows, together with (the analogue of) Proposition II.3.14, that $\varphi(x^n)$ converges to $\varphi(y)$. ∎

It is useful to introduce a shorthand. We denote by

$$\mathcal{L}(E_1, \ldots, E_m; F)$$

the set of all continuous multilinear maps from $E_1 \times \cdots \times E_m$ to F. Evidently $\mathcal{L}(E_1, \ldots, E_m; F)$ is a vector subspace of $C(E_1 \times \cdots \times E_m, F)$. Often, all the E_j are the same, and we use the notation

$$\mathcal{L}^m(E, F) := \mathcal{L}(E, \ldots, E; F) .$$

In addition, we set $\mathcal{L}^1(E, F) := \mathcal{L}(E, F)$ and $\mathcal{L}^0(E, F) := F$. Finally, let

$$\|\varphi\| := \inf\big\{ \alpha \geq 0 \; ; \; \|\varphi(x_1, \ldots, x_m)\| \leq \alpha \|x_1\| \cdot \, \cdots \, \cdot \|x_m\|, \; x_j \in E_j \big\} \qquad (4.1)$$

for $\varphi \in \mathcal{L}(E_1, \ldots, E_m; F)$.

4.2 Theorem

(i) For $\varphi \in \mathcal{L}(E_1, \ldots, E_m; F)$, we find

$$\|\varphi\| = \sup\big\{ \|\varphi(x_1, \ldots, x_m)\| \; ; \; \|x_j\| \leq 1, \; 1 \leq j \leq m \big\}$$

and

$$\|\varphi(x_1, \ldots, x_m)\| \leq \|\varphi\| \, \|x_1\| \cdot \, \cdots \, \cdot \|x_m\|$$

for $(x_1, \ldots, x_m) \in E_1 \times \cdots \times E_m$.

(ii) The relation (4.1) defines a norm and

$$\mathcal{L}(E_1, \ldots, E_m; F) := \big(\mathcal{L}(E_1, \ldots, E_m; F), \|\cdot\|\big)$$

is a Banach space.

(iii) When $\dim E_j < \infty$ for $1 \leq j \leq m$, every m-linear map is continuous.

Proof For $m = 1$, these are implied by Proposition VI.2.3, Conclusion VI.2.4(e), and Theorems 1.1 and 1.6. The general case can be verified through the obvious modification of these proofs, and we leave this to you as an exercise. ∎

The canonical isomorphism

The norm on $\mathcal{L}(E_1, \ldots, E_m; F)$ is a natural extension of the operator norm on $\mathcal{L}(E, F)$. The next theorem shows this norm is also natural in another way.

4.3 Theorem   The spaces $\mathcal{L}(E_1, \ldots, E_m; F)$ and $\mathcal{L}\big(E_1, \mathcal{L}(E_2, \ldots, \mathcal{L}(E_m, F) \cdots)\big)$ are isometrically isomorphic.

Proof We verify the statement for $m = 2$. The general case obtains via a simple induction argument.

(i) For $T \in \mathcal{L}\big(E_1, \mathcal{L}(E_2, F)\big)$ we set

$$\varphi_T(x_1, x_2) := (Tx_1)x_2 \quad \text{for } (x_1, x_2) \in E_1 \times E_2 \, .$$

Then $\varphi_T : E_1 \times E_2 \to F$ is bilinear, and

$$\|\varphi_T(x_1, x_2)\| \leq \|T\| \, \|x_1\| \, \|x_2\| \quad \text{for } (x_1, x_2) \in E_1 \times E_2 \, .$$

Therefore φ_T belongs to $\mathcal{L}(E_1, E_2; F)$, and $\|\varphi_T\| \leq \|T\|$.

(ii) Suppose $\varphi \in \mathcal{L}(E_1, E_2; F)$. Then we set

$$T_\varphi(x_1)x_2 := \varphi(x_1, x_2) \quad \text{for } (x_1, x_2) \in E_1 \times E_2 .$$

Because

$$\|T_\varphi(x_1)x_2\| = \|\varphi(x_1, x_2)\| \le \|\varphi\| \, \|x_1\| \, \|x_2\| \quad \text{for } (x_1, x_2) \in E_1 \times E_2 ,$$

we get

$$T_\varphi(x_1) \in \mathcal{L}(E_2, F) \quad \text{for } \|T_\varphi(x_1)\| \le \|\varphi\| \, \|x_1\|$$

for every $x_1 \in E_1$. Therefore

$$T_\varphi := [x_1 \mapsto T_\varphi(x_1)] \in \mathcal{L}\big(E_1, \mathcal{L}(E_2, F)\big) \text{ and } \|T_\varphi\| \le \|\varphi\| .$$

(iii) Altogether, we have proved that the maps

$$T \mapsto \varphi_T : \mathcal{L}\big(E_1, \mathcal{L}(E_2, F)\big) \to \mathcal{L}(E_1, E_2; F)$$

and

$$\varphi \mapsto T_\varphi : \mathcal{L}(E_1, E_2; F) \to \mathcal{L}\big(E_1, \mathcal{L}(E_2, F)\big) \tag{4.2}$$

are linear, continuous, and inverses of each other. Thus they are topological isomorphisms. In particular, $T_{\varphi_T} = T$, and we get

$$\|T\| = \|T_{\varphi_T}\| \le \|\varphi_T\| \le \|T\| .$$

Thus the map $T \mapsto \varphi_T$ is an isometry. ∎

Convention $\mathcal{L}(E_1, \ldots, E_m; F)$ and $\mathcal{L}\big(E_1, \mathcal{L}\big(E_2, \ldots, \mathcal{L}(E_m, F) \cdots\big)\big)$ are identified using the isometric isomorphism (4.2).

4.4 Conclusions (a) For $m \in \mathbb{N}$, we have

$$\mathcal{L}\big(E, \mathcal{L}^m(E, F)\big) = \mathcal{L}^{m+1}(E, F) .$$

Proof This follows immediately from Theorem 4.3 and the above convention. ∎

(b) $\mathcal{L}^m(E, F)$ is a Banach space.

Proof This follows from Theorem 4.3 and Remark 1.3(c). ∎

Symmetric multilinear maps

Suppose $m \ge 2$ and $\varphi : E^m \to F$ is m-linear. We say φ is **symmetric** if

$$\varphi(x_{\sigma(1)}, \ldots, x_{\sigma(m)}) = \varphi(x_1, \ldots, x_m)$$

for every $(x_1, \ldots, x_m)$ and every permutation σ of $\{1, \ldots, m\}$. We set

$$\mathcal{L}^m_{\mathrm{sym}}(E, F) := \big\{ \varphi \in \mathcal{L}^m(E, F) \, ; \, \varphi \text{ is symmetric} \big\} .$$

4.5 Proposition $\mathcal{L}^m_{\mathrm{sym}}(E, F)$ *is a closed vector subspace of* $\mathcal{L}^m(E, F)$ *and is therefore itself a Banach space.*

Proof Suppose (φ_k) is a sequence in $\mathcal{L}^m_{\mathrm{sym}}(E, F)$ that converges in $\mathcal{L}^m(E, F)$ to φ. For every $(x_1, \ldots, x_m) \in E^m$ and every permutation[3] $\sigma \in \mathsf{S}_m$, we then have

$$\varphi(x_{\sigma(1)}, \ldots, x_{\sigma(m)}) = \lim_k \varphi_k(x_{\sigma(1)}, \ldots, x_{\sigma(m)}) = \lim_k \varphi_k(x_1, \ldots, x_m)$$
$$= \varphi(x_1, \ldots, x_m) \ .$$

Therefore φ is symmetric. $\blacksquare$

The derivative of multilinear maps

The next theorem shows that m-linear maps are actually continuously differentiable and that the derivatives are sums of $(m-1)$-linear functions.

4.6 Proposition $\mathcal{L}(E_1, \ldots, E_m; F)$ *is a vector subspace of* $C^1(E_1 \times \cdots \times E_m, F)$. *And, for* $\varphi \in \mathcal{L}(E_1, \ldots, E_m; F)$ *and* $(x_1, \ldots, x_m) \in E_1 \times \cdots \times E_m$, *we have*

$$\partial\varphi(x_1, \ldots, x_m)(h_1, \ldots, h_m) = \sum_{j=1}^m \varphi(x_1, \ldots, x_{j-1}, h_j, x_{j+1}, \ldots, x_m)$$

for $(h_1, \ldots, h_m) \in E_1 \times \cdots \times E_m$.

Proof We denote by A_x the map

$$(h_1, \ldots, h_m) \mapsto \sum_{j=1}^m \varphi(x_1, \ldots, x_{j-1}, h_j, x_{j+1}, \ldots, x_m)$$

from $E_1 \times \cdots \times E_m$ to F. Then it is not hard to verify that

$$(x \mapsto A_x) \in C\big(E_1 \times \cdots \times E_m, \mathcal{L}(E_1 \times \cdots \times E_m, F)\big) \ .$$

Letting $y_k := x_k + h_k$, the equation

$$\varphi(y_1, \ldots, y_m) = \varphi(x_1, \ldots, x_m) + \sum_{k=1}^m \varphi(x_1, \ldots, x_{k-1}, h_k, y_{k+1}, \ldots, y_m)$$

follows from the multilinearity of φ. Because each of the maps

$$E_{k+1} \times \cdots \times E_m \to F \ , \quad (z_{k+1}, \ldots, z_m) \mapsto \varphi(x_1, \ldots, x_{k-1}, h_k, z_{k+1}, \ldots, z_m)$$

---

[3]Recall that S_n is the group of permutations of the set $\{1, \ldots, n\}$ (see the end of Section I.7).

is $(m-k)$-linear, we likewise get

$$\varphi(x_1, \ldots, x_{k-1}, h_k, y_{k+1}, \ldots, y_m)$$
$$= \varphi(x_1, \ldots, x_{k-1}, h_k, x_{k+1}, \ldots, x_m)$$
$$+ \sum_{j=1}^{m-k} \varphi(x_1, \ldots, x_{k-1}, h_k, x_{k+1}, \ldots, x_{k+j-1}, h_{k+j}, y_{k+j+1}, \ldots, y_m) .$$

Consequently, we find

$$\varphi(x_1 + h_1, \ldots, x_m + h_m) - \varphi(x_1, \ldots, x_m) - A_x h = r(x, h) ,$$

where $r(x, h)$ is a sum of function-valued multilinear maps, in which every summand has at least two different h_i for suitable $i \in \{1, \ldots, m\}$. Thus Theorem 4.2(i) implies that $r(x, h) = o(\|h\|)$ as $h \to 0$. ∎

4.7 Corollary   *Suppose X is open in $\mathbb{K}$ and $\varphi \in \mathcal{L}(E_1, \ldots, E_m; F)$ and $f_j \in C^1(X, E_j)$ for $1 \le j \le m$. Then the map*

$$\varphi(f_1, \ldots, f_m) : X \to F , \quad x \mapsto \varphi(f_1(x), \ldots, f_m(x))$$

belongs to $C^1(X, F)$, and we have

$$\partial\big(\varphi(f_1, \ldots, f_m)\big) = \sum_{j=1}^{m} \varphi(f_1, \ldots, f_{j-1}, f_j', f_{j+1}, \ldots, f_m) .$$

Proof   This follows from $f_j'(x) \in E_j$ (Proposition 4.6) and the chain rule. ∎

The signature formula

$$\det[a_k^j] = \sum_{\sigma \in S_m} \text{sign}(\sigma) a_{\sigma(1)}^1 \cdot \cdots \cdot a_{\sigma(m)}^m$$

shows that $\det[a_k^j]$ is an m-linear function of the row vectors $a_\bullet^j := (a_1^j, \ldots, a_m^j)$ (or, alternatively, of the column vectors $a_k^\bullet := (a_k^1, \ldots, a_k^m)$).

For $a_1, \ldots, a_m \in \mathbb{K}^m$ with $a_k = (a_k^1, \ldots, a_k^m)$, we set

$$[a_1, \ldots, a_m] := [a_k^j] \in \mathbb{K}^{m \times m} .$$

In other words, $[a_1, \ldots, a_m]$ is the square matrix with the column vectors $a_k^\bullet := a_k$ for $1 \le k \le m$. Therefore, the determinant function

$$\det : \underbrace{\mathbb{K}^m \times \cdots \times \mathbb{K}^m}_{m} \to \mathbb{K} , \quad (a_1, \ldots, a_m) \mapsto \det[a_1, \ldots, a_m]$$

is a well-defined m-linear map.

4.8 Examples   Suppose X is open in $\mathbb{K}$.

(a) For $a_1, \ldots, a_m \in C^1(X, \mathbb{K}^m)$, we have $\det[a_1, \ldots, a_m] \in C^1(X, \mathbb{K})$ and

$$
\left(\det[a_1, \ldots, a_m]\right)' = \sum_{j=1}^{m} \det[a_1, \ldots, a_{j-1}, a'_j, a_{j+1}, \ldots, a_m] .
$$

(b) Suppose $\varphi \in \mathcal{L}(E_1, E_2; F)$ and $(f, g) \in C^1(X, E_1 \times E_2)$. Then the (**generalized**) **product rule** is

$$
\partial \varphi(f, g) = \varphi(\partial f, g) + \varphi(f, \partial g) .
$$

Exercises

1   Let $H := \bigl(H, (\cdot \,|\, \cdot)\bigr)$ be a finite-dimensional Hilbert space. Show these facts:

(a) For every $a \in \mathcal{L}^2(H, \mathbb{K})$, there is exactly one $A \in \mathcal{L}(H)$, the **linear operator induced** by a, satisfying

$$
a(x, y) = (Ax \,|\, y) \quad \text{for } x, y \in H .
$$

The map

$$
\mathcal{L}^2(H, \mathbb{K}) \to \mathcal{L}(H) , \quad a \mapsto A \tag{4.3}
$$

thus induced is an isometric isomorphism.

(b) In the real case, we have $a \in \mathcal{L}^2_{\mathrm{sym}}(H, \mathbb{R}) \Longleftrightarrow A = A^*$.

(c) Suppose $H = \mathbb{R}^m$ and $(\cdot \,|\, \cdot)$ is the Euclidean scalar product. Then show

$$
a(x, y) = \sum_{j,k=1}^{m} a_{jk} x^j y^k \quad \text{for } x = (x^1, \ldots, x^m) , \quad y = (y^1, \ldots, y^m) , \quad x, y \in \mathbb{K}^m ,
$$

where $[a_{jk}]$ is the matrix representation of the linear operator induced by a.

(Hint for (a): the Riesz representation theorem.[4])

2   Suppose X is open in E and $f \in C^1\bigl(X, \mathcal{L}^m(E, F)\bigr)$ satisfies $f(X) \subset \mathcal{L}^m_{\mathrm{sym}}(E, F)$. Then show $\partial f(x) h \in \mathcal{L}^m_{\mathrm{sym}}(E, F)$ for $x \in X$ and $h \in E$.

3   For $A \in \mathbb{K}^{n \times n}$, show $\det e^A = e^{\mathrm{tr}(A)}$. (Hint: Let $h(t) := \det e^{tA} - e^{t\,\mathrm{tr}(A)}$ for $t \in \mathbb{R}$. With the help of Example 4.8(a), conclude $h' = 0$.)

4   For $T_1, T_2 \in \mathcal{L}(F, E)$ and $S \in \mathcal{L}(E, F)$, let $g(T_1, T_2)(S) := T_1 S T_2$. Verify

(a) $g(T_1, T_2) \in \mathcal{L}\bigl(\mathcal{L}(E, F), \mathcal{L}(F, E)\bigr)$;

(b) $g \in \mathcal{L}^2\bigl(\mathcal{L}(F, E); \mathcal{L}(\mathcal{L}(E, F), \mathcal{L}(F, E))\bigr)$.

5   Suppose H is a Hilbert space $A \in \mathcal{L}(H)$ and $a(x) := (Ax \,|\, x)$ for $x \in H$. Show

(a) $\partial a(x) y = (Ax \,|\, y) + (Ay \,|\, x)$ for $x, y \in H$;

(b) if $H = \mathbb{R}^n$, then $\nabla a(x) = (A + A^*)x$ for $x \in \mathbb{R}^n$.

---

[4]Because the Riesz representation theorem also holds for infinite-dimensional Hilbert spaces, (4.3) is also true in such spaces.

5 Higher derivatives

We can now define higher derivatives. We will see that if a function is continuously differentiable, its higher derivatives are symmetric multilinear maps.

Primarily, this section generalizes the Taylor expansion to functions of more than one variable. As one of its consequences, it will give, as in the one-dimensional case, a sufficient criterion of the existence of local extrema.

In the following, suppose

- E and F are Banach spaces and X is an open subset of E.

Definitions

As mentioned previously, we will define higher derivatives inductively, as we did in the one variable case.

Suppose $f \colon X \to F$ and $x_0 \in X$. We then set $\partial^0 f := f$. Therefore $\partial^0 f(x_0)$ belongs to $F = \mathcal{L}^0(E, F)$. Suppose now that $m \in \mathbb{N}^\times$ and $\partial^{m-1} f \colon X \to \mathcal{L}^{m-1}(E, F)$ is already defined. If

$$\partial^m f(x_0) := \partial(\partial^{m-1} f)(x_0) \in \mathcal{L}\big(E, \mathcal{L}^{m-1}(E, F)\big) = \mathcal{L}^m(E, F)$$

exists, we call $\partial^m f(x_0)$ the **m-th derivative of f at x_0**, and f is **m-times differentiable at x_0**. When $\partial^m f(x)$ exists for every $x \in X$, we call

$$\partial^m f \colon X \to \mathcal{L}^m(E, F)$$

the **m-th derivative** of f and say f is **m-times differentiable**. If $\partial^m f$ is continuous as well, we say f is **m-times continuously differentiable**. We set

$$C^m(X, F) := \{\, f \colon X \to F \;;\; f \text{ is } m\text{-times continuously differentiable} \,\}$$

and

$$C^\infty(X, F) := \bigcap_{m \in \mathbb{N}} C^m(X, F) \,.$$

Evidently $C^0(X, F) = C(X, F)$, and f is **smooth** or **infinitely continuously differentiable** if f belongs to $C^\infty(X, F)$.

Instead of $\partial^m f$, we sometimes write $D^m f$ or $f^{(m)}$; we also set $f' := f^{(1)}$, $f'' := f^{(2)}$, $f''' := f^{(3)}$ etc.

5.1 Remarks Suppose $m \in \mathbb{N}$.

(a) The m-th derivative is linear, that is,

$$\partial^m (f + \alpha g)(x_0) = \partial^m f(x_0) + \alpha \partial^m g(x_0)$$

for $\alpha \in \mathbb{K}$ and $f, g \colon X \to F$ if f and g are m-times differentiable at x_0.

Proof This follows by induction from Proposition 3.1. ∎

(b) $C^m(X,F) = \{\, f : X \to F \;;\; \partial^j f \in C\big(X, \mathcal{L}^j(E,F)\big),\; 0 \leq j \leq m \,\}$.

Proof This is implied by Proposition 2.1(ii) and the definition of the m-th derivative. ∎

(c) One can now easily verify that $C^{m+1}(X,F)$ is a vector subspace of $C^m(X,F)$. Also $C^\infty(X,F)$ is a vector subspace of $C^m(X,F)$. In particular, we have the chain of inclusions

$$C^\infty(X,F) \subset \cdots \subset C^{m+1}(X,F) \subset C^m(X,F) \subset \cdots \subset C(X,F) \;.$$

In addition, for $k \in \mathbb{N}$ the map

$$\partial^k : C^{m+k}(X,F) \to C^m(X,F)$$

is defined and linear.

Proof We leave the simple check to you. ∎

5.2 Theorem *Suppose $f \in C^2(X,F)$. Then*

$$\partial^2 f(x) \in \mathcal{L}^2_{\mathrm{sym}}(E,F) \quad \text{for } x \in X \;,$$

that is,[1]

$$\partial^2 f(x)[h,k] = \partial^2 f(x)[k,h] \quad \text{for } x \in X \text{ and } h,k \in E \;.$$

Proof (i) Suppose $x \in X$ and $r > 0$ such that $\mathbb{B}(x,2r) \subset X$ and $h,k \in r\mathbb{B}$. We set $g(y) := f(y+h) - f(y)$ and

$$r(h,k) := \int_0^1 \left[\int_0^1 \{\partial^2 f(x+sh+tk) - \partial^2 f(x)\} h\, ds \right] k\, dt \;.$$

From the mean value theorem (Theorem 3.10) and from the linearity of the derivative, we get

$$f(x+h+k) - f(x+k) - f(x+h) + f(x) = g(x+k) - g(x)$$

$$= \int_0^1 \partial g(x+tk)k\, dt = \int_0^1 \big(\partial f(x+h+tk) - \partial f(x+tk)\big)k\, dt$$

$$= \int_0^1 \left(\int_0^1 \partial^2 f(x+sh+tk)h\, ds \right) k\, dt = \partial^2 f(x)[h,k] + r(h,k) \;,$$

because $x+sh+tk$ belongs to $\mathbb{B}(x,2r)$ for $s,t \in [0,1]$.

(ii) Setting $\widetilde{g}(y) := f(y+k) - f(y)$ and

$$\widetilde{r}(k,h) := \int_0^1 \left(\int_0^1 \{\partial^2 f(x+sk+th) - \partial^2 f(x)\} k\, ds \right) h\, dt \;,$$

---

[1] For clarity, we have enclosed the arguments of multilinear maps in square brackets.

it follows analogously that

$$f(x + h + k) - f(x + h) - f(x + k) + f(x) = \partial^2 f(x)[k, h] + \widetilde{r}(k, h) \ .$$

Thus we get

$$\partial^2 f(x)[h, k] - \partial^2 f(x)[k, h] = \widetilde{r}(k, h) - r(h, k) \ .$$

Noting also the estimate

$$\|\widetilde{r}(k, h)\| \vee \|r(h, k)\| \leq \sup_{0 \leq s,t \leq 1} \|\partial^2 f(x + sh + tk) - \partial^2 f(x)\| \, \|h\| \, \|k\| \ ,$$

we find

$$\|\partial^2 f(x)[h, k] - \partial^2 f(x)[k, h]\|$$
$$\leq 2 \sup_{0 \leq s,t \leq 1} \|\partial^2 f(x + sh + tk) - \partial^2 f(x)\| \, \|h\| \, \|k\| \ .$$

In this inequality, we can replace h and k by τh and τk with $\tau \in (0, 1]$. Then we have

$$\|\partial^2 f(x)[h, k] - \partial^2 f(x)[k, h]\|$$
$$\leq 2 \sup_{0 \leq s,t \leq 1} \|\partial^2 f\big(x + \tau(sh + tk)\big) - \partial^2 f(x)\| \, \|h\| \, \|k\| \ .$$

Finally, the continuity of $\partial^2 f$ implies

$$\sup_{0 \leq s,t \leq 1} \|\partial^2 f\big(x + \tau(sh + tk)\big) - \partial^2 f(x)\| \to 0 \quad (\tau \to 0) \ .$$

Therefore $\partial^2 f(x)[h, k] = \partial^2 f(x)[k, h]$ for $h, k \in r\mathbb{B}$, and we are done. ∎

5.3 Corollary For $f \in C^m(X, F)$ such that $m \geq 2$, we have

$$\partial^m f(x) \in \mathcal{L}^m_{\mathrm{sym}}(E, F) \quad \text{for } x \in X \ .$$

Proof Taking cues from Theorem 5.2, we now construct an induction proof in m. The induction step $m \to m + 1$ goes as follows. Because $\mathcal{L}^m_{\mathrm{sym}}(E, F)$ is a closed vector subspace or $\mathcal{L}(E, F)$, it follows from the induction hypothesis that

$$\partial^{m+1} f(x)h_1 \in \mathcal{L}^m_{\mathrm{sym}}(E, F) \quad \text{for } h_1 \in E \ ,$$

(see Exercise 4.2). In particular, we get

$$\partial^{m+1} f(x)[h_1, h_{\sigma(2)}, h_{\sigma(3)}, \dots, h_{\sigma(m+1)}] = \partial^{m+1} f(x)[h_1, h_2, h_3, \dots, h_{m+1}]$$

for every $(h_1, \dots, h_{m+1}) \in E^{m+1}$ and every permutation σ of $\{2, \dots, m + 1\}$. Because every $\tau \in \mathsf{S}_{m+1}$ can be represented as a composition of transpositions, it suffices to verify that

$$\partial^{m+1} f(x)[h_1, h_2, h_3, \dots, h_{m+1}] = \partial^{m+1} f(x)[h_2, h_1, h_3, \dots, h_{m+1}] \ .$$

Because $\partial^{m+1} f(x) = \partial^2(\partial^{m-1} f)(x)$, this follows from Theorem 5.2. ∎

Higher order partial derivatives

We consider first the case $E = \mathbb{R}^n$ with $n \geq 2$.

For $q \in \mathbb{N}^\times$ and indices $j_1, \ldots, j_q \in \{1, \ldots, n\}$, we call

$$\frac{\partial^q f(x)}{\partial x^{j_1} \partial x^{j_2} \cdots \partial x^{j_q}} := \partial_{j_1} \partial_{j_2} \cdots \partial_{j_q} f(x) \quad \text{for } x \in X ,$$

the **q-th order partial derivative**[2] for $f : X \to F$ at x. The map f is said to be **q-times [continuously] partially differentiable** if its derivatives up to and including order q exist [and are continuous].

5.4 Theorem   *Suppose X is open in $\mathbb{R}^n$, $f : X \to F$, and $m \in \mathbb{N}^\times$. Then the following statements hold:*

(i) *f belongs to $C^m(X, F)$ if and only if f is m-times continuously partially differentiable.*

(ii) *For $f \in C^m(X, F)$, we have*

$$\frac{\partial^q f}{\partial x^{j_1} \cdots \partial x^{j_q}} = \frac{\partial^q f}{\partial x^{j_{\sigma(1)}} \cdots \partial x^{j_{\sigma(q)}}} \quad \text{for } 1 \leq q \leq m ,$$

for every permutation $\sigma \in \mathsf{S}_q$, that is, the partial derivatives are independent of the order of differentiation.[3]

Proof   First, we observe that for $h_1, \ldots, h_q \in \mathbb{R}^n$ such that $q \leq m$, we have

$$\partial^q f(x)[h_1, \ldots, h_q] = \partial(\cdots \partial(\partial f(x) h_1) h_2 \cdots) h_q \quad \text{for } x \in X .$$

In particular, it follows that

$$\partial^q f(x)[e_{j_1}, \ldots, e_{j_q}] = \partial_{j_q} \partial_{j_{q-1}} \cdots \partial_{j_1} f(x) \quad \text{for } x \in X . \tag{5.1}$$

Consequently, we find

$$\partial^q f(x)[h_1, \ldots, h_q] = \sum_{j_1, \ldots, j_q = 1}^{n} \partial_{j_q} \partial_{j_{q-1}} \cdots \partial_{j_1} f(x) h_1^{j_1} \cdots h_q^{j_q} \tag{5.2}$$

for $x \in X$ and $h_i = (h_i^1, \ldots, h_i^n) \in \mathbb{R}^n$ with $1 \leq i \leq q$.

---

[2]In an obvious notation, we simplify writing multiple successive derivatives of the same variable as, for example,

$$\frac{\partial^3 f}{\partial x^1 \partial x^4 \partial x^4} = \frac{\partial^3 f}{\partial x^1 (\partial x^4)^2} , \quad \frac{\partial^4 f}{\partial x^2 \partial x^3 \partial x^3 \partial x^2} = \frac{\partial^4 f}{\partial x^2 (\partial x^3)^2 \partial x^2} ,$$

etc. Note though, that one should generally maintain the differentiation order between different variables, because generally

$$\frac{\partial^2 f}{\partial x^1 \partial x^2} \neq \frac{\partial^2 f}{\partial x^2 \partial x^1}$$

(see Remark 5.6).

[3]One can show that the partial derivatives are already independent of that order when $f : X \subset \mathbb{R}^n \to F$ is m-times totally differentiable (see Exercise 17). In general, the sequence of partial derivatives can not be exchanged when f is only m-times partially differentiable (see Remark 5.6).

(i) Formula (5.1) immediately implies that every element of $C^m(X, F)$ is m-times continuously partially differentiable.

Conversely, suppose that f is m-times continuously partially differentiable. In the case $m = 1$, we know from Theorem 2.10 that f belongs to $C^1(X, F)$. We assume that the statement is correct for $m-1 \geq 1$. From this assumption and (5.2), we see that $\partial_j(\partial^{m-1} f)$ exists and is continuous for $j \in \{1, \ldots, n\}$. Theorem 2.10 then says

$$\partial(\partial^{m-1} f) \colon X \to \mathcal{L}\big(\mathbb{R}^n, \mathcal{L}^{m-1}(\mathbb{R}^n, F)\big) = \mathcal{L}^m(\mathbb{R}^n, F)$$

exists and is continuous. Therefore $f \in C^m(X, F)$.

(ii) This follows immediately from Corollary 5.3 and (5.1). ∎

5.5 Corollary *Suppose $f \colon X \subset \mathbb{R}^n \to \mathbb{K}$. Then we have these:*
 (i) *f belongs to $C^2(X, \mathbb{K})$ if and only if*

$$f, \partial_j f, \partial_j \partial_k f \in C(X, \mathbb{K}) \quad \text{for } 1 \leq j, k \leq n .$$

(ii) *(Schwarz theorem) If f belongs to $C^2(X, \mathbb{K})$, then*

$$\partial_j \partial_k f(x) = \partial_k \partial_j f(x) \quad \text{for } x \in X , \quad 1 \leq j, k \leq n .$$

5.6 Remark The Schwarz theorem is false when f is only twice partially differentiable.

Proof We consider $f \colon \mathbb{R}^2 \to \mathbb{R}$ with

$$f(x, y) := \begin{cases} \dfrac{xy(x^2 - y^2)}{x^2 + y^2} , & (x, y) \neq (0, 0) , \\[2mm] 0 , & (x, y) = (0, 0) . \end{cases}$$

Then we have

$$\partial_1 f(x, y) = \begin{cases} \dfrac{y(x^4 + 4x^2 y^2 - y^4)}{(x^2 + y^2)^2} , & (x, y) \neq (0, 0) , \\[2mm] 0 , & (x, y) = (0, 0) , \end{cases}$$

and it follows that

$$\partial_2 \partial_1 f(0, 0) = \lim_{h \to 0} \frac{\partial_1 f(0, h) - \partial_1 f(0, 0)}{h} = -1 .$$

Using $f(y, x) = -f(x, y)$, we find

$$\partial_2 f(y, x) = \lim_{h \to 0} \frac{f(y, x + h) - f(y, x)}{h} = -\lim_{h \to 0} \frac{f(x + h, y) - f(x, y)}{h} = -\partial_1 f(x, y) .$$

Therefore $\partial_1 \partial_2 f(y, x) = -\partial_2 \partial_1 f(x, y)$ for $(x, y) \in \mathbb{R}^2$. Then $\partial_2 \partial_1 f(0, 0) \neq 0$ gives $\partial_1 \partial_2 f(0, 0) \neq -\partial_2 \partial_1 f(0, 0)$. ∎

The chain rule

The chain rule is also important for higher derivatives, though there is generally no simple representation for higher derivatives of composed functions.

5.7 Theorem (chain rule) *Suppose Y is open in F and G is a Banach space. Also suppose $m \in \mathbb{N}^\times$ and $f \in C^m(X, F)$ with $f(X) \subset Y$ and $g \in C^m(Y, G)$. Then we have $g \circ f \in C^m(X, G)$.*

Proof Using Theorem 3.3, the statement follows from an induction in m. We leave the details to you as an exercise. ∎

Taylor's formula

We now extend Taylor's theorem to functions of multiple variables. For that, we use the following notation:

$$
\partial^k f(x)[h]^k := \begin{cases} \partial^k f(x)[\underbrace{h, \ldots, h}_{k\text{-times}}] \,, & 1 \le k \le q \,, \\[2mm] f(x) \,, & k = 0 \,, \end{cases}
$$

for $x \in X$, $h \in E$, and $f \in C^q(X, F)$.

5.8 Theorem (Taylor's formula) *Suppose X is open in E, $q \in \mathbb{N}^\times$, and f belongs to $C^q(X, F)$. Then*

$$
f(x + h) = \sum_{k=0}^{q} \frac{1}{k!} \partial^k f(x)[h]^k + R_q(f, x; h)
$$

for $x \in X$ and $h \in E$ such that $[\![x, x + h]\!] \subset X$. Here

$$
R_q(f, x; h) := \int_0^1 \frac{(1 - t)^{q-1}}{(q - 1)!} \left[\partial^q f(x + th) - \partial^q f(x) \right] [h]^q \, dt \in F
$$

*is the **q-th order remainder** of f at the point x.*

Proof When $q = 1$, the formula reduces to the mean value theorem in integral form (Theorem 3.10):

$$
f(x + h) = f(x) + \int_0^1 \partial f(x + th) h \, dt
$$

$$
= f(x) + \partial f(x) h + \int_0^1 \left[\partial f(x + th) - \partial f(x) \right] h \, dt \,.
$$

Suppose now $q = 2$. Letting $u(t) := \partial f(x+th)h$ and $v(t) := t-1$ for $t \in [0,1]$, we have $u'(t) = \partial^2 f(x+th)[h]^2$ and $v' = 1$. Therefore, by the generalized product rule (Example 4.8(b)) and by integrating the above formula, we get[4]

$$f(x+h) = f(x) + \partial f(x)h + \int_0^1 (1-t)\partial^2 f(x+th)[h]^2 \, dt$$

$$= \sum_{k=0}^{2} \frac{1}{k!} \partial^k f(x)[h]^k + R_2(f,x;h) \ .$$

For $q = 3$, we set $u(t) := \partial^2 f(x+th)[h]^2$ and $v(t) := -(1-t)^2/2$. We then get in similar fashion that

$$f(x+h) = f(x) + \partial f(x)h + \frac{1}{2}\partial^2 f(x)[h]^2 + \int_0^1 \frac{(1-t)^2}{2} \partial^3 f(x+th)[h]^3 \, dt \ .$$

In the general case $q > 3$, the claim follows using simple induction arguments. ∎

5.9 Remark   For $f \in C^q(X,F)$ and $x, h \in E$ such that $[\![x, x+h]\!] \subset X$, we have

$$\|R_q(f,x;h)\| \le \frac{1}{q!} \max_{0 \le t \le 1} \|\partial^q f(x+th) - \partial^q f(x)\| \, \|h\|^q \ .$$

In particular, we get

$$R_q(f,x;h) = o(\|h\|^q) \quad (h \to 0) \ .$$

Proof   This follows from Proposition VI.4.3 and the continuity of $\partial^q f$. ∎

5.10 Corollary   Suppose $f \in C^q(X,F)$ with $q \in \mathbb{N}^\times$ and $x \in X$. Then we have[5]

$$f(x+h) = \sum_{k=0}^{q} \frac{1}{k!} \partial^k f(x)[h]^k + o(\|h\|^q) \quad (h \to 0) \ .$$

Functions of m variables

In the remainder of this section, we consider functions of m real variables, that is, we set $E = \mathbb{R}^m$. In this case, it is convenient to represent partial derivatives using multiindices.

---

[4]See the proof of Proposition VI.5.4.
[5]For sufficiently small h, $[\![x, x+h]\!]$ lies in X.

Suppose $\alpha = (\alpha_1, \ldots, \alpha_m) \in \mathbb{N}^m$ is a multiindex of length[6] $|\alpha| = \sum_{j=1}^m \alpha_j$, and let $f \in C^{|\alpha|}(X, F)$. Then we write

$$\partial^\alpha f := \partial_1^{\alpha_1} \partial_2^{\alpha_2} \cdots \partial_m^{\alpha_m} f = \frac{\partial^{|\alpha|} f}{(\partial x^1)^{\alpha_1} (\partial x^2)^{\alpha_2} \cdots (\partial x^m)^{\alpha_m}} \quad \text{for } \alpha \neq 0$$

and $\partial^0 f := f$.

5.11 Theorem (Taylor's theorem) *Suppose $f \in C^q(X, F)$ with $q \in \mathbb{N}^\times$ and $x \in X$. Then*

$$f(y) = \sum_{|\alpha| \leq q} \frac{\partial^\alpha f(x)}{\alpha!} (y - x)^\alpha + o(|x - y|^q) \quad (x \to y) .$$

Proof We set $h := y - x$ and write $h = \sum_{j=1}^m h_j e_j$. Then

$$\partial^k f(x)[h]^k = \partial^k f(x) \left[\sum_{j_1=1}^m h_{j_1} e_{j_1}, \ldots, \sum_{j_k=1}^m h_{j_k} e_{j_k} \right]$$

$$= \sum_{j_1=1}^m \cdots \sum_{j_k=1}^m \partial^k f(x)[e_{j_1}, \ldots, e_{j_k}] h_{j_1} \cdot \ldots \cdot h_{j_k}$$

(5.3)

for $1 \leq k \leq q$. The number of k-tuples $(j_1, \ldots, j_k)$ of numbers $1 \leq j_i \leq m$ for which all of the numbers $\ell \in \{1, \ldots, m\}$ occur exactly α_ℓ-times is equal to

$$\frac{k!}{(\alpha_1)! (\alpha_2)! \cdot \ldots \cdot (\alpha_m)!} = \frac{k!}{\alpha!} .$$

(5.4)

From (5.1) and Corollary 5.3, the claim follows from

$$\partial^k f(x)[e_{j_1}, c_{j_2}, \ldots, e_{j_k}] = \partial_m^{\alpha_m} \partial_{m-1}^{\alpha_{m-1}} \cdots \partial_1^{\alpha_1} f(x) = \partial^\alpha f(x) .$$

(5.5)

Summarizing, we get from (5.3)–(5.5) that

$$\partial^k f(x)[h]^k = \sum_{|\alpha|=k} \frac{k!}{\alpha!} \partial^\alpha f(x) h^\alpha \quad \text{for } |\alpha| = k ,$$

and the theorem follows from Corollary 5.10. ∎

5.12 Remarks Suppose $f \in C^q(X, \mathbb{R})$ for some $q \in \mathbb{N}^\times$.

(a) If $q = 1$, we have

$$f(y) = f(x) + df(x)(y - x) + o(|y - x|) = f(x) + \left(\nabla f(x) \mid y - x \right) + o(|y - x|)$$

as $y \to x$.

---

[6]See Section I.8.

(b) If $q = 2$, we have

$$f(y) = f(x) + df(x)(y - x) + \frac{1}{2}\partial^2 f(x)[y - x]^2 + o(|y - x|^2)$$
$$= f(x) + \left(\nabla f(x)\,\middle|\,y - x\right) + \frac{1}{2}\left(H_f(x)(y - x)\,\middle|\,y - x\right) + o(|y - x|^2)$$

as $y \to x$. Here,[7]

$$H_f(x) := [\partial_j \partial_k f(x)] \in \mathbb{R}^{m \times m}_{\text{sym}}$$

denotes the **Hessian matrix** of f at x. In other words, $H_f(x)$ is the representation matrix of the linear operator induced by the bilinear form $\partial^2 f(x)[\cdot\,\cdot]$ at $\mathbb{R}^m$ (see Exercise 4.1). ∎

Sufficient criterion for local extrema

In the following remarks, we gather several results from linear algebra.

5.13 Remarks   **(a)** Let $H := (H, (\cdot|\cdot))$ be a finite-dimensional real Hilbert space. The bilinear form $b \in \mathcal{L}^2(H, \mathbb{R})$ is said to be **positive [semi]definite** if it is symmetric and

$$b(x, x) > 0 \quad [b(x, x) \geq 0] \quad \text{for } x \in H \backslash \{0\} \ .$$

It is **negative [semi]definite** if $-b$ is **positive [semi]definite**. If b is neither positive nor negative semidefinite, then b is **indefinite**. The form b induces the linear operator[8] $B \in \mathcal{L}(H)$ with $(Bx|y) = b(x, y)$ for $x, y \in H$, which is said to be **positive** or **negative [semi]definite** if b has the corresponding property.

Suppose $A \in \mathcal{L}(H)$. Then these statements are equivalent:

 (i) A is positive [semi]definite.

 (ii) $A = A^*$ and $(Ax|x) > 0 \ [\geq 0]$ for $x \in H \backslash \{0\}$.

 (iii) $A = A^*$ and there is an $\alpha > 0 \ [\alpha \geq 0]$ such that $(Ax|x) \geq \alpha |x|^2$ for $x \in H$.

Proof   This follows from Exercises 4.1 and III.4.11. ∎

(b) Suppose $H = (H, (\cdot|\cdot))$ is a Hilbert space of dimension m and $A \in \mathcal{L}(H)$ is self adjoint, that is, $A = A^*$. Then all eigenvalues of A are real and semisimple. In addition, there is an ONB of eigenvectors $h_1, \ldots, h_m$ which give A the **spectral representation**

$$A = \sum_{j=1}^{m} \lambda_j (\cdot | h_j) h_j \ , \tag{5.6}$$

---

[7]We write $\mathbb{R}^{m \times m}_{\text{sym}}$ for the vector subspace of $\mathbb{R}^{m \times m}$ that consists of all symmetric $(m \times m)$ matrices.

[8]See Exercise 4.1.

where, in $\lambda_1, \ldots, \lambda_m$, the eigenvalues of A appear according to their multiplicity and $Ah_j = \lambda_j h_j$ for $1 \leq j \leq m$. This means in particular that the matrix of A is diagonal in this basis, that is, $[A] = \mathrm{diag}(\lambda_1, \ldots, \lambda_m)$.

Finally, A is positive definite if and only if all its eigenvalues are positive; A is positive semidefinite if and only if none of its eigenvalues are negative.

Proof The statements that $\sigma(A) \subset \mathbb{R}$ and that the spectral representation exists are standard in linear algebra (for example, [Art93, Section VII.5]).[9]

From (5.6), we read off

$$(Ax \,|\, x) = \sum_{j=1}^{m} \lambda_j \,|(x \,|\, h_j)|^2 \quad \text{for } x \in H \ .$$

The claimed characterization of A is now an easy consequence of the assumed positive [semi]definiteness of the eigenvalues. ∎

We can now derive — in a (partial) generalization of the results of Application IV.3.9 — a sufficient criteria for maxima and minima of real-valued functions of multiple real variables.

5.14 Theorem *Suppose X is open in $\mathbb{R}^m$ and $x_0 \in X$ is a critical point of $f \in C^2(X, \mathbb{R})$. Then*

 (i) *if $\partial^2 f(x_0)$ is positive definite, then f has an isolated local minimum at x_0;*

 (ii) *if $\partial^2 f(x_0)$ is negative definite, then f has an isolated local maximum at x_0;*

(iii) *if $\partial^2 f(x_0)$ is indefinite, then f does not have an local extremum at x_0.*

Proof Because x_0 is a critical point of f, according to Remark 5.12(b), we have

$$f(x_0 + \xi) = f(x_0) + \frac{1}{2} \partial^2 f(x_0)[\xi]^2 + o(|\xi|^2) \quad (\xi \to 0) \ .$$

(i) Because $\partial^2 f(x_0)$ is positive definite, there is an $\alpha > 0$ such that

$$\partial^2 f(x_0)[\xi]^2 \geq \alpha \,|\xi|^2 \quad \text{for } \xi \in \mathbb{R}^m \ .$$

We fix an $\delta > 0$ such that $\left| o(|\xi|^2) \right| \leq \alpha \,|\xi|^2 /4$ for $\xi \in \bar{\mathbb{B}}^m(0, \delta)$ and then find the estimate

$$f(x_0 + \xi) \geq f(x_0) + \frac{\alpha}{2} |\xi|^2 - \frac{\alpha}{4} |\xi|^2 = f(x_0) + \frac{\alpha}{4} |\xi|^2 \quad \text{for } |\xi| \leq \delta \ .$$

Therefore f has a local minimum at x_0.

(ii) The statement follows by applying (i) to $-f$.

(iii) Because $\partial^2 f(x_0)$ is indefinite, there are $\xi_1, \xi_2 \in \mathbb{R}^m \backslash \{0\}$ such that

$$\alpha := \partial^2 f(x_0)[\xi_1]^2 > 0 \quad \text{and} \quad \beta := \partial^2 f(x_0)[\xi_2]^2 < 0 \ .$$

---

[9] See also Example 10.17(b).

Also, we find $t_j > 0$ such that $[\![x_0, x_0 + t_j\xi_j]\!] \subset X$ and

$$\frac{\alpha}{2} + \frac{o(t^2 |\xi_1|^2)}{t^2 |\xi_1|^2} |\xi_1|^2 > 0 \quad \text{and} \quad \frac{\beta}{2} + \frac{o(t^2 |\xi_2|^2)}{t^2 |\xi_2|^2} |\xi_2|^2 < 0$$

for $0 < t < t_1$ and $0 < t < t_2$, respectively. Consequently,

$$f(x_0 + t\xi_1) = f(x_0) + t^2\left(\frac{\alpha}{2} + \frac{o(t^2 |\xi_1|^2)}{t^2 |\xi_1|^2} |\xi_1|^2\right) > f(x_0) \quad \text{for } 0 < t < t_1 ,$$

and

$$f(x_0 + t\xi_2) = f(x_0) + t^2\left(\frac{\beta}{2} + \frac{o(t^2 |\xi_2|^2)}{t^2 |\xi_2|^2} |\xi_2|^2\right) < f(x_0) \quad \text{for } 0 < t < t_2 .$$

Therefore x_0 is not a local extremal point of f. ∎

5.15 Examples We consider

$$f\colon \mathbb{R}^2 \to \mathbb{R} , \quad (x,y) \mapsto c + \delta x^2 + \varepsilon y^2$$

with $c \in \mathbb{R}$ and $\delta, \varepsilon \in \{-1, 1\}$. Then

$$\nabla f(x,y) = 2(\delta x, \varepsilon y) \quad \text{and} \quad H_f(x,y) = 2\begin{bmatrix} \delta & 0 \\ 0 & \varepsilon \end{bmatrix} .$$

Thus $(0,0)$ is the only critical point of f.

(a) Taking $f(x,y) = c + x^2 + y^2$ so that $\delta = \varepsilon = 1$, we find $H_f(0,0)$ is positive definite. Thus f has an isolated (absolute) minimum at $(0,0)$.

(b) Taking $f(x,y) = c - x^2 - y^2$, we find $H_f(0,0)$ is negative definite, and f has an isolated (absolute) maximum at $(0,0)$.

(c) Taking $f(x,y) = c + x^2 - y^2$ makes $H_f(0,0)$ indefinite. Therefore f does not have an extremum at $(0,0)$; rather $(0,0)$ is a "saddle point".

      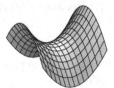

Minimum   Maximum   Saddle

(d) When $\partial^2 f$ is semidefinite at a critical point of f, we cannot learn anything more from studying the second derivative. For consider the maps $f_j\colon \mathbb{R}^2 \to \mathbb{R}$, $j = 1, 2, 3$, with

$$f_1(x,y) := x^2 + y^4 , \quad f_2(x,y) := x^2 , \quad f_3(x,y) := x^2 + y^3 .$$

In each case $(0,0)$ is a critical point of f_j, and the Hessian matrix $H_{f_j}(0,0)$ is positive semidefinite.

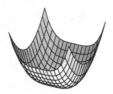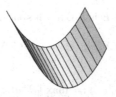

However, we can easily see that $(0,0)$ is an isolated minimum for f_1, *not* an isolated minimum for f_2, and not a local extremal point for f_3. ∎

Exercises

1 If $A \in \mathcal{L}(E,F)$, show $A \in C^\infty(E,F)$ and $\partial^2 A = 0$.

2 If $\varphi \in \mathcal{L}(E_1,\ldots,E_m;F)$, show $\varphi \in C^\infty(E_1 \times \cdots \times E_m, F)$ and $\partial^{m+1}\varphi = 0$.

3 Suppose X is open in $\mathbb{R}^m$. The (m-dimensional) **Laplacian** Δ is defined by

$$\Delta : C^2(X,\mathbb{R}) \to C(X,\mathbb{R}) , \quad u \mapsto \Delta u := \sum_{j=1}^m \partial_j^2 u .$$

We say the function $u \in C^2(X,\mathbb{R})$ is **harmonic** in X if $\Delta u = 0$. The harmonic functions thus comprise the kernel of the linear map Δ. Verify that $g_m : \mathbb{R}^m \setminus \{0\} \to \mathbb{R}$ defined by

$$g_m(x) := \begin{cases} \log|x| , & m = 2 , \\ |x|^{2-m} , & m > 2 , \end{cases}$$

is harmonic in $\mathbb{R}^m \setminus \{0\}$.

4 For $f,g \in C^2(X,\mathbb{R})$, show

$$\Delta(fg) = g\Delta f + 2(\nabla f \mid \nabla g) + f\Delta g .$$

5 Suppose $g : [0,\infty) \times \mathbb{R} \to \mathbb{R}^2$, $(r,\varphi) \mapsto (r\cos\varphi, r\sin\varphi)$. Verify these:

(a) $g \mid (0,\infty) \times (-\pi,\pi) \to \mathbb{R}^2 \setminus H$ for $H := \{ (x,y) \in \mathbb{R}^2 ; x \leq 0, y = 0 \} = (-\mathbb{R}^+) \times \{0\}$ is topological.

(b) $\operatorname{im}(g) = \mathbb{R}^2$.

(c) If X is open in $\mathbb{R}^2 \setminus H$ and $f \in C^2(X,\mathbb{R})$, then

$$(\Delta f) \circ g = \frac{\partial^2(f \circ g)}{\partial r^2} + \frac{1}{r}\frac{\partial(f \circ g)}{\partial r} + \frac{1}{r^2}\frac{\partial^2(f \circ g)}{\partial \varphi^2}$$

on $g^{-1}(X)$.

6 Let $X := \mathbb{R}^n \times (0,\infty)$ and $p(x,y) := y/(|x|^2 + y^2)$ for $(x,y) \in X$. Calculate Δp.

7 Verify

$$(\Delta u) \circ A = \Delta(u \circ A) \quad \text{for } u \in C^2(\mathbb{R}^n,\mathbb{R})$$

if $A \in \mathcal{L}(\mathbb{R}^n)$ is **orthogonal**, that is, if $A^*A = 1$.

8 Suppose X is open in $\mathbb{R}^m$ and E is a Banach space. For $k \in \mathbb{N}$, let

$$BC^k(X, E) := \Big(\big\{ u \in BC(X, E) \; ; \; \partial^\alpha u \in BC(X, E), \ |\alpha| \leq k \big\}, \ \|\cdot\|_{k,\infty} \Big) \,,$$

where

$$\|u\|_{k,\infty} := \max_{|\alpha| \leq k} \|\partial^\alpha u\|_\infty \,.$$

Verify that

(a) $BC^k(X, E)$ is a Banach space;

(b) $BUC^k(X, E) := \big\{ u \in BC(X, E) \; ; \; \partial^\alpha u \in BUC(X, E), \ |\alpha| \leq k \big\}$ is a closed vector subspace of $BC^k(X, E)$ and is therefore itself a Banach space.

9 Let $q \in \mathbb{N}$ and $a_\alpha \in \mathbb{K}$ for $\alpha \in \mathbb{N}^m$ with $|\alpha| \leq q$. Also suppose $a_\alpha \neq 0$ for some $\alpha \in \mathbb{N}^m$ with $|\alpha| = q$. Then we call

$$\mathcal{A}(\partial) : C^q(X, \mathbb{K}) \to C(X, \mathbb{K}) \,, \quad u \mapsto \mathcal{A}(\partial)u := \sum_{|\alpha| \leq q} a_\alpha \partial^\alpha u$$

a **linear differential operator** of order q with constant coefficients.

Show for $k \in \mathbb{N}$ that

(a) $\mathcal{A}(\partial) \in \mathcal{L}\big(BC^{k+q}(X, \mathbb{K}), BC^k(X, \mathbb{K}) \big)$;

(b) $\mathcal{A}(\partial) \in \mathcal{L}\big(BUC^{k+q}(X, \mathbb{K}), BUC^k(X, \mathbb{K}) \big)$.

10 For $u \in C^2(\mathbb{R} \times \mathbb{R}^m, \mathbb{R})$ and $(t, x) \in \mathbb{R} \times \mathbb{R}^m$ we set

$$\square u := \partial_t^2 u - \Delta_x u$$

and

$$(\partial_t - \Delta)u := \partial_t u - \Delta_x u \,,$$

where Δ_x is the Laplacian in $x \in \mathbb{R}^m$. We call $\square$ the **wave operator** (or **d'Alembertian**). We call $\partial_t - \Delta$ the **heat operator**.

(a) Calculate $(\partial_t - \Delta)k$ in $(0, \infty) \times \mathbb{R}^m$ when

$$k(t, x) := t^{-m/2} \exp(-|x|^2/4t) \quad \text{for } (t, x) \in (0, \infty) \times \mathbb{R}^m \,.$$

(b) Suppose $g \in C^2(\mathbb{R}, \mathbb{R})$ $c > 0$, and $v \in S^{m-1}$. Calculate $\square w$ when

$$w(t, x) := g(v \cdot x - tc) \quad \text{for } (t, x) \in \mathbb{R} \times \mathbb{R}^m \,.$$

11 Suppose X is an open convex subset of a Banach space E and $f \in C^2(X, \mathbb{R})$. Show these statements are equivalent:

(i) f is convex;

(ii) $f(x) \geq f(a) + \partial f(a)(x - a)$, for $a, x \in X$;

(iii) $\partial^2 f(a)[h]^2 \geq 0$, for $a \in X$ and $h \in E$.

If $E = \mathbb{R}^m$, these statements are also equivalent to

(iv) $H_f(a)$ is positive semidefinite for $a \in X$.

12   For all $\alpha \in \mathbb{R}$, classify the critical points of the function

$$f_\alpha : \mathbb{R}^2 \to \mathbb{R} , \quad (x,y) \mapsto x^3 - y^3 + 3\alpha xy ,$$

as maxima, minima, or saddle points.

13   Suppose $f : \mathbb{R}^2 \to \mathbb{R}$, $(x,y) \mapsto (y - x^2)(y - 2x^2)$. Prove these:
(a) $(0,0)$ is not a local minimum for f.
(b) For every $(x_0, y_0) \in \mathbb{R}^2 \setminus \{(0,0)\}$, $\mathbb{R} \to \mathbb{R}$ $t \mapsto f(tx_0, ty_0)$ has an isolated local minimum at 0.

14   Determine, up to (including) second order, the Taylor expansion of

$$(0,\infty) \times (0,\infty) \to \mathbb{R} , \quad (x,y) \mapsto (x - y)/(x + y)$$

at the point $(1,1)$.

15   Suppose X is open in $\mathbb{R}^m$. For $f, g \in C^1(X, \mathbb{R}^m)$, define $[f,g] \in C(X, \mathbb{R}^m)$ by

$$[f,g](x) := \partial f(x)g(x) - \partial g(x)f(x) \quad \text{for } x \in X .$$

We call $[f,g]$ the **Lie bracket** of f and g.
Verify these statements:
 (i) $[f,g] = -[g,f]$;
 (ii) $[\alpha f + \beta g, h] = \alpha[f,h] + \beta[g,h]$ for $\alpha, \beta \in \mathbb{R}$ and $h \in C^1(X, \mathbb{R}^m)$;
 (iii) $[\varphi f, \psi g] = \varphi\psi[f,g] + (\nabla\varphi \mid f)\psi f - (\nabla\psi \mid g)\varphi g$ for $\varphi, \psi \in C^1(X, \mathbb{R})$;
 (iv) (**Jacobi identity**)
     If $f, g, h \in C^2(X, \mathbb{R}^m)$, then $[[f,g],h] + [[g,h],f] + [[h,f],g] = 0$.

16   Show that

$$\mathbb{R}^n \to \mathbb{R} , \quad x \mapsto \begin{cases} \exp(1/(|x|^2 - 1)) , & |x| < 1 , \\ 0 , & |x| \geq 1 , \end{cases}$$

is smooth.

17   Suppose X is open in $\mathbb{R}^m$ and $f : X \to F$ is m-times differentiable. Then show

$$\partial^q f(x) \in \mathcal{L}_{\text{sym}}^q(X, F) \quad \text{for } x \in X \text{ and } 1 \leq q \leq m .$$

In particular, show for every permutation $\sigma \in S_q$ that

$$\frac{\partial^q f(x)}{\partial x^{j_1} \cdot \ldots \cdot \partial x^{j_q}} = \frac{\partial^q f(x)}{\partial x^{j_{\sigma(1)}} \cdot \ldots \cdot \partial x^{j_{\sigma(q)}}} \quad \text{for } x \in X .$$

(Hint: It suffices to consider the case $q = m = 2$. Apply the mean value theorem on

$$[0,1] \to F , \quad t \mapsto f(x + tsh_1 + tsh_2) - f(x + tsh_1) ,$$

and verify that

$$\lim_{\substack{s \to 0 \\ s > 0}} \frac{f(x + sh_1 + sh_2) - f(x + sh_1) - f(x + sh_2) + f(x)}{s^2} = \partial^2 f(x)[h_1, h_2] \ .)$$

18 Suppose H is a finite-dimensional Hilbert space, and $A \in \mathcal{L}(H)$ is positive [or negative] definite. Show then A that belongs to $\mathcal{L}\mathrm{aut}(H)$, and A^{-1} is positive [or negative] definite.

19 Suppose H is a finite-dimensional Hilbert space and $A \in C^1([0,T], \mathcal{L}(H))$. Also suppose $A(t)$ symmetric for every $t \in [0,T]$. Prove these:

(a) $\partial A(t)$ is symmetric for $t \in [0,T]$.

(b) If $\partial A(t)$ is positive definite for every $t \in [0,T]$, then, for $0 \le \sigma < \tau \le T$, we have

  (i) $A(\tau) - A(\sigma)$ is positive definite;

  (ii) $A^{-1}(\tau) - A^{-1}(\sigma)$ is negative definite if $A(t)$ is an automorphism for every $t \in [0,T]$.

(Hint for (ii): Differentiate $t \mapsto A^{-1}(t)A(t)$.)

20 Let H be a finite-dimensional Hilbert space, and let $A, B \in \mathcal{L}(H)$. Show that if A, B, and $A - B$ are positive definite, then $A^{-1} - B^{-1}$ is negative definite.
(Hint: Consider $t \mapsto B + t(A - B)$ for $t \in [0,1]$, and study Exercise 19.)

21 Suppose X is open in $\mathbb{R}^m$ and $f, g \in C^q(X, \mathbb{K})$. Show that fg belongs to $C^q(X, \mathbb{K})$ and prove the **Leibniz rule**:

$$\partial^\alpha(fg) = \sum_{\beta \le \alpha} \binom{\alpha}{\beta} \partial^\beta f \, \partial^{\alpha - \beta} g \quad \text{for } \alpha \in \mathbb{N}^m \text{ and } |\alpha| \le q \ .$$

Here $\beta \le \alpha$ means that $\beta_j \le \alpha_j$ for $1 \le j \le m$, and

$$\binom{\alpha}{\beta} := \binom{\alpha_1}{\beta_1} \cdot \ldots \cdot \binom{\alpha_m}{\beta_m}$$

for $\alpha, \beta \in \mathbb{N}^m$ with $\alpha = (\alpha_1, \ldots, \alpha_m)$ and $\beta = (\beta_1, \ldots, \beta_m)$. (Hint: Theorem IV.1.12(ii) and induction.)

6   Nemytskii operators and the calculus of variations

Although we developed the differential calculus for general Banach spaces, we have so far almost exclusively demonstrated it with finite-dimensional examples. In this section, we remedy that by giving you a glimpse of the scope of the general theory. To that end, we consider first the simplest nonlinear map between function spaces, namely, the "covering operator", which connects the two function spaces using a fixed nonlinear map. We restrict to covering operators in Banach spaces of continuous functions, and seek their continuity and differentiability properties.

As an application, we will study the basic problem of the calculus of variations, namely, the problem of characterizing and finding the extrema of real-valued functions of "infinitely many variables". In particular, we derive the Euler–Lagrange equation, whose satisfaction is necessary for a variational problem to have a local extremum.

Nemytskii operators

Suppose T, X, and Y are nonempty sets and φ is a map from $T \times X$ to Y. Then we define the **induced Nemytskii** or **covering operator** $\varphi^\natural$ through

$$\varphi^\natural : X^T \to Y^T , \quad u \mapsto \varphi\big(\cdot, u(\cdot)\big) .$$

This means $\varphi^\natural$ is the map that associates to every function $u : T \to X$ the function $\varphi^\natural(u) : T \to Y, \ t \mapsto \varphi\big(t, u(t)\big)$.

In the following suppose

- T is a compact metric space;
  E and F are Banach spaces;
  X is open in E.

When we do not expect any confusion, we denote both norms on E and F by $|\cdot|$.

6.1 Lemma $C(T, X)$ is open in the Banach space $C(T, E)$.

Proof   From Theorem V.2.6, we know that $C(T, E)$ is a Banach space.

It suffices to consider the case $X \neq E$. Hence let $u \in C(T, X)$. Then, because $u(T)$ is the continuous image of a compact metric space, it is a compact subset of X (see Theorem III.3.6). Then according to Example III.3.9(c), there is an $r > 0$ such that $d(u(T), X^c) \geq 2r$. Then, for $v \in u + r\mathbb{B}_{C(T,E)}$, that $|u(t) - v(t)| < r$ for $t \in T$ implies that $v(T)$ lies in the r-neighborhood of $u(T)$ in E and, therefore, in X. Thus v belongs to $C(T, X)$. ∎

The continuity of Nemytskii operators

We first show that continuous maps induce continuous Nemytskii operators in spaces of continuous functions.

6.2 Theorem *Suppose $\varphi \in C(T \times X, F)$. Then*

(i) $\varphi^\natural \in C\big(C(T, X), C(T, F)\big)$;

(ii) *if φ is bounded on bounded sets, this is also true of $\varphi^\natural$.*

Proof For $u \in C(T, X)$, suppose $\widetilde{u}(t) := \big(t, u(t)\big)$ for $t \in T$. Then $\widetilde{u}$ clearly belongs to $C(T, T \times X)$ (see Proposition III.1.10). Thus Theorem III.1.8 implies that $\varphi^\natural(u) = \varphi \circ \widetilde{u}$ belongs to $C(T, F)$.

(i) Suppose (u_j) is a sequence in $C(T, X)$ with $u_j \to u_0$ in $C(T, X)$. Then the set $M := \{\, u_j \;;\; j \in \mathbb{N}^\times \,\} \cup \{u_0\}$ is compact in $C(T, X)$, according to Example III.3.1(a).

We set $M(T) := \bigcup\{\, m(T) \;;\; m \in M \,\}$ and want to show $M(T)$ is compact in X. To see this, let (x_j) be a sequence in $M(T)$. Then there are $t_j \in T$ and $m_j \in M$ such that $x_j = m_j(t_j)$. Because T and M are sequentially compact and because of Exercise III.3.1, we find $(t, m) \in T \times M$ and a subsequence $\big((t_{j_k}, m_{j_k})\big)_{k \in \mathbb{N}}$ that converges in $T \times M$ to (t, m). From this, we get

$$|x_{j_k} - m(t)| \leq |m_{j_k}(t_{j_k}) - m(t_{j_k})| + |m(t_{j_k}) - m(t)|$$
$$\leq \|m_{j_k} - m\|_\infty + |m(t_{j_k}) - m(t)| \to 0$$

as $k \to \infty$ because m is continuous. Thus (x_{j_k}) converges to the element $m(t)$ in $M(T)$, which proves the sequential compactness, and therefore the compactness, of $M(T)$.

Thus, according to Exercise III.3.1, $T \times M(T)$ is compact. Now Theorem III.3.13 shows that the restriction of φ to $T \times M(T)$ is uniformly continuous. From this, we get

$$\|\varphi^\natural(u_j) - \varphi^\natural(u_0)\|_\infty = \max_{t \in T}\big|\varphi\big(t, u_j(t)\big) - \varphi\big(t, u_0(t)\big)\big| \to 0$$

as $j \to \infty$ because $\|u_j - u_0\|_\infty \to 0$. We also find that $\big(t, u_j(t)\big)$ and $\big(t, u_0(t)\big)$ belong to $T \times M(T)$ for $t \in T$ and $j \in \mathbb{N}$. Therefore $\varphi^\natural$ is sequentially continuous and hence, because of Theorem III.1.4, continuous.

(ii) Suppose $B \subset C(T, X)$ and there is an $R > 0$ such that $\|u\|_\infty \leq R$ for $u \in B$. Then

$$B(T) := \bigcup\{\, u(T) \;;\; u \in B \,\} \subset X \,,$$

and $|x| \leq R$ for $x \in B(T)$. Therefore $B(T)$ is bounded in X. Thus $T \times B(T)$ is bounded in $T \times X$. Because φ is bounded on bounded sets, there is an $r > 0$ such that $|\varphi(t, x)| \leq r$ for $(t, x) \in T \times B(T)$. This implies $\|\varphi^\natural(u)\|_\infty \leq r$ for $u \in B$. Therefore $\varphi^\natural$ is bounded on bounded sets because this is the case for φ. ∎

6.3 Remark When E is finite-dimensional, every $\varphi \in C(T \times X, F)$ is bounded on sets of the form $T \times B$, where B is bounded in E and where $\overline{B}$, the closure of B in E, lies in X.

Proof This follows from Theorem III.3.5, Corollary III.3.7 and Theorem 1.4. ∎

The differentiability of Nemytskii operators

Suppose $p \in \mathbb{N} \cup \{\infty\}$. Then we write

$$\varphi \in C^{0,p}(T \times X, F)$$

if, for every $t \in T$, the function $\varphi(t, \cdot) \colon X \to F$ is p-times differentiable and if its derivatives, which we denote by $\partial_2^q \varphi$, satisfy

$$\partial_2^q \varphi \in C(T \times X, \mathcal{L}^q(E, F)) \quad \text{for } q \in \mathbb{N} \text{ and } q \leq p \,.$$

Consequently, $C^{0,0}(T \times X, F) = C(T \times X, F)$.

6.4 Theorem For $\varphi \in C^{0,p}(T \times X, F)$, $\varphi^\natural$ belongs to $C^p(C(T, X), C(T, F))$, and[1]

$$[\partial \varphi^\natural(u)h](t) = \partial_2 \varphi(t, u(t))h(t) \quad \text{for } t \in T \,,$$

$u \in C(T, X)$, and $h \in C(T, E)$.

Proof Because of Theorem 6.2, we may assume $p \geq 1$. First we make

$$G \colon C(T, X) \times C(T, E) \to F^T \,, \quad (u, h) \mapsto \partial_2 \varphi(\cdot, u(\cdot))h(\cdot)$$

the Nemytskii operator induced by the map[2]

$$[(t, (x, \xi)) \mapsto \partial_2 \varphi(t, x)\xi] \in C(T \times (X \times E), F) \,.$$

Then Theorem 6.2 implies that $G(u, h)$ belongs to $C(T, F)$. Obviously

$$\|G(u, h)\|_{C(T,F)} \leq \max_{t \in T} \|\partial_2 \varphi(t, u(t))\|_{\mathcal{L}(E,F)} \|h\|_{C(T,E)} \,, \tag{6.1}$$

and $G(u, \cdot)$ is linear. Consequently, we see that

$$A(u) := G(u, \cdot) \in \mathcal{L}(C(T, E), C(T, F)) \quad \text{for } u \in C(T, X) \,.$$

The function $\partial_2 \varphi \in C(T \times X, \mathcal{L}(E, F))$ induces a Nemytskii operator $\widetilde{A}$. Then, from Theorems 6.2 and 1.1,

$$\widetilde{A} \in C(C(T, X), C(T, \mathcal{L}(E, F))) \,. \tag{6.2}$$

Suppose now $u \in C(T, X)$. We choose $\varepsilon > 0$ such that $u(T) + \varepsilon \mathbb{B}_E \subset X$. Then $u + h$ belongs to $C(T, X)$ for every $h \in C(T, E)$ such that $\|h\|_\infty < \varepsilon$, and

---

[1] Here and in like situations, all statements about derivatives apply to the natural case $p \geq 1$.
[2] $C(T, X) \times C(T, E)$ and $C(T, X \times E)$ are identified in the obvious way.

the mean value theorem in integral form (Theorem 3.10) implies

$$
\begin{aligned}
\big|\varphi^{\natural}&(u+h)(t) - \varphi^{\natural}(u)(t) - \big(A(u)h\big)(t)\big| \\
&= \big|\varphi\big(t,u(t)+h(t)\big) - \varphi\big(t,u(t)\big) - \partial_2\varphi\big(t,u(t)\big)h(t)\big| \\
&= \left|\int_0^1 \big[\partial_2\varphi\big(t,u(t)+\tau h(t)\big) - \partial_2\varphi\big(t,u(t)\big)\big]h(t)\,d\tau\right| \\
&\leq \|h\|_{C(T,E)}\int_0^1 \big\|\widetilde{A}(u+\tau h) - \widetilde{A}(u)\big\|_{C(T,\mathcal{L}(E,F))}\,d\tau \ .
\end{aligned}
$$

Consequently, it follows from (6.2) that

$$
\varphi^{\natural}(u+h) - \varphi^{\natural}(u) - A(u)h = o(\|h\|_{C(T,E)}) \quad (h\to 0) \ .
$$

Thus, $\partial\varphi^{\natural}(u)$ exists and equals $A(u)$. Analogously to (6.1), we consider

$$
\|\partial\varphi^{\natural}(u) - \partial\varphi^{\natural}(v)\|_{\mathcal{L}(C(T,E),C(T,F))} \leq \big\|\widetilde{A}(u) - \widetilde{A}(v)\big\|_{C(T,\mathcal{L}(E,F))} \ ,
$$

which, because of (6.2), implies

$$
\partial\varphi^{\natural} \in C\big(C(T,X),\mathcal{L}\big(C(T,E),C(T,F)\big)\big) \ .
$$

This proves the theorem for $p=1$.

The general case now follows simply by induction, which we leave to you. ∎

6.5 Corollary *If $\partial_2\varphi$ is bounded on bounded sets, so is $\partial\varphi^{\natural}$.*

Proof This follows from (6.1) and Theorem 6.2(ii). ∎

6.6 Examples **(a)** Let $\varphi(\xi) := \sin\xi$ for $\xi \in \mathbb{R}$ and $T := [0,1]$. Then the Nemytskii operator

$$
\varphi^{\natural} : C(T) \to C(T) \ , \quad u \mapsto \sin u(\cdot)
$$

induced by φ is a C^{∞} map,

$$
\big(\partial\varphi^{\natural}(u)h\big)(t) = \big[\cos u(t)\big]h(t) \quad \text{for } t \in T
$$

and $u,h \in C(T)$. This means that in this case the linear map $\partial\varphi^{\natural}(u) \in \mathcal{L}\big(C(T)\big)$ is the function $\cos u(\cdot)$ understood as a **multiplication operator**, that is,

$$
C(T) \to C(T) \ , \quad h \mapsto \big[\cos u(\cdot)\big]h \ .
$$

(b) Suppose $-\infty < \alpha < \beta < \infty$ and $\varphi \in C^{0,p}\big([\alpha,\beta]\times X,F\big)$. Also let

$$
f(u) := \int_\alpha^\beta \varphi\big(t,u(t)\big)\,dt \quad \text{for } u \in C\big([\alpha,\beta],X\big) \ . \tag{6.3}
$$

Then $f \in C^p\big(C([\alpha,\beta],X),F\big)$ and

$$\partial f(u)h = \int_\alpha^\beta \partial_2\varphi\big(t,u(t)\big)h(t)\,dt \quad \text{for } u \in C([\alpha,\beta],X)\;,\quad h \in C([\alpha,\beta],E)\;.$$

Proof   With $T := [\alpha,\beta]$, Theorem 6.4 implies that $\varphi^\natural$ belongs to $C^p\big(C(T,X),C(T,F)\big)$ and that $\partial\varphi^\natural(u)$ is the multiplication operator

$$C(T,E) \to C(T,F)\;,\quad h \mapsto \partial_2\varphi\big(\cdot,u(\cdot)\big)h\;.$$

From Section VI.3, we know that $\int_\alpha^\beta$ belongs to $\mathcal{L}\big(C(T,F),F\big)$. Therefore, it follows from chain rule (Theorem 3.3) and Example 2.3(a) that

$$f = \int_\alpha^\beta \circ\, \varphi^\natural \in C^p\big(C(T,X),F\big)$$

and

$$\partial f(u) = \int_\alpha^\beta \circ\, \partial\varphi^\natural(u) \quad \text{for } u \in C(T,X)\;. \;\blacksquare$$

(c) Suppose $-\infty < \alpha < \beta < \infty$ and $\varphi \in C^{0,p}\big([\alpha,\beta]\times(X\times E),F\big)$. Then $C^1([\alpha,\beta],X)$ is open in $C^1([\alpha,\beta],E)$, and, for the map defined through

$$\Phi(u)(t) := \varphi\big(t,u(t),\dot u(t)\big) \quad \text{for } u \in C^1([\alpha,\beta],X) \text{ and } \alpha \le t \le \beta\;,$$

we have

$$\Phi \in C^p\big(C^1([\alpha,\beta],X),C([\alpha,\beta],F)\big)$$

and

$$\big[\partial\Phi(u)h\big](t) = \partial_2\varphi\big(t,u(t),\dot u(t)\big)h(t) + \partial_3\varphi\big(t,u(t),\dot u(t)\big)\dot h(t)$$

for $t \in [\alpha,\beta]$, $u \in C^1([\alpha,\beta],X)$, and $h \in C^1([\alpha,\beta],E)$.

Proof   The canonical inclusion

$$i: C^1([\alpha,\beta],E) \to C([\alpha,\beta],E)\;,\quad u \mapsto u \tag{6.4}$$

is linear and continuous. This is because the maximum norm that $C([\alpha,\beta],E)$ induces on $C^1([\alpha,\beta],E)$ is weaker than the C^1 norm. Therefore, because of Lemma 6.1 and Theorem III.2.20,

$$C^1([\alpha,\beta],X) = i^{-1}\big(C([\alpha,\beta],X)\big)$$

is open in $C^1([\alpha,\beta],E)$.

From (6.4), it follows that

$$\Psi := \big[u \mapsto (u,\dot u)\big] \in \mathcal{L}\big(C^1(T,X),C(T,X)\times C(T,E)\big)$$

for $T := [\alpha,\beta]$. Because $\Phi = \varphi^\natural \circ \Psi$, we conclude using Theorem 6.4 and the chain rule. $\blacksquare$

(d) Suppose the assumptions of (c) are satisfied and

$$f(u) := \int_\alpha^\beta \varphi\big(t, u(t), \dot{u}(t)\big)\, dt \quad \text{for } u \in C^1\big([\alpha, \beta], X\big) .$$

Then

$$f \in C^p\big(C^1\big([\alpha, \beta], X\big), F\big)$$

and

$$\partial f(u)h = \int_\alpha^\beta \big\{ \partial_2\varphi\big(t, u(t), \dot{u}(t)\big)h(t) + \partial_3\varphi\big(t, u(t), \dot{u}(t)\big)\dot{h}(t) \big\}\, dt$$

for $u \in C^1\big([\alpha, \beta], X\big)$ and $h \in C^1\big([\alpha, \beta], E\big)$.

Proof Because $f = \int_\alpha^\beta \circ \Phi$, this follows from (c) and the chain rule. ∎

The differentiability of parameter-dependent integrals

As a simple consequence of the examples above, we will study the **parameter-dependent integral**

$$\int_\alpha^\beta \varphi(t, x)\, dt .$$

In Volume III, we will prove, in the framework of Lebesgue integration theory, a general version of the following theorem about the continuity and the differentiability of parameter-dependent integrals.

6.7 Proposition *Suppose* $-\infty < \alpha < \beta < \infty$ *and* $p \in \mathbb{N} \cup \{\infty\}$. *Also assume* φ *belongs to* $C^{0,p}\big([\alpha, \beta] \times X, F\big)$, *and let*

$$\Phi(x) := \int_\alpha^\beta \varphi(t, x)\, dt \quad \text{for } x \in X .$$

Then we have $\Phi \in C^p(X, F)$ *and*

$$\partial\Phi = \int_\alpha^\beta \partial_2\varphi(t, \cdot)\, dt . \tag{6.5}$$

Proof[3] For $x \in X$, let u_x be the constant map $[\alpha, \beta] \to E$, $t \mapsto x$. Then, for the function ψ defined through $x \mapsto u_x$, we clearly have

$$\psi \in C^\infty\big(X, C\big([\alpha, \beta], X\big)\big) ,$$

---

[3] We recommend also that you devise a direct proof that does not make use of Examples 6.6.

where $\partial\psi(x)\xi = u_\xi$ for $\xi \in E$. With (6.3), we get $\Phi(x) = f(u_x) = f \circ \psi(x)$. Therefore, from Example 6.6(b) and the chain rule, it follows that $\Phi \in C^p(X, F)$ and[4]

$$\partial\Phi(x)\xi = \partial f\big(\psi(x)\big)\partial\psi(x)\xi = \partial f(u_x)u_\xi = \int_\alpha^\beta \partial_2\varphi(t,x)\xi \, dt$$

for $x \in X$ and $\xi \in E$. Because $\partial_2\varphi(\cdot,x) \in C\big([\alpha,\beta], \mathcal{L}(E,F)\big)$ for $x \in X$, we have

$$\int_\alpha^\beta \partial_2\varphi(t,x) \, dt \in \mathcal{L}(E,F) \ ,$$

and thus

$$\int_\alpha^\beta \partial_2\varphi(t,x) \, dt \, \xi = \int_\alpha^\beta \partial_2\varphi(t,x)\xi \, dt \quad \text{for } \xi \in E$$

(see Exercise VI.3.3). From this, we get

$$\partial\Phi(x)\xi = \int_\alpha^\beta \partial\varphi(t,x) \, dt \, \xi \quad \text{for } x \in X \text{ and } \xi \in E$$

and therefore (6.5). ∎

6.8 Corollary *Suppose the assumptions of Proposition 6.7 are satisfied with $p \le 1$ and $E := \mathbb{R}^m$ and also*

$$\Psi(x) := \int_{a(x)}^{b(x)} \varphi(t,x) \, dt \quad \text{for } x \in X$$

and $a,b \in C\big(X,(\alpha,\beta)\big)$. Then Ψ belongs to $C^p(X,F)$, and for $x \in X$ we have

$$\partial\Psi(x) = \int_{a(x)}^{b(x)} \partial_2\varphi(t,x) \, dt + \varphi\big(b(x),x\big)\partial b(x) - \varphi\big(a(x),x\big)\partial a(x) \ .$$

Proof We set

$$\Phi(x,y,z) := \int_y^z \varphi(t,x) \, dt \quad \text{for } (x,y,z) \in X \times (\alpha,\beta) \times (\alpha,\beta) \ .$$

Proposition 6.7 secures a $\Phi(\cdot,y,z) \in C^p(X,F)$ such that

$$\partial_1\Phi(x,y,z) := \int_y^z \partial_2\varphi(t,x) \, dt$$

for every choice of $(x,y,z) \in X \times (\alpha,\beta) \times (\alpha,\beta)$, if $p = 1$.

---

[4]Here and in the following statements we understand that statements about derivatives apply to the case $p \ge 1$.

Suppose $(x, y) \in X \times (\alpha, \beta)$. Then it follows from Theorem VI.4.12 that $\Phi(x, y, \cdot)$ belongs to $C^p\big((\alpha, \beta), F\big)$ and

$$\partial_3\Phi(x, y, z) = \varphi(z, x) \quad \text{for } z \in (\alpha, \beta) \ .$$

Analogously, by noting that $(x, z) \in X \times (\alpha, \beta)$ because of (VI.4.3), we find $\Phi(x, \cdot, z)$ belongs to $C^p\big((\alpha, \beta), F\big)$ and

$$\partial_2\Phi(x, y, z) = -\varphi(y, x) \quad \text{for } y \in (\alpha, \beta) \ .$$

From this, we easily derive

$$\partial_k\Phi \in C\big(X \times (\alpha, \beta) \times (\alpha, \beta), F\big) \quad \text{for } 1 \leq k \leq 3$$

and get $\Phi \in C^p\big(X \times (\alpha, \beta) \times (\alpha, \beta), F\big)$ from Theorem 2.10. The chain rule now implies

$$\Psi = \Phi\big(\cdot, a(\cdot), b(\cdot)\big) \in C^p(X, F)$$

and

$$\begin{aligned}
\partial\Psi(x) &= \partial_1\Phi\big(x, a(x), b(x)\big) + \partial_2\Phi\big(x, a(x), b(x)\big)\partial a(x) \\
&\quad + \partial_3\Phi\big(x, a(x), b(x)\big)\partial b(x) \\
&= \int_{a(x)}^{b(x)} \partial_2\varphi(t, x)\, dt + \varphi\big(b(x), x\big)\partial b(x) - \varphi\big(a(x), x\big)\partial a(x)
\end{aligned}$$

for $x \in X$. $\blacksquare$

From the formula for $\partial\Psi$, an easy induction shows $\Psi \in C^p(X, F)$ for $p \geq 2$ if φ belongs to $C^{p-1,p}\big([0, 1] \times X, F\big)$, where $C^{p-1,p}$ is defined in the obvious way.

Variational problems

In the remaining part of this section, we assume that[5]

- $-\infty < \alpha < \beta < \infty$;
  X is open in $\mathbb{R}^m$;
  $a, b \in X$;
  $L \in C^{0,1}\big([\alpha, \beta] \times (X \times \mathbb{R}^m), \mathbb{R}\big)$.

Then the integral

$$f(u) := \int_\alpha^\beta L\big(t, u(t), \dot{u}(t)\big)\, dt \tag{6.6}$$

exists for every $u \in C^1\big([\alpha, \beta], X\big)$. The definition (6.6) gives rise to a **variational problem with free boundary conditions** when we seek to minimize the integral over

---

[5]In most applications L is only defined on a set of the form $[\alpha, \beta] \times X \times Y$, where Y is open in $\mathbb{R}^m$ (see Exercise 10). We leave you the simple modifications needed to include this case.

all C^1 functions u in X. We write that problem as

$$\int_\alpha^\beta L(t, u, \dot{u})\, dt \Rightarrow \text{Min} \quad \text{for } u \in C^1\big([\alpha, \beta], X\big) . \qquad (6.7)$$

Thus, problem (6.7) becomes one of minimizing f over the set $U := C^1\big([\alpha, \beta], X\big)$.

The definition (6.6) can also be a part of a **variational problem with fixed boundary conditions**, when we minimize the function f over a constrained set

$$U_{a,b} := \big\{ u \in C^1\big([\alpha, \beta], X\big) \ ; \ u(\alpha) = a, \ u(\beta) = b \big\} .$$

We write this problem as

$$\int_\alpha^\beta L(t, u, \dot{u})\, dt \Rightarrow \text{Min} \quad \text{for } u \in C^1\big([\alpha, \beta], X\big) , \quad u(\alpha) = a , \quad u(\beta) = b . \quad (6.8)$$

Every solution of (6.7) and (6.8) is said to be **extremal**[6] for the variational problems (6.7) and (6.8), respectively.

To treat this problem efficiently, we need this:

6.9 Lemma Let $E := C^1\big([\alpha, \beta], \mathbb{R}^m\big)$ and

$$E_0 := \big\{ u \in E \ ; \ u(\alpha) = u(\beta) = 0 \big\} =: C_0^1\big([\alpha, \beta], \mathbb{R}^m\big) .$$

Then E_0 is a closed vector subspace of E and is therefore itself a Banach space. For $\bar{u} \in U_{a,b}$, $U_{a,b} - \bar{u}$ is open in E_0.

Proof It is clear that E_0 is a closed vector subspace of E. Because, according to Example 6.6(c), U is open in E and because $U_{a,b} = U \cap (\bar{u} + E_0)$, we know $U_{a,b}$ is open in the affine space $\bar{u} + E_0$. Now it is clear that $U_{a,b} - \bar{u}$ is open in E_0. ∎

6.10 Lemma

(i) $f \in C^1(U, \mathbb{R})$ and

$$\partial f(u)h = \int_\alpha^\beta \big\{ \partial_2 L\big(t, u(t), \dot{u}(t)\big) h(t) + \partial_3 L\big(t, u(t), \dot{u}(t)\big) \dot{h}(t) \big\}\, dt$$

for $u \in U$ and $h \in E$.

(ii) Suppose $\bar{u} \in U_{a,b}$, and let $g(v) := f(\bar{u} + v)$ for $v \in V := U_{a,b} - \bar{u}$. Then g belongs to $C^1(V, \mathbb{R})$ with

$$\partial g(v)h = \partial f(\bar{u} + v)h \quad \text{for } v \in V \text{ and } h \in E_0 . \qquad (6.9)$$

Proof (i) is a consequence of Example 6.6(d), and (ii) follows from (i). ∎

---

[6]We will also use the term extremal in the calculus of variations to describe solutions to the Euler–Lagrange equation.

6.11 Lemma   *For* $u \in C([\alpha, \beta], \mathbb{R}^m)$, *suppose*

$$\int_\alpha^\beta \big(u(t) \big| v(t) \big)\, dt = 0 \quad \text{for } v \in C_0^1([\alpha, \beta], \mathbb{R}^m) . \tag{6.10}$$

Then $u = 0$.

Proof   Suppose $u \neq 0$. Then there exists a $k \in \{1, \dots, m\}$ and a $t_0 \in (\alpha, \beta)$ such that $u_k(t_0) \neq 0$. It suffices to consider the case $u_k(t_0) > 0$. From the continuity of u_k, there are $\alpha < \alpha' < \beta' < \beta$ such that $u_k(t) > 0$ for $t \in (\alpha', \beta')$ (see Example III.1.3(d)).

Choosing $v \in C_0^1([\alpha, \beta], \mathbb{R}^m)$ such that $v_j = 0$ for $j \neq k$, it follows from (6.10) that

$$\int_\alpha^\beta u_k(t) w(t)\, dt = 0 \quad \text{for } w \in C_0^1([\alpha, \beta], \mathbb{R}) . \tag{6.11}$$

Now suppose $w \in C_0^1([\alpha, \beta], \mathbb{R})$ such that $w(t) > 0$ for $t \in (\alpha', \beta')$ and $w(t) = 0$ for t outside of the interval (α', β') (see Exercise 7). Then we have

$$\int_\alpha^\beta u_k(t) w(t)\, dt = \int_{\alpha'}^{\beta'} u_k(t) w(t)\, dt > 0 ,$$

which contradicts (6.11).   ∎

The Euler–Lagrange equation

We can now prove the fundamental result of the calculus of variations:

6.12 Theorem   *Suppose* u *is extremal for the variational problem* (6.7) *with free boundary conditions or (alternatively)* (6.8) *with fixed boundary conditions. Also suppose*

$$\big[t \mapsto \partial_3 L\big(t, u(t), \dot{u}(t) \big) \big] \in C^1([\alpha, \beta], \mathbb{R}) . \tag{6.12}$$

Then u *satisfies the* **Euler–Lagrange equation**

$$\big[\partial_3 L(\cdot, u, \dot{u}) \big]^{\boldsymbol{\cdot}} = \partial_2 L(\cdot, u, \dot{u}) . \tag{6.13}$$

In the problem with free boundary conditions, u *also satisfies the* **natural boundary conditions**

$$\partial_3 L\big(\alpha, u(\alpha), \dot{u}(\alpha) \big) = \partial_3 L\big(\beta, u(\beta), \dot{u}(\beta) \big) = 0 . \tag{6.14}$$

For fixed boundary conditions, u *satisfies*

$$u(\alpha) = a \quad \text{and} \quad u(\beta) = b . \tag{6.15}$$

Proof (i) We consider first the problem (6.7). By assumption, the function f defined in (6.6) assumes its minimum at $u \in U$. Thus it follows from Lemma 6.10(i), Theorem 3.13, and Proposition 2.5 that

$$\int_\alpha^\beta \{\partial_2 L(t, u(t), \dot{u}(t))h(t) + \partial_3 L(t, u(t), \dot{u}(t))\dot{h}(t)\} \, dt = 0 \qquad (6.16)$$

for $h \in E = C^1([\alpha, \beta], \mathbb{R}^m)$. Using the assumption (6.12), we can integrate by parts to get

$$\int_\alpha^\beta \partial_3 L(\cdot, u, \dot{u})\dot{h} \, dt = \partial_3 L(\cdot, u, \dot{u})h\big|_\alpha^\beta - \int_\alpha^\beta [\partial_3 L(\cdot, u, \dot{u})]^{\cdot} h \, dt .$$

We then get from (6.16) that

$$\int_\alpha^\beta \{\partial_2 L(\cdot, u, \dot{u}) - [\partial_3 L(\cdot, u, \dot{u})]^{\cdot}\}h \, dt + \partial_3 L(\cdot, u, \dot{u})h\big|_\alpha^\beta = 0 \qquad (6.17)$$

for $h \in E$. In particular, we have

$$\int_\alpha^\beta \{\partial_2 L(\cdot, u, \dot{u}) - [\partial_3 L(\cdot, u, \dot{u})]^{\cdot}\}h \, dt = 0 \quad \text{for } h \in E_0 , \qquad (6.18)$$

because

$$\partial_3 L(\cdot, u, \dot{u})h\big|_\alpha^\beta = 0 \quad \text{for } h \in E_0 . \qquad (6.19)$$

Now the Euler–Lagrange equations (6.13) is implied by (6.18) and Lemma 6.11. Consequently, it follows from (6.17) that

$$\partial_3 L(\cdot, u, \dot{u})h\big|_\alpha^\beta = 0 \quad \text{for } h \in E_0 . \qquad (6.20)$$

For $\xi, \eta \in \mathbb{R}^m$, we set

$$h_{\xi,\eta}(t) := \frac{\beta - t}{\beta - \alpha}\xi + \frac{t - \alpha}{\beta - \alpha}\eta \quad \text{for } \alpha \leq t \leq \beta .$$

Then $h_{\xi,\eta}$ belongs to E, and from (6.20) we get for $h = h_{\xi,\eta}$ that

$$\partial_3 L(\beta, u(\beta), \dot{u}(\beta))\eta - \partial_3 L(\alpha, u(\alpha), \dot{u}(\alpha))\xi = 0 . \qquad (6.21)$$

Because (6.21) holds for every choice of $\xi, \eta \in \mathbb{R}^m$, the natural boundary conditions (6.14) are satisfied.

(ii) Now we consider the problem (6.8). We fix a $\bar{u} \in U_{a,b}$ and set $v := u - \bar{u}$. Then it follows from Lemma 6.10(ii) that g assumes its minimum at the point v of the open set $V := U_{a,b} - \bar{u}$ of the Banach space E_0. It therefore follows from Lemma 6.10(ii), Theorem 3.13, and Proposition 2.5 that (6.16) also holds in this case, though only for $h \in E_0$. On the other hand, because of (6.19), we get from (6.17) the validity of (6.18) and, consequently, the Euler–Lagrange equation. The statement (6.15) is trivial. ∎

6.13 Remarks (a) In Theorem 6.12, it is assumed that the variational problem has a solution, that is, that there is an extremal u. In (6.12) we made an additional assumption placing an implicit regularity condition on the extremal u. Only under this extra assumption is the Euler–Lagrange equation a necessary condition for the presence of an extremal u.

The **indirect method of the calculus of variations** gives the Euler–Lagrange equation a central role. In it, we seek to verify that there is a function u satisfying equation (6.13) and the boundary conditions (6.14) or (6.15). Because the Euler–Lagrange equation is a (generally nonlinear) differential equation, this leads to the subject of **boundary value problems** for differential equations. If it can be proved, using methods of boundary value problem, that there are solutions, we then select in a second step any that are actually extremal.

In contrast is the **direct method of the calculus of variations**, in which one seeks to directly show that the function f has a minimum at a point $u \in U$ (or the function g at a point $v \in V$). Then Theorem 3.13 and Proposition 2.5 says that $\partial f(u) = 0$ (or $\partial g(v) = 0$) and therefore that the relation (6.16) is satisfied for all $h \in E_1$ (or all $h \in E_0$). If the auxiliary "regularity problem" can be solved — that is, if it can be proved that the extremal u satisfies the condition (6.12) — then it also satisfies the Euler–Lagrange differential equation.

Thus Theorem 6.12 has two possible applications. In the first case, we may use the theory of boundary value problems for differential equations to solve the Euler–Lagrange equation, that is, to minimize the integral. This is the classical method of calculus of variations.

In general, however, it is very difficult to prove existence results for boundary value problems. The second case thus takes another point of view: We take a boundary value problem as given and then try to find the variational problem that gives rise to it, so that the differential equation can be interpreted as an Euler–Lagrange equation. If such a variational problem exists, then we try to prove it has an extremal solution that satisfies the regularity condition (6.12). Such a proof then implies that the original boundary value problem also has a solution. This turns out to be a powerful method for studying boundary value problems.

Unfortunately, the details surpass the scope of this text. For more, you should consult the literature or take courses on variational methods, differential equations, and (nonlinear) functional analysis.

(b) Clearly Theorem 6.12 also holds if u is only a critical point of f (or g), that is, if $\partial f(u) = 0$ (or $\partial g(v) = 0$ with $v := u - \bar{u}$). In this case, one says that u is a **stationary value** for the integral (6.6).

(c) Suppose E is a Banach space, Z is open in E, and $F: Z \to \mathbb{R}$. If the directional derivative of F exists for some $h \in E \setminus \{0\}$ in $z_0 \in Z$, then (in the calculus of variations) we call $D_h F(z_0)$ the **first variation of F in the direction h** and write it as $\delta F(z_0; h)$. If the first variation of F exists in every direction, we call

$$\delta F(z_0) := \delta F(z_0; \cdot) : E \to \mathbb{R}$$

the **first variation** of F at z_0.

If F has a local extremum at $z_0 \in Z$ at which the first variation exists, then $\delta F(z_0) = 0$. In other words: the vanishing of the first variation is a necessary condition for the presence of a local extremum.

Proof This is a reformulation of Theorem 3.13. ∎

(d) Under our assumptions, f (or g) is continuously differentiable. Hence, the first variation exists and agrees with ∂f (or ∂g). In many cases, however, the Euler–Lagrange equation can already be derived from weaker assumptions which only guarantee the existence of the first variation.

(e) In the calculus of variations, it is usual to use write $\partial_2 L$ as L_u and $\partial_3 L$ as $L_{\dot{u}}$, respectively. Then the Euler–Lagrange equation takes the abbreviated form

$$\frac{d}{dt}(L_{\dot{u}}) = L_u \, , \tag{6.22}$$

that is, it loses its arguments. If also we assume L and the extremal u are twice differentiable, we can write (6.22) as $L_{t,\dot{u}} + L_{u,\dot{u}}\dot{u} + L_{\dot{u},\dot{u}}\ddot{u} = L_u$, where $L_{t,\dot{u}} = \partial_1 \partial_3 L$ etc. ∎

Classical mechanics

The calculus of variations is important in physics. To demonstrate this, we consider a system of massive point particles in m degrees of freedom, which are specified by the (generalized) **coordinates** $q - (q^1, \ldots, q^m)$. The fundamental problem in classical mechanics is to determine the state $q(t)$ of the system at a given time t, given knowledge of its initial position q_0 at time t_0.

We denote by $\dot{q} := dq/dt$ the (generalized) **velocity coordinates**, by $T(t, q, \dot{q})$ the **kinetic energy**, and by $U(t, q)$ the **potential energy**. Then we make the fundamental assumption of **Hamilton's principle of least action**. This says that between each pair of times t_0 and t_1, the system goes from q_0 to q_1 in a way that minimizes

$$\int_{t_0}^{t_1} [T(t, q, \dot{q}) - U(t, q)] \, dt \, ,$$

that is, the actual path chosen by the system is the one that minimizes this integral over all (virtual) paths starting from q_0 at t_0 and arriving at q_1 at t_1. In this context, $L := T - U$ is called the **Lagrangian** of the system.

Hamilton's principle means that the path $\{ q(t) \; ; \; t_0 \le t \le t_1 \}$ that the system takes is an extremum of the variational problem with fixed boundary conditions

$$\int_{t_0}^{t_1} L(t, q, \dot{q}) \, dt \Rightarrow \text{Min} \quad \text{for } q \in C^1\big([t_0, t_1], \mathbb{R}^m\big) , \quad q(t_0) = q_0 , \quad q(t_1) = q_1 .$$

It then follows from Theorem 6.12 that q satisfies the Euler–Lagrange equation

$$\frac{d}{dt}(L_{\dot{q}}) = L_q$$

if the appropriate regularity conditions are also satisfied.

In physics, engineering, and other fields that use variational methods (for example, economics), it is usual to presume the regularity conditions are satisfied and to postulate the validity of the Euler–Lagrange equation (and to see the existence of an extremal path as physically obvious). The Euler–Lagrange equation is then used to extract the form of the extremal path and to subsequently understand the behavior of the system.

6.14 Examples   (a) We consider the motion of an unconstrained point particle of (positive) mass m moving in three dimensions acted on by a time-independent potential field $U(t, q) = U(q)$. We assume the kinetic energy depends only on $\dot{q}$ and is given by

$$T(\dot{q}) = m \, |\dot{q}|^2 / 2 .$$

The Euler–Lagrange differential equation takes the form

$$m\ddot{q} = -\nabla U(q) . \tag{6.23}$$

Because $\ddot{q}$ is the **acceleration** of the system, (6.23) is the same as **Newton's equation of motion** for the motion of a particle acted on by a **conservative**[7] **force** $-\nabla U$.

Proof   We identify $(\mathbb{R}^3)' = \mathcal{L}(\mathbb{R}^3, \mathbb{R})$ with $\mathbb{R}^3$ using the Riesz representation theorem. We then get from $L(t, q, \dot{q}) = T(\dot{q}) - U(q)$ the relations

$$\partial_2 L(t, q, \dot{q}) = -\nabla U(q) \text{ and } \partial_3 L(t, q, \dot{q}) = \nabla T(\dot{q}) = m\dot{q} ,$$

which prove the claim. ∎

(b) In a generalization of (a), we consider the motion of N unconstrained, massive point particles in a potential field. We write $x = (x_1, \ldots, x_N)$ for the coordinates, where $x_j \in \mathbb{R}^3$ specifies the position of the j-th particle. Further suppose X is open in $\mathbb{R}^{3N}$ and $U \in C^1(X, \mathbb{R})$. Letting m_j be the mass of the j-th particle, the kinetic energy of the system is

$$T(\dot{x}) := \sum_{j=1}^{N} \frac{m_j}{2} \, |\dot{x}_j|^2 .$$

---

[7]A force field is called **conservative** if it has a potential (see Remark VIII.4.10(c)).

Then we get a system of N Euler–Lagrange equations:

$$-m_j \ddot{x}_j = \nabla_{x_j} U(x) \quad \text{for } 1 \le j \le N .$$

Here, ∇_{x_j} denotes the gradient of U in the variables $x_j \in \mathbb{R}^3$. ∎

Exercises

1 Suppose $I = [\alpha, \beta]$, where $-\infty < \alpha < \beta < \infty$, $k \in C(I \times I, \mathbb{R})$, $\varphi \in C^{0,1}(I \times E, F)$. Also let

$$\Phi(u)(t) := \int_\alpha^\beta k(t, s) \varphi(s, u(s)) \, ds \quad \text{for } t \in I$$

and $u \in C(I, E)$. Prove that $\Phi \in C^1(C(I, E), C(I, F))$ and

$$(\partial \Phi(u)h)(t) = \int_\alpha^\beta k(t, s) \partial_2 \varphi(s, u(s)) h(s) \, ds \quad \text{for } t \in I$$

and $u, h \in C(I, E)$.

2 Suppose I and $J := [\alpha, \beta]$ are compact perfect intervals and $f \in C(I \times J, E)$. Show that

$$\int_J \left(\int_I f(s, t) \, ds \right) dt = \int_I \left(\int_J f(s, t) \, dt \right) ds .$$

(Hint: Consider

$$J \to E , \quad y \mapsto \int_I \left(\int_\alpha^y f(s, t) \, dt \right) ds$$

and apply Proposition 6.7 and Corollary VI.4.14.)

3 For $f \in C([\alpha, \beta], E)$, show

$$\int_\alpha^s \left(\int_\alpha^t f(\tau) \, d\tau \right) dt = \int_\alpha^s (s - t) f(t) \, dt \quad \text{for } s \in [\alpha, \beta] .$$

4 Suppose I and J are compact perfect intervals and $\rho \in C(I \times J, \mathbb{R})$. Verify that

$$u(x, y) := \int_I \left(\int_J \log(((x - s)^2 + (y - t)^2) \rho(s, t) \, dt \right) ds \quad \text{for } (x, y) \in \mathbb{R}^2 \setminus (I \times J)$$

is harmonic.

5 Define

$$h : \mathbb{R} \to [0, 1) , \quad s \mapsto \begin{cases} e^{-1/s^2} , & s > 0 , \\ 0 , & s \le 0 , \end{cases}$$

and, for $-\infty < \alpha < \beta < \infty$, define $k : \mathbb{R} \to \mathbb{R}$, $s \mapsto h(s - \alpha)h(\beta - s)$ and

$$\ell : \mathbb{R} \to \mathbb{R} , \quad t \mapsto \int_\alpha^t k(s) \, ds \Big/ \int_\alpha^\beta k(s) \, ds .$$

Then show that
(a) $\ell \in C^\infty(\mathbb{R}, \mathbb{R})$;

(b) ℓ is increasing;

(c) $\ell(t) = 0$ for $t \leq \alpha$ and $\ell(t) = 1$ for $t \geq \beta$.

6 Suppose $-\infty < \alpha_j < \beta_j < \infty$ for $j = 1, \ldots, m$ and $A := \prod_{j=1}^{m} (\alpha_j, \beta_j) \subset \mathbb{R}^m$. Show that there is a $g \in C^\infty(\mathbb{R}^m, \mathbb{R})$ such that $g(x) > 0$ for $x \in A$ and $g(x) = 0$ for $x \in A^c$.
(Hint: Consider

$$g(x_1, \ldots, x_m) := k_1(x_1) \cdot \cdots \cdot k_m(x_m) \quad \text{for } (x_1, \ldots, x_m) \in \mathbb{R}^m \ ,$$

where $k_j(t) := h(t - \alpha_j) h(\beta_j - t)$ for $t \in [\alpha_j, \beta_j]$ and $j = 1, \ldots, m$.)

7 Suppose $K \subset \mathbb{R}^m$ is compact and U is an open neighborhood of K. Then show there is an $f \in C^\infty(\mathbb{R}^m, \mathbb{R})$ such that

(a) $f(x) \in [0, 1]$ for $x \in \mathbb{R}^m$;

(b) $f(x) = 1$ for $x \in K$;

(c) $f(x) = 0$ for $x \in U^c$.

(Hint: For $x \in K$, there exists an $\varepsilon_x > 0$ such that $A_x := \overline{\mathbb{B}}(x, \varepsilon_x) \subset U$. Then choose $g_x \in C^\infty(\mathbb{R}^m, \mathbb{R})$ with $g_x(y) > 0$ for $y \in A_x$ (see Exercise 6) and choose $x_0, \ldots, x_n \in K$ such that $K \subset \bigcup_{j=0}^{n} A_{x_j}$. Then $G := g_{x_0} + \cdots + g_{x_n}$ belongs to $C^\infty(\mathbb{R}^m, \mathbb{R})$, and the number $\delta := \min_{x \in K} G(x) > 0$. Finally, let ℓ be as in Exercise 5 with $\alpha = 0$, $\beta = \delta$, and $f := \ell \circ G$.)

8 Define $E_0 := \{ u \in C^1([0, 1], \mathbb{R}) \ ; \ u(0) = 0, \ u(1) = 1 \}$ and $f(u) := \int_0^1 [t\dot{u}(t)]^2 \, dt$ for $u \in E_0$. Then show $\inf\{ f(u) \ ; \ u \in E_0 \} = 0$, although there is no $u_0 \in E_0$ such that $f(u_0) = 0$.

9 For $a, b \in \mathbb{R}^m$, consider the variational problem with fixed boundary conditions

$$\int_\alpha^\beta |\dot{u}(t)| \, dt \Rightarrow \text{Min} \quad \text{for } u \in C^1([\alpha, \beta], \mathbb{R}^m) \ , \quad u(\alpha) = a \ , \quad u(\beta) = b \ . \tag{6.24}$$

(a) Find all solutions of the Euler–Lagrange equations for (6.24) with the property that $|\dot{u}(t)| = \text{const}$.

(b) How do the Euler–Lagrange equations for (6.24) read when $m = 2$ under the assumption $\dot{u}_1(t) \neq 0$ and $\dot{u}_2(t) \neq 0$ for $t \in [\alpha, \beta]$?

10 Find the Euler–Lagrange equations when

(a) $\int_\alpha^\beta u\sqrt{1 + \dot{u}^2} \, dt \Rightarrow \text{Min}$ for $u \in C^1([\alpha, \beta], \mathbb{R})$;

(b) $\int_\alpha^\beta \sqrt{(1 + \dot{u}^2)/u} \, dt \Rightarrow \text{Min}$ for $u \in C^1([\alpha, \beta], (0, \infty))$.

11 Suppose f is defined through (6.6) with $L \in C^{0,2}([\alpha, \beta] \times (X \times \mathbb{R}^m), \mathbb{R})$. Show $f \in C^2(U, \mathbb{R})$ and

$$\partial^2 f(u)[h, k] = \int_\alpha^\beta \{ \partial_2^2 L(\cdot, u, \dot{u})[h, k] + \partial_2 \partial_3 L(\cdot, u, \dot{u})[h, \dot{k}]$$
$$+ \partial_2 \partial_3 L(\cdot, u, \dot{u})[\dot{h}, k] + \partial_3^2 L(\cdot, u, \dot{u})[\dot{h}, \dot{k}] \} \, dt$$

for $h, k \in E_1$. Derive from this a condition sufficient to guarantee that a solution u to Euler–Lagrange equation is a local minimum of f.

12 Suppose $T(\dot{q}) := m |\dot{q}|^2 /2$ and $U(t, q) = U(q)$ for $q \in \mathbb{R}^3$. Prove that the **total energy** $E := T + U$ is constant along every solution q of the Euler–Lagrange equation for the variational problem

$$\int_{t_0}^{t_1} [T(\dot{q}) - U(q)]\, dt \Rightarrow \text{Min} \quad \text{for } q \in C^2([t_0, t_1], \mathbb{R}^3) .$$

7 Inverse maps

Suppose J is an open interval in $\mathbb{R}$ and $f \colon J \to \mathbb{R}$ is differentiable. Also suppose $a \in J$ with $f'(a) \neq 0$. Then the linear approximation

$$\mathbb{R} \to \mathbb{R} \,, \quad x \mapsto f(a) + f'(a)(x - a)$$

to f is invertible at a. This also stays true locally, that is, there is an $\varepsilon > 0$ such that f is invertible on $X := (a - \varepsilon, a + \varepsilon)$. Additionally, $Y = f(X)$ is an open interval, and the inverse map $f^{-1} \colon Y \to \mathbb{R}$ is differentiable with

$$(f^{-1})'\big(f(x)\big) = \big(f'(x)\big)^{-1} \quad \text{for } x \in X$$

(Compare with Theorem IV.1.8).

  In this section, we develop the natural generalization of this for functions of multiple variables.

The derivative of the inverse of linear maps

In the following, suppose

- E and F are Banach spaces over the field $\mathbb{K}$.

We want to determine the differentiability of the map

$$\operatorname{inv} \colon \mathcal{L}\mathrm{is}(E, F) \to \mathcal{L}(F, E) \,, \quad A \mapsto A^{-1}$$

and, if necessary, find its derivative. To make sense of this question, we must first verify that $\mathcal{L}\mathrm{is}(E, F)$ is an open subset of the Banach space $\mathcal{L}(E, F)$. To show this, we first prove a theorem about "geometric series" in the Banach algebra $\mathcal{L}(E)$. Here we let $I := 1_E$.

7.1 Proposition *Suppose $A \in \mathcal{L}(E)$ with $\|A\| < 1$. Then $I - A$ belongs to $\mathcal{L}\mathrm{aut}(E)$, $(I - A)^{-1} = \sum_{k=0}^{\infty} A^k$, and $\|(I - A)^{-1}\| \leq (1 - \|A\|)^{-1}$.*

Proof We consider the geometric series $\sum A^k$ in the Banach algebra $\mathcal{L}(E)$. Because $\|A^k\| \leq \|A\|^k$ and

$$\sum_{k=0}^{\infty} \|A\|^k = 1/(1 - \|A\|) \,, \tag{7.1}$$

it follows from the majorant criterion (Theorem II.8.3) that $\sum A^k$ converges absolutely in $\mathcal{L}(E)$. In particular, the value of the series

$$B := \sum_{k=0}^{\infty} A^k$$

is a well-defined element of $\mathcal{L}(E)$. Clearly $AB = BA = \sum_{k=1}^{\infty} A^k$. Therefore $(I - A)B = B(I - A) = I$, and consequently $(I - A)^{-1} = B$. Finally, Remark II.8.2(c) and (7.1) gives the estimate

$$\|(I - A)^{-1}\| = \|B\| \leq (1 - \|A\|)^{-1} . \blacksquare$$

7.2 Proposition

(i) $\mathcal{L}\mathrm{is}(E, F)$ is open in $\mathcal{L}(E, F)$.

(ii) The map

$$\mathrm{inv} : \mathcal{L}\mathrm{is}(E, F) \to \mathcal{L}(F, E) , \quad A \mapsto A^{-1}$$

is infinitely many times continuously differentiable, and[1]

$$\partial\,\mathrm{inv}(A)B = -A^{-1}BA^{-1} \quad \text{for } A \in \mathcal{L}\mathrm{is}(E, F) \text{ and } B \in \mathcal{L}(E, F) . \quad (7.2)$$

Proof   (i) Suppose $A_0 \in \mathcal{L}\mathrm{is}(E, F)$ and $A \in \mathcal{L}(E, F)$ such that $\|A - A_0\| < 1/\|A_0^{-1}\|$. Because

$$A = A_0 + A - A_0 = A_0\left[I + A_0^{-1}(A - A_0)\right] , \quad (7.3)$$

it suffices to verify that $I + A_0^{-1}(A - A_0)$ belongs to $\mathcal{L}\mathrm{aut}(E)$. Since

$$\| - A_0^{-1}(A - A_0)\| \leq \|A_0^{-1}\|\,\|A - A_0\| < 1 ,$$

this follows from Proposition 7.1. Therefore the open ball in $\mathcal{L}(E, F)$ with center A_0 and radius $1/\|A_0^{-1}\|$ belongs to $\mathcal{L}\mathrm{is}(E, F)$.

(ii) We have $2\,\|A - A_0\| < 1/\|A_0^{-1}\|$. From (7.3), we get

$$A^{-1} = \left[I + A_0^{-1}(A - A_0)\right]^{-1}A_0^{-1} ,$$

and thus

$$A^{-1} - A_0^{-1} = \left[I + A_0^{-1}(A - A_0)\right]^{-1}\left(I - \left[I + A_0^{-1}(A - A_0)\right]\right)A_0^{-1} ,$$

that is,

$$\mathrm{inv}(A) - \mathrm{inv}(A_0) = -\left[I + A_0^{-1}(A - A_0)\right]^{-1}A_0^{-1}(A - A_0)A_0^{-1} . \quad (7.4)$$

From this and Proposition 7.1, we derive

$$\| \mathrm{inv}(A) - \mathrm{inv}(A_0)\| \leq \frac{\|A_0^{-1}\|^2\,\|A - A_0\|}{1 - \|A_0^{-1}(A - A_0)\|} < 2\,\|A_0^{-1}\|^2\,\|A - A_0\| . \quad (7.5)$$

---

[1]Because of Remark 2.2(e), this formula reduces when $E = F = \mathbb{K}$ to $(1/z)' = -1/z^2$.

Therefore inv is continuous.

(iii) We next prove that inv is differentiable. So let $A \in \mathcal{L}is(E, F)$. For $B \in \mathcal{L}(E, F)$ such that $\|B\| < 1/\|A^{-1}\|$ we have according to (i) that $A + B$ belongs to $\mathcal{L}is(E, F)$ and

$$(A + B)^{-1} - A^{-1} = (A + B)^{-1}[A - (A + B)]A^{-1} = -(A + B)^{-1}BA^{-1} .$$

This implies

$$\text{inv}(A + B) - \text{inv}(A) + A^{-1}BA^{-1} = \big[\text{inv}(A) - \text{inv}(A + B)\big]BA^{-1} .$$

Consequently,

$$\text{inv}(A + B) - \text{inv}(A) + A^{-1}BA^{-1} = o(\|B\|) \ \ (B \to 0)$$

follows from the continuity of inv. Therefore $\text{inv} \colon \mathcal{L}is(E, F) \to \mathcal{L}(F, E)$ is differentiable, and (7.2) holds.

(iv) We express

$$g \colon \mathcal{L}(F, E)^2 \to \mathcal{L}\big(\mathcal{L}(E, F), \mathcal{L}(F, E)\big)$$

through

$$g(T_1, T_2)(S) := -T_1 S T_2 \quad \text{for } T_1, T_2 \in \mathcal{L}(F, E) \text{ and } S \in \mathcal{L}(E, F) .$$

Then is it easy to see that g is bilinear and continuous, and therefore

$$g \in C^\infty\big(\mathcal{L}(F, E)^2, \mathcal{L}\big(\mathcal{L}(E, F), \mathcal{L}(F, E)\big)\big)$$

(see Exercises 4.4 and 5.2). In addition,

$$\partial \text{inv}(A) = g\big(\text{inv}(A), \text{inv}(A)\big) \quad \text{for } A \in \mathcal{L}is(E, F) . \tag{7.6}$$

Therefore we get from (ii) the continuity of ∂inv. Thus the map inv belongs to $C^1\big(\mathcal{L}is(E, F), \mathcal{L}(F, E)\big)$. Finally, it follows from (7.6) with the help of the chain rule and a simple induction argument that this map is smooth. ∎

The inverse function theorem

We can now prove a central result about the local behavior of differentiable maps, which — as we shall see later — has extremely far-reaching consequences.

7.3 Theorem (inverse function) *Suppose X is open in E and $x_0 \in X$. Also suppose for $q \in \mathbb{N}^\times \cup \{\infty\}$ that $f \in C^q(X, F)$. Finally, suppose*

$$\partial f(x_0) \in \mathcal{L}\mathrm{is}(E, F) \ .$$

Then there is an open neighborhood U of x_0 in X and an open neighborhood V of $y_0 := f(x_0)$ with these properties:

(i) *$f : U \to V$ is bijective.*

(ii) *$f^{-1} \in C^q(V, E)$, and for every $x \in U$, we have*

$$\partial f(x) \in \mathcal{L}\mathrm{is}(E, F) \quad \text{and} \quad \partial f^{-1}(f(x)) = \left[\partial f(x)\right]^{-1} \ .$$

Proof (i) We set $A := \partial f(x_0)$ and $h := A^{-1} f : X \to E$. Then it follows from Exercise 5.1 and the chain rule (Theorem 5.7), that h belongs $C^q(X, E)$ and $\partial h(x_0) = A^{-1} \partial f(x_0) = I$. Thus we lose no generality in considering the case $E = F$ and $\partial f(x_0) = I$ (see Exercise 5.1).

(ii) We make another simplification by setting

$$h(x) := f(x + x_0) - f(x_0) \quad \text{for } x \in X_1 := X - x_0 \ .$$

Then X_1 is open in E, and $h \in C^q(X_1, E)$ with $h(0) = 0$ and $\partial h(0) = \partial f(x_0) = I$. Thus it suffices to consider the case

$$E = F \ , \quad x_0 = 0 \ , \quad f(0) = 0 \ , \quad \partial f(0) = I \ .$$

(iii) We show that f is locally bijective around 0. Precisely, we prove that there are null neighborhoods[2] U and V such that the equation $f(x) = y$ has a unique solution in U for every $y \in V$ and such that $f(U) \subset V$. This problem is clearly equivalent to determining the null neighborhoods U and V such that the map

$$g_y : U \to E \ , \quad x \mapsto x - f(x) + y$$

has exactly one fixed point for every $y \in V$. Starting, we set $g := g_0$. Because $\partial f(0) = I$, we have $\partial g(0) = 0$, and we find using the continuity of ∂g an $r > 0$ such that

$$\|\partial g(x)\| \leq 1/2 \quad \text{for } x \in 2r\overline{\mathbb{B}} \ . \tag{7.7}$$

Because $g(0) = 0$, we get $g(x) = \int_0^1 \partial g(tx) x \, dt$ by applying the mean value theorem in integral form (Theorem 3.10). In turn, we get

$$\|g(x)\| \leq \int_0^1 \|\partial g(tx)\| \, \|x\| \, dt \leq \|x\|/2 \quad \text{for } x \in 2r\overline{\mathbb{B}} \ . \tag{7.8}$$

---

[2]That is, neighborhoods of 0.

Thus for every $y \in r\overline{\mathbb{B}}$, we have the estimate

$$\|g_y(x)\| \leq \|y\| + \|g(x)\| \leq 2r \quad \text{for } x \in 2r\overline{\mathbb{B}} \ ,$$

that is, g_y maps $2r\overline{\mathbb{B}}$ within itself for every $y \in r\overline{\mathbb{B}}$.

For $x_1, x_2 \in 2r\overline{\mathbb{B}}$, we find with the help of the mean value theorem that

$$\|g_y(x_1) - g_y(x_2)\| = \left\| \int_0^1 \partial g\big(x_2 + t(x_1 - x_2)\big)(x_1 - x_2)\, dt \right\| \leq \frac{1}{2} \|x_1 - x_2\| \ .$$

Consequently, g_y is a contraction on $2r\overline{\mathbb{B}}$ for every $y \in r\overline{\mathbb{B}}$. From this, the Banach fixed point theorem (Theorem IV.4.3) secures the existence of a unique $x \in 2r\overline{\mathbb{B}}$ such that $g_y(x) = x$ and therefore such that $f(x) = y$.

We set $V := r\mathbb{B}$ and $U := f^{-1}(V) \cap 2r\mathbb{B}$. Then U is an open null neighborhood, and $f|U : U \to V$ is bijective.

(iv) Now we show that $f^{-1} : V \to E$ is continuous. So we consider

$$x = x - f(x) + f(x) = g(x) + f(x) \quad \text{for } x \in U \ .$$

Therefore

$$\|x_1 - x_2\| \leq \frac{1}{2} \|x_1 - x_2\| + \|f(x_1) - f(x_2)\| \quad \text{for } x_1, x_2 \in U \ ,$$

and consequently

$$\|f^{-1}(y_1) - f^{-1}(y_2)\| \leq 2 \|y_1 - y_2\| \quad \text{for } y_1, y_2 \in V \ . \tag{7.9}$$

Thus $f^{-1} : V \to E$ is Lipschitz continuous.

(v) We show the differentiability of $f^{-1} : V \to E$ and prove that the derivative is given by

$$\partial f^{-1}(y) = \big[\partial f(x)\big]^{-1} \quad \text{with} \quad x := f^{-1}(y) \ . \tag{7.10}$$

First we want to show $\partial f(x)$ belongs to $\mathcal{L}\mathrm{aut}(E)$ for $x \in U$. From $f(x) = x - g(x)$, it follows that $\partial f(x) = I - \partial g(x)$ for $x \in U$. Noting (7.7), we see $\partial f(x) \in \mathcal{L}\mathrm{aut}(E)$ follows from Proposition 7.1.

Suppose now $y, y_0 \in V$, and let $x := f^{-1}(y)$ and $x_0 := f^{-1}(y_0)$. Then

$$f(x) - f(x_0) = \partial f(x_0)(x - x_0) + o(\|x - x_0\|) \quad (x \to x_0) \ ,$$

and we get as $x \to x_0$ that

$$\|f^{-1}(y) - f^{-1}(y_0) - \big[\partial f(x_0)\big]^{-1}(y - y_0)\|$$
$$= \|x - x_0 - \big[\partial f(x_0)\big]^{-1}\big(f(x) - f(x_0)\big)\| \leq c \|o(\|x - x_0\|)\|$$

where $c = \left\| \left[\partial f^{-1}(x_0) \right]^{-1} \right\|$. Because (7.9) gives the estimate $2\left\| y - y_0 \right\| \geq \left\| x - x_0 \right\|$, we finally have

$$\frac{\left\| f^{-1}(y) - f^{-1}(y_0) - \left[\partial f(x_0) \right]^{-1}(y - y_0) \right\|}{\left\| y - y_0 \right\|} \leq \frac{2c\left\| o(\left\| x - x_0 \right\|) \right\|}{\left\| x - x_0 \right\|} \tag{7.11}$$

as $x \to x_0$. We now pass the limit $y \to y_0$ so that $x \to x_0$ follows from (iv), and (7.11) shows that f^{-1} is differentiable at y_0, with derivative $\left[\partial f(x_0) \right]^{-1}$.

(vi) It remains to verify that f^{-1} belongs to $C^q(V, E)$. So consider what (7.10) shows, that is,

$$\partial f^{-1} = (\partial f \circ f^{-1})^{-1} = \mathrm{inv} \circ \partial f \circ f^{-1} . \tag{7.12}$$

We know from (iv) that f^{-1} belongs to $C(V, E)$ with $f^{-1}(V) \cap \mathbb{B}(x_0, 2r) = U$ and, by assumption, that $\partial f \in C(U, \mathcal{L}(E))$. Using Proposition 7.2 it follows that ∂f^{-1} belongs to $C(V, \mathcal{L}(E))$, which proves $f^{-1} \in C^1(V, E)$. For $q > 1$, one proves $f^{-1} \in C^p(V, E)$ by induction from (7.12) using the chain rule. ∎

Diffeomorphisms

Suppose X is open in E, Y is open in F, and $q \in \mathbb{N} \cup \{\infty\}$. We call the map $f : X \to Y$ a C^q **diffeomorphism** from X to Y if it is bijective,

$$f \in C^q(X, Y), \quad \text{and} \quad f^{-1} \in C^q(Y, E) .$$

We may call a C^0 diffeomorphism a **homeomorphism** or a **topological map**. We set

$$\mathrm{Diff}^q(X, Y) := \{ f : X \to Y \; ; \; f \text{ is a } C^q \text{ diffeomorphism} \} .$$

The map $g : X \to F$ is a **locally C^q diffeomorphism**[3] if every $x_0 \in X$ has open neighborhoods $U \in \mathfrak{U}_X(x_0)$ and $V \in \mathfrak{U}_F\big(g(x_0)\big)$ such that $g|U$ belongs to $\mathrm{Diff}^q(U, V)$. We denote the set of all locally C^q diffeomorphisms from X to F by $\mathrm{Diff}^q_{\mathrm{loc}}(X, F)$.

7.4 Remarks Suppose X is open in E and Y is open in F.

(a) For every $q \in \mathbb{N}$, we have the inclusions

$$\mathrm{Diff}^\infty(X, Y) \subset \mathrm{Diff}^{q+1}(X, Y) \subset \mathrm{Diff}^q(X, Y) \subset \mathrm{Diff}^q_{\mathrm{loc}}(X, F) ,$$

$$\mathrm{Diff}^\infty_{\mathrm{loc}}(X, F) \subset \mathrm{Diff}^{q+1}_{\mathrm{loc}}(X, F) \subset \mathrm{Diff}^q_{\mathrm{loc}}(X, F) .$$

(b) Suppose $f \in C^{q+1}(X, Y)$ and $f \in \mathrm{Diff}^q(X, Y)$. Then it does not follow generally that f belongs to $\mathrm{Diff}^{q+1}(X, Y)$.

---

[3]When $q = 0$, one speaks of a **local homeomorphism** or a **locally topological map**.

Proof We set $E := F := \mathbb{R}$ and $X := Y := \mathbb{R}$ and consider $f(x) := x^3$. Then f is a smooth topological map, that is, $f \in C^\infty(X,Y) \cap \mathrm{Diff}^0(X,Y)$. However, f^{-1} is not differentiable at 0. ∎

(c) Local topological maps are **open**, that is, they map open sets to open sets (see Exercise III.2.14).

Proof This follows easily from Theorems I.3.8(ii) and III.2.4(ii). ∎

(d) For $f \in \mathrm{Diff}^1_{\mathrm{loc}}(X,Y)$, we have $\partial f(x) \in \mathcal{L}\mathrm{is}(E,F)$ for $x \in X$.

Proof This is a consequence of the chain rule (see Exercise 1). ∎

7.5 Corollary *Suppose X is open in E, $q \in \mathbb{N}^\times \cup \{\infty\}$, and $f \in C^q(X,F)$. Then*

$$f \in \mathrm{Diff}^q_{\mathrm{loc}}(X,F) \Longleftrightarrow \partial f(x) \in \mathcal{L}\mathrm{is}(E,F) \text{ for } x \in X \ .$$

Proof "⟹" From Remark 7.4(a), it follows by induction that f belongs to $\mathrm{Diff}^1_{\mathrm{loc}}(X,F)$. The claim then follows from Remark 7.4(d).

"⟸" This follows from the inverse function theorem ∎

7.6 Remark Under the assumptions of Corollary 7.5, $\partial f(x)$ belongs to $\mathcal{L}\mathrm{is}(E,F)$ for $x \in X$. Then f is locally topological, and thus $Y := f(X)$ is open. In general, however, f is not a C^q diffeomorphism from X to Y.

Proof Let $X := E := F := \mathbb{C}$ and $f(z) := e^z$. Then f is smooth and $\partial f(z) = e^z \neq 0$ for $z \in \mathbb{C}$. Because of its $2\pi i$-periodicity, f is not injective. ∎

The solvability of nonlinear systems of equations

We will now formulate the inverse function theorem for the case $E = F = \mathbb{R}^m$. We will exploit that every linear map on $\mathbb{R}^m$ is continuous and that a linear map is invertible if its determinant does not vanish.

In the rest of this section, suppose X is open in $\mathbb{R}^m$ and $x_0 \in X$. Also suppose $q \in \mathbb{N}^\times \cup \{\infty\}$ and $f = (f^1, \ldots, f^m) \in C^q(X, \mathbb{R}^m)$.

7.7 Theorem *If $\det\big(\partial f(x_0)\big) \neq 0$, then there are open neighborhoods U of x_0 and V of $f(x_0)$ such that $f \,|\, U$ belongs to $\mathrm{Diff}^q(U,V)$.*

Proof From Theorem 1.6 we know that $\mathrm{Hom}(\mathbb{R}^m, \mathbb{R}^m) = \mathcal{L}(\mathbb{R}^m)$. Linear algebra teaches us that

$$\partial f(x_0) \in \mathcal{L}\mathrm{aut}(\mathbb{R}^m) \Longleftrightarrow \det\big(\partial f(x_0)\big) \neq 0$$

(for example, [Gab96, A3.6(a)]). The claim then follows from Theorem 7.3. ∎

7.8 Corollary *If* $\det(\partial f(x_0)) \neq 0$, *then there are open neighborhoods* U *of* x_0 *and* V *of* $f(x_0)$ *such that the system of equations*

$$f^1(x^1, \ldots, x^m) = y^1 \ ,$$

$$\vdots \qquad\qquad\qquad (7.13)$$

$$f^m(x^1, \ldots, x^m) = y^m$$

has exactly one solution

$$x^1 = \boldsymbol{x}^1(y^1, \ldots, y^m), \ldots, x^m = \boldsymbol{x}^m(y^1, \ldots, y^m)$$

for every m-*tuple* $(y^1, \ldots, y^m) \in V$ *in* U. *The functions* $\boldsymbol{x}^1, \ldots, \boldsymbol{x}^m$ *belong to* $C^q(V, \mathbb{R})$.

7.9 Remarks **(a)** According to Remark 1.18(b) and Corollary 2.9, the determinant of the linear map $\partial f(x)$ can be calculated using the Jacobi matrix $[\partial_k f^j(x)]$, that is, $\det(\partial f(x)) = \det[\partial_k f^j(x)]$. This is called the **Jacobian determinant** of f (at the point x). We may also simply call it the **Jacobian** and write it as

$$\frac{\partial(f^1, \ldots, f^m)}{\partial(x^1, \ldots, x^m)}(x) \ .$$

(b) Because the proof of the inverse function theorem is constructive, it can in principle be used to calculate (approximately) $f^{-1}(x)$. In particular, we can approximate the solution in the finite-dimensional case described in Corollary 7.8 by looking at the equation system (7.13) in the neighborhood of x_0.

Proof This follows because the proof of Theorem 7.3 is based on the contraction theorem and therefore on the method of successive approximation. ∎

Exercises

1 Suppose X is open in E, Y is open in F, and $f \in \text{Diff}^1_{\text{loc}}(X, Y)$. Show that $\partial f(x_0)$ belongs to $\mathcal{L}\text{is}(E, F)$ for $x_0 \in X$, and $\partial f^{-1}(y_0) = [\partial f(x_0)]^{-1}$, where $y_0 := f(x_0)$.

2 Suppose $m, n \in \mathbb{N}^\times$, and X is open in $\mathbb{R}^n$. Show that if $\text{Diff}^1_{\text{loc}}(X, \mathbb{R}^m)$ is not empty, then $m = n$.

3 For the maps defined in (a)–(d), find $Y := f(X)$ and f^{-1}. Also decide whether $f \in \text{Diff}^q_{\text{loc}}(X, \mathbb{R}^2)$ or $f \in \text{Diff}^q(X, Y)$.
(a) $X := \mathbb{R}^2$ and $f(x, y) := (x + a, y + b)$ for $(a, b) \in \mathbb{R}^2$;
(b) $X := \mathbb{R}^2$ and $f(x, y) := (x^2 - x - 2, 3y)$;
(c) $X := \mathbb{R}^2 \setminus \{(0, 0)\}$ and $f(x, y) := (x^2 - y^2, 2xy)$;
(d) $X := \{ (x, y) \in \mathbb{R}^2 \ ; \ 0 < y < x \}$ and $f(x, y) := (\log xy, 1/(x^2 + y^2))$.
(Hint for (c): $\mathbb{R}^2 \longleftrightarrow \mathbb{C}$.)

4 Suppose
$$f : \mathbb{R}^2 \to \mathbb{R}^2 , \quad (x, y) \mapsto (\cosh x \cos y, \sinh x \sin y)$$
and $X := \{ (x, y) \in \mathbb{R}^2 ; x > 0 \}$ and $Y := f(X)$. Show that

(a) $f | X \in \mathrm{Diff}^\infty_{\mathrm{loc}}(X, \mathbb{R}^2)$;

(b) $f | X \notin \mathrm{Diff}^\infty(X, Y)$;

(c) for $U := \{ (x, y) \in X ; 0 < y < 2\pi \}$ and $V := Y \setminus ([0, \infty) \times \{0\})$, $f | U$ belongs to $\mathrm{Diff}^\infty(U, V)$;

(d) $Y = \mathbb{R}^2 \setminus ([-1, 1] \times \{0\})$.

5 Suppose X is open in $\mathbb{R}^m$ and $f \in C^1(X, \mathbb{R}^m)$. Show that

(a) if $\partial f(x) \in \mathcal{L}\mathrm{is}(\mathbb{R}^m)$ for $x \in X$, then $x \mapsto |f(x)|$ does not have a maximum in X;

(b) if $\partial f(x) \in \mathcal{L}\mathrm{is}(\mathbb{R}^m)$ and $f(x) \neq 0$ for $x \in X$, then $x \mapsto |f(x)|$ has no minimum.

6 Suppose H is a real Hilbert space and
$$f : H \to H , \quad x \mapsto x / \sqrt{1 + |x|^2} .$$

Letting $Y := \mathrm{im}(f)$, show $f \in \mathrm{Diff}^\infty(H, Y)$. What are f^{-1} and ∂f?

7 Suppose X is open in $\mathbb{R}^m$ and $f \in C^1(X, \mathbb{R}^m)$. Also suppose there is an $\alpha > 0$ such that
$$|f(x) - f(y)| \geq \alpha |x - y| \quad \text{for } x, y \in X . \tag{7.14}$$
Show that $Y := f(X)$ is open in $\mathbb{R}^m$ and $f \in \mathrm{Diff}^1(X, Y)$. Also show for $X = \mathbb{R}^m$ that $Y = \mathbb{R}^m$.

(Hint: From (7.14), it follows that $\partial f(x) \in \mathcal{L}\mathrm{is}(X, Y)$ for $x \in X$. If $X = \mathbb{R}^m$, then (7.14) implies that Y is closed in $\mathbb{R}^m$.)

8 Suppose $f \in \mathrm{Diff}^1(\mathbb{R}^m, \mathbb{R}^m)$ and $g \in C^1(\mathbb{R}^m, \mathbb{R}^m)$, and either of these assumptions is satisfied:

(a) f^{-1} and g is Lipschitz continuous;

(b) g vanishes outside of a bounded subset of $\mathbb{R}^m$.

Show then there is an $\varepsilon_0 > 0$ such that $f + \varepsilon g \in \mathrm{Diff}^1(\mathbb{R}^m, \mathbb{R}^m)$ for $\varepsilon \in (-\varepsilon_0, \varepsilon_0)$.
(Hint: Consider $\mathrm{id}_{\mathbb{R}^m} + f^{-1} \circ (\varepsilon g)$ and apply Exercise 7.)

8 Implicit functions

In the preceding sections, we have studied (in the finite-dimensional case) the solvability of nonlinear systems of equations. We concentrated on the case in which the number of equations equals the number of variables. Now, we explore the solvability of nonlinear system having more unknowns than equations.

We will prove the main result, the implicit function theorem, without troubling ourselves with the essentially auxiliary considerations needed for proving it on general Banach spaces. To illustrate the fundamental theorem, we prove the fundamental existence and continuity theorems for ordinary differential equations. Finite-dimensional applications of the implicit function theorem will be treated in the two subsequent sections.

For motivation, we consider the function $f \colon \mathbb{R}^2 \to \mathbb{R}$, $(x,y) \mapsto x^2 + y^2 - 1$. Suppose $(a,b) \in \mathbb{R}^2$ where $a \neq \pm 1$, $b > 0$ and $f(a,b) = 0$. Then there are open intervals A and B with $a \in A$ and $b \in B$ such that for every $x \in A$ there exists exactly one $y \in B$ such that $f(x,y) = 0$. By the assignment $x \mapsto y$, we define a map $g \colon A \to B$ with $f\big(x,g(x)\big) = 0$ for $x \in A$. Clearly, we have $g(x) = \sqrt{1-x^2}$. In addition, there is an open interval $\widetilde{B}$ such that $-b \in \widetilde{B}$ and a $\widetilde{g} \colon A \to \widetilde{B}$ such that $f\big(x,\widetilde{g}(x)\big) = 0$ for $x \in A$. Here, we have $\widetilde{g}(x) = -\sqrt{1-x^2}$. The function g is uniquely determined through f and (a,b) on A, and the function $\widetilde{g}$ is defined by f and $(a,-b)$. We say therefore that g and $\widetilde{g}$ are implicitly defined by f near (a,b) and $(a,-b)$, respectively. The functions g and $\widetilde{g}$ are **solutions** for y (as a function of x) of the equation $f(x,y) = 0$ near (a,b) and $(a,-b)$, respectively. We say $f(x,y) = 0$ defines g **implicitly** near (a,b).

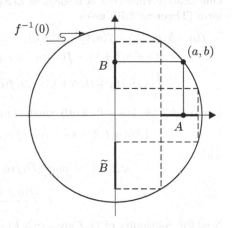

Such solutions can clearly not be given in neighborhoods of $(1,0)$ or $(-1,0)$. To hint at the obstruction, we note that for $a = \pm 1$, we have $\partial_2 f(a,b) = 0$, whereas for $a \neq \pm 1$ this quantity does not vanish, as $\partial_2 f(a,b) = 2b$.

Differentiable maps on product spaces

Here, suppose

- E_1, E_2 and F are Banach spaces over $\mathbb{K}$;
  $q \in \mathbb{N}^\times \cup \{\infty\}$.

Suppose X_j is open in E_j for $j = 1,2$, and $f \colon X_1 \times X_2 \to F$ is differentiable at (a,b). Then the functions $f(\cdot,b) \colon X_1 \to F$ and $f(a,\cdot) \colon X_2 \to F$ are also differentiable at a and b, respectively. To avoid confusion with the standard partial derivatives, we write $D_1 f(a, b)$ for the derivative of $f(\cdot, b)$ at a, and we write $D_2 f(a, b)$ for the derivative of $f(a, \cdot)$ at b,

8.1 Remarks (a) Obviously $D_j f \colon X_1 \times X_2 \to \mathcal{L}(E_j, F)$ for $j = 1, 2$.

(b) The statements
  (i) $f \in C^q(X_1 \times X_2, F)$ and
  (ii) $D_j f \in C^{q-1}\big(X_1 \times X_2, \mathcal{L}(E_j, F)\big)$ for $j = 1, 2$
are equivalent. Because of them, we have

$$\partial f(a, b)(h, k) = D_1 f(a, b)h + D_2 f(a, b)k$$

for $(a, b) \in X_1 \times X_2$ and $(h, k) \in E_1 \times E_2$ (see Theorem 5.4 and Proposition 2.8).

Proof The implication "(i)$\Rightarrow$(ii)" is clear.

"(ii)$\Rightarrow$(i)" Suppose $(a, b) \in X_1 \times X_2$ and

$$A(h, k) := D_1 f(a, b)h + D_2 f(a, b)k \quad \text{for } (h, k) \in E_1 \times E_2 .$$

One easily verifies that A belongs to $\mathcal{L}(E_1 \times E_2, F)$. The mean value theorem in integral form (Theorem 3.10) gives

$$
\begin{aligned}
f(a &+ h, b + k) - f(a, b) - A(h, k) \\
&= f(a + h, b + k) - f(a, b + k) + f(a, b + k) - f(a, b) - A(h, k) \\
&= \int_0^1 \big[D_1 f(a + th, b + k) - D_1 f(a, b)\big] h \, dt + f(a, b + k) - f(a, b) - D_2 f(a, b)k
\end{aligned}
$$

if $\max\{\|h\|, \|k\|\}$ is sufficiently small. From this, we get the estimate

$$\|f(a + h, b + k) - f(a, b) - A(h, k)\| \le \varphi(h, k) \max\{\|h\|, \|k\|\}$$

with

$$
\begin{aligned}
\varphi(h, k) := \ &\max_{0 \le t \le 1} \|D_1 f(a + th, b + k) - D_1 f(a, b)\| \\
&+ \frac{\|f(a, b + k) - f(a, b) - D_2 f(a, b)k\|}{\|k\|} .
\end{aligned}
$$

Now the continuity of $D_1 f$ gives $\varphi(h, k) \to 0$ for $(h, k) \to (0, 0)$. Thus we see that

$$f(a + h, b + k) - f(a, b) - A(h, k) = o(\|(h, k)\|) \quad \big((h, k) \to (0, 0)\big) .$$

Therefore f is differentiable in (a, b) with $\partial f(a, b) = A$. Finally, the regularity assumption on $D_j f$ and the definition of A implies that $\partial f \in C^{q-1}(X_1 \times X_2, F)$ and therefore that $f \in C^q(X_1 \times X_2, F)$. ∎

(c) In the special case $E_1 = \mathbb{R}^m$, $E_2 = \mathbb{R}^n$, and $F = \mathbb{R}^\ell$, we have

$$
[D_1 f] = \begin{bmatrix} \partial_1 f^1 & \cdots & \partial_m f^1 \\ \vdots & & \vdots \\ \partial_1 f^\ell & \cdots & \partial_m f^\ell \end{bmatrix}, \quad
[D_2 f] = \begin{bmatrix} \partial_{m+1} f^1 & \cdots & \partial_{m+n} f^1 \\ \vdots & & \vdots \\ \partial_{m+1} f^\ell & \cdots & \partial_{m+n} f^\ell \end{bmatrix}
$$

with $f = (f^1, \ldots, f^\ell)$.

Proof   This follows from (b) and Corollary 2.9. ∎

The implicit function theorem

With what we just learned, we can turn to the solvability of nonlinear equations. The result is fundamental.

8.2 Theorem (implicit function) *Suppose W is open in $E_1 \times E_2$ and $f \in C^q(W, F)$. Further, suppose $(x_0, y_0) \in W$ such that*

$$f(x_0, y_0) = 0 \quad \text{and} \quad D_2 f(x_0, y_0) \in \mathcal{L}\mathrm{is}(E_2, F) .$$

Then there are open neighborhoods $U \in \mathfrak{U}_W(x_0, y_0)$ and $V \in \mathfrak{U}_{E_1}(x_0)$, and also a unique $g \in C^q(V, E_2)$ such that

$$\big((x, y) \in U \text{ and } f(x, y) = 0 \big) \iff \big(x \in V \text{ and } y = g(x) \big) . \tag{8.1}$$

In addition,

$$\partial g(x) = -\big[D_2 f\big(x, g(x)\big) \big]^{-1} D_1 f\big(x, g(x)\big) \quad \text{for } x \in V . \tag{8.2}$$

Proof   (i) Let $A := D_2 f(x_0, y_0) \in \mathcal{L}\mathrm{is}(E_2, F)$ and $\widetilde{f} := A^{-1} f \in C^q(W, E_2)$. Then

$$\widetilde{f}(x_0, y_0) = 0 \quad \text{and} \quad D_2 \widetilde{f}(x_0, y_0) = I_{E_2} .$$

Thus because of Exercise 5.1 we lose no generality in considering the case $F = E_2$ and $D_2 f(x_0, y_0) = I_{E_2}$. Also, we can assume $W = W_1 \times W_2$ for open neighborhoods $W_1 \in \mathfrak{U}_{E_1}(x_0)$ and $W_2 \in \mathfrak{U}_{E_2}(y_0)$.

(ii) For the map

$$\varphi : W_1 \times W_2 \to E_1 \times E_2 , \quad (x, y) \mapsto \big(x, f(x, y) \big)$$

we have $\varphi \in C^q(W_1 \times W_2, E_1 \times E_2)$ with[1]

$$\partial \varphi(x_0, y_0) = \begin{bmatrix} I_{E_1} & 0 \\ D_1 f(x_0, y_0) & I_{E_2} \end{bmatrix} \in \mathcal{L}(E_1 \times E_2) .$$

We verify at once that

$$\begin{bmatrix} I_{E_1} & 0 \\ -D_1 f(x_0, y_0) & I_{E_2} \end{bmatrix} \in \mathcal{L}(E_1 \times E_2)$$

is the inverse of $\partial \varphi(x_0, y_0)$. Consequently, we have $\partial \varphi(x_0, y_0) \in \mathcal{L}\mathrm{aut}(E_1 \times E_2)$, and because $\varphi(x_0, y_0) = (x_0, 0)$, the inverse function theorem (Theorem 7.3) guarantees

---

[1] Here and in similar situations, we use the natural matrix notation.

the existence of open neighborhoods $U \in \mathfrak{U}_{E_1 \times E_2}(x_0, y_0)$ and $X \in \mathfrak{U}_{E_1 \times E_2}(x_0, 0)$ on which $\varphi | U \in \mathrm{Diff}^q(U, X)$. We set $\psi := (\varphi | U)^{-1} \in \mathrm{Diff}^q(X, U)$ and write ψ in the form

$$\psi(\xi, \eta) = \big(\psi_1(\xi, \eta), \psi_2(\xi, \eta)\big) \quad \text{for } (\xi, \eta) \in X .$$

Then $\psi_j \in C^q(X, E_2)$ for $j = 1, 2$, and the definition of φ shows

$$(\xi, \eta) = \varphi(\psi(\xi, \eta)) = \big(\psi_1(\xi, \eta), f(\psi_1(\xi, \eta), \psi_2(\xi, \eta))\big) \quad \text{for } (\xi, \eta) \in X .$$

Therefore, we recognize

$$\psi_1(\xi, \eta) = \xi \text{ and } \eta = f(\xi, \psi_2(\xi, \eta)) \quad \text{for } (\xi, \eta) \in X . \tag{8.3}$$

Furthermore, $V := \big\{ x \in E_1 \ ; \ (x, 0) \in X \big\}$ is an open neighborhood of x_0 in E_1, and for $g(x) := \psi_2(x, 0)$ with $x \in V$, we have $g \in C^q(V, E_2)$ and

$$\big(x, f(x, g(x))\big) = \big(\psi_1(x, 0), f(\psi_1(x, 0), \psi_2(x, 0))\big)$$
$$= \varphi(\psi_1(x, 0), \psi_2(x, 0)) = \varphi \circ \psi(x, 0) = (x, 0)$$

for $x \in V$. This together with (8.3) implies (8.1). The uniqueness of g is clear.

(iii) We set $h(x) := f(x, g(x))$ for $x \in V$. Then $h = 0$ and the chain rule together with Remark 8.1(b), imply

$$\partial h(x) = D_1 f(x, g(x)) I_{E_1} + D_2 f(x, g(x)) \partial g(x) = 0 \quad \text{for } x \in V . \tag{8.4}$$

From $q \geq 1$, it follows that

$$D_2 f \in C^{q-1}(U, \mathcal{L}(E_2)) \subset C(U, \mathcal{L}(E_2)) .$$

According to Proposition 7.2(i), $\mathcal{L}\mathrm{aut}(E_2)$ is open in $\mathcal{L}(E_2)$. Therefore

$$\big((D_2 f)^{-1}(\mathcal{L}\mathrm{aut}(E_2))\big) \cap U = \big\{ (x, y) \in U \ ; \ D_2 f(x, y) \in \mathcal{L}\mathrm{aut}(E_2) \big\}$$

is an open neighborhood of (x_0, y_0) in U. By making U smaller, we can therefore assume $D_2 f(x, y) \in \mathcal{L}\mathrm{aut}(E_2)$ for $(x, y) \in U$. Then from (8.4), we get (8.2). $\blacksquare$

8.3 Remark Theorem 8.2 says that near (x_0, y_0) the fiber $f^{-1}(0)$ is the graph of a C^q function. $\blacksquare$

By formulating the implicit function theorem in the special case $E_1 = \mathbb{R}^m$ and $E_2 = F = \mathbb{R}^n$, we get a statement about the "local solvability of nonlinear systems of equations depending on parameters".

8.4 Corollary   *Suppose W is open in $\mathbb{R}^{m+n}$ and $f \in C^q(W, \mathbb{R}^n)$. Also suppose $(a, b) \in W$ such that $f(a, b) = 0$, that is*

$$f^1(a^1, \ldots, a^m, b^1, \ldots, b^n) = 0 ,$$

$$\vdots$$

$$f^n(a^1, \ldots, a^m, b^1, \ldots, b^n) = 0 .$$

Then if

$$\frac{\partial(f^1, \ldots, f^n)}{\partial(x^{m+1}, \ldots, x^{m+n})}(a, b) := \det\left[\partial_{m+k} f^j(a, b)\right]_{1 \le j, k \le n} \ne 0 ,$$

there is an open neighborhood U of (a, b) in W, an open neighborhood V of a in $\mathbb{R}^m$, and a $g \in C^q(V, \mathbb{R}^n)$ such that

$$\big((x, y) \in U \text{ and } f(x, y) = 0 \big) \Longleftrightarrow \big(x \in V \text{ and } y = g(x) \big) .$$

That is, there is a neighborhood V of a in $\mathbb{R}^m$ such that the system of equations

$$f^1(x^1, \ldots, x^m, y^1, \ldots, y^n) = 0 ,$$

$$\vdots$$

$$f^n(x^1, \ldots, x^m, y^1, \ldots, y^n) = 0$$

has exactly one solution

$$y^1 = g^1(x^1, \ldots, x^m) ,$$

$$\vdots$$

$$y^n = g^n(x^1, \ldots, x^m)$$

near $b = (b^1, \ldots, b^n)$ for every m-tuple $(x^1, \ldots, x^m) \in V$. Also, the solutions $g^1, \ldots, g^n$ are C^q functions in the "parameters" $(x^1, \ldots, x^m)$ from V, and

$$[\partial g(x)] = - \begin{bmatrix} \partial_{m+1} f^1 & \cdots & \partial_{m+n} f^1 \\ \vdots & & \vdots \\ \partial_{m+1} f^n & \cdots & \partial_{m+n} f^n \end{bmatrix}^{-1} \begin{bmatrix} \partial_1 f^1 & \cdots & \partial_m f^1 \\ \vdots & & \vdots \\ \partial_1 f^n & \cdots & \partial_m f^n \end{bmatrix}$$

for $x \in V$, where the derivatives $\partial_k f^j$ are evaluated at $(x, g(x))$.

Proof   Because $D_2 f(x, y)$ belongs to $\mathcal{L}aut(\mathbb{R}^n)$ if and only if

$$\frac{\partial(f^1, \ldots, f^n)}{\partial(x^{m+1}, \ldots, x^{m+n})}(x, y) \ne 0 ,$$

we verify all the claims from Theorem 8.2 and Remark 8.1(c). ∎

Regular values

Suppose X is open in $\mathbb{R}^m$ and $f: X \to \mathbb{R}^n$ is differentiable. Then $x \in X$ is called a **regular point** of f if $\partial f(x) \in \mathcal{L}(\mathbb{R}^m, \mathbb{R}^n)$ is surjective. The map is called **regular** or a **submersion** if every point in X is regular. Finally, we say $y \in \mathbb{R}^n$ is a **regular value** of f if the fiber $f^{-1}(y)$ consists only of regular points.

8.5 Remarks   (a) If $m < n$, then f has no regular points.

Proof   This follows from $\operatorname{rank}(\partial f(x)) \leq m$. ∎

(b) If $n \leq m$, then $x \in X$ is a regular point of f if and only if $\partial f(x)$ has rank[2] n.

(c) If $n = 1$, then x is a regular point of f if and only if $\nabla f(x) \neq 0$.

(d) Every $y \in \mathbb{R}^n \setminus \operatorname{im}(f)$ is a regular value of f.

(e) Suppose $x_0 \in X$ is a regular point of f with $f(x_0) = 0$. Then there are n variables that uniquely solve the system of equations

$$f^1(x^1, \ldots, x^m) = 0 \,,$$

$$\vdots$$

$$f^n(x^1, \ldots, x^m) = 0$$

in a neighborhood of x_0 as functions of then $m - n$ other variables. When f is in the class C^q, the solutions are also C^q functions.

Proof   From (a), we have $m \geq n$. Also, we can make

$$\frac{\partial(f^1, \ldots, f^n)}{\partial(x^{m-n+1}, \ldots, x^m)}(x_0) \neq 0$$

by suitably permuting the coordinates in $\mathbb{R}^m$, that is, by applying an appropriate orthogonal transformation of $\mathbb{R}^m$. The remark then follows from Corollary 8.4. ∎

(f) Suppose $0 \in \operatorname{im}(f)$ is a regular value of $f \in C^q(X, \mathbb{R}^n)$. Then for every $x_0 \in f^{-1}(0)$, there is a neighborhood U in $\mathbb{R}^m$ such that $f^{-1}(0) \cap U$ is the graph of a C^q function of $m - n$ variables.

Proof   This follows from (e) and Remark 8.3. ∎

Ordinary differential equations

In Section 1, we used the exponential map, to help address existence and uniqueness questions for *linear* differential equations. We next treat differential equations of the form $\dot{x} = f(t, x)$, where now f may be nonlinear.

---

[2]Suppose E and F are finite-dimensional Banach spaces and $A \in \mathcal{L}(E, F)$. Then the **rank** of A is defined by $\operatorname{rank}(A) := \dim(\operatorname{im}(A))$. Obviously, $\operatorname{rank}(A)$ is same as the linear algebraic rank of the representation matrix $[A]_{\mathcal{E}, \mathcal{F}}$ in any bases $\mathcal{E}$ of E and $\mathcal{F}$ of F.

In the rest of this section, suppose

- J is an open interval in $\mathbb{R}$;
  E is a Banach space;
  D is an open subset of E;
  $f \in C(J \times D, E)$.

The function $u : J_u \to D$ is said to be the **solution of the differential equation**
$\dot{x} = f(t, x)$ in E if J_u is a perfect subinterval of J, $u \in C^1(J_u, D)$, and

$$\dot{u}(t) = f\big(t, u(t)\big) \quad \text{for } t \in J_u .$$

By giving $(t_0, x_0) \in J \times D$, the pair

$$\dot{x} = f(t, x) \text{ and } x(t_0) = x_0 \qquad (8.5)_{(t_0, x_0)}$$

becomes an **initial value problem** for $\dot{x} = f(t, x)$. The map $u : J_u \to D$ is a **solution**
to $(8.5)_{(t_0, x_0)}$ if u satisfies the differential equation $\dot{x} = f(t, x)$ and $u(t_0) = x_0$. It
is a **noncontinuable** (or **maximal**), if there is no solution $v : J_v \to D$ of $(8.5)_{(t_0, x_0)}$
such that $v \supset u$ and $v \neq u$. In this case J_u is a **maximal existence interval** for
$(8.5)_{(t_0, x_0)}$. When $J_u = J$, we call u a **global** solution of $(8.5)_{(t_0, x_0)}$.

8.6 Remarks   (a) Global solutions are noncontinuable.

(b) Suppose J_u is a perfect subinterval of J. Then $u \in C(J_u, D)$ is a solution of
$(8.5)_{(t_0, x_0)}$ if and only if u satisfies the integral equation

$$u(t) = x_0 + \int_{t_0}^{t} f\big(s, u(s)\big)\, ds \quad \text{for } t \in J_u .$$

Proof   Because $s \mapsto f\big(s, u(s)\big)$ is continuous, the remark follows easily from the fundamental theorem of calculus. ∎

(c) Suppose $A \in \mathcal{L}(E)$ and $g \in C(\mathbb{R}, E)$. Then the initial value problem

$$\dot{x} = Ax + g(t) \text{ and } x(t_0) = x_0$$

has the unique global solution

$$u(t) = e^{(t-t_0)A} x_0 + \int_{t_0}^{t} e^{(t-s)A} g(s)\, ds \quad \text{for } t \in \mathbb{R} .$$

Proof   This is a consequence of Theorem 1.17. ∎

(d) When $E = \mathbb{R}$, one says that $\dot{x} = f(t, x)$ is a **scalar** differential equation.

(e) The initial value problem

$$\dot{x} = x^2 \text{ and } x(t_0) = x_0 \qquad (8.6)$$

has at most one solution for $(t_0, x_0) \in \mathbb{R}^2$.

Proof Suppose $u \in C^1(J_u, \mathbb{R})$ and $v \in C^1(J_v, \mathbb{R})$ are solutions of (8.6). For $t \in J_u \cap J_v$, let $w(t) := u(t) - v(t)$, and let I be a compact subinterval of $J_u \cap J_v$ with $t_0 \in I$. Then it suffices to verify that w vanishes on I. From (b), we have

$$w(t) = u(t) - v(t) = \int_{t_0}^t \left(u^2(s) - v^2(s) \right) ds = \int_{t_0}^t \left(u(s) + v(s) \right) w(s) \, ds \quad \text{for } t \in I \ .$$

Letting $\alpha := \max_{s \in I} |u(s) + v(s)|$, it follows that

$$|w(t)| \leq \alpha \left| \int_{t_0}^t |w(s)| \, ds \right| \quad \text{for } t \in I \ ,$$

and the claim is implied by Gronwall's lemma. ∎

(f) The initial value problem (8.6) has no global solution for $x_0 \neq 0$.

Proof Suppose $u \in C^1(J_u, \mathbb{R})$ is a solution of (8.6) with $x_0 \neq 0$. Because 0 is the unique global solution for the initial value $(t_0, 0)$, it follows from (e) that $u(t) \neq 0$ for $t \in J_u$. Therefore (8.6) implies

$$\frac{\dot{u}}{u^2} = \left(-\frac{1}{u} \right)^{\cdot} = 1$$

on J_u. By integration, we find

$$\frac{1}{x_0} - \frac{1}{u(t)} = t - t_0 \quad \text{for } t \in J_u \ ,$$

and therefore

$$u(t) = 1/(t_0 - t + 1/x_0) \quad \text{for } t \in J_u \ .$$

Thus we get $J_u \neq J = \mathbb{R}$. ∎

(g) The initial value problem (8.6) has, for every $(t_0, x_0) \in \mathbb{R}^2$, the unique non-continuable solution $u(\cdot, t_0, x_0) \in C^1(J(t_0, x_0), \mathbb{R})$ with

$$J(t_0, x_0) = \begin{cases} (t_0 + 1/x_0, \infty) \ , & x_0 < 0 \ , \\ \mathbb{R} \ , & x_0 = 0 \ , \\ (-\infty, t_0 + 1/x_0) \ , & x_0 > 0 \ , \end{cases}$$

and

$$u(t, t_0, x_0) = \begin{cases} 0 \ , & x_0 = 0 \ , \\ 1/(t_0 - t + 1/x_0) \ , & x_0 \neq 0 \ . \end{cases}$$

Proof This follows from (e) and the proof of (f). ∎

(h) The scalar initial value problem

$$\dot{x} = 2\sqrt{|x|} \ , \quad x(0) = 0 \tag{8.7}$$

has uncountably many global solutions.

Proof   Clearly $u_0 \equiv 0$ is a global solution.
For $\alpha < 0 < \beta$, let

$$u_{\alpha,\beta}(t) := \begin{cases} -(t-\alpha)^2 \,, & t \in (-\infty, \alpha] \,, \\ 0 \,, & t \in (\alpha, \beta) \,, \\ (t-\beta)^2 \,, & t \in [\beta, \infty) \,. \end{cases}$$

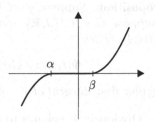

One verifies easily that $u_{\alpha,\beta}$ is a global solution of (8.7). ∎

Separation of variables

It is generally not possible to solve a given differential equation "explicitly". For certain scalar differential equations, though, a unique local solution to an initial value problem can be given with the help of the implicit function theorem.

The function $\Phi \colon J \times D \to \mathbb{R}$ is said to a **first integral** of $\dot{x} = f(t,x)$ if, for every solution $u \colon J_u \to D$ of $\dot{x} = f(t,x)$, the map

$$J_u \to \mathbb{R} \,, \quad t \mapsto \Phi\big(t, u(t)\big)$$

is constant. If in addition

$$\Phi \in C^1(J \times D, \mathbb{R}) \quad \text{and} \quad \partial_2 \Phi(t,x) \neq 0 \quad \text{for } (t,x) \in J \times D \,,$$

then Φ is a **regular first integral**.

8.7 Proposition   *Suppose Φ is a regular first integral of the scalar differential equation $\dot{x} = f(t,x)$ and*

$$f(t,x) = -\partial_1 \Phi(t,x) / \partial_2 \Phi(t,x) \quad \text{for } (t,x) \in J \times D \,.$$

Then for every $(t_0, x_0) \in J \times D$, there is an open interval I of J and U from D with $(t_0, x_0) \in I \times U$ such that the initial value problem

$$\dot{x} = f(t,x) \,, \quad x(t_0) = x_0$$

on I has exactly one solution u with $u(I) \subset U$. It is achieved by solving the equation $\Phi(t,x) = \Phi(t_0, x_0)$ for x.

Proof   This follows immediately from Theorem 8.2. ∎

By suitably choosing Φ, we get an important class of scalar differential equations with "separated variables". As the next corollary shows, we can solve these equations by quadrature.

8.8 Proposition *Suppose $g \in C(J, \mathbb{R})$ and $h \in C(D, \mathbb{R})$ with $h(x) \neq 0$ for $x \in D$. Also suppose $G \in C^1(J, \mathbb{R})$ and $H \in C^1(D, \mathbb{R})$ are antiderivatives of g and h, respectively. Then*

$$\Phi(t, x) := G(t) - H(x) \quad \text{for } (t, x) \in J \times D ,$$

is a regular first integral of $\dot{x} = g(t)/h(x)$.

Proof Obviously Φ belongs to $C^1(J \times D, \mathbb{R})$, and $\partial_2 \Phi(t, x) = -h(x) \neq 0$. Now suppose $u \in C^1(J_u, D)$ is a solution of $\dot{x} = g(t)/h(x)$. Then it follows from the chain rule that

$$\left[\Phi(t, u(t)) \right]^{\cdot} = \partial_1 \Phi(t, u(t)) + \partial_2 \Phi(t, u(t)) \dot{u}(t)$$

$$= g(t) - h(u(t)) \cdot \frac{g(t)}{h(u(t))} = 0 .$$

Therefore $t \mapsto \Phi(t, u(t))$ is constant on J_u. ∎

8.9 Corollary (separation of variables) *Suppose $g \in C(J, \mathbb{R})$ and $h \in C(D, \mathbb{R})$ with $h(x) \neq 0$ for $x \in D$. Then the initial value problem*

$$\dot{x} = g(t)/h(x) , \quad x(t_0) = x_0$$

has a unique local solution for every $(t_0, x_0) \in J \times D$. That solution can be obtained by solving the equation

$$\int_{x_0}^{x} h(\xi) \, d\xi = \int_{t_0}^{t} g(\tau) \, d\tau \tag{8.8}$$

for x.

Proof This is an immediate consequence of Theorems 8.7 and 8.8. ∎

8.10 Remark Under the assumptions of Corollary 8.9,

$$\frac{dx}{dt} = \frac{g(t)}{h(x)} , \quad x(t_0) = x_0$$

implies formally $h(x) \, dx = g(t) \, dt$, the "equation with separated variables". Formal integration then gives (8.8). ∎

8.11 Examples (a) We consider the initial value problem

$$\dot{x} = 1 + x^2 , \quad x(t_0) = x_0 . \tag{8.9}$$

Letting $g := 1$ and $h := 1/(1 + X^2)$, it follows from Corollary 8.9 that

$$\arctan x - \arctan x_0 = \int_{x_0}^{x} \frac{d\xi}{1 + \xi^2} = \int_{t_0}^{t} d\tau = t - t_0 .$$

Therefore
$$x(t) = \tan(t - \alpha) \quad \text{for } t \in (\alpha - \pi/2, \alpha + \pi/2)$$

and $\alpha := t_0 - \arctan x_0$ is the unique maximal solution of (8.9). In particular, (8.9) has no global solutions.

(b) Suppose $\alpha > 0$, $D := (0, \infty)$ and $x_0 > 0$. For

$$\dot{x} = x^{1+\alpha} , \quad x(0) = x_0 , \tag{8.10}$$

Corollary 8.9 implies

$$\int_{x_0}^{x} \xi^{-1-\alpha} \, d\xi = t .$$

Therefore
$$x(t) = (x_0^{-\alpha} - \alpha t)^{-1/\alpha} \quad \text{for } -\infty < t < x_0^{-\alpha}/\alpha$$

is the unique maximal solution of (8.10), which is therefore not globally solvable.

(c) On $D := (1, \infty)$, we consider the initial value problem

$$\dot{x} = x \log x , \quad x(0) = x_0 . \tag{8.11}$$

Because $\log \circ \log$ is an antiderivative on D of $x \mapsto 1/(x \log x)$, we get from Corollary 8.9 that

$$\log(\log x) - \log(\log x_0) = \int_{x_0}^{x} \frac{d\xi}{\xi \log \xi} = \int_0^t d\tau = t .$$

From this we derive that
$$x(t) = x_0^{(e^t)} \quad \text{for } t \in \mathbb{R} ,$$

is the unique global solution of (8.11).

(d) Let $-\infty < a < b < \infty$ and $f \in C\big((a, \infty), \mathbb{R}\big)$ such that $f(x) > 0$ for $x \in (a, b)$ and $f(b) = 0$. For $x_0 \in (a, b)$ and $t_0 \in \mathbb{R}$, we consider the initial value problem

$$\dot{x} = f(x) , \quad x(t_0) = x_0 .$$

According to Corollary 8.9, we get the unique local solution $u \colon J_u \to (a, b)$ by solving
$$t = t_0 + \int_{x_0}^{x} \frac{d\xi}{f(\xi)} =: H(x)$$

for x. Because $H' = 1/f$, H is strictly increasing. Therefore $u = H^{-1}$, and J_u agrees with $H\big((a, b)\big)$. Also

$$T^* := \lim_{x \to b-0} H(x) = \sup J_u$$

exists in $\overline{\mathbb{R}}$, and

$$T^* < \infty \iff \int_{x_0}^{b} \frac{d\xi}{f(\xi)} < \infty .$$

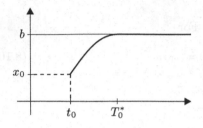
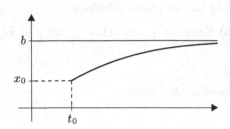

Now $\lim_{t \to T^* - 0} u(t) = b$ implies

$$\lim_{t \to T^* - 0} \dot{u}(t) = \lim_{t \to T^* - 0} f\big(u(t)\big) = 0 .$$

In the case $T^* < \infty$, it then follows from Proposition IV.1.16 that

$$v(t) := \begin{cases} u(t) , & t \in J_u , \\ b , & t \in [T^*, \infty) , \end{cases}$$

gives a continuation of u. Analogous considerations for

$$T_* := \lim_{x \to a + 0} H(x) = \inf J_u$$

are left to you.

(e) For $a \in C(J, \mathbb{R})$ and $(t_0, x_0) \in J \times \mathbb{R}$, we consider the initial value problem

$$\dot{x} = a(t)x , \quad x(t_0) = x_0 \tag{8.12}$$

for the **scalar linear homogeneous differential equation** $\dot{x} = a(t)x$ with **time-dependent coefficients** a. Therefore (8.12) *has the unique global solution*

$$x(t) = x_0 e^{\int_{t_0}^{t} a(\tau)\, d\tau} \quad \text{for } t \in J .$$

Proof From Gronwall's lemma, it follows that (8.12) has at most one solution. If $u \in C^1(J_u, \mathbb{R})$ is a solution and there is a $t_1 \in J_u$ such that $u(t_1) = 0$, then the uniqueness result for the initial value problem $\dot{x} = a(t)x$, $x(t_1) = 0$ implies that $u = 0$.

Suppose therefore $x_0 \neq 0$. Then we get by separation of variables that

$$\log|x| - \log|x_0| = \int_{x_0}^{x} \frac{d\xi}{\xi} = \int_{t_0}^{t} a(\tau)\, d\tau$$

and thus

$$|x(t)| = |x_0|\, e^{\int_{t_0}^{t} a(\tau)\, d\tau} \quad \text{for } t \in J .$$

Because $x(t) \neq 0$ for $t \in J$, the claim follows. ∎

Lipschitz continuity and uniqueness

Remark 8.6(h) shows that the initial value problem $(8.5)_{(t_0,x_0)}$ is generally not uniquely solvable. We will see next that we can guarantee uniqueness if we require that f is somewhat regular than merely continuous.

Suppose E and F are Banach spaces, $X \subset E$ and $I \subset \mathbb{R}$. Then $f \in C(I \times X, F)$ is said to be **locally Lipschitz continuous** in $x \in X$, if every (t_0, x_0) in $I \times X$ has a neighborhood $U \times V$ such that

$$\| f(t,x) - f(t,y) \| \le L \, \| x - y \| \quad \text{for } t \in U \text{ and } x,y \in V$$

for some $L \ge 0$. We set

$$C^{0,1^-}(I \times X, F) := \left\{ f \in C(I \times X, F) \, ; \, f \text{ is locally Lipschitz continuous in } x \in X \right\} .$$

If I is a single point, and thus f has the form $f \colon X \to F$, we call f **locally Lipschitzcontinuous**, and

$$C^{1^-}(X, F) := \left\{ f \colon X \to F \, ; \, f \text{ is locally Lipschitz continuous} \right\} .$$

8.12 Remarks (a) Obviously, we have $C^{0,1^-}(I \times X, F) \subset C(I \times X, F)$.

(b) Suppose X is open in E, and $f \in C(I \times X, F)$. Assume $\partial_2 f$ exists and belongs to $C\big(I \times X, \mathcal{L}(E, F)\big)$. Then $f \in C^{0,1^-}(I \times X, F)$.

Proof Let $(t_0, x_0) \in I \times X$. Because $\partial_2 f$ belongs to $C\big(I \times X, \mathcal{L}(E, F)\big)$ and X open is in E, there is an $\varepsilon > 0$ such that $\mathbb{B}(x_0, \varepsilon) \subset X$ and

$$\| \partial_2 f(t_0, x_0) - \partial_2 f(t, x) \| \le 1 \quad \text{for } (t,x) \in U \times V , \tag{8.13}$$

where we set $U := (t_0 - \varepsilon, t_0 + \varepsilon) \cap I$ and $V := \mathbb{B}(x_0, \varepsilon)$. Putting $L := 1 + \| \partial_2 f(t_0, x_0) \|$, it follows from the mean value theorem (Theorem 3.9) and (8.13) that

$$\| f(t,x) - f(t,y) \| \le \sup_{0 \le s \le 1} \| \partial_2 f\big(t, x + s(y - x)\big) \| \, \| x - y \| \le L \, \| x - y \|$$

for $(t,x), (t,y) \in U \times V$. Therefore f is locally Lipschitz continuous in x. ∎

(c) Polynomials of degree ≥ 2 are locally Lipschitz continuous but not Lipschitz continuous.

(d) Suppose $I \times X$ is compact and $f \in C^{0,1^-}(I \times X, F)$. Then f is **uniformly Lipschitz continuous** in $x \in X$, that is, there is an $L \ge 0$ such that

$$\| f(t,x) - f(t,y) \| \le L \, \| x - y \| \quad \text{for } x,y \in X \text{ and } t \in I .$$

Proof For every $(t,x) \in I \times X$, there are $\varepsilon_t > 0$, $\varepsilon_x > 0$, and $L(t,x) \ge 0$ such that

$$\| f(s,y) - f(s,z) \| \le L(t,x) \, \| y - z \| \quad \text{for } (s,y), (s,z) \in \mathbb{B}(t, \varepsilon_t) \times \mathbb{B}(x, \varepsilon_x) .$$

Because $I \times X$ is compact, there exist $(t_0, x_0), \ldots, (t_m, x_m) \in I \times X$ such that

$$I \times X \subset \bigcup_{j=0}^{m} \mathbb{B}(t_j, \varepsilon_{t_j}) \times \mathbb{B}(x_j, \varepsilon_{x_j}/2) \ .$$

Because $f(I \times X)$ is compact, there is a $R > 0$ such that $f(I \times X) \subset R\mathbb{B}$. We set

$$\delta := \min\{\varepsilon_{x_0}, \ldots, \varepsilon_{x_m}\}/2 > 0$$

and

$$L := \max\{L(t_0, x_0), \ldots, L(t_m, x_m), 2R/\delta\} > 0 \ .$$

Suppose now $(t, x), (t, y) \in I \times X$. Then there is an $k \in \{0, \ldots, m\}$ such that

$$(t, x) \in \mathbb{B}(t_k, \varepsilon_{t_k}) \times \mathbb{B}(x_k, \varepsilon_{x_k}/2) \ .$$

If $\|x - y\| < \delta$, we find (t, y) lies in $\mathbb{B}(t_k, \varepsilon_{t_k}) \times \mathbb{B}(x_k, \varepsilon_{x_k})$, and we get

$$\|f(t, x) - f(t, y)\| \leq L(t_k, x_k) \|x - y\| \leq L \|x - y\| \ .$$

On the other hand, if $\|x - y\| \geq \delta$, it follows that

$$\|f(t, x) - f(t, y)\| \leq \frac{2R}{\delta} \cdot \delta \leq L \|x - y\| \ .$$

Therefore f is uniformly Lipschitz continuous in $x \in X$. ∎

(e) Suppose X is compact in E and $f \in C^{1-}(X, F)$. Then f is Lipschitz continuous.
Proof This is a special case of (d). ∎

8.13 Theorem Let $f \in C^{0,1-}(J \times D, E)$, and suppose $u \colon J_u \to D$ and $v \colon J_v \to D$ are solutions of $\dot{x} = f(t, x)$ with $u(t_0) = v(t_0)$ for some $t_0 \in J_u \cap J_v$. Then $u(t) = v(t)$ for every $t \in J_u \cap J_v$.

Proof Suppose $I \subset J_u \cap J_v$ is a compact interval with $t_0 \in I$ and $w := u - v$. It suffices to verify that $w \,|\, I = 0$.

Because $K := u(I) \cup v(I) \subset D$ is compact, Remark 8.12(d) guarantees the existence of an $L \geq 0$ such that

$$\big\| f\big(s, u(s)\big) - f\big(s, v(s)\big) \big\| \leq L \|u(s) - v(s)\| = L \|w(s)\| \quad \text{for } s \in I \ .$$

Therefore from

$$w(t) = u(t) - v(t) = \int_{t_0}^{t} \big(f\big(s, u(s)\big) - f\big(s, v(s)\big) \big) \, ds \quad \text{for } t \in J_u \cap J_v \ ,$$

(see Remark 8.6(b)) the inequality

$$\|w(t)\| \leq L \left| \int_{t_0}^{t} \|w(s)\| \, ds \right| \quad \text{for } t \in I$$

follows, and Gronwall's lemma implies $w \,|\, I = 0$. ∎

The Picard–Lindelöf theorem

In the following, suppose

- $J \subset \mathbb{R}$ is an open interval;
  E is a finite-dimensional Banach space;
  D is open in E;
  $f \in C^{0,1-}(J \times D, E)$.

We now prove the fundamental local existence and uniqueness theorem for of ordinary differential equations.

8.14 Theorem (Picard–Lindelöf)   *Suppose* $(t_0, x_0) \in J \times D$. *Then there is an* $\alpha > 0$ *such that the initial value problem*

$$\dot{x} = f(t,x) , \quad x(t_0) = x_0 \tag{8.14}$$

has a unique solution on $I := [t_0 - \alpha, t_0 + \alpha]$.

Proof   (i) Because $J \times D \subset \mathbb{R} \times E$ is open, there are $a, b > 0$ such that

$$R := [t_0 - a, t_0 + a] \times \overline{\mathbb{B}}(x_0, b) \subset J \times D .$$

From the local Lipschitz continuity of f in x, we find an $L > 0$ such that

$$\|f(t,x) - f(t,y)\| \leq L \|x - y\| \quad \text{for } (t,x), (t,y) \in R .$$

Finally, it follows from the compactness of R and Remark 8.12(a) that there is an $M > 0$ such that

$$\|f(t,x)\| \leq M \quad \text{for } (t,x) \in R .$$

(ii) Suppose $\alpha := \min\{a, b/M, 1/(2L)\} > 0$ and $I := [t_0 - \alpha, t_0 + \alpha]$. We set

$$T(y)(t) := x_0 + \int_{t_0}^{t} f(\tau, y(\tau)) \, d\tau \quad \text{for } y \in C(I, D) , \quad t \in I .$$

If we can show that the map $T \colon C(I,D) \to C(I,E)$ has exactly one fixed point u, it will follows from Remark 8.6(b) that u is the unique solution of (8.14).

(iii) Let

$$X := \left\{ y \in C(I, E) \; ; \; y(t_0) = x_0, \ \max_{t \in I} \|y(t) - x_0\| \leq b \right\} .$$

Then X is a closed subset of the Banach space $C(I, E)$ and therefore a complete metric space (see Exercise II.6.4).

For $y \in X$, we have $y(t) \in \overline{\mathbb{B}}(x_0, b) \subset D$ for $t \in I$. Therefore $T(y)$ is defined, and $T(y)(t_0) = x_0$. Also

$$\|T(y)(t) - x_0\| = \left\| \int_{t_0}^{t} f(\tau, y(\tau)) \, d\tau \right\| \leq \alpha \sup_{(t,\xi) \in R} \|f(t,\xi)\| \leq \frac{b}{M} \cdot M = b .$$

This shows that T maps the space X into itself.

(iv) For $y, z \in X$, we find

$$\|T(y)(t) - T(z)(t)\| = \left\| \int_{t_0}^{t} \Big(f(\tau, y(\tau)) - f(\tau, z(\tau)) \Big) \, d\tau \right\|$$

$$\leq \alpha \max_{t \in I} \| f(t, y(t)) - f(t, z(t)) \| \leq \alpha L \max_{t \in I} \| y(t) - z(t) \|$$

for $t \in I$. Using the definition of α, it follows that

$$\|T(y) - T(z)\|_{C(I,E)} \leq \frac{1}{2} \| y - z \|_{C(I,E)} \; .$$

Therefore $T \colon X \to X$ is a contraction. Then the contraction theorem (Theorem IV.4.3) gives a unique fixed point u in X. ∎

8.15 Remarks (a) The solution of (8.14) on I can be calculated using the method of successive approximation (or iteratively) by putting

$$u_{m+1}(t) := x_0 + \int_{t_0}^{t} f(\tau, u_m(\tau)) \, d\tau \quad \text{for } m \in \mathbb{N} \text{ and } t \in I \; ,$$

with $u_0(t) = x_0$ for $t \in I$. The sequence (u_m) converges uniformly on I to u, and we have the error estimate

$$\|u_m - u\|_{C(I,E)} \leq \alpha M / 2^{m-1} \quad \text{for } m \in \mathbb{N}^{\times} \; . \tag{8.15}$$

Proof The first statement follows immediately from the last proof and from Theorem IV.4.3(ii). Statement (iii) of that theorem gives also the error estimate

$$\|u_m - u\|_{C(I,E)} \leq 2^{-m+1} \|u_1 - u_0\|_{C(I,E)} \; .$$

The claim now follows because for $t \in I$ we have

$$\|u_1(t) - u_0(t)\| = \left\| \int_{t_0}^{t} f(\tau, u_0(\tau)) \, d\tau \right\| \leq \alpha M \; . ∎$$

(b) The error estimate (8.15) be improved to

$$\|u_m - u\|_{C(I,E)} < \frac{M \alpha \sqrt{e}}{2^m (m+1)!} \quad \text{for } m \in \mathbb{N} \; ,$$

(see Exercise 11).

(c) The assumption that E is finite-dimensional was only used to prove the existence of the bound M on $f(R)$. Therefore Theorem 8.14 and its proof stays correct if the assumption that "E is finite-dimensional" is replaced with "f is bounded on bounded sets".

(d) Although the continuity of f is not enough to guarantee the uniqueness of the solution (8.14), it is enough to prove the existence of a solution (see [Ama95, Theorem II.7.3]). ∎

Finally we show that the local solution of Theorem 8.14 can be extended to a unique noncontinuable solution.

8.16 Theorem *For every $(t_0, x_0) \in J \times D$, there is exactly one noncontinuable solution $u(\cdot, t_0, x_0) \colon J(t_0, x_0) \to D$ of the initial value problem*

$$\dot{x} = f(t, x) , \quad x(t_0) = x_0 .$$

The maximal existence interval is open, that is, $J(t_0, x_0) = \big(t^-(t_0, x_0), t^+(t_0, x_0)\big)$.

Proof Suppose $(t_0, x_0) \in J \times D$. According to Theorem 8.14, there is an α_0 and a unique solution u of $(8.5)_{(t_0, x_0)}$ on $I_0 := [t_0 - \alpha_0, t_0 + \alpha_0]$. We set $x_1 := u(t_0 + \alpha_0)$ and $t_1 := t_0 + \alpha_0$ and apply Theorem 8.14 to the initial value problem $(8.5)_{(t_1, x_1)}$. Then there is an $\alpha_1 > 0$ and a unique solution v of $(8.5)_{(t_1, x_1)}$ on $I_1 := [t_1 - \alpha_1, t_1 + \alpha_1]$. Theorem 8.13 shows that $u(t) = v(t)$ for $t \in I_0 \cap I_1$. Therefore

$$u_1(t) := \begin{cases} u(t) , & t \in I_0 , \\ v(t) , & t \in I_1 , \end{cases}$$

is a solution of $(8.5)_{(t_0, x_0)}$ on $I_0 \cup I_1$. An analogous argument shows that u can be continued to the left past $t_0 - \alpha_0$. Suppose now

$$t^+ := t^+(t_0, x_0) := \sup\big\{ \beta \in \mathbb{R} ; \ (8.5)_{(t_0, x_0)} \text{ has a solution on } [t_0, \beta] \big\}$$

and

$$t^- := t^-(t_0, x_0) := \inf\big\{ \gamma \in \mathbb{R} ; \ (8.5)_{(t_0, x_0)} \text{ has a solution on } [\gamma, t_0] \big\} .$$

The considerations above show that $t^+ \in (t_0, \infty]$ and $t^- \in [-\infty, t_0)$ are well defined and that $(8.5)_{(t_0, x_0)}$ has a noncontinuable solution u on (t^-, t^+). Now it follows from Theorem 8.13 that u is unique. ∎

8.17 Examples **(a)** (ordinary differential equations of m-th order) Suppose $E := \mathbb{R}^m$ and $g \in C^{0,1-}(J \times D, \mathbb{R})$. Then

$$x^{(m)} = g(t, x, \dot{x}, \dots, x^{(m-1)}) \tag{8.16}$$

is an **m-th order ordinary differential equation**. The function $u \colon J_u \to \mathbb{R}$ is a **solution** of (8.14) if $J_u \subset J$ is a perfect interval, u belongs to $C^m(J_u, \mathbb{R})$,

$$\big(u(t), \dot{u}(t), \dots, u^{(m-1)}(t)\big) \in D \quad \text{for } t \in J_u ,$$

and

$$u^{(m)}(t) = g\big(t, u(t), \dot{u}(t), \dots, u^{(m-1)}(t)\big) \quad \text{for } t \in J_u .$$

For $t_0 \in J$ and $x_0 := (x_0^0, \ldots, x_0^{m-1}) \in D$, the pair

$$x^{(m)} = g\big(t, x, \dot{x}, \ldots, x^{(m-1)}\big)$$
$$x(t_0) = x_0^0, \ldots, \ x^{(m-1)}(t_0) = x_0^{m-1} \qquad\qquad (8.17)_{(t_0, x_0)}$$

is called an **initial value problem** for (8.16) with **initial value** x_0 and **initial time** t_0.

In this language we have, *For every $(t_0, x_0) \in J \times D$, there is a unique maximal solution $u\colon J(t_0, x_0) \to \mathbb{R}$ of $(8.17)_{(t_0, x_0)}$. The maximal existence interval $J(t_0, x_0)$ is open.*

Proof We define $f\colon J \times D \to \mathbb{R}^m$ by

$$f(t, y) := \big(y^2, y^3, \ldots, y^m, g(t, y)\big) \quad \text{for } t \in J \text{ and } y = (y^1, \ldots, y^m) \in D . \qquad (8.18)$$

Then f belongs to $C^{0,1^-}(J \times D, \mathbb{R}^m)$. Using Theorem 8.16, there is a unique noncontinuable solution $z\colon J(t_0, x_0) \to \mathbb{R}^m$ of $(8.5)_{(t_0, x_0)}$. One can then verify easily that $u := \mathrm{pr}_1 \circ z\colon J(t_0, x_0) \to \mathbb{R}$ is a solution of $(8.17)_{(t_0, x_0)}$.

Conversely, if $v\colon J(t_0, x_0) \to \mathbb{R}$ is a solution of $(8.17)_{(t_0, x_0)}$, then the vector function $(v, \dot{v}, \ldots, v^{(m-1)})$ solves the initial value problem $(8.5)_{(t_0, x_0)}$ on $J(t_0, x_0)$, where f is defined through (8.18). From this follows $v = u$. An analogous argument shows that u is noncontinuable. ∎

(b) (Newton's equation of motion in one dimension) Suppose V is open in $\mathbb{R}$ and $f \in C^1(V, \mathbb{R})$. For some $x_0 \in V$, suppose

$$U(x) := \int_{x_0}^x f(\xi)\, d\xi \text{ for } x \in V \quad \text{and} \quad T(y) := y^2/2 \text{ for } y \in \mathbb{R} .$$

Finally let $D := V \times \mathbb{R}$ and $L := T - U$. According to Example 6.14(a), the differential equation

$$-\ddot{x} = f(x) \qquad\qquad (8.19)$$

is Newton's equation of motion for the (one-dimensional) motion of a massive particle acted on by the conservative force $-\nabla U = f$. From (a), we know that (8.19) is equivalent to the system

$$\dot{x} = y , \quad \dot{y} = -f(x) \qquad\qquad (8.20)$$

and therefore to a first order differential equation

$$\dot{u} = F(u) , \qquad\qquad (8.21)$$

where $F(u) := \big(y, -f(x)\big)$ for $u = (x, y) \in D$.

The function $E := T + U\colon D \to \mathbb{R}$, called the **total energy**, is a first integral of (8.21). This means that the motion of the particle **conserves energy**[3], that is, every solution $x \in C^2(J, V)$ of (8.21) satisfies $E\big(x(t), \dot{x}(t)\big) = E\big(x(t_0), \dot{x}(t_0)\big)$ for every $t \in J$ and $t_0 \in J$.

---

[3]See Exercise 6.12.

Proof Obviously E belongs to $C^1(D, \mathbb{R})$. Every solution $u : J_u \to D$ of (8.20) has

$$[E(u(t))]^{\cdot} = [y^2(t)/2 + U(x(t))]^{\cdot} = y(t)\dot{y}(t) + f(x(t))\dot{x}(t) = 0 \quad \text{for } t \in J_u \;.$$

Therefore $E(u)$ is constant. ∎

(c) From (b), every solution of (8.20) lies in a level set $E^{-1}(c)$ of the total energy.

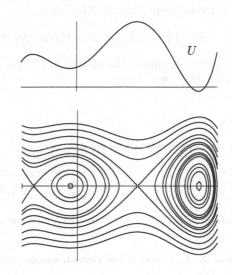

Conversely, the existence statement (a) means that for every point (x_0, y_0) on a level set of E, there is a solution of (8.20) with $u(0) = (x_0, y_0)$. Therefore, the set of level sets of E is the same as the set of all (maximal) solution curves of (8.20); this set is called the **phase portrait** of (8.19).

The phase portrait has these properties:

(i) The critical points of E are exactly the points $(x, 0) \in D$ with $f(x) = 0$. This means that the stationary points of (8.20) lie on the x-axis and are exactly the critical points of the potential U.

(ii) Every level set is symmetric about the x-axis.

(iii) If c is a regular value of E, then $E^{-1}(c)$ admits a local representation as the graph of a C^2 function. More precisely, for every $u_0 \in E^{-1}(c)$, there are positive numbers b and ε and a $\varphi \in C^2\big((-\varepsilon, \varepsilon), \mathbb{R}\big)$ with $\varphi(0) = 0$ such that

$$E^{-1}(c) \cap \mathbb{B}(u_0, b) = \operatorname{tr}(g) \;, \tag{8.22}$$

where $g \in C^2\big((-\varepsilon, \varepsilon), \mathbb{R}^2\big)$ is defined by

$$g(s) := \begin{cases} (s, \varphi(s)) + u_0 \;, & \partial_2 E(u_0) \neq 0 \;, \\ (\varphi(s), s) + u_0 \;, & \partial_1 E(u_0) \neq 0 \;. \end{cases}$$

(iv) Suppose $(x_0, 0)$ is a regular point of E and $c := U(x_0)$. Then the level set $E^{-1}(c)$ slices the x-axis orthogonally.

Proof (i) This follows from $\nabla E(x, y) = (y, f(x))$.

(ii) This statement is a consequence of $E(x, -y) = E(x, y)$ for $(x, y) \in D$.

(iii) This follows from Remarks 8.4(e) and (f).

(iv) We apply the notation of (iii). Because

$$\nabla E(x_0, 0) = \big(f(x_0), 0\big) \neq (0, 0) \ ,$$

we see (8.22) is satisfied with $u_0 := (x_0, 0)$ and $g(s) := \big(\varphi(s) + x_0, s\big)$ for $s \in (-\varepsilon, \varepsilon)$. Further, it follows from Example 3.6(c) that

$$0 = \big(\nabla E(x_0, 0) \,\big|\, \dot{g}(0)\big) = \big((f(x_0), 0) \,\big|\, (\dot{\varphi}(0), 1)\big) = f(x_0)\dot{\varphi}(0) \ .$$

Therefore $\dot{\varphi}(0) = 0$, that is, the tangent to $E^{-1}(c)$ at $(x_0, 0)$ is parallel to the y-axis (see Remark IV.1.4(a)). ∎

Exercises

1 Show that the system of equations

$$x^2 + y^2 - u^2 - v = 0 \ ,$$
$$x^2 + 2y^2 + 3u^2 + 4v^2 = 1$$

can be solved near $(1/2, 0, 1/2, 0)$ for (u, v). What are the first derivatives of u and v in (x, y)?

2 Suppose E, F, G and H are Banach spaces and X is open in H. Further suppose

$$A \in C^1\big(X, \mathcal{L}(E, F)\big) \text{ and } B \in C^1\big(X, \mathcal{L}(E, G)\big)$$

and

$$\big(A(x), B(x)\big) \in \mathcal{L}\mathrm{is}(E, F \times G) \quad \text{for } x \in X \ .$$

Finally, suppose $(f, g) \in F \times G$ and

$$\varphi: X \to E \ , \quad x \mapsto \big(A(x), B(x)\big)^{-1}(f, g) \ .$$

Show that

$$\partial\varphi(x)h = -S(x)\partial A(x)\big[h, \varphi(x)\big] - T(x)\partial B(x)\big[h, \varphi(x)\big] \quad \text{for } (x, h) \in X \times H \ ,$$

where, for $x \in X$,

$$S(x) := \big(A(x), B(x)\big)^{-1}\big|\,(F \times \{0\}) \quad \text{and} \quad T(x) := \big(A(x), B(x)\big)^{-1}\big|\,(\{0\} \times G) \ .$$

3 Determine the general solution of the **scalar linear inhomogeneous differential equation**

$$\dot{x} = a(t)x + b(t)$$

with **time-dependent coefficients** $a, b \in C(J, \mathbb{R})$.

4 Suppose D is open in $\mathbb{R}$ and $f \in C(D, \mathbb{R})$. Show that the **similarity differential equation**

$$\dot{x} = f(x/t)$$

is equivalent to the separable differential equation

$$\dot{y} = \big(f(y) - y\big)/t$$

by using the transformation $y := x/t$.

5 What is the general solution of $\dot{x} = (x^2 + tx + t^2)/t^2$? (Hint: Exercise 4.)

6 Determine the solution of the initial value problem $\dot{x} + tx^2 = 0$, $x(0) = 2$.

7 Suppose $a, b \in C(J, \mathbb{R})$ and $\alpha \neq 1$. Show that the **Bernoulli differential equation**

$$\dot{x} = a(t)x + b(t)x^\alpha$$

becomes linear differential equation

$$\dot{y} = (1 - \alpha)\big(a(t)y + b(t)\big)$$

through the transformation $y := x^{1-\alpha}$.

8 Calculate the general solution u of the **logistic differential equation**

$$\dot{x} = (\alpha - \beta x)x \quad \text{for } \alpha, \beta > 0 .$$

Also find $\lim_{t \to t^+} u(t)$ and the turning points of u.

9 Using the substitution $y = x/t$, determine the general solution to $\dot{x} = (t+x)/(t-x)$.

10 Suppose $f \in C^{1-}(D, E)$, and let $u : J(\xi) \to \mathbb{R}$ denote the noncontinuable solution of

$$\dot{x} = f(x) , \quad x(0) = \xi .$$

Determine $\mathcal{D}(f) := \big\{ (t, \xi) \in \mathbb{R} \times D ; \ t \in J(\xi) \big\}$ if f is given through

$$\mathbb{R} \to \mathbb{R} , \quad x \mapsto x^{1+\alpha} , \quad \alpha \geq 0 ;$$
$$E \to E , \quad x \mapsto Ax , \quad A \in \mathcal{L}(E) ;$$
$$\mathbb{R} \to \mathbb{R} , \quad x \mapsto 1 + x^2 .$$

11 Prove the error estimate of Remark 8.14(b). (Hint: With the notation of the proof Theorem 8.13, it follows by induction that

$$\|u_{m+1}(t) - u_m(t)\| \leq ML^m \frac{|t - t_0|^{m+1}}{(m+1)!} \quad \text{for } m \in \mathbb{N} \text{ and } t \in I ,$$

and consequently $\|u_{m+1} - u_m\|_{C(I,E)} < M\alpha 2^{-m}/(m+1)!$ for $m \in \mathbb{N}$.)

12 Suppose $A \in C\big(J, \mathcal{L}(E)\big)$ and $b \in C(J, E)$. Show that the initial value problem

$$\dot{x} = A(t)x + b(t) , \quad x(t_0) = x_0$$

has a unique global solution for every $(t_0, x_0) \in J \times E$. (Hints: Iterate an integral equation, and use Exercise IV.4.1 and $\int_{t_0}^t \int_{t_0}^s d\sigma\, ds = (t - t_0)^2/2$.)

9 Manifolds

From Remark 8.5(e), we know that the solution set of nonlinear equations near regular points can be described by graphs. In this section, we will more precisely study the subsets of $\mathbb{R}^n$ that can be represented locally through graphs; namely, we study submanifolds of $\mathbb{R}^n$. We introduce the important concepts of regular parametrization and local charts, which allow us to locally represent submanifolds using functions.

- In this entire section, q belongs to $\mathbb{N}^\times \cup \{\infty\}$.

Submanifolds of $\mathbb{R}^n$

A subset M of $\mathbb{R}^n$ is said to be an **m-dimensional C^q submanifold** of $\mathbb{R}^n$ if, for every $x_0 \in M$, there is in $\mathbb{R}^n$ an open neighborhood U of x_0, an open set V in $\mathbb{R}^n$, and a $\varphi \in \mathrm{Diff}^q(U,V)$ such that $\varphi(U \cap M) = V \cap \left(\mathbb{R}^m \times \{0\}\right)$.

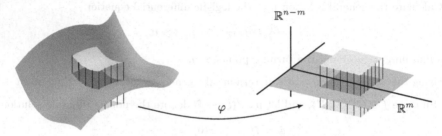

One- and two-dimensional submanifolds of $\mathbb{R}^n$ are called (**imbedded**) **curves** in $\mathbb{R}^n$ and (**imbedded**) **surfaces** in $\mathbb{R}^n$, respectively. Submanifolds of $\mathbb{R}^n$ of dimension $n-1$ (or **codimension** 1) are called (imbedded) **hypersurfaces** (in $\mathbb{R}^n$). Instead of C^q submanifold of $\mathbb{R}^n$ we will often — especially when the "surrounding space" $\mathbb{R}^n$ is unimportant — simply say C^q **manifold**.[1]

An m-dimensional submanifold of $\mathbb{R}^n$ is defined locally in $\mathbb{R}^n$, that is, in terms of open neighborhoods. Up to small deformations, each of these neighborhood lies in $\mathbb{R}^n$ in the same way that $\mathbb{R}^m$ is contained within $\mathbb{R}^n$ as a vector subspace.

9.1 Examples (a) A subset X of $\mathbb{R}^n$ is an n-dimensional C^∞ submanifold of $\mathbb{R}^n$ if and only if X is open in $\mathbb{R}^n$.

Proof If X is an n-dimensional C^∞ submanifold of $\mathbb{R}^n$ and $x_0 \in X$, then there is an open neighborhood U of x_0, an open set V in $\mathbb{R}^n$, and a $\varphi \in \mathrm{Diff}^\infty(U,V)$ such that $\varphi(U \cap X) = V$. Therefore $U \cap X = \varphi^{-1}(V) = U$. Thus U belongs to X, which shows that X is open.

Suppose now X is open in $\mathbb{R}^n$. We set $U := X$, $V := X$, and $\varphi := \mathrm{id}_X$. We then see X is an n-dimensional C^∞ submanifold of $\mathbb{R}^n$. ∎

---

[1]Note that the empty set is a manifold of every dimension $\leq n$. The dimension of a nonempty manifold is uniquely determined, as Remark 9.16(a) will show.

(b) Let $M := \{x_0, \ldots, x_k\} \subset \mathbb{R}^n$. Then M is a 0-dimensional C^∞ submanifold of $\mathbb{R}^n$.

Proof We set $\alpha := \min\{|x_i - x_j| \ ; \ 0 \le i, j \le k, \ i \ne j\}$ and choose $y \in M$. Then $\mathbb{B}(y, \alpha)$ is an open neighborhood of y in $\mathbb{R}^n$, and for $\varphi(x) := x - y$ such that $x \in \mathbb{B}(y, \alpha)$, we have $\varphi \in \mathrm{Diff}^\infty(\mathbb{B}(y, \alpha), \alpha\mathbb{B})$ and $\varphi(\mathbb{B}(y, \alpha) \cap M) = \{0\}$. ∎

(c) Suppose $\psi \in \mathrm{Diff}^q(\mathbb{R}^n, \mathbb{R}^n)$ and M is an m-dimensional C^q submanifold of $\mathbb{R}^n$. Then $\psi(M)$ is an m-dimensional C^q submanifold of $\mathbb{R}^n$.

Proof We leave this to you as an exercise. ∎

(d) Every C^q submanifold of $\mathbb{R}^n$ is also a C^r submanifold of $\mathbb{R}^n$ for $1 \le r \le q$. ∎

(e) The diffeomorphism φ and the open set V of $\mathbb{R}^n$ in the above definition can be chosen so that $\varphi(x_0) = 0$.

Proof It suffices to combine the given φ with the C^∞ diffeomorphism $y \mapsto y - \varphi(x_0)$. ∎

Graphs

The next theorem shows that graphs are manifolds and thus furnish a large class of examples.

9.2 Proposition *Suppose X is open in $\mathbb{R}^m$ and $f \in C^q(X, \mathbb{R}^n)$. Then $\mathrm{graph}(f)$ is an m-dimensional C^q submanifold of $\mathbb{R}^{m+n}$.*

Proof We set $U := X \times \mathbb{R}^n$ and consider

$$\varphi : U \to \mathbb{R}^{m+n} = \mathbb{R}^m \times \mathbb{R}^n , \quad (x, y) \mapsto (x, y - f(x)) .$$

Then we have $\varphi \in C^q(U, \mathbb{R}^m \times \mathbb{R}^n)$ with $\mathrm{im}(\varphi) = U$. In addition, $\varphi : U \to U$ is bijective with $\varphi^{-1}(x, z) = (x, z + f(x))$. Therefore φ is a C^q diffeomorphism of U onto itself and $\varphi(U \cap \mathrm{graph}(f)) = X \times \{0\} = U \cap (\mathbb{R}^m \times \{0\})$. ∎

The regular value theorem

The next theorem gives a new interpretation of Remark 8.5(f). It provides one of the most natural ways to understand submanifolds.

9.3 Theorem (regular value) *Suppose X is open in $\mathbb{R}^m$ and c is a regular value of $f \in C^q(X, \mathbb{R}^n)$. Then $f^{-1}(c)$ is an $(m - n)$-dimensional C^q submanifold of $\mathbb{R}^m$.*

Proof This follows immediately from Remark 8.5(f) and Proposition 9.2. ∎

9.4 Corollary *Suppose X is open in $\mathbb{R}^n$ and $f \in C^q(X, \mathbb{R})$. If $\nabla f(x) \ne 0$ for $x \in f^{-1}(c)$, the **level set** $f^{-1}(c)$ of f is a C^q hyperplane of $\mathbb{R}^n$.*

Proof   See Remark 8.5(c). ∎

9.5 Examples   (a) The function

$$f : \mathbb{R}^2 \to \mathbb{R} , \quad (x,y) \mapsto x^2 - y^2$$

has $(0,0)$ as its only critical point. Its level
sets are hyperbolas.

(b) The (**Euclidean**) n-**sphere** $S^n := \{\, x \in \mathbb{R}^{n+1} \;;\; |x| = 1 \,\}$ is a C^∞ hypersurface
in $\mathbb{R}^{n+1}$ of dimension n.

Proof   The function $f : \mathbb{R}^{n+1} \to \mathbb{R}$, $x \mapsto |x|^2$ is smooth, and $S^n = f^{-1}(1)$. Because
$\nabla f(x) = 2x$, we know 1 is a regular value of f. The claim then follows from Corol-
lary 9.4 ∎

(c) The **orthogonal group**[2] $O(n) := \{\, A \in \mathbb{R}^{n \times n} \;;\; A^\top A = 1_n \,\}$ is a C^∞ submanifold of $\mathbb{R}^{n \times n}$ of dimension $n(n-1)/2$.

Proof   (i) From Exercise 1.5, we know (because $A^\top = A^*$) that $A^\top A$ is symmetric for
every $A \in \mathbb{R}^{n \times n}$. It is easy to verify that $\mathbb{R}^{n \times n}_{\mathrm{sym}}$ has dimension $n(n+1)/2$ (which is the
number of entries on and above the diagonal of an $(n \times n)$ matrix). For the map

$$f : \mathbb{R}^{n \times n} \to \mathbb{R}^{n \times n}_{\mathrm{sym}} , \quad A \mapsto A^\top A ,$$

we have $O(n) = f^{-1}(1_n)$. First, we observe that

$$g : \mathbb{R}^{n \times n} \times \mathbb{R}^{n \times n} \to \mathbb{R}^{n \times n} , \quad (A,B) \mapsto A^\top B$$

is bilinear and consequently smooth. Then because $f(A) = g(A,A)$, the map f is also
smooth. Further, we have (see Proposition 4.6)

$$\partial f(A)B = A^\top B + B^\top A \quad \text{for } A,B \in \mathbb{R}^{n \times n} .$$

(ii) Suppose $A \in f^{-1}(1_n) = O(n)$ and $S \in \mathbb{R}^{n \times n}_{\mathrm{sym}}$. For $B := AS/2$, we then have

$$\partial f(A)B = \frac{1}{2} A^\top A S + \frac{1}{2} S A^\top A = S$$

because $A^\top A = 1_n$. Therefore $\partial f(A)$ is surjective for every $A \in f^{-1}(1_n)$. Thus 1_n is
a regular value of f. Because $\dim(\mathbb{R}^{n \times n}) = n^2$ and $n^2 - n(n+1)/2 = n(n-1)/2$, the
claim follows from Theorem 9.3. ∎

The immersion theorem

Suppose X is open in $\mathbb{R}^m$. The map $f \in C^1(X, \mathbb{R}^n)$ is called an **immersion**
(of X in $\mathbb{R}^n$) if $\partial f(x) \in \mathcal{L}(\mathbb{R}^m, \mathbb{R}^n)$ is injective for every $x \in X$. Then f is a
regular parametrization of $F := f(X)$. Finally, F is an m-**dimensional** (**regular**)
parametrized hypersurface, and X is its **parameter domain**. A 1-dimensional
or 2-dimensional parametrized hypersurface is a (**regular**) **parametrized curve** or
(**regular**) **parametrized surface**, respectively.

---

[2]Here 1_n denotes the identity matrix in $\mathbb{R}^{n \times n}$. For more on $O(n)$, see also Exercises 1 and 2.

9.6 Remarks and Examples (a) If $f \in C^1(X, \mathbb{R}^n)$ is an immersion, then $m \le n$.

Proof For $n < m$, that there is no injection $A \in \mathcal{L}(\mathbb{R}^m, \mathbb{R}^n)$ follows immediately from the rank formula (2.4). ∎

(b) For every $\ell \in \mathbb{N}^\times$, the restriction of $\mathbb{R} \to \mathbb{R}^2$, $t \mapsto \big(\cos(\ell t), \sin(\ell t)\big)$ to $(0, 2\pi)$ is a C^∞ immersion. The image of $[0, 2\pi)$ is the unit circle S^1, which is traversed ℓ times.

(c) The map $(-\pi, \pi) \to \mathbb{R}^2$, $t \mapsto (1 + 2\cos t)(\cos t, \sin t)$ is a smooth immersion. The closure of its image is called the **limaçon of Pascal**.

(d) It is easy to see that $(-\pi/4, \pi/2) \to \mathbb{R}^2$, $t \mapsto \sin 2t(-\sin t, \cos t)$ is an injective C^∞ immersion. ∎

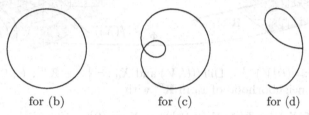

for (b) for (c) for (d)

The next theorem shows that m-dimensional parametrized hypersurfaces are represented locally by manifolds.

9.7 Theorem (immersion) *Suppose X is open in $\mathbb{R}^m$ and $f \in C^q(X, \mathbb{R}^n)$ is an immersion. Then there is for every $x_0 \in X$ an open neighborhood X_0 in X such that $f(X_0)$ is an m-dimensional C^q submanifold of $\mathbb{R}^n$.*

Proof (i) By permuting the coordinates, we can assume without loss of generality that the first m rows of the Jacobi matrix $[\partial f(x_0)]$ are linearly independent. Therefore

$$\frac{\partial(f^1, \ldots, f^m)}{\partial(x^1, \ldots, x^m)}(x_0) \ne 0 \ .$$

(ii) We consider the open set $X \times \mathbb{R}^{n-m}$ in $\mathbb{R}^n$ and the map

$$\psi \colon X \times \mathbb{R}^{n-m} \to \mathbb{R}^n \ , \quad (x, y) \mapsto f(x) + (0, y) \ .$$

Obviously, ψ belongs to the class C^q, and $\partial\psi(x_0, 0)$ has the matrix

$$[\partial\psi(x_0, 0)] = \begin{bmatrix} A & 0 \\ B & 1_{n-m} \end{bmatrix}$$

with

$$A := \begin{bmatrix} \partial_1 f^1 & \cdots & \partial_m f^1 \\ \vdots & & \vdots \\ \partial_1 f^m & \cdots & \partial_m f^m \end{bmatrix}(x_0) \text{ and } B := \begin{bmatrix} \partial_1 f^{m+1} & \cdots & \partial_m f^{m+1} \\ \vdots & & \vdots \\ \partial_1 f^n & \cdots & \partial_m f^n \end{bmatrix}(x_0) \ .$$

Therefore $\partial\psi(x_0,0)$ belongs to $\mathcal{L}\mathrm{aut}(\mathbb{R}^n)$, and thus

$$\det\left[\partial\psi(x_0,0)\right] = \det A = \frac{\partial(f^1,\ldots,f^m)}{\partial(x^1,\ldots,x^m)}(x_0) \neq 0 \ .$$

Consequently the inverse function theorem (Theorem 7.3) says there are open neighborhoods $V \in \mathfrak{U}_{\mathbb{R}^n}(x_0,0)$ and $U \in \mathfrak{U}_{\mathbb{R}^n}(\psi(x_0,0))$ with $\psi|V \in \mathrm{Diff}^q(V,U)$.

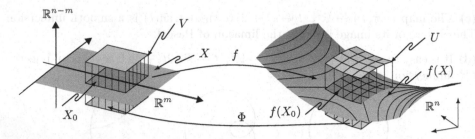

We now set $\Phi := (\psi|V)^{-1} \in \mathrm{Diff}^q(U,V)$ and $X_0 := \left\{\, x \in \mathbb{R}^m \ ; \ (x,0) \in V \,\right\}$. Then X_0 is an open neighborhood of x_0 in $\mathbb{R}^m$ with

$$\Phi\big(U \cap f(X_0)\big) = \Phi\big(\psi(X_0 \times \{0\})\big) = X_0 \times \{0\} = V \cap \big(\mathbb{R}^m \times \{0\}\big) \ . \ \blacksquare$$

9.8 Corollary *Suppose I is an open interval and $\gamma \in C^q(I,\mathbb{R}^n)$. Further let $t_0 \in I$ and $\dot\gamma(t_0) \neq 0$. Then there is an open subinterval I_0 of I with $t_0 \in I_0$ such that the trace of $\gamma|I_0$ is an imbedded C^q curve in $\mathbb{R}^n$.*

Proof This follows immediately Theorem 9.7. $\blacksquare$

9.9 Remarks (a) The smooth path

$$\gamma \colon \mathbb{R} \to \mathbb{R}^2 \ , \quad t \mapsto (t^3,t^2)$$

satisfies $\dot\gamma(t) \neq 0$ for $t \neq 0$. At $t = 0$, the derivative of γ vanishes. The trace of γ, the **Neil parabola**, is "not smooth" there but rather comes to a point.

(b) Let $f \in C^q(X,\mathbb{R}^n)$ be an immersion. Then $f(X)$ is not generally a C^q submanifold of $\mathbb{R}^n$ because $f(X)$ can have "self intersections"; see Example 9.6(c).

(c) The immersion given in Example 9.6(c) is not injective. Also, images of injective immersions are generally not submanifolds; see Example 9.6(d) and Exercise 16. On the other hand, Example 9.6(b) shows that images of noninjective immersions can indeed be submanifolds.

(d) Suppose X is open in $\mathbb{R}^m = \mathbb{R}^m \times \{0\} \subset \mathbb{R}^n$, and $f \in C^q(X,\mathbb{R}^n)$ is an immersion. Then for every $x_0 \in X$, there is an open neighborhood V of $(x_0,0)$

in $\mathbb{R}^n$, an open set U in $\mathbb{R}^n$, and a $\psi \in \mathrm{Diff}^q(V, U)$ such that $\psi(x, 0) = f(x)$ for $x \in X$ with $(x, 0) \in V$.

Proof This follows immediately from the proof of Theorem 9.7. ∎

Embeddings

Suppose $g : I \to \mathbb{R}^2$ is the injective C^∞ immersion of Example 9.6(d). We have already determined that $S = \mathrm{im}(g)$ represents an embedded curve in $\mathbb{R}^2$. This raises the question, What other properties must an injective immersion have so that its image is a submanifold? If one analyzes the example above, it is easy to see that $g^{-1} : S \to I$ is not continuous. Therefore the map $g : I \to S$ is not topological. Indeed, the next theorem shows that if an injective immersion has a continuous inverse, its image is a submanifold. We say a (C^q) immersion $f : X \to \mathbb{R}^n$ is a (C^q) **embedding** of X in $\mathbb{R}^n$ if $f : X \to f(X)$ is topological.[3]

9.10 Proposition *Suppose X is open in $\mathbb{R}^m$ and $f \in C^q(X, \mathbb{R}^n)$ is an embedding. Then $f(X)$ is an m-dimensional C^q submanifold of $\mathbb{R}^n$.*

Proof We set $M := f(X)$ and choose $y_0 \in M$. According to Theorem 9.7, $x_0 := f^{-1}(y_0)$ has an open neighborhood X_0 in X such that $M_0 := f(X_0)$ is an m-dimensional submanifold of $\mathbb{R}^n$.

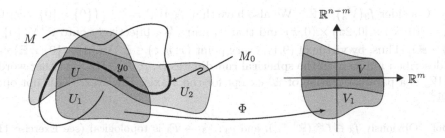

Hence there are open neighborhoods U_1 of y_0 and V_1 of 0 in $\mathbb{R}^n$ as well as a C^q diffeomorphism Φ from U_1 to V_1 such that $\Phi(M_0 \cap U_1) = V_1 \cap (\mathbb{R}^m \times \{0\})$. Because f is topological, M_0 is open in M. Hence there is an open set U_2 in $\mathbb{R}^n$ such that $M_0 = M \cap U_2$ (see Proposition III.2.26). Therefore $U := U_1 \cap U_2$ is an open neighborhood of y_0 in $\mathbb{R}^n$, and $V := \Phi(U)$ is an open neighborhood of 0 in $\mathbb{R}^n$ with $\Phi(M \cap U) = V \cap (\mathbb{R}^m \times \{0\})$. The claim follows because this holds for every $y_0 \in M$. ∎

---

[3]Naturally, $f(X)$ carries with it the induced topology from $\mathbb{R}^n$. If the context makes the meaning of X and $\mathbb{R}^n$ unambiguous, we speak in short of an **embedding**.

9.11 Examples   **(a)** (spherical coordinates)
Let

$$f_3 \colon \mathbb{R}^3 \to \mathbb{R}^3 \,, \quad (r, \varphi, \vartheta) \mapsto (x, y, z)$$

be defined through

$$
\begin{aligned}
x &= r \cos\varphi \sin\vartheta \,,\\
y &= r \sin\varphi \sin\vartheta \,,\\
z &= r \cos\vartheta \,,
\end{aligned}
\qquad (9.1)
$$

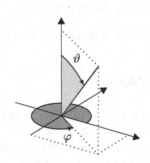

and let $V_3 := (0, \infty) \times (0, 2\pi) \times (0, \pi)$. Then $g_3 := f_3 | V_3$ is a C^∞ embedding of V_3 in $\mathbb{R}^3$, and $F_3 = g_3(V_3) = \mathbb{R}^3 \setminus H_3$, where H_3 denotes the closed half plane $\mathbb{R}^+ \times \{0\} \times \mathbb{R}$.

If one restricts (r, φ, ϑ) to a subset of V_3 of the form

$$(r_0, r_1) \times (\varphi_0, \varphi_1) \times (\vartheta_0, \vartheta_1) \,,$$

one gets a regular parametrization of a "spherical shell".

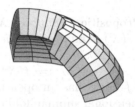

Consider $f_3(\overline{V}_3) = \mathbb{R}^3$. We also have that $f_3(W_3) = \mathbb{R}^3 \setminus (\{0\} \times \{0\} \times \mathbb{R})$ for $W_3 := (0, \infty) \times [0, 2\pi) \times (0, \pi)$ and that f_3 maps W_3 bijectively onto $\mathbb{R}^3 \setminus (\{0\} \times \{0\} \times \mathbb{R})$. Thus, by virtue of (9.1), every point $(x, y, z) \in \mathbb{R}^3 \setminus (\{0\} \times \{0\} \times \mathbb{R})$ can be described uniquely in the **spherical coordinates** $(r, \varphi, \vartheta) \in W_3$. In other words, $f_3 | W_3$ is a parametrization of $\mathbb{R}^3$ except for the z-axis. However, W_3 is not open in $\mathbb{R}^3$.

Proof   Obviously $f_3 \in C^\infty(\mathbb{R}^3, \mathbb{R}^3)$, and $g_3 \colon V_3 \to F_3$ is topological (see Exercise 11). Also, we find

$$
[\partial f_3(r, \varphi, \vartheta)] =
\begin{bmatrix}
\cos\varphi \sin\vartheta & -r \sin\varphi \sin\vartheta & r \cos\varphi \cos\vartheta \\
\sin\varphi \sin\vartheta & r \cos\varphi \sin\vartheta & r \sin\varphi \cos\vartheta \\
\cos\vartheta & 0 & -r \sin\vartheta
\end{bmatrix} \,.
$$

The determinant of ∂f_3 can be calculated by expanding along the last row, giving the value $-r^2 \sin\vartheta \neq 0$ for $(r, \varphi, \vartheta) \in V_3$. Hence g_3 is an immersion and therefore an embedding. The remaining statements are clear. $\blacksquare$

(b) (spherical coordinates[4])
We define

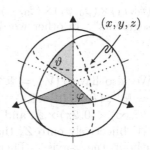

$$f_2 : \mathbb{R}^2 \to \mathbb{R}^3 , \quad (\varphi, \vartheta) \mapsto (x, y, z)$$

through

$$x = \cos\varphi \sin\vartheta ,$$
$$y = \sin\varphi \sin\vartheta , \qquad (9.2)$$
$$z = \cos\vartheta$$

and set $V_2 := (0, 2\pi) \times (0, \pi)$. Then the restriction $g_2 := f_2 | V_2$ is a C^∞ embedding of V_2 in $\mathbb{R}^3$ and $F_2 := g_2(V_2) = S^2 \backslash H_3$. In other words, F_2 is obtained from S^2 by removing the half circle where the half plane H_3 intersects with S^2. Restricting (φ, ϑ) to a subset V_2 of the form $(\varphi_0, \varphi_1) \times (\vartheta_0, \vartheta_1)$, one gets a regular parametrization of a section of S^2, which means that the rectangle $\overline{V}_2$ is "curved into (part of) a sphere" by f_2. Note also that for $W_2 := [0, 2\pi) \times (0, \pi)$, we have $f_2(W_2) = S^2 \backslash \{\pm e_3\}$, where e_3 is the **north pole** and $-e_3$ is the **south pole**. Also f_2 maps the parameter domain W_2 bijectively onto $S^2 \backslash \{\pm e_3\}$. Using (9.2), $S^2 \backslash \{\pm e_3\}$ can therefore be described through **spherical coordinates** $(\varphi, \vartheta) \in W_2$, although $f_2 | W_2$ is not a regular parametrization of $S^2 \backslash \{\pm e_3\}$ because W_2 is not open in $\mathbb{R}^2$.

Proof It is clear that $f_2 = f_3(1, \cdot, \cdot) \in C^\infty(\mathbb{R}^2, \mathbb{R}^3)$ and $F_2 = S^2 \cap F_3$. Since F_3 is open in $\mathbb{R}^3$, it follows that F_2 is open in S^2. Further, $g_2 = g_3(1, \cdot, \cdot)$ maps V_2 bijectively onto F_2, and $g_2^{-1} = g_3^{-1} | F_2$. Therefore $g_2^{-1} : F_2 \to V_2$ is continuous. Because

$$[\partial f_2(\varphi, \vartheta)] = \begin{bmatrix} -\sin\varphi \sin\vartheta & \cos\varphi \cos\vartheta \\ \cos\varphi \sin\vartheta & \sin\varphi \cos\vartheta \\ 0 & -\sin\vartheta \end{bmatrix} \qquad (9.3)$$

consists of the last two columns of the regular matrix $[\partial f_3(1, \varphi, \vartheta)]$, we see that $\partial g_2(\varphi, \vartheta)$ is injective for $(\varphi, \vartheta) \in V_2$. Thus g_2 is a regular C^∞ parametrization. ∎

(c) (cylindrical coordinates) Define

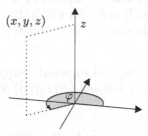

$$f : \mathbb{R}^3 \to \mathbb{R}^3 , \quad (r, \varphi, z) \mapsto (x, y, z)$$

through

$$x = r\cos\varphi ,$$
$$y = r\sin\varphi , \qquad (9.4)$$
$$z = z ,$$

and let $V := (0, \infty) \times (0, 2\pi) \times \mathbb{R}$. Then $g := f | V$ is a C^∞ embedding of V in $\mathbb{R}^3$ with $g(V) = F_3 = \mathbb{R}^3 \backslash H_3$. By further restricting g to a subset of V of the form

---

[4]Note the reuse of this terminology.

$R := (r_0, r_1) \times (\varphi_0, \varphi_1) \times (z_0, z_1)$, one gets a regular parametrization of "cylindrical shell segments". In other words, f "bends" the rectangular volume R into this shape.

We also have $f(\overline{V}) = \mathbb{R}^3$ and
$f(W) = \mathbb{R}^3 \setminus (\{0\} \times \{0\} \times \mathbb{R}) =: Z$ for
$W := (0, \infty) \times [0, 2\pi) \times \mathbb{R}$, and $f \mid W$
maps W bijectively onto Z, that is,
$\mathbb{R}^3$ without the z-axis. Therefore
(9.4) describes Z through **cylindrical
coordinates**, although $f \mid W$ is not a
regular parametrization of Z because
W is not open.

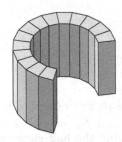

Proof It is obvious that g maps bijectively onto F_3, and it is easy to see that g^{-1} is smooth (see Exercise 12). Also, we have

$$[\partial f(r, \varphi, z)] = \begin{bmatrix} \cos\varphi & -r\sin\varphi & 0 \\ \sin\varphi & r\cos\varphi & 0 \\ 0 & 0 & 1 \end{bmatrix},$$

which follows from $\det \partial f(r, \varphi, z) = r$. Therefore $f \mid (0, \infty) \times \mathbb{R} \times \mathbb{R}$ is a C^∞ immersion of the open half space $(0, \infty) \times \mathbb{R} \times \mathbb{R}$ in $\mathbb{R}^3$, and g is a C^∞ embedding of V in $\mathbb{R}^3$. The remaining statements are also straightforward. ∎

(d) (cylindrical coordinates)
We set $r = 1$ in (9.4). Then
$g(1, \cdot, \cdot)$ is a C^∞ embedding of
$(0, 2\pi) \times \mathbb{R}$ in $\mathbb{R}^3$, for which the
obvious analogue of (c) holds.

(e) (surfaces of rotation) Suppose J is an open interval in $\mathbb{R}$ and $\rho, \sigma \in C^q(J, \mathbb{R})$ with $\rho(t) > 0$ and $\big(\dot\rho(t), \dot\sigma(t)\big) \neq (0, 0)$ for $t \in J$. Then

$$r : J \times \mathbb{R} \to \mathbb{R}^3 , \quad (t, \varphi) \mapsto \big(\rho(t)\cos\varphi, \rho(t)\sin\varphi, \sigma(t)\big)$$

is a C^q immersion of $J \times \mathbb{R}$ in $\mathbb{R}^3$.

The map

$$\gamma \colon J \to \mathbb{R}^3 , \quad t \mapsto \bigl(\rho(t), 0, \sigma(t)\bigr)$$

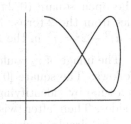

is a C^q immersion of J in $\mathbb{R}^3$. The image Γ of γ is a regularly parametrized curve that lies in the x-z plane, and $R := r(J \times \mathbb{R})$ is generated by rotating Γ about the z-axis. Thus R is called a **surface of revolution**, and Γ is a **meridian curve** of R. If γ is not injective, then R contains circles in planes parallel to the x-y-axis, at which R intersects itself.

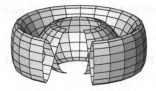
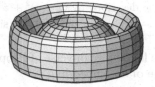

Suppose I is an open subinterval of J such that $\gamma \,|\, I$ is an embedding, and K is an open subinterval of $[0, 2\pi]$. Then $r \,|\, I \times K$ is also an embedding.

Proof   The Jacobi matrix of r is

$$\begin{bmatrix} \dot{\rho}(t) \cos\varphi & -\rho(t) \sin\varphi \\ \dot{\rho}(t) \sin\varphi & \rho(t) \cos\varphi \\ \dot{\sigma}(t) & 0 \end{bmatrix} . \tag{9.5}$$

The determinant of the (2×2)-matrix obtained by discarding the last row has the value $\rho(t)\dot{\rho}(t)$. Therefore, the matrix (9.5) has rank 2 if $\dot{\rho}(t) \neq 0$. If $\dot{\rho}(t) = 0$, then $\dot{\sigma}(t) \neq 0$, and at one least of the two determinants obtained by striking either the first or second row from (9.5) is different from 0. This proves that r is an immersion. We leave remaining statements to you. ∎

(f) (tori)   Suppose $0 < r < a$. The equation $(x - a)^2 + z^2 = r^2$ defines a circle in the x-z plane, and, by rotating it about the z-axis,

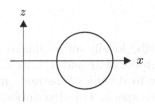
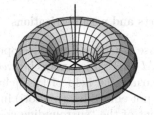

we get a **(2-)torus**, $\mathsf{T}^2_{a,r}$. For $\tau_2 \in C^\infty(\mathbb{R}^2, \mathbb{R}^3)$ with

$$\tau_2(t, \varphi) := \bigl((a + r \cos t) \cos\varphi, (a + r \cos t) \sin\varphi, r \sin t\bigr) ,$$

we find that $\tau_2\big([0,2\pi]^2\big) = \mathsf{T}^2_{a,r}$ and that $\tau_2|(0,2\pi)^2$ is an embedding. The image of the open square $(0,2\pi)^2$ under τ_2 is the surface obtained by "cutting" two circles from the 2-torus: first, $x^2 + y^2 = (r+a)^2$ in the x-y plane, and, second, $(x-a)^2 + z^2 = r^2$ in the x-z plane.

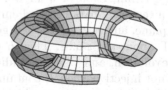

The image of τ_2 could also be described as follows: The square $[0,2\pi]^2$ is first made into a tube by "identifying" two of its opposite sides. Then, after twisting this tube into a ring, its circular ends are also identified.

Proof The map $t \mapsto (a + r\cos t, 0, r\sin t)$ is a C^∞ immersion of $\mathbb{R}$ in $\mathbb{R}^3$, whose image is the circle described by $(x-a)^2 + z^2 = r^2$. Therefore the claim follows easily from (e) with $\rho(t) := a + r\cos t$ and $\sigma(t) := r\sin t$. ∎

The following converse of Proposition 9.10 shows every submanifold is locally the image of an embedding.

9.12 Theorem *Suppose M is an m-dimensional C^q submanifold of $\mathbb{R}^n$. Then every point p of M has a neighborhood U in M such that U is the image of an open set in $\mathbb{R}^m$ under a C^q embedding.*

Proof For every $p \in M$, there are open neighborhoods $\widetilde{U}$ of p and $\widetilde{V}$ of 0 in $\mathbb{R}^n$, as well as a C^q diffeomorphism $\Phi : \widetilde{U} \to \widetilde{V}$, such that $\Phi(M \cap \widetilde{U}) = \widetilde{V} \cap \big(\mathbb{R}^m \times \{0\}\big)$. We set $U := M \cap \widetilde{U}$, $V := \big\{x \in \mathbb{R}^m \; ; \; (x,0) \in \widetilde{V}\big\}$, and

$$g : V \to \mathbb{R}^n, \quad x \mapsto \Phi^{-1}\big((x,0)\big).$$

Then U is open in M, and V is open in $\mathbb{R}^m$; also g belongs to $C^q(V, \mathbb{R}^n)$. In addition, g maps the set V bijectively onto U, and rank $\partial g(x) = m$ for $x \in V$ because $\big[\partial g(x)\big]$ consists of the first m column of the regular matrix $\big[\partial \Phi^{-1}\big((x,0)\big)\big]$. Because $g^{-1} = \Phi | U$, we clearly see that g, interpreted as a map in the topological subspace U of $\mathbb{R}^n$, is a topological map from V to U. ∎

Local charts and parametrizations

We now use what we have developed to describe locally an m-dimensional submanifold M of $\mathbb{R}^n$ using maps between open subsets of M and $\mathbb{R}^m$. In Volume III, we will learn — the help of local charts — how to describe "abstract" manifolds, which are not embedded (a priori) in a Euclidean space. This description is largely independent of the "surrounding space".

Suppose M is a subset of $\mathbb{R}^n$ and $p \in M$. We denote by

$$i_M : M \to \mathbb{R}^n, \quad x \mapsto x$$

the canonical injection of M into $\mathbb{R}^n$. The map φ is called an m-dimensional (**local**) $\pmb{C^q}$ **chart of** $\pmb{M}$ **around** $\pmb{p}$ if

- $U := \operatorname{dom}(\varphi)$ is an open neighborhood of p in M;
- φ is a homeomorphism of U onto the open set $V := \varphi(U)$ of $\mathbb{R}^m$;
- $g := i_M \circ \varphi^{-1}$ is a C^q immersion.

The set U is the **charted territory** of φ, V is the **parameter range**, and g is the **parametrization** of U in φ. Occasionally, we write (φ, U) for φ and (g, V) for g. An m-dimensional $\pmb{C^q}$ **atlas** for M is a family $\{\, \varphi_\alpha \;;\; \alpha \in \mathsf{A} \,\}$ of m-dimensional C^q charts of M whose charted territories $U_\alpha := \operatorname{dom}(\varphi_\alpha)$ cover the set M, that is, $M = \bigcup_\alpha U_\alpha$. Finally, the $(x^1, \ldots, x^m) := \varphi(p)$ are the **local coordinates** of $p \in U$ in the chart φ.

The next theorem shows that a manifold can be described through charts. This means in particular that an m-dimensional submanifold of $\mathbb{R}^n$ locally "looks like $\mathbb{R}^m$", that is, it is locally homeomorphic to an open subset of $\mathbb{R}^m$.

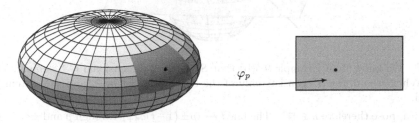

9.13 Theorem *Suppose M is an m-dimensional C^q submanifold of $\mathbb{R}^n$. Then, for every $p \in M$, there is an m-dimensional C^q chart of M around p. Therefore M has an m-dimensional C^q atlas. If M is compact, then M has a finite atlas.*

Proof Theorem 9.12 guarantees that every $p \in M$ has an m-dimensional C^q chart φ_p around it. Therefore $\{\, \varphi_p \;;\; p \in M \,\}$ is an atlas. Because the charted territories of this atlas form an open cover of M, the last statement follows from the definition of compactness. $\blacksquare$

9.14 Examples (a) Suppose X is open in $\mathbb{R}^n$. Then X has a C^∞ atlas with only one chart, namely, the **trivial chart** id_X.

(b) Suppose X is open in $\mathbb{R}^m$ and $f \in C^q(X, \mathbb{R}^n)$. Then, according to Proposition 9.2, $\operatorname{graph}(f)$ is m-dimensional C^q submanifold of $\mathbb{R}^{m+n}$, and it has an atlas consisting only of the chart φ with $\varphi\big((x, f(x))\big) = x$ for $x \in X$.

(c) The sphere S^2 has a C^∞ atlas with exactly two charts.

Proof In the notation of Example 9.11(b), let $U_2 := g_2(V_2)$ and $\varphi_2 : U_2 \to V_2$, and let $(x, y, z) \mapsto (\varphi, \vartheta)$ be the map that inverts g_2. Then φ_2 is a C^∞ chart of S^2.

Now we define $\widetilde{g}_2 : \widetilde{V}_2 \to S^2$ through $\widetilde{V}_2 := (\pi, 3\pi) \times (0, \pi)$ and

$$\widetilde{g}_2(\varphi, \vartheta) := (\cos\varphi \sin\vartheta, \cos\vartheta, \sin\varphi \sin\vartheta) \ .$$

Then $\widetilde{g}_2(\widetilde{V}_2)$ is obtained from S^2 by removing the half circle where it intersects with the coordinate half plane $(-\mathbb{R}^+) \times \mathbb{R} \times \{0\}$. Obviously $U_2 \cup \widetilde{U}_2 = S^2$ with $\widetilde{U}_2 := \widetilde{g}_2(\widetilde{V}_2)$, and the proof of Example 9.11(b) shows that $\widetilde{\varphi}_2 := g_2^{-1}$ is a C^∞ chart of S^2. ∎

(d) For every $n \in \mathbb{N}$, the n-sphere S^n has an atlas with exactly two smooth charts. For $n \geq 1$, these charts are the **stereographic projections** $\varphi_\pm$, where φ_+ $[\varphi_-]$ assigns to every $p \in S^n \backslash \{e_{n+1}\}$ $[\, p \in S^n \backslash \{-e_{n+1}\}\,]$ the "puncture point", that is, the point where the line connecting the **north pole** e_{n+1} [**south pole** $-e_{n+1}$] to p intersects the **equatorial hyperplane** $\mathbb{R}^n \times \{0\}$.

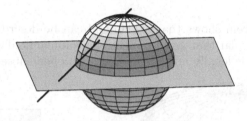

Proof We know from Example 9.5(b) that S^n is an n-dimensional C^∞ submanifold of $\mathbb{R}^n$. When $n = 0$, the S^n consists only of the two points ± 1 in $\mathbb{R}$, and the statement is trivial.

Suppose therefore $n \in \mathbb{N}^\times$. The line $t \mapsto tp \pm (1-t)e_{n+1}$ through p and $\pm e_{n+1} \in S^n$ intersects the hyperplane $x^{n+1} = 0$ when $t = 1/(1 \mp p^{n+1})$. Therefore the puncture point has the coordinates $x = p'/(1 \mp p^{n+1}) \in \mathbb{R}^n$, where $p = (p', p^{n+1}) \in \mathbb{R}^n \times \mathbb{R}$. These define the maps $\varphi_\pm : S^n \backslash \{\pm e_{n+1}\} \to \mathbb{R}^n$, $p \mapsto x$, and they are obviously continuous.

To calculate their inverses, we consider the lines

$$t \mapsto t(x, 0) \pm (1 - t)e_{n+1}$$

through $(x, 0) \in \mathbb{R}^n \times \mathbb{R}$ and $\pm e_{n+1} \in S^n$. They intersect $S^n \backslash \{\pm e_{n+1}\}$ when $t > 0$ and $t^2 |x|^2 + (1 - t)^2 = 1$ and, therefore, when $t = 2/(1 + |x|^2)$. With this we get

$$\varphi_\pm^{-1}(x) = \frac{(2x, \pm(|x|^2 - 1))}{1 + |x|^2} \quad \text{for } x \in \mathbb{R}^n \ .$$

Thus $g_\pm := i_{S^n} \circ \varphi_\pm^{-1}$ belongs to $C^\infty(\mathbb{R}^n, \mathbb{R}^{n+1})$. That $\operatorname{rank} \partial g_\pm(x) = n$ for $x \in \mathbb{R}^n$ is checked easily. ∎

(e) The torus $\mathsf{T}_{a,r}^2$ is a C^∞ hypersurface in $\mathbb{R}^3$ and has an atlas with three charts.

Proof Suppose $X := \mathbb{R}^3 \backslash \{0\} \times \{0\} \times \mathbb{R}$ and

$$f : X \to \mathbb{R} , \quad (x, y, z) \mapsto \left(\sqrt{x^2 + y^2} - a\right)^2 + z^2 - r^2 \ .$$

Then $f \in C^\infty(X, \mathbb{R})$, and 0 is a regular value of f. One verifies easily that $f^{-1}(0) = \mathsf{T}_{a,r}^2$. Therefore, according to Theorem 9.3, $\mathsf{T}_{a,r}^2$ is a smooth surface in $\mathbb{R}^3$. The map φ that inverts $\tau_2 \,|\, (0, 2\pi)^2$ is a two-dimensional C^∞ chart of $\mathsf{T}^2_{a,r}$. A second such chart $\widetilde{\varphi}$ serves as the map that inverts $\tau_2 \,|\, (\pi, 3\pi)^2$. Finally, we define a third chart $\widehat{\varphi}$ as the inverse of $\tau_2 \,|\, (\pi/2, 5\pi/2)^2$, and then $\{\varphi, \widetilde{\varphi}, \widehat{\varphi}\}$ is an atlas for $\mathsf{T}^2_{a,r}$. ∎

Change of charts

The local geometric meaning of a curve or a surface (generally, a submanifold of $\mathbb{R}^n$) is independent of its description via local charts. For concrete calculations, it is necessary to work using a specific chart. As we move around the manifold, our calculation may take us to the boundary of the chart we are using, forcing a "change of charts". Thus, we should understand how that goes. In addition to this practical justification, understanding changes of charts will help us understand how the local description of a manifold is "put together" to form the global description.

Suppose $\big\{ (\varphi_\alpha, U_\alpha) \,;\, \alpha \in \mathsf{A} \big\}$ is an m-dimensional C^q atlas for $M \subset \mathbb{R}^n$. We call the maps[5]

$$\varphi_\beta \circ \varphi_\alpha^{-1} : \varphi_\alpha(U_\alpha \cap U_\beta) \to \varphi_\beta(U_\alpha \cap U_\beta) \quad \text{for } \alpha, \beta \in \mathsf{A} \,,$$

transition functions. They specify how the charts in the atlas $\{\varphi_\alpha \,;\, \alpha \in \mathsf{A}\}$ are "stitched together".

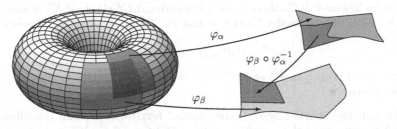

9.15 Proposition If $(\varphi_\alpha, U_\alpha)$ and (φ_β, U_β) are m-dimensional C^q charts of a C^q manifold of dimension m, then

$$\varphi_\beta \circ \varphi_\alpha^{-1} \in \mathrm{Diff}^q\big(\varphi_\alpha(U_\alpha \cap U_\beta), \varphi_\beta(U_\alpha \cap U_\beta)\big) \,,$$

where $(\varphi_\beta \circ \varphi_\alpha^{-1})^{-1} = \varphi_\alpha \circ \varphi_\beta^{-1}$.

Proof (i) It is clear that $\varphi_\beta \circ \varphi_\alpha^{-1}$ is bijective and its inverse is $\varphi_\alpha \circ \varphi_\beta^{-1}$. Thus it suffices to verify that $\varphi_\beta \circ \varphi_\alpha^{-1}$ belongs to $C^q\big(\varphi_\alpha(U_\alpha \cap U_\beta), \mathbb{R}^m\big)$.

(ii) We set $V_\gamma := \varphi_\gamma(U_\gamma)$, and g_γ is the parametrization belonging to $(\varphi_\gamma, U_\gamma)$ for $\gamma \in \{\alpha, \beta\}$. Further suppose $x_\gamma \in \varphi_\gamma(U_\alpha \cap U_\beta)$ with $g_\alpha(x_\alpha) = g_\beta(x_\beta) =: p$. Because g_γ is an injective C^q immersion, Remark 9.9(d) gives open neighborhoods

---

[5]Here and nearby, we always assume that $U_\alpha \cap U_\beta \neq \emptyset$ when working with the transition function $\varphi_\beta \circ \varphi_\alpha^{-1}$.

$\widetilde{U}_\gamma$ of p and $\widetilde{V}_\gamma$ of $(x_\gamma, 0)$ in $\mathbb{R}^n$, as well as a $\psi_\gamma \in \mathrm{Diff}^q(\widetilde{V}_\gamma, \widetilde{U}_\gamma)$ such that $\psi_\gamma(y, 0) = g_\gamma(y)$ for all $y \in V_\gamma$ with $(y, 0) \in \widetilde{V}_\gamma$. We now set

$$V := \big\{ x \in V_\alpha \ ; \ (x, 0) \in \widetilde{V}_\alpha \big\} \cap \varphi_\alpha(U_\alpha \cap U_\beta) \ .$$

Clearly V is an open neighborhood of x_α in $\mathbb{R}^m$. Using the continuity of g_α, we can (possibly by shrinking V) assume that $g_\alpha(V)$ is contained in $\widetilde{U}_\beta$. Therefore

$$\varphi_\beta \circ \varphi_\alpha^{-1}(x) = \psi_\beta^{-1} \circ g_\alpha(x) \quad \text{for } x \in V \ .$$

Because $g_\alpha \in C^q(V_\alpha, \mathbb{R}^n)$, the chain rule shows that $\psi_\beta^{-1} \circ g_\alpha$ belongs to $C^q(V, \mathbb{R}^n)$.[6] It therefore follows that $\varphi_\beta \circ \varphi_\alpha^{-1} \in C^q(V, \mathbb{R}^m)$, because the image of $\psi_\beta^{-1} \circ g_\alpha$ coincides with that of $\varphi_\beta \circ \varphi_\alpha^{-1}$ and therefore lies in $\mathbb{R}^m$. Because this holds for every $x_\alpha \in \varphi_\alpha(U_\alpha \cap U_\beta)$ and because belonging to the class C^q is a local property, the claim is proved. ∎

9.16 Remarks (a) The dimension of a submanifold of $\mathbb{R}^n$ is unique. Thus for a given m-dimensional manifold, it makes sense to speak simply of its "charts" instead of its "m-dimensional charts".

Proof Suppose M is an m-dimensional C^q submanifold of $\mathbb{R}^n$ and $p \in M$. Then, according to Theorem 9.13, there is an m-dimensional C^q chart (φ, U) around p. Let (ψ, V) be an m'-dimensional C^q chart around p. Then the proof of Proposition 9.15 shows that

$$\psi \circ \varphi^{-1} \in \mathrm{Diff}^q\big(\varphi(U \cap V), \psi(U \cap V)\big) \ ,$$

where $\varphi(U \cap V)$ is open in $\mathbb{R}^m$ and $\psi(U \cap V)$ is open in $\mathbb{R}^{m'}$. This implies $m = m'$ (see (7.2)), as desired. ∎

(b) Through the charts (φ_1, U_1), the charted territories U_1 can described using the local coordinates $(x^1, \ldots, x^m) = \varphi_1(q) \in \mathbb{R}^m$ for $q \in U$. If (φ_2, U_2) is a second chart, then U_2 has its own local coordinates $(y^1, \ldots, y^m) = \varphi_2(q) \in \mathbb{R}^m$. Thus $U_1 \cap U_2$ has a description in two coordinate systems $(x^1, \ldots, x^m)$ and $(y^1, \ldots, y^m)$. The transition function $\varphi_2 \circ \varphi_1^{-1}$ is nothing other than the **coordinate transformation** $(x^1, \ldots, x^m) \mapsto (y^1, \ldots, y^m)$ that converts the coordinates x into the coordinates y.[7] ∎

Exercises

1 Let H be a finite-dimensional real Hilbert space, and let $\varphi : H \to H$ be an isometry with $\varphi(0) = 0$. Verify that

(a) φ is linear;

(b) $\varphi^* \varphi = \mathrm{id}_H$;

---

[6]Here and elsewhere, we often apply the same symbol for a map and its restriction to a subset of its domain of definition.

[7]Coordinate transformations will be discussed in detail in Volume III.

(c) $\varphi \in \mathcal{L}\mathrm{aut}(H)$, and φ^{-1} is an isometry.

2 (a) For $A \in \mathbb{R}^{n \times n}$, show these statements are equivalent:

(i) A is an isometry;

(ii) $A \in O(n)$;

(iii) $|\det A| = 1$;

(iv) the column vectors $a_k^\bullet = (a_k^1, \ldots, a_k^n)$ for $k = 1, \ldots, n$ form an ONB of $\mathbb{R}^n$;

(v) the row vectors $a_\bullet^j = (a_1^j, \ldots, a_n^j)$ for $k = 1, \ldots, n$ form an ONB of $\mathbb{R}^n$.

(b) Show that $O(n)$ is a matrix group.

3 Prove the statement of Example 9.1(c).

4 Suppose M and N are respectively m and n-dimensional C^q submanifolds of $\mathbb{R}^k$ and $\mathbb{R}^\ell$. Prove that $M \times N$ is an $(m + n)$-dimensional C^q submanifold of $\mathbb{R}^{k+\ell}$.

5 Decide whether $\mathcal{L}\mathrm{aut}(\mathbb{R}^n)$ is a submanifold of $\mathcal{L}(\mathbb{R}^n)$.

6 Define $f : \mathbb{R}^3 \to \mathbb{R}^2$ through

$$f(x,y,z) := (x^2 + xy - y - z, 2x^2 + 3xy - 2y - 3z) .$$

Show that $f^{-1}(0)$ is an embedded curve in $\mathbb{R}^3$.

7 Suppose $f, g : \mathbb{R}^4 \to \mathbb{R}^3$ is given by

$$f(x,y,z,u) := (xz - y^2, yu - z^2, xu - yz) ,$$
$$g(x,y,z,u) := \left(2(xz + yu), 2(xu - yz), z^2 + u^2 - x^2 - y^2\right) .$$

Verify that

(a) $f^{-1}(0) \backslash \{0\}$ is an embedded surface in $\mathbb{R}^4$;

(b) for every $a \in \mathbb{R}^3 \backslash \{0\}$, $g^{-1}(a)$ is an embedded curve in $\mathbb{R}^4$.

8 Which of the sets

(a) $K := \left\{ (x,y) \in \mathbb{R}^n \times \mathbb{R} ; |x|^2 = y^2 \right\}$,

(b) $\left\{ (x,y) \in K ; y > 0 \right\}$,

(c) $K \backslash \{(0,0)\}$

are submanifolds of $\mathbb{R}^{n+1}$?

9 For $A \in \mathbb{R}_{\mathrm{sym}}^{n \times n}$, show $\left\{ x \in \mathbb{R}^n ; (x \,|\, Ax) = 1 \right\}$ is a smooth hypersurface of $\mathbb{R}^n$. Sketch the curve for $n = 2$ and the surface for $n = 3$. (Hint: Exercise 4.5.)

10 Show for the **special orthogonal group** $SO(n) := \left\{ A \in O(n) ; \det A = 1 \right\}$ that

(a) $SO(n)$ is a subgroup of $O(n)$;

(b) $SO(n)$ is a smooth submanifold of $\mathbb{R}^{n \times n}$. What is its dimension?

11 Recursively define $f_n \in C^\infty(\mathbb{R}^n, \mathbb{R}^n)$ through

$$f_2(y) := (y^1 \cos y^2, y^1 \sin y^2) \qquad \text{for } y = (y^1, y^2) \in \mathbb{R}^2 ,$$

and
$$f_{n+1}(y) := \left(f_n(y')\sin y^{n+1}, y^1\cos y^{n+1}\right) \quad \text{for } y = (y', y^{n+1}) \in \mathbb{R}^n \times \mathbb{R}.$$
Further let $V_2 := (0,\infty) \times (0, 2\pi)$ and $V_n := V_2 \times (0,\pi)^{n-2}$ for $n \geq 3$.

(a) Find f_n explicitly.

(b) Show that

   (i) $|f_n(1, y^2, \ldots, y^n)| = 1$ and $|f_n(y)| = |y^1|$;

   (ii) $f_n(V_n) = \mathbb{R}^n \backslash H_n$ with $H_n := \mathbb{R}^+ \times \{0\} \times \mathbb{R}^{n-2}$;

   (iii) $f_n(\overline{V}_n) = \mathbb{R}^n$;

   (iv) $g_n := f_n|V_n : V_n \to \mathbb{R}^n \backslash H_n$ is topological;

   (v) $\det \partial f_n(y) = (-1)^n (y^1)^{n-1} \sin^{n-2}(y^n) \cdots \sin(y^3)$ for $y \in V_n$ and $n \geq 3$.

Thus g_n is a C^∞ embedding of V_n in $\mathbb{R}^n$. The coordinates induced by f_n are called **n-dimensional polar** (or **spherical**) **coordinates** (see Example 9.11(a)).

12   Denote by $f : \mathbb{R}^3 \to \mathbb{R}^3$ the cylindrical coordinates in $\mathbb{R}^3$ (see Example 9.11(c)). Further let $V := (0,\infty) \times (0, 2\pi) \times \mathbb{R}$ and $g := f|V$. Prove that

(a) g is an C^∞ embedding of V in $\mathbb{R}^3$;

(b) $g(1, \cdot, \cdot)$ is a C^∞ embedding of $(0, 2\pi) \times \mathbb{R}$ in $\mathbb{R}^3$;

(c) $f(V) = \mathbb{R}^3 \backslash H_3$ (see Exercise 11);

(d) $f(\overline{V}) = \mathbb{R}^3$.

13   (a) Show that the **elliptical cylinder**[8]
$$M_{a,b} := \left\{ (x, y, z) \in \mathbb{R}^3 \;;\; x^2/a^2 + y^2/b^2 = 1 \right\} \quad \text{for } a, b \in (0, \infty),$$
is a C^∞ hypersurface in $\mathbb{R}^3$.

(b) Suppose $W := (0, 2\pi) \times \mathbb{R}$ and
$$f_1 : W \to \mathbb{R}^3 \quad \text{for } (\varphi, z) \mapsto (a\cos\varphi, b\sin\varphi, z),$$
$$f_2 : W \to \mathbb{R}^3 \quad \text{for } (\varphi, z) \mapsto (-a\sin\varphi, b\cos\varphi, z).$$
Further let $U_j := f_j|W$ and $\varphi_j := (f_j|U_j)^{-1}$ for $j = 1, 2$. Show that $\{(\varphi_1, \varphi_2)\}$ is an atlas of M, and calculate the transition function $\varphi_1 \circ \varphi_2^{-1}$.

14   Suppose M is a nonempty compact m-dimensional C^1 submanifold of $\mathbb{R}^n$ with $m \geq 1$. Prove that M does not have an atlas with only one chart.

15   Show that the surface given by the last map of Example 9.11(e) is not a submanifold of $\mathbb{R}^3$.

16   For $g : (-\pi/4, \pi/2) \to \mathbb{R}^2$, $t \mapsto \sin(2t)(-\sin t, \cos t)$ verify that

(a) g is an injective C^∞ immersion;

(b) $\operatorname{im}(g)$ is not an embedded curve in $\mathbb{R}^2$.

17   Calculate the transition function $\varphi_- \circ \varphi_+^{-1}$ for the atlas $\{\varphi_-, \varphi_+\}$, where $\varphi_\pm$ are the stereographic projections of S^n.

---

[8] $M_{a,a}$ is the usual **cylinder**.

18 Suppose M and N are respectively submanifolds of $\mathbb{R}^m$ and $\mathbb{R}^n$, and suppose that $\big\{ (\varphi_\alpha, U_\alpha) \; ; \; \alpha \in \mathsf{A} \big\}$ and $\big\{ (\psi_\beta, V_\beta) \; ; \; \beta \in \mathsf{B} \big\}$ are atlases of M and N. Verify that

$$\big\{ (\varphi_\alpha \times \psi_\beta, U_\alpha \times V_\beta) \; ; \; (\alpha, \beta) \in \mathsf{A} \times \mathsf{B} \big\} ,$$

where $\varphi_\alpha \times \psi_\beta(p, q) := \big(\varphi_\alpha(p), \psi_\beta(q)\big)$, is an atlas of $M \times N$.

10 Tangents and normals

We now introduce linear structures, which make it possible to assign derivatives to maps between submanifolds. These structures will be described using local coordinates, which are of course very helpful in concrete calculations.

To illustrate why we need such a structure, we consider a real function $f : S^2 \to \mathbb{R}$ on the unit sphere S^2 in $\mathbb{R}^3$. The naïve attempt to define the derivative of f through a limit of difference quotients is immediately doomed to fail: For $p \in S^2$ and $h \in \mathbb{R}^3$ with $h \neq 0$, $p + h$ does not generally lie in S^2, and thus the "increment" $f(p + h) - f(p)$ of f is not even defined at the point p.

The tangential in $\mathbb{R}^n$

We begin with the simple situation of an n-dimensional submanifold of $\mathbb{R}^n$ (see Example 9.1(a)). Suppose X is open in $\mathbb{R}^n$ and $p \in X$. The **tangential space** T_pX of X at point p is the set $\{p\} \times \mathbb{R}^n$ equipped with the induced Euclidean vector space structure of $\mathbb{R}^n = \big(\mathbb{R}^n, (\cdot\,|\,\cdot)\big)$, that is,

$$(p, v) + \lambda(p, w) := (p, v + \lambda w) \text{ and } \big((p, v)\,|\,(p, w)\big)_p := (v\,|\,w)$$

for $(p, v), (p, w) \in T_pX$ and $\lambda \in \mathbb{R}$. The element $(p, v) \in T_pX$ is called a **tangential vector** of X at p and will also be denoted by $(v)_p$. We call v the **tangent part** of $(v)_p$.[1]

10.1 Remark The tangential space T_pX and $\mathbb{R}^n$ are isometric isomorphic Hilbert spaces. One can clearly express the isometric isomorphism by "attaching" $\mathbb{R}^n$ at the point $p \in X$. ∎

Suppose Y is open in $\mathbb{R}^\ell$ and $f \in C^1(X, Y)$. Then the linear map

$$T_pf : T_pX \to T_{f(p)}Y , \quad (p, v) \mapsto \big(f(p), \partial f(p)v\big)$$

is called the **tangential** of f at the point p.

10.2 Remarks (a) Obviously

$$T_pf \in \mathcal{L}(T_pX, T_{f(p)}Y) .$$

Because $v \mapsto f(p) + \partial f(p)v$ approximates (up to first order) the map f at the point p when v is in a null neighborhood of $\mathbb{R}^n$, we know $\mathrm{im}(T_pf)$ is a vector subspace of $T_{f(p)}Y$, which approximates (to first order) $f(X)$ at the point $f(p)$.

---

[1]We distinguish "tangential" from "tangent": a tangential vector contains the "base point" p, whereas its tangent part does not.

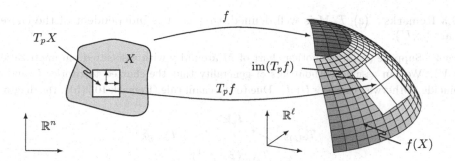

(b) If Z is open in $\mathbb{R}^s$ and $g \in C^1(Y, Z)$, the **chain rule** reads

$$T_p(g \circ f) = T_{f(p)}g \circ T_p f .$$

In other words, the following are commutative diagrams.

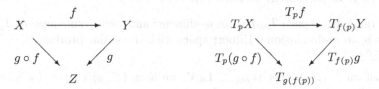

Proof This follows from the chain rule for C^1 maps (see Theorem 3.3). ∎

(c) For $f \in \mathrm{Diff}^1(X, Y)$, we have

$$T_p f \in \mathcal{L}is(T_p X, T_{f(p)}Y) \quad \text{and} \quad (T_p f)^{-1} = T_{f(p)}f^{-1} \quad \text{for } p \in X .$$

Proof This is a consequence of (b). ∎

The tangential space

In the rest of this section, suppose

- M is an m-dimensional C^q submanifold of $\mathbb{R}^n$;
  $q \in \mathbb{N}^\times \cup \{\infty\}$;
  $p \in M$, and (φ, U) is a chart of M around p;
  (g, V) is the parametrization belonging to (φ, U).

The **tangential space** $T_p M$ of M at the point p is the image of $T_{\varphi(p)}V$ under $T_{\varphi(p)}g$, and therefore $T_p M = \mathrm{im}(T_{\varphi(p)}g)$. The elements of $T_p M$ are called **tangential vectors** of M at p, and $TM := \bigcup_{p \in M} T_p M$ is the **tangential bundle**[2] of M.

---

[2]For $p \in M$, we have the inclusion $T_p M \subset M \times \mathbb{R}^n$. Therefore TM is a subset of $M \times \mathbb{R}^n$. We use the simpler term "tangent" here because it is more standard and because there is no need for the "bundle" of tangent parts.

10.3 Remarks   (a) T_pM is well defined, that is, it is independent of the chosen chart (φ, U).

Proof Suppose $(\widetilde{\varphi}, \widetilde{U})$ is another chart of M around p with associated parametrization $(\widetilde{g}, \widetilde{V})$. We can assume without loss of generality that the charted territories U and $\widetilde{U}$ coincide. Otherwise consider $U \cap \widetilde{U}$. Due to the chain rule (Remark 10.2(b)), the diagram

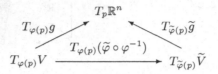

commutes. Because of Proposition 9.15 and Remark 10.2(c), $T_{\varphi(p)}(\widetilde{\varphi} \circ \varphi^{-1})$ is an isomorphism, which then easily implies the claim. ∎

(b) If M is an subset of $\mathbb{R}^n$, then the above definition of T_pM agrees with that of Remark 10.1. In particular, we have $TM = M \times \mathbb{R}^n$.

(c) The tangential space T_pM is an m-dimensional vector subspace of $T_p\mathbb{R}^n$ and therefore is an m-dimensional Hilbert space with the scalar product $(\cdot | \cdot)_p$ induced from $T_p\mathbb{R}^n$.

Proof Let $x_0 := \varphi(p)$. For $(v)_{x_0} \in T_{x_0}V$, we have $(T_{x_0}g)(v)_{x_0} = (p, \partial g(x_0)v)$. It therefore follows that

$$\operatorname{rank} T_{x_0}g = \operatorname{rank} \partial g(x_0) = m \ ,$$

which proves the claim. ∎

(d) Because $T_{\varphi(p)}g : T_{\varphi(p)}V \to T_p\mathbb{R}^n$ is injective and $T_pM = \operatorname{im}(T_{\varphi(p)}g)$, there is exactly one $A \in \mathcal{L}\mathrm{is}(T_pM, T_{\varphi(p)}V)$ such that $(T_{\varphi(p)}g)(A) = i_{T_pM}$, where i_{T_pM} is the canonical injection of T_pM into $T_p\mathbb{R}^n$. In other words, A is the inverse of $T_{\varphi(p)}g$, when $T_{\varphi(p)}g$ is understood as a map of $T_{\varphi(p)}V$ onto its image T_pM. We call $T_p\varphi := A$ the **tangential of the chart** φ at point p. Further, $(T_p\varphi)v \in T_{\varphi(p)}V$ is the **representation** of the tangential vector $v \in T_pM$ in the **local coordinates** induced by φ. If $(\widetilde{\varphi}, \widetilde{U})$ is another chart of M around p, then the diagram

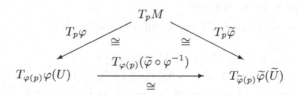

commutes, where $\cong$ means "isomorphic".

Proof Without loss of generality, assume $U = \widetilde{U}$. For $\widetilde{g} := i_M \circ \widetilde{\varphi}^{-1}$, it follows from $\widetilde{g} = (i_M \circ \varphi^{-1}) \circ (\varphi \circ \widetilde{\varphi}^{-1}) = g \circ (\varphi \circ \widetilde{\varphi}^{-1})$ and the chain rule of Remark 10.2(b) that

$$T_{\widetilde{\varphi}(p)}\widetilde{g} = T_{\varphi(p)}g \, T_{\widetilde{\varphi}(p)}(\varphi \circ \widetilde{\varphi}^{-1}) \ .$$

From this, the definition of $T_p\varphi$ and $T_p\widetilde{\varphi}$, Remark 10.2(c), and Proposition 9.15, we get the relation

$$T_p\widetilde{\varphi} = \left(T_{\widetilde{\varphi}(p)}(\varphi \circ \widetilde{\varphi}^{-1})\right)^{-1} T_p\varphi = T_{\varphi(p)}(\widetilde{\varphi} \circ \varphi^{-1})\, T_p\varphi \;,$$

which proves the claim. ∎

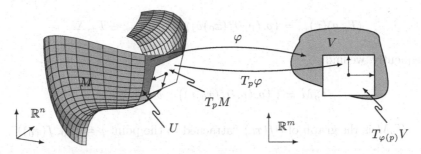

(e) (the scalar product in local coordinates) Suppose $x_0 := \varphi(p)$ and

$$g_{jk}(x_0) := \left(\partial_j g(x_0)\,\middle|\,\partial_k g(x_0)\right) \quad \text{for } 1 \le j, k \le m \;.$$

Then $[g_{jk}] \in \mathbb{R}^{m\times m}_{\mathrm{sym}}$ is called the (**first**) **fundamental matrix** of M with respect to the chart φ at p (or the parametrization g at x_0). It is positive definite.

For $v, w \in T_pM$, we have[3]

$$(v\,|\,w)_p = \sum_{j,k=1}^{m} g_{jk}(x_0) v^j w^k \;, \tag{10.1}$$

where v^j and w^k are respectively the components of the local representation of the tangent part of $(T_p\varphi)v$ and $(T_p\varphi)w$ with respect to the standard basis.

Proof For $\xi, \eta \in \mathbb{R}^m$, we have

$$\sum_{j,k=1}^{m} g_{jk}(x_0)\xi^j\eta^k = \left(\partial g(x_0)\xi \,\middle|\, \partial g(x_0)\eta\right) \;. \tag{10.2}$$

In particular, because $\partial g(x_0)$ is injective, we have

$$\left([g_{jk}](x_0)\xi \,\middle|\, \xi\right) = |\partial g(x_0)\xi|^2 > 0 \quad \text{for } \xi \in \mathbb{R}^m \setminus \{0\} \;.$$

Thus the fundamental matrix is positive definite.

Because $v = T_{x_0}g\,(T_p\varphi)v$ and $(T_p\varphi)v = \sum_{j=1}^{m} v^j e_j$ it follows from the definition of $(\cdot\,|\,\cdot)_p$ that

$$(v\,|\,w)_p = \left((T_{x_0}g)(T_p\varphi)v \,\middle|\, (T_{x_0}g)(T_p\varphi)w\right)_p = \left(\partial g(x_0)\sum_{j=1}^{m} v^j e_j \,\middle|\, \partial g(x_0)\sum_{k=1}^{m} w^k e_k\right) .$$

Consequently (10.1) implies (10.2). ∎

---

[3]See also Remark 2.17(c).

10.4 Example (the tangential space of a graph) Suppose X is open in $\mathbb{R}^m$, and let $f \in C^q(X, \mathbb{R}^\ell)$. According to Proposition 9.2, $M := \text{graph}(f)$ is an m-dimensional C^q submanifold of $\mathbb{R}^m \times \mathbb{R}^\ell = \mathbb{R}^{m+\ell}$. Then $g(x) := \big(x, f(x)\big)$ for $x \in X$ is a C^q parametrization of M, and with $p = \big(x_0, f(x_0)\big) \in M$, we have

$$(T_{x_0}g)(v)_{x_0} = \big(p, (v, \partial f(x_0)v)\big) \quad \text{for } (v)_{x_0} \in T_{x_0}X .$$

Consequently, we find

$$T_p M = \big\{ \, \big(p, (v, \partial f(x_0)v)\big) \, ; \, v \in \mathbb{R}^m \, \big\} ,$$

that is, $T_p M$ is the graph of $\partial f(x_0)$ "attached to the point $p = \big(x_0, f(x_0)\big)$".

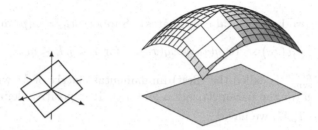

The representation of $T_p M$ in $\mathbb{R}^n$ for $n = m + \ell$ is obtained by identifying $(\eta)_p = (p, \eta) \in T_p \mathbb{R}^n$ with $p + \eta \in \mathbb{R}^n$. Then it follows that

$$T_p M \ni \Big(\big(x_0, f(x_0)\big), (v, \partial f(x_0)v) \Big)$$
$$= \big(x_0 + v, f(x_0) + \partial f(x_0)v\big) \in \mathbb{R}^m \times \mathbb{R}^\ell = \mathbb{R}^n ,$$

and we get

$$T_p M = \big\{ \, \big(x, f(x_0) + \partial f(x_0)(x - x_0)\big) \, ; \, x \in \mathbb{R}^m \, \big\}$$
$$= \text{graph}\big(x \mapsto f(x_0) + \partial f(x_0)(x - x_0)\big) .$$

This representation shows once more that $T_p M$ is the higher-dimensional generalization of the concept of the tangent to a curve (see Remark IV.1.4). ∎

Suppose $\varepsilon > 0$ with $\varphi(p) + te_j \in V$ for $t \in (-\varepsilon, \varepsilon)$ and $j \in \{1, \dots, m\}$. Then

$$\gamma_j(t) := g\big(\varphi(p) + te_j\big) \quad \text{for } t \in (-\varepsilon, \varepsilon) ,$$

is called the j-th **coordinate path** through p.

10.5 Remark   For $x_0 := \varphi(p)$, we have

$$T_pM = \mathrm{span}\big\{ \big(\partial_1 g(x_0)\big)_p, \ldots, \big(\partial_m g(x_0)\big)_p \big\} ,$$

that is, *the tangential vectors at p on the coordinate
paths γ_j form a basis of T_pM.*

Proof   For the j-th column of $\big[\partial g(x_0)\big]$, we have

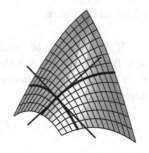

$$\partial_j g(x_0) = \partial g(x_0)e_j = \dot{\gamma}_j(0) .$$

The remark follows because g is an immersion, and consequently the column vectors of
$\big[\partial g(x_0)\big]$ are linearly independent, and $\dim T_pM = m$. ∎

Characterization of the tangential space

Tangential vectors of M at p can be described as tangential vectors of regular paths
in M. The next theorem provides a geometric interpretation of the tangential
space.

10.6 Theorem   *For every $p \in M$, we have*

$$T_pM = \Big\{ (v)_p \in T_p\mathbb{R}^n \ ;$$

$$\exists\, \varepsilon > 0, \ \exists\, \gamma \in C^1\big((-\varepsilon,\varepsilon),\mathbb{R}^n\big) \text{ such that } \mathrm{im}(\gamma) \subset M, \ \gamma(0) = p, \ \dot{\gamma}(0) = v \Big\}.$$

In other words, for every $(v)_p \in T_pM \subset T_p\mathbb{R}^n$, there is a C^1 path in $\mathbb{R}^n$ passing
through p that is entirely contained in M and has $(v)_p$ as its tangential vector at
p. Every tangential vector of such a path belongs to T_pM.

Proof   (i) Suppose $(v)_p \in T_pM$ and $x_0 := \varphi(p)$. Then there exists a $\xi \in \mathbb{R}^m$ such
that $v = \partial g(x_0)\xi$. Because $V = \varphi(U)$ at $\mathbb{R}^m$ is open, there is an $\varepsilon > 0$ such that
$x_0 + t\xi \in V$ for $t \in (-\varepsilon,\varepsilon)$. We now set $\gamma(t) := g(x_0 + t\xi)$ for $t \in (-\varepsilon,\varepsilon)$, so that
γ is a C^1 path in M with $\gamma(0) = p$ and $\dot{\gamma}(0) = \partial g(x_0)\xi = v$.

(ii) Suppose $\gamma \in C^1\big((-\varepsilon,\varepsilon),\mathbb{R}^n\big)$ for $\mathrm{im}(\gamma) \subset M$ and $\gamma(0) = p$. According
to Remark 9.9(d), there is an open neighborhood $\widetilde{V}$ of $(x_0,0)$ in $\mathbb{R}^n$, an open
neighborhood $\widetilde{U}$ in $\mathbb{R}^n$, and a $\psi \in \mathrm{Diff}^q(\widetilde{V},\widetilde{U})$ such that $\psi(x,0) = g(x)$ for $x \in V$.
By shrinking ε, we can assume that $\mathrm{im}(\gamma) \subset U \cap \widetilde{U}$. From this, it follows that

$$\gamma(t) = (g \circ \varphi \circ \gamma)(t) = (g \circ \mathrm{pr}_{\mathbb{R}^m} \circ \psi^{-1} \circ \gamma)(t) ,$$

and we get from the chain rule that

$$\dot{\gamma}(0) = \partial g(x_0)(\mathrm{pr}_{\mathbb{R}^m} \circ \psi^{-1} \circ \gamma)^{\cdot}(0) .$$

For $\xi := (\mathrm{pr}_{\mathbb{R}^m} \circ \psi^{-1} \circ \gamma)\dot{}(0) \in \mathbb{R}^m$ and $v := \partial g(x_0)\xi \in \mathbb{R}^n$, we have $(v)_p \in T_pM$, and we are done. ∎

If X is open in $\mathbb{R}^n$ and the point $c \in \mathbb{R}^\ell$ is a regular value of $f \in C^q(X, \mathbb{R}^\ell)$, we know from Theorem 9.3 that $M = f^{-1}(c)$ is a C^q submanifold of $\mathbb{R}^n$ of dimension $(n - \ell)$. In the next theorem, we show that an analogous statement holds for the "linearization", that is,

$$T_pM = (T_pf)^{-1}(c) = \ker(T_pf) \ .$$

10.7 Theorem (regular value) *Suppose X is open in $\mathbb{R}^n$, and $c \in \mathbb{R}^\ell$ is a regular value of $f \in C^q(X, \mathbb{R}^\ell)$. For the $(n - \ell)$-dimensional C^q submanifold $M := f^{-1}(c)$ of $\mathbb{R}^n$, we then have $T_pM = \ker(T_pf)$ for $p \in M$.*

Proof Suppose $(v)_p \in T_pM \subset T_p\mathbb{R}^n$. According to Theorem 10.6, there is an $\varepsilon > 0$ and a path $\gamma \in C^1\big((-\varepsilon, \varepsilon), \mathbb{R}^n\big)$ such that $\mathrm{im}(\gamma) \subset M$, with $\gamma(0) = p$ and $\dot{\gamma}(0) = v$. In particular, we have $f(\gamma(t)) = c$ for every $t \in (-\varepsilon, \varepsilon)$, and we find by differentiating this relation that

$$\partial f\big(\gamma(0)\big)\dot{\gamma}(0) = \partial f(p)v = 0 \ .$$

It follows that $T_pM \subset \ker(T_pf)$.

Because p is a regular point of f, we have

$$\dim\big(\mathrm{im}(T_pf)\big) = \dim\big(\mathrm{im}(\partial f(p))\big) = \ell \ .$$

Consequently, the rank formula (2.4) gives

$$\dim\big(\ker(T_pf)\big) = n - \ell = \dim(T_pM) \ .$$

Therefore T_pM is not a proper vector subspace of $\ker(T_pf)$. ∎

Differentiable maps

Suppose N is a C^r submanifold of $\mathbb{R}^\ell$ and $1 \le s \le \min\{q, r\}$. Also suppose $f \in C(M, N)$ and (ψ, W) is a chart of N around $f(p)$. Then $U \cap f^{-1}(W)$ is an open neighborhood of p in M. Therefore by shrinking U, we can assume without loss of generality that $f(U) \subset W$. The function f is said to be (**s-times**) [**continuously**] **differentiable** at p if the map

$$f_{\varphi,\psi} := \psi \circ f \circ \varphi^{-1} : \varphi(U) \to \psi(W)$$

at point $\varphi(p)$ is (s-times) [continuously] differentiable.

We say the map $f \in C(M, N)$ is (**s-times**) [**continuously**] **differentiable** if f is (s-times) [continuously] differentiable at every point M. We denote the set of all s-times continuously differentiable functions from M to N by $C^s(M, N)$, and

$$\text{Diff}^s(M, N) := \left\{ f \in C^s(M, N) \; ; \; f \text{ is bijective}, \; f^{-1} \in C^s(N, M) \right\}$$

is the set of all C^s **diffeomorphisms** from M to N. Finally, we say that M and N are C^s-**diffeomorphic** if $\text{Diff}^s(M, N)$ is nonempty.

10.8 Remarks (a) The previous definitions are independent of the choice of charts.

Proof If $(\widetilde{\varphi}, \widetilde{U})$ and $(\widetilde{\psi}, \widetilde{W})$ are other charts of M around p and of N around $f(p)$, respectively, such that $f(\widetilde{U}) \subset \widetilde{W}$, then

$$f_{\widetilde{\varphi},\widetilde{\psi}} = \widetilde{\psi} \circ f \circ \widetilde{\varphi}^{-1} = (\widetilde{\psi} \circ \psi^{-1}) \circ f_{\varphi,\psi} \circ (\varphi \circ \widetilde{\varphi}^{-1}) . \tag{10.3}$$

This, with Proposition 9.15 and the chain rule, then gives the theorem. ∎

(b) If M and N respectively have dimension n and ℓ, that is, M is open in $\mathbb{R}^n$ and N is open in $\mathbb{R}^\ell$, then the above definition agrees with the one in Section 5.

(c) The function $f_{\varphi,\psi}$ is the **local representation** of f in the charts φ and ψ, or the **local coordinate representation**. In contrast with the function f, which maps between the "curved sets" M and N, $f_{\varphi,\psi}$ is a map between open subsets of Euclidean spaces.

(d) It is generally not possible to define sensibly the concept of a C^s map between M and N that is coordinate independent if $s > \min\{q, r\}$.

Proof This follows from (10.3), because the transition functions only belong to C^q or C^r, respectively. ∎

Suppose $f : M \to N$ is differentiable at p, and (ψ, W) is a chart of N around $f(p)$ such that $f(U) \subset W$. Then the diagram

$$
\begin{array}{ccc}
M \supset U & \xrightarrow{\quad f \quad} & W \subset N \\
{\scriptstyle \varphi} \downarrow {\scriptstyle \cong} & & {\scriptstyle \cong} \downarrow {\scriptstyle \psi} \\
\mathbb{R}^m \supset \varphi(U) & \xrightarrow[\quad f_{\varphi,\psi} \quad]{} & \psi(W) \subset \mathbb{R}^{\bar{n}}
\end{array}
\tag{10.4}
$$

commutes, where $\cong$ means C^1-diffeomorphic[4] and $\bar{n}$ is the dimension of N. We

---

[4]Note the local representation $\varphi_{\varphi,\text{id}} = \text{id}_{\varphi(U)}$.

now define the **tangential $T_p f$ of f at p** by requiring the diagram

$$
\begin{array}{ccc}
T_p M & \xrightarrow{\ \ T_p f\ \ } & T_{f(p)} N \\[2pt]
T_p \varphi \Big\downarrow \cong & & \cong \Big\downarrow T_{f(p)} \psi \\[2pt]
T_{\varphi(p)}\varphi(U) & \xrightarrow{\ T_{\varphi(p)} f_{\varphi,\psi}\ } & T_{\psi(f(p))}\psi(W)
\end{array}
\tag{10.5}
$$

commutes, where now $\cong$ means "isomorphic".

10.9 Remarks (a) The tangential $T_p f$ is coordinate independent, and $T_p f \in \mathcal{L}(T_p M, T_{f(p)} N)$.

Proof Suppose $(\widetilde{\varphi}, \widetilde{U})$ is a chart of M around p and $(\widetilde{\psi}, \widetilde{W})$ is a chart around $f(p)$ such that $f(\widetilde{U}) \subset \widetilde{W}$. Then (10.3) and the chain rule of Remark 10.2(b) gives

$$
T_{\widetilde{\varphi}(p)} f_{\widetilde{\varphi},\widetilde{\psi}} = T_{\widetilde{\varphi}(p)}\big((\widetilde{\psi}\circ\psi^{-1})\circ f_{\varphi,\psi}\circ(\varphi\circ\widetilde{\varphi}^{-1})\big)
$$
$$
= T_{\psi(f(p))}(\widetilde{\psi}\circ\psi^{-1})\circ T_{\varphi(p)} f_{\varphi,\psi}\circ T_{\widetilde{\varphi}(p)}(\varphi\circ\widetilde{\varphi}^{-1})\ .
$$

The remark then follows from Remark 10.3(d). ∎

(b) Suppose O is another manifold, and $g: N \to O$ is differentiable at $f(p)$. Then we have the **chain rule**

$$
T_p(g\circ f) = T_{f(p)} g\, T_p f
\tag{10.6}
$$

and

$$
T_p \mathrm{id}_M = \mathrm{id}_{T_p M}\ .
\tag{10.7}
$$

If f belongs to $\mathrm{Diff}^1(M,N)$, then

$$
T_p f \in \mathcal{L}\mathrm{is}(T_p M, T_{f(p)} N) \quad\text{and}\quad (T_p f)^{-1} = T_{f(p)} f^{-1} \quad\text{for } p \in M\ .
$$

Proof The statements (10.6) and (10.7) follow easily from the commutativity of diagram (10.5) and the chain rule of Remark 10.2(b). The remaining claims are immediate consequences of (10.6) and (10.7). ∎

10.10 Examples (a) The canonical injection i_M of M into $\mathbb{R}^n$ is in $C^q(M, \mathbb{R}^n)$, and for $\psi := \mathrm{id}_{\mathbb{R}^n}$, we have

$$
T_{\varphi(p)}(i_M)_{\varphi,\psi} = T_{\varphi(p)} g\ .
$$

Thus $T_p i_M$ is the canonical injection of $T_p M$ into $T_p \mathbb{R}^n$.

Proof This follows obviously from $(i_M)_{\varphi,\psi} = i_M \circ \varphi^{-1} = g$. ∎

(b) Suppose X is an open neighborhood of M and $\widetilde{f} \in C^s(X, \mathbb{R}^\ell)$. Further suppose $\widetilde{f}(M) \subset N$. Then $f := \widetilde{f}\,|\,M$ belongs to $C^s(M,N)$ and

$$
T_{f(p)} i_N\, T_p f = T_p \widetilde{f}\, T_p i_M \quad\text{for } p \in M\ ,
$$

that is, the diagram

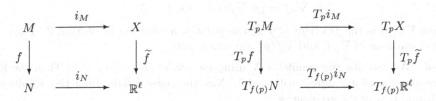

commutes.

Proof Because N is a C^r submanifold of $\mathbb{R}^\ell$, we can assume, possibly by shrinking W, that there exists an open neighborhood $\widetilde{W}$ of W in $\mathbb{R}^\ell$ and a C^r diffeomorphism Ψ of $\widetilde{W}$ on an open subset of $\mathbb{R}^\ell$ such that $\Psi \supset \psi$. Thus we find

$$f_{\varphi,\psi} = \psi \circ f \circ \varphi^{-1} = \Psi \circ \widetilde{f} \circ g \in C^s(V, \mathbb{R}^\ell) \ .$$

Because this holds for every pair of charts (φ, U) of M and (ψ, W) of N such that $f(U) \subset W$, we can conclude that f belongs to $C^s(M, N)$. Because $i_N \circ f = \widetilde{f} \circ i_M$, the last part of the claim is a consequence of the chain rule. ∎

(c) Suppose X is open in $\mathbb{R}^n$, Y is open in $\mathbb{R}^\ell$, and $f \in C^q(X, \mathbb{R}^\ell)$ with $f(X) \subset Y$. Then f belongs to $C^q(X, Y)$, where X and Y are respectively understood as n- and ℓ-dimensional submanifolds of $\mathbb{R}^n$ and $\mathbb{R}^\ell$. In addition, we have

$$T_p f = \big(f(p), \partial f(p)\big) \quad \text{for } p \in X \ .$$

Proof This is a special case of (b). ∎

The differential and the gradient

If $f : M \to \mathbb{R}$ is differentiable at p, then

$$T_p f : T_p M \to T_{f(p)}\mathbb{R} = \{f(p)\} \times \mathbb{R} \subset \mathbb{R} \times \mathbb{R} \ .$$

With the canonical projection $\mathrm{pr}_2 : \mathbb{R} \times \mathbb{R} \to \mathbb{R}$ onto the second factor, we set

$$d_p f := \mathrm{pr}_2 \circ T_p f \in \mathcal{L}(T_p M, \mathbb{R}) = (T_p M)' \ ,$$

and call $d_p f$ **differential** of f at the point p. Therefore $d_p f$ is a continuous linear form on $T_p M$. Because the tangential space is an m-dimensional Hilbert space, there is, using the Riesz representation theorem, a unique vector $\nabla_p f := \nabla_p^M f \in T_p M$ such that

$$(d_p f)v = (\nabla_p f | v)_p \quad \text{for } v \in T_p M \ ,$$

which we call the **gradient of f at the point p**.

10.11 Remarks (a) Suppose X is open in $\mathbb{R}^n$ and $f \in C^1(X, \mathbb{R})$. Then

$$\nabla_p f = \big(p, \nabla f(p)\big) \quad \text{for } p \in X ,$$

where $\nabla f(p)$ is the gradient of f at the point p, as defined in Section 2. Thus, the two definitions of $\nabla_p f$ and $\nabla f(p)$ are consistent.

Proof We describe the manifolds X using the trivial chart (id_X, X). Then the local representation of $d_p f$ agrees with $\partial f(p)$. Now the claim follows from the definition of $(\cdot|\cdot)_p$ and that of the gradient. ∎

(b) (representation in local coordinates) Suppose $f \in C^1(M, \mathbb{R})$, and $f_\varphi := f \circ \varphi^{-1}$ is the local representation of f in the charts φ of M and $\mathrm{id}_\mathbb{R}$ of $\mathbb{R}$, that is, $f_\varphi = f_{\varphi, \mathrm{id}_\mathbb{R}}$. Further suppose $[g^{jk}]$ is the inverse of the fundamental matrix $[g_{jk}]$ with respect to φ at p. Then, in the local coordinates induced by φ, the local representation $(T_p\varphi)\nabla_p f$ of the gradient $\nabla_p f \in T_p M^5$ has tangent part

$$\Big(\sum_{k=1}^m g^{1k}(x_0)\partial_k f_\varphi(x_0), \ldots, \sum_{k=1}^m g^{mk}(x_0)\partial_k f_\varphi(x_0) \Big)$$

where $x_0 := \varphi(p)$.

Proof Using the definitions of $\nabla_p f$ and $d_p f$, it follows from Proposition 2.8(i) that

$$(\nabla_p f \,|\, v)_p = (d_p f)v = \partial(f \circ \varphi^{-1})(x_0)(T_p\varphi)v = \sum_{j=1}^m \partial_j f_\varphi(x_0)v^j \tag{10.8}$$

for $v \in T_p M$ and $(T_p\varphi)v = \sum_{j=1}^m v^j e_j$. For $(T_p\varphi)\nabla_p f = \sum_{j=1}^m w^j e_j$, we get using Remark 10.3(e) that

$$(\nabla_p f \,|\, v)_p = \sum_{j,k=1}^m g_{jk}(x_0)w^j v^k \quad \text{for } v \in T_p M . \tag{10.9}$$

Now (10.8), (10.9), and the symmetry of $[g_{jk}]$ imply

$$\sum_{k=1}^m g_{jk}(x_0)w^k = \partial_j f_\varphi(x_0) \quad \text{for } 1 \le j \le m ,$$

and therefore, after multiplication by $[g_{jk}]^{-1} = [g^{jk}]$, we are done. ∎

The next theorem gives a necessary condition for $p \in M$ to be an extremal point of a differentiable real-valued function on M. This generalizes Theorem 3.13.

10.12 Theorem If $p \in M$ is a local extremal point of $f \in C^1(M, \mathbb{R})$, then $\nabla_p f = 0$.

Proof Because $f_\varphi \in C^1(V, \mathbb{R})$ has a local extremum at $x_0 = \varphi(p)$, we have $\partial f_\varphi(x_0) = 0$ (see Proposition 2.5 and Theorem 3.13). Thus, for $v \in T_p M$ and the tangent part ξ of $(T_p\varphi)v$, we have

$$(\nabla_p f \,|\, v)_p = (d_p f)v = \partial f_\varphi(x_0)\xi = 0 . \quad ∎$$

---

[5] Compare this to formula (2.6).

Normals

The orthogonal complement of T_pM at $T_p\mathbb{R}^n$ is called the **normal space** of M at p and will be denoted by $T_p^\perp M$. The vectors in $T_p^\perp M$ are the **normals** of M at p, and $T^\perp M := \bigcup_{p \in M} T_p^\perp M$ is the **normal bundle** of M.

10.13 Proposition *Suppose X is open in $\mathbb{R}^n$, and c is a regular value of $f \in C^q(X, \mathbb{R}^\ell)$. If $M := f^{-1}(c)$ is not empty, $\{\nabla_p f^1, \ldots, \nabla_p f^\ell\}$ is a basis of $T_p^\perp M$.*

Proof (i) According to Remark 10.11(a), $\nabla_p f^j$ has the tangent part $\nabla f^j(p)$. From the surjectivity of $\partial f(p)$, it follows that the vectors $\nabla f^1(p), \ldots, \nabla f^\ell(p)$ in $\mathbb{R}^n$ — and thus $\nabla_p f^1, \ldots, \nabla_p f^\ell$ in $T_p\mathbb{R}^n$ — are linearly independent.

(ii) Suppose $v \in T_pM$. From Theorem 10.6, we know that there is an $\varepsilon > 0$ and a $\gamma \in C^1\big((-\varepsilon, \varepsilon), \mathbb{R}^n\big)$ such that $\mathrm{im}(\gamma) \subset M$, $\gamma(0) = p$, and $\gamma(0) = v$. Because $f^j\big(\gamma(t)\big) = c^j$ for $t \in (-\varepsilon, \varepsilon)$, it follows that

$$0 = (f^j \circ \gamma)^{\cdot}(0) = \big(\nabla f^j\big(\gamma(0)\big)\,\big|\,\dot\gamma(0)\big) = (\nabla_p f^j \,|\, v)_p \quad \text{for } 1 \le j \le \ell .$$

Because this is true for every $v \in T_pM$, we see $\nabla_p f^1, \ldots, \nabla_p f^\ell$ belongs to $T_p^\perp M$. Then $\dim(T_p^\perp M) = n - \dim(T_pM) = \ell$. $\blacksquare$

10.14 Examples
(a) For the sphere S^{n-1} in $\mathbb{R}^n$, we have $S^{n-1} = f^{-1}(1)$ when

$$f : \mathbb{R}^n \to \mathbb{R}, \quad x \mapsto |x|^2 .$$

Because $\nabla_p f = (p, 2p)$, we have

$$T_p^\perp S^{n-1} = (p, \mathbb{R}p) .$$

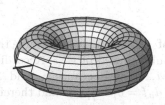

(b) For $X := \mathbb{R}^3 \setminus (\{0\} \times \{0\} \times \mathbb{R})$ and

$$f(x_1, x_2, x_3) := \left(\sqrt{x_1^2 + x_2^2} - 2\right)^2 + x_3^2 ,$$

we have $f \in C^\infty(X, \mathbb{R})$, and[6] $f^{-1}(1) = \mathsf{T}_{2,1}^2$. Then normal vector at the point $p = (p_1, p_2, p_3)$ is given by $\nabla_p f = (p, 2(p-k))$, where

$$k := \left(\frac{2p_1}{\sqrt{p_1^2 + p_2^2}}, \frac{2p_2}{\sqrt{p_1^2 + p_2^2}}, 0\right) .$$

Proof This follows from Example 9.14(e) and recalculation. $\blacksquare$

---

[6] See Example 9.11(f).

(c) Suppose X is open in $\mathbb{R}^n$ and $f \in C^q(X, \mathbb{R})$. Then a **unit normal**, that is, a normal vector of length 1, is given on $M := \operatorname{graph}(f)$ at $p := (x, f(x))$ by $\nu_p := (p, \nu(x)) \in T_p\mathbb{R}^{n+1}$, where

$$\nu(x) := \frac{(-\nabla f(x), 1)}{\sqrt{1 + |\nabla f(x)|^2}} \in \mathbb{R}^{n+1} .$$

Proof For the parametrization g defined by $g(x) := (x, f(x))$ for $x \in X$, we have

$$\partial_j g(x) := (e_j, \partial_j f(x)) \quad \text{for } x \in X , \quad 1 \le j \le n .$$

Clearly $\nu(x)$ is a vector of length 1 in $\mathbb{R}^{n+1}$ and is orthogonal to every vector $\partial_j g(x)$. Therefore the claim follows from Remark 10.5. ∎

Constrained extrema

In many applications, we seek to find extremal points of a function $F: \mathbb{R}^n \to \mathbb{R}$ subject to some constraint. In other words, the constraints mean that not every point of $\mathbb{R}^n$ is allowed to extremize F, and instead the sought-for point must belong to a subset M. Very often these constraints are described by equations of the form $h^1(p) = 0, \ldots, h^\ell(p) = 0$, and the solution set of these equations is a submanifold, namely, the set M. If $F|M$ has a local extremum at $p \in M$, then p is called an **extremal point** of F **under the constraints** $h^1(p) = 0, \ldots, h^\ell(p) = 0$.

The next theorem, which is important in practice, specifies a necessary condition for a point to be a constrained extremal point.

10.15 Theorem (Lagrange multipliers) *Suppose X is open in $\mathbb{R}^n$ and we are given functions $F, h^1, \ldots, h^\ell \in C^1(X, \mathbb{R})$ for $\ell < n$. Also suppose 0 is a regular value of the map $h := (h^1, \ldots, h^\ell)$, and suppose $M := h^{-1}(0)$ is not empty. If $p \in M$ is an extremal point of F under the constraints $h^1(p) = 0, \ldots, h^\ell(p) = 0$, there are unique real numbers $\lambda_1, \ldots, \lambda_\ell$, the **Lagrange multipliers**, for which p is a critical point of*

$$F - \sum_{j=1}^{\ell} \lambda_j h^j \in C^1(X, \mathbb{R}) .$$

Proof From the regular value theorem, we know that M is an $(n-\ell)$-dimensional C^1 submanifold of $\mathbb{R}^n$. According to Example 10.10(b), $f := F|M$ belongs to $C^1(M, \mathbb{R})$, and because $i_\mathbb{R} = \operatorname{id}_\mathbb{R}$, we have $T_p f = T_p F\, T_p i_M$. From this, it follows that $d_p f = d_p F\, T_p i_M$, and therefore

$$\left(\nabla_p f\,|\,(v)_p\right)_p = d_p F T_p i_M(v)_p = \left(\nabla_p F\,|\,T_p i_M(v)_p\right)_p = \left(\nabla F(p)\,|\,v\right) \qquad (10.10)$$

for $(v)_p \in T_p M \subset T_p\mathbb{R}^n$.

If p is a critical point of f, Theorem 10.12 says $\nabla_p f = 0$. Now $\nabla F(p) \in T_p^\perp M$ follows from (10.10). Therefore Proposition 10.13 shows there are unique real numbers $\lambda_1, \ldots, \lambda_\ell$ such that

$$\nabla F(p) = \sum_{j=1}^{\ell} \lambda_j \nabla h^j(p) \, ,$$

which, because of Remark 3.14(a), finishes the proof. ∎

10.16 Remark The Lagrange multiplier method turns the problem of finding the extrema of F under the constraints $h^1(p) = 0, \ldots, h^\ell(p) = 0$ into the problem of finding the critical points of the function

$$F - \sum_{j=1}^{\ell} \lambda_j h^j \in C^1(X, \mathbb{R})$$

(without constraints). We determine the critical points and the Lagrange multipliers by solving the $\ell + n$ equations

$$h^j(p) = 0 \quad \text{for } 1 \leq j \leq \ell \, ,$$

$$\partial_k \Big(F - \sum_{j=1}^{\ell} \lambda_j h^j \Big)(p) = 0 \quad \text{for } 1 \leq k \leq n \, ,$$

for the $n+\ell$ unknowns $p_1, \ldots, p_n, \lambda_1, \ldots, \lambda_\ell$, where $p = (p^1, \ldots, p^n)$. Subsequently, we must find which of these critical points are actually extrema. ∎

Applications of Lagrange multipliers

In the following examples, which are of independent interest, we demonstrate nontrivial applications of the Lagrange multipliers. In particular, we give a short proof of the principal axis transformation theorem, which is shown by other means in linear algebra.

10.17 Examples (a) Given arbitrary vectors $a_j \in \mathbb{R}^n$ for $1 \leq j \leq n$, **Hadamard's inequality** says

$$\big| \det[a_1, \ldots, a_n] \big| \leq \prod_{j=1}^{n} |a_j| \, .$$

Proof (i) From linear algebra, it is known that the determinant is an n-linear function of column vectors. It therefore suffices to verify

$$-1 \leq \det[a_1, \ldots, a_n] \leq 1 \quad \text{for } a_j \in S^{n-1} \text{ and } 1 \leq j \leq n \, .$$

(ii) We set

$$F(x) := \det[x_1, \ldots, x_n] \quad \text{for } h^j(x) := |x_j|^2 - 1 , \quad h = (h^1, \ldots, h^n) ,$$

and $x = (x_1, \ldots, x_n) \in \mathbb{R}^n \times \cdots \times \mathbb{R}^n = \mathbb{R}^{n^2}$. Then F belongs to $C^\infty(\mathbb{R}^{n^2}, \mathbb{R})$, h belongs to $C^\infty(\mathbb{R}^{n^2}, \mathbb{R}^n)$, and[7]

$$[\partial h(x)] = 2 \begin{bmatrix} x_1^\top & & & & \\ & x_2^\top & & 0 & \\ & & \ddots & & \\ & 0 & & x_{n-1}^\top & \\ & & & & x_n^\top \end{bmatrix} \in \mathbb{R}^{n \times n^2} .$$

Clearly the rank of $\partial h(x)$ is maximal for every $x \in h^{-1}(0)$. Therefore 0 is a regular value of h, and $M := h^{-1}(0)$ is an $n(n-1)$-dimensional C^∞ submanifold of $\mathbb{R}^{n^2}$. Further, M is compact because $M = S^{n-1} \times \cdots \times S^{n-1}$. Therefore $f := F \,|\, M \in C^\infty(M, \mathbb{R})$ assumes a minimum and a maximum.

(iii) Suppose $p = (p_1, \ldots, p_n) \in M$ is a extremal point of f. Using Lagrange multipliers, there are $\lambda_1, \ldots, \lambda_n \in \mathbb{R}$ such that

$$\nabla F(p) = \sum_{j=1}^n \lambda_j \nabla h^j(p) = 2 \sum_{j=1}^n \lambda_j (0, \ldots, 0, p_j, 0, \ldots, 0) \tag{10.11}$$

$$= 2(\lambda_1 p_1, \ldots, \lambda_n p_n) \in \mathbb{R}^{n^2} .$$

In addition, Example 4.8(a) implies

$$\frac{\partial F}{\partial x_j^k}(p) = \det[p_1, \ldots, p_{j-1}, e_k, p_{j+1}, \ldots, p_n] \quad \text{for } 1 \le j, k \le n . \tag{10.12}$$

We set $B := [p_1, \ldots, p_n]$ and denote by $B^\sharp := [b_{jk}^\sharp]_{1 \le j,k \le n}$ the **cofactor matrix** to B whose entries are $b_{jk}^\sharp := (-1)^{j+k} \det B_{jk}$, where B_{jk} is obtained from B by removing the k-th row and the j-th column (see [Gab96, § A.3.7]). Then (10.11) and (10.12) result in

$$\frac{\partial F}{\partial x_j^k}(p) = b_{jk}^\sharp = 2\lambda_j p_j^k .$$

Because $B^\sharp B = (\det B) 1_n$, we have

$$\delta_{ij} \det B = \sum_{k=1}^n b_{ik}^\sharp p_j^k = 2\lambda_i (p_i \,|\, p_j) \quad \text{for } 1 \le i, j \le n . \tag{10.13}$$

By choosing $i = j$ in (10.13), we find

$$2\lambda_1 = \cdots = 2\lambda_n = \det B . \tag{10.14}$$

---

[7]$x_j^\top$ means that x_j is understood as a row vector.

In the case $\det B = 0$, the claim is obviously true. If $\det B \neq 0$, then (10.13) and (10.14) show that p_i and p_j are orthogonal for $i \neq j$. Therefore B belongs to $O(n)$, and we get $|\det B| = 1$ (see Exercise 9.2). Now the claim follows because $F(p) = \det B$. ∎

(b) (principal axis transformation) Suppose $A \in \mathcal{L}_{\mathrm{sym}}(\mathbb{R}^n)$. Then there are real numbers $\lambda_1 \geq \lambda_2 \geq \cdots \geq \lambda_n$ and $x_1, \ldots, x_n \in S^{n-1}$ such that $Ax_k = \lambda_k x_k$ for $1 \leq k \leq n$, that is, x_k is an eigenvector with eigenvalue λ_k. The $x_1, \ldots, x_n$ form an ONB of $\mathbb{R}^n$. In this basis, A has the matrix $[A] = \mathrm{diag}(\lambda_1, \ldots, \lambda_n)$. We also have

$$\lambda_k = \max\{ (Ax \,|\, x) \;;\; x \in S^{n-1} \cap E_k \} \quad \text{for } k = 1, \ldots, n ,$$

where $E_1 := \mathbb{R}^n$ and $E_k := \big(\mathrm{span}\{x_1, \ldots, x_{k-1}\}\big)^{\perp}$ for $k = 2, \ldots, n$.

Proof (i) We set $h^0(x) := |x|^2 - 1$ and $F(x) := (Ax \,|\, x)$ for $x \in \mathbb{R}^n$. Then 0 is a regular value of $h^0 \in C^\infty(\mathbb{R}^n, \mathbb{R})$, and F has a maximum on $S^{n-1} = h^{-1}(0)$. Suppose $x_1 \in S^{n-1}$ is a maximal point of $f := F \,|\, S^{n-1}$. Using Lagrange multipliers, there is a $\lambda_1 \in \mathbb{R}$ such that $\nabla F(x_1) = 2Ax_1 = 2\lambda_1 x_1$ (see Exercise 4.5). Therefore x_1 is an eigenvector of A with eigenvalue λ_1. We also have

$$\lambda_1 = \lambda_1(x_1 \,|\, x_1) = (Ax_1 \,|\, x_1) = f(x_1) ,$$

because $x_1 \in S^{n-1}$.

(ii) We now construct $x_2, \ldots, x_n$ recursively. Supposing $x_1, \ldots, x_{k-1}$ are already known for $k \geq 2$, we set $h := (h^0, h^1, \ldots, h^{k-1})$ with $h^j(x) := 2(x_j \,|\, x)$ for $1 \leq j \leq k-1$. Then $h^{-1}(0) = S^{n-1} \cap E_k$ is compact, and there exists an $x_k \in S^{n-1} \cap E_k$ such that $f(x) \leq f(x_k)$ for $x \in S^{n-1} \cap E_k$. In addition, one verifies (because $\mathrm{rank}\, B = \mathrm{rank}\, B^{\top}$ for $B \in \mathbb{R}^{k \times n}$) that

$$\mathrm{rank}\, \partial h(x) = \mathrm{rank}[x, x_1, \ldots, x_{k-1}] = k \quad \text{for } x \in S^{n-1} \cap E_k .$$

Therefore 0 is a regular value of h, and we use Theorem 10.15 to find real numbers $\mu_0, \ldots, \mu_{k-1}$ such that

$$2Ax_k = \nabla F(x_k) = \sum_{j=0}^{k-1} \mu_j \nabla h^j(x_k) = 2\mu_0 x_k + 2 \sum_{j=1}^{k-1} \mu_j x_j . \tag{10.15}$$

Because $(x_k \,|\, x_j) = 0$ for $1 \leq j \leq k-1$, we have

$$(Ax_k \,|\, x_j) = (x_k \,|\, Ax_j) = \lambda_j(x_k \,|\, x_j) = 0 \quad \text{for } 1 \leq j \leq k-1 .$$

It therefore follows from (10.15) that

$$0 = (Ax_k \,|\, x_j) = \mu_j \quad \text{for } j = 1, \ldots, k-1 .$$

We also see from (10.15) that x_k is a eigenvector of A with eigenvalue μ_0. Finally we get

$$\mu_0 = \mu_0(x_k \,|\, x_k) = (Ax_k \,|\, x_k) = f(x_k) .$$

Thus we are finished (see Remark 5.13(b)). ∎

10.18 Remark   Suppose $A \in \mathcal{L}_{\mathrm{sym}}(\mathbb{R}^n)$ with $n \geq 2$, and let $\lambda_1 \geq \lambda_2 \geq \cdots \geq \lambda_n$ be the eigenvalues of A. Further let $x_1, \ldots, x_n$ be an ONB of $\mathbb{R}^n$, where x_k is an eigenvector of A with eigenvalue λ_k. For $x = \sum_{j=1}^n \xi^j x_j \in \mathbb{R}^n$, we then have

$$(Ax|x) = \sum_{j=1}^n \lambda_j (\xi^j)^2 . \tag{10.16}$$

We now assume that

$$\lambda_1 \geq \cdots \geq \lambda_k > 0 > \lambda_{k+1} \geq \cdots \geq \lambda_m$$

for some $k \in \{0, \ldots, m\}$. Then $\gamma \in \{-1, 1\}$ is a regular value of the C^∞ map

$$a: \mathbb{R}^n \to \mathbb{R} , \quad x \mapsto (Ax|x)$$

(see Exercise 4.5). Therefore, due to the regular value theorem,

$$a^{-1}(\gamma) = \big\{ x \in \mathbb{R}^n ;\ (Ax|x) = \gamma \big\}$$

is a C^∞ hypersurface in $\mathbb{R}^n$. Letting $\alpha_j := 1/\sqrt{|\lambda_j|}$, it follows from (10.16) that

$$\sum_{j=1}^k \Big(\frac{\xi^j}{\alpha_j}\Big)^2 - \sum_{j=k+1}^n \Big(\frac{\xi^j}{\alpha_j}\Big)^2 = \gamma \tag{10.17}$$

for $x = \sum_{j=1}^n \xi^j x_j \in a^{-1}(\gamma)$.

If A is positive definite, then $\lambda_1 \geq \cdots \geq \lambda_n > 0$ follows from Remark 5.13(b). In this case, we read off from (10.17) that $a^{-1}(1)$ is an $(n-1)$-dimensional **ellipsoid** with **principal axes** $\alpha_1 x_1, \ldots, \alpha_n x_n$. If A is indefinite, then (10.17) shows that $a^{-1}(\pm 1)$ is (in general) a **hyperboloid** with **principal axes** $\alpha_1 x_1, \ldots, \alpha_n x_n$.

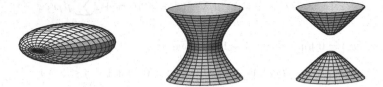

This extends the known even-dimensional case to higher dimensions.

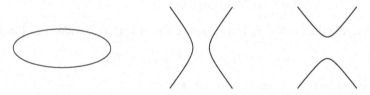

These considerations clarify the name "principal axis transformation". If one or more eigenvalues of A vanishes, $a^{-1}(\gamma)$ is a cylinder with an ellipsoidal or hyperboloidal profile. ∎

Exercises

1 Denote by (g_n, V_n) the regular parametrization of $F_n := \mathbb{R}^n \setminus H_n$ by n-dimensional polar coordinates (see Exercise 9.11).

(a) Show that the first fundamental matrix $[(g_n)_{jk}]$ of F_n with respect to g_n is given by

$$\begin{bmatrix} 1 & 0 \\ 0 & r^2 \end{bmatrix} \quad \text{for } n = 2 \,,$$

or, in the case $n \geq 3$, by

$$\text{diag}\left(1, r^2 \sin^2(y^3) \cdot \cdots \cdot \sin^2(y^n), r^2 \sin^2(y^3) \cdot \cdots \cdot r^2 \sin^2(y^{n-1}), \ldots, r^2 \sin^2(y^3), r^2\right)$$

for $(r, y^2, \ldots, y^n) \in V_n$.

(b) Suppose $f \in C^1(F_n, \mathbb{R})$, and let φ_n denote the chart belonging to g_n. Calculate the representation of $\nabla_p f$ in n-dimensional polar coordinates, that is, in the local coordinates induced by φ_n.

2 Let (g, V) be a parametrization of an m-dimensional C^1 submanifold M of $\mathbb{R}^n$. Also denote by

$$\sqrt{g}(y) := \sqrt{\det[g_{jk}(y)]} \quad \text{for } y \in V$$

the **Gram determinant** of M with respect to g. Verify these statements:

(a) If $m = 1$, then $\sqrt{g} = |\dot{g}|$.

(b) If $m = 2$ and $n = 3$, then

$$\sqrt{g} = \sqrt{\left(\frac{\partial(g^2, g^3)}{\partial(x,y)}\right)^2 + \left(\frac{\partial(g^3, g^1)}{\partial(x,y)}\right)^2 + \left(\frac{\partial(g^1, g^2)}{\partial(x,y)}\right)^2} = |\partial_1 g \times \partial_2 g| \,,$$

where $\times$ denotes the cross (or vector) product (see also Section VIII.2).

(c) Suppose V is open in $\mathbb{R}^n$ and $f \in C^1(V, \mathbb{R})$. For the parametrization $g : V \to \mathbb{R}^{n+1}$, $x \mapsto (x, f(x))$ of the graph of f, show that $\sqrt{g} = \sqrt{1 + |\nabla f|^2}$.

3 Determine $T_p S^2$ at $p = (0, 1, 0)$ in

(a) spherical coordinates (see Example 9.11(b));

(b) the coordinates coming from the stereographic projection.

4 Determine $T_p T_{2,1}^2$ at $p = (\sqrt{2}, \sqrt{2}, 1)$.

5 Suppose M is an m-dimensional C^q submanifold of $\mathbb{R}^n$ and (φ, U) is a chart around $p \in M$. Show

(a) every open subset of M is an m-dimensional C^q submanifold of $\mathbb{R}^n$;

(b) if U is understood as a manifold, then φ belongs to $\text{Diff}^q(U, \varphi(U))$, and the tangential of the chart $T_p \varphi$ agrees with that of $\varphi \in C^q(U, \varphi(U))$ at the point p.

6 Find $T_A GL(n)$ for $A \in GL(n) := \mathcal{L}\mathrm{aut}(\mathbb{R}^n)$.

7 Show the tangential space of the orthogonal group $O(n)$ at 1_n is a vector space of skew-symmetric $(n \times n)$-matrices, that is,

$$T_{1_n} O(n) = \left(1_n, \, \{\, A \in \mathbb{R}^{n \times n} \; ; \; A + A^\top = 0 \,\}\right) .$$

(Hint: Use Example 9.5(c) and Theorem 10.7.)

8 Show that the tangential space of the special orthogonal group $SO(n)$ at 1_n is

$$T_{1_n} SO(n) = \left(1_n, \, \{\, A \in \mathbb{R}^{n \times n} \; ; \; \mathrm{tr}(A) = 0 \,\}\right) .$$

(Hint: For $A \in \mathbb{R}^{n \times n}$ with $\mathrm{tr}(A) = 0$, note $\gamma(t) = e^{tA}$ for $t \in \mathbb{R}$, and see Theorem 10.6.)

9 Show

(a) for $\psi \in \mathrm{Diff}^q(M, N)$ and $p \in M$, we have $T_p \psi \in \mathcal{L}\mathrm{is}(T_p M, T_{\psi(p)} N)$;

(b) if M and N are diffeomorphic C^q submanifolds of $\mathbb{R}^n$, their dimensions coincide.

10 Show

(a) $S^1 \times \mathbb{R}$ (see Exercise 9.4) and $\{\, (x, y, z) \in \mathbb{R}^3 \; ; \; x^2 + y^2 = 1 \,\}$ are diffeomorphic;

(b) $S^1 \times S^1$ and $\mathsf{T}_{2,1}^2$ are diffeomorphic;

(c) S^n and $r S^n$ for $r > 0$ are diffeomorphic.

11 Suppose $M := \{\, (x, y, z) \in \mathbb{R}^3 \; ; \; x^2 + y^2 = 1 \,\}$ and

$$\nu : M \to S^2 , \quad (x, y, z) \mapsto (x, y, 0) .$$

Then $\nu \in C^\infty(M, S^2)$. Also $T_p \nu$ is symmetric for $p \in M$ and has the eigenvalues 0 and 1.

12 Suppose X is open in $\mathbb{R}^n$ and $f \in C^1(X, \mathbb{R})$. Also let $\nu : M \to S^n$ be a unit normal on $M := \mathrm{graph}(f)$. Show that ν belongs to $C^\infty(M, S^n)$ and that $T_p \nu$ is symmetric for $p \in M$.

13 Suppose M and ν are as in Exercise 12. Also suppose

$$\varphi_\alpha : M \to \mathbb{R}^{n+1} , \quad p \mapsto p + \alpha \nu(p)$$

for $\alpha \in \mathbb{R}$. Show that there is an $\alpha_0 > 0$ such that $\varphi_\alpha(M)$ is a smooth hypersurface diffeomorphic to M for every $\alpha \in (-\alpha_0, \alpha_0)$.

14 Find the volume of the largest rectangular box that is contained within the ellipsoid $\{\, (x, y, z) \in \mathbb{R}^3 \; ; \; (x/a)^2 + (y/b)^2 + (z/c)^2 = 1 \,\}$ with $a, b, c > 0$.

Chapter VIII

Line integrals

In this chapter, we return to the theory of integrating functions of a real variable. We will now consider integrals which are not only over intervals but also over continuously differentiable maps of intervals, namely, curves. We will see that this generalization of the integral has important and profound consequences.

Of course, we must first make precise the notion of a curve, which we do in Section 1. In addition, we will introduce the concept of arc length and derive an integral formula for calculating it.

In Section 2, we discuss the differential geometry of curves. In particular, we prove the existence of an associated n-frame, a certain set of n vectors defined along the curve. For plane curves, we study curvature, and for space curves we will also study the torsion. The material in this section contributes mostly to your general mathematical knowledge, and it will not be needed for the remainder of this chapter.

Section 3 treats differential forms of first order. Here we make rigorous the ad hoc definition of differentials introduced in Chapter VI. We will also derive several simple rules for dealing with such differential forms. These rules represent the foundation of the theory of line integrals, as we will learn in Section 4 when differential forms emerge as the integrands of line integrals. We will prove the fundamental theorem of line integrals, which characterizes the vector fields that can be obtained as gradients of potentials.

Sections 5 and 6 find a particularly important application of line integrals in complex analysis, also known as the theory of functions of a complex variable. In Section 5, we derive the fundamental properties of holomorphic functions, in particular, Cauchy's integral theorem and formula. With these as aids, we prove the fundamental result, which says that a function is holomorphic if and only if it is analytic. We apply to general theory to the Fresnel integral, thus showing how the Cauchy integral theorem can be used to calculate real integrals.

In Section 6, we study meromorphic functions and prove the important

residue theorem and its version in homology theory. To this end, we introduce
the concept of winding number and derive its most important properties. To conclude this volume, we show the scope of the residue theorem by calculating several
Fourier integrals.

1 Curves and their lengths

This section mainly proves that a continuously differentiable compact curve Γ has a finite length $L(\Gamma)$ given by

$$L(\Gamma) = \int_a^b |\dot\gamma(t)| \, dt .$$

Here γ is an arbitrary parametrization of Γ.

In the following suppose

- $E = (E, |\cdot|)$ is a Banach space over the field $\mathbb{K}$ and $I = [a, b]$ is a compact interval.

The total variation

Suppose $f : I \to E$ and $\mathfrak{Z} := (t_0, \ldots, t_n)$ is a partition of I. Then

$$L_{\mathfrak{Z}}(f) := \sum_{j=1}^n |f(t_j) - f(t_{j-1})|$$

is the **length of the piecewise straight path** $\big(f(t_0), \ldots, f(t_n)\big)$ in E, and

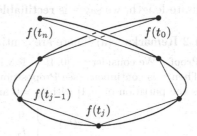

$$\mathrm{Var}(f, I) := \sup\{\, L_{\mathfrak{Z}}(f) \;;\; \mathfrak{Z} = (t_0, \ldots, t_n) \text{ is a partition of } I \,\}$$

is called the **total variation** (or simply the **variation**) of f over I. We say f is of **bounded variation** if $\mathrm{Var}(f, I) < \infty$.

1.1 Lemma *For $f : [a, b] \to E$ and $c \in [a, b]$, we have*

$$\mathrm{Var}\big(f, [a, b]\big) = \mathrm{Var}\big(f, [a, c]\big) + \mathrm{Var}\big(f, [c, b]\big) . \qquad (1.1)$$

Proof (i) Suppose $c \in [a, b]$. Without loss of generality, we may assume that $\mathrm{Var}\big(f, [a, c]\big)$ and $\mathrm{Var}\big(f, [c, b]\big)$ are finite, for otherwise the function $f : [a, b] \to E$ would not be of bounded variation, and the claim would be already obvious.

(ii) Suppose $\mathfrak{Z}$ is a partition of $[a, b]$ and $\widetilde{\mathfrak{Z}}$ is a refinement of $\mathfrak{Z}$ that contains the point c. In addition, set $\widetilde{\mathfrak{Z}}_1 := \widetilde{\mathfrak{Z}} \cap [a, c]$ and $\widetilde{\mathfrak{Z}}_2 := \widetilde{\mathfrak{Z}} \cap [c, b]$. Then

$$L_{\mathfrak{Z}}(f) \leq L_{\widetilde{\mathfrak{Z}}}(f) = L_{\widetilde{\mathfrak{Z}}_1}(f) + L_{\widetilde{\mathfrak{Z}}_2}(f) \leq \mathrm{Var}\big(f, [a, c]\big) + \mathrm{Var}\big(f, [c, b]\big) .$$

By forming the supremum with respect to $\mathfrak{Z}$, we find

$$\mathrm{Var}\big(f, [a, b]\big) \leq \mathrm{Var}\big(f, [a, c]\big) + \mathrm{Var}\big(f, [c, b]\big) .$$

(iii) For $\varepsilon > 0$, there are partitions $\mathfrak{Z}_1$ of $[a,c]$ and $\mathfrak{Z}_2$ of $[c,b]$ such that

$$L_{\mathfrak{Z}_1}(f) \geq \text{Var}(f,[a,c]) - \varepsilon/2 \quad \text{and} \quad L_{\mathfrak{Z}_2}(f) \geq \text{Var}(f,[c,b]) - \varepsilon/2 \ .$$

For $\mathfrak{Z} := \mathfrak{Z}_1 \vee \mathfrak{Z}_2$, we have

$$L_{\mathfrak{Z}_1}(f) + L_{\mathfrak{Z}_2}(f) = L_{\mathfrak{Z}}(f) \leq \text{Var}(f,[a,b]) \ ,$$

and thus

$$\text{Var}(f,[a,c]) + \text{Var}(f,[c,b]) \leq L_{\mathfrak{Z}_1}(f) + L_{\mathfrak{Z}_2}(f) + \varepsilon \leq \text{Var}(f,[a,b]) + \varepsilon \ .$$

The claim is now implied by (ii). ∎

Rectifiable paths

Interpreting $\gamma \in C(I,E)$ as a continuous path in E, we also call $\text{Var}(\gamma,I)$ the **length** (or **arc length**) of γ and write it as $L(\gamma)$. If $L(\gamma) < \infty$, that is, if γ has a finite length, we say γ is **rectifiable**.

1.2 Remarks (a) There are continuous paths that are not rectifiable.[1]

Proof We consider $\gamma \colon [0,1] \to \mathbb{R}$ with $\gamma(0) := 0$ and $\gamma(t) := t \cos^2(\pi/2t)$ for $t \in (0,1]$. Then γ is continuous (see Proposition III.2.24). For $n \in \mathbb{N}^\times$, suppose $\mathfrak{Z}_n = (t_0,\dots,t_{2n})$ is the partition of $[0,1]$ with $t_0 = 0$ and $t_j = 2/(2n+1-j)$ for $1 \leq j \leq 2n$. Because

$$\gamma(t_j) = \begin{cases} 0\,, & j = 2k, \quad 0 \leq k \leq n\,, \\ t_j\,, & j = 2k+1, \quad 0 \leq k < n\,, \end{cases}$$

we have

$$L_{\mathfrak{Z}_n}(\gamma) = \sum_{j=1}^{2n} |\gamma(t_j) - \gamma(t_{j-1})| = \sum_{k=0}^{n-1} t_{2k+1} = \frac{1}{2} \sum_{k=1}^{n} \frac{1}{k} \to \infty$$

as $n \to \infty$. Therefore γ is not rectifiable. ∎

(b) Suppose $\gamma \colon [a,b] \to E$ is Lipschitz continuous with Lipschitz constant λ. Then γ is rectifiable, and $L(\gamma) \leq \lambda(b-a)$.

Proof For every partition $\mathfrak{Z} = (t_0,\dots,t_n)$ of $[a,b]$, we have

$$L_{\mathfrak{Z}}(\gamma) = \sum_{j=1}^{n} |\gamma(t_j) - \gamma(t_{j-1})| \leq \lambda \sum_{j=1}^{n} |t_j - t_{j-1}| = \lambda(b-a) \ . ∎$$

(c) Of course, the length of a path γ depends on the norm on E. However, that it is rectifiable does not change when going to a different but equivalent norm. ∎

The path we considered in Remark 1.2(a) is indeed continuous but not differentiable at 0. The next result shows that continuously differentiable paths are always rectifiable.

---

[1] One can also show that there are continuous paths in $\mathbb{R}^2$ whose image fills out the entire unit disc $\overline{\mathbb{B}}$ (see Exercise 8). Such "space filling" paths are called **Peano curves**.

1.3 Theorem *Suppose $\gamma \in C^1(I, E)$. Then γ is rectifiable, and we have*

$$L(\gamma) = \int_a^b |\dot{\gamma}(t)| \, dt \ .$$

Proof (i) It suffices to consider the case $a < b$, in which I is not just a single point.

(ii) The rectifiability of γ follows immediately from the fundamental theorem of calculus. That is, if $\mathfrak{Z} = (t_0, \ldots, t_n)$ is partition of $[a, b]$, then

$$L_{\mathfrak{Z}}(\gamma) = \sum_{j=1}^n |\gamma(t_j) - \gamma(t_{j-1})| = \sum_{j=1}^n \left| \int_{t_{j-1}}^{t_j} \dot{\gamma}(t) \, dt \right|$$

$$\leq \sum_{j=1}^n \int_{t_{j-1}}^{t_j} |\dot{\gamma}(t)| \, dt = \int_a^b |\dot{\gamma}(t)| \, dt \ .$$

Therefore

$$L(\gamma) = \mathrm{Var}(\gamma, [a, b]) \leq \int_a^b |\dot{\gamma}(t)| \, dt \ . \tag{1.2}$$

(iii) Suppose now $s_0 \in [a, b)$. From Lemma 1.1, we know for every $s \in (s_0, b)$ that

$$\mathrm{Var}(\gamma, [a, s]) - \mathrm{Var}(\gamma, [a, s_0]) = \mathrm{Var}(\gamma, [s_0, s]) \ .$$

We also find

$$|\gamma(s) - \gamma(s_0)| \leq \mathrm{Var}(\gamma, [s_0, s]) \leq \int_{s_0}^s |\dot{\gamma}(t)| \, dt \ .$$

Here the first inequality follows because (s_0, s) is a partition of $[s_0, s]$, and the second comes from (1.2).

Thus because $s_0 < s$, we have

$$\left| \frac{\gamma(s) - \gamma(s_0)}{s - s_0} \right| \leq \frac{\mathrm{Var}(\gamma, [a, s]) - \mathrm{Var}(\gamma, [a, s_0])}{s - s_0} \leq \frac{1}{s - s_0} \int_{s_0}^s |\dot{\gamma}(t)| \, dt \ . \tag{1.3}$$

Because γ has continuous derivatives, it follows from the mean value theorem in integral form and from Theorem VI.4.12 that

$$|\dot{\gamma}(s_0)| = \lim_{s \to s_0} \left| \frac{\gamma(s) - \gamma(s_0)}{s - s_0} \right| \leq \lim_{s \to s_0} \left[\frac{1}{s - s_0} \int_{s_0}^s |\dot{\gamma}(t)| \, dt \right] = |\dot{\gamma}(s_0)| \ .$$

Thus (1.3) and like considerations for $s < s_0$ show that $s \mapsto \mathrm{Var}(\gamma, [a, s])$ is differentiable with

$$\frac{d}{ds} \mathrm{Var}(\gamma, [a, s]) = |\dot{\gamma}(s)| \quad \text{for } s \in [a, b] \ .$$

Therefore $s \mapsto \mathrm{Var}(\gamma, [a, s])$ belongs to $C^1(I, \mathbb{R})$. In addition, the fundamental theorem of calculus gives

$$\mathrm{Var}(\gamma, [a, b]) = \int_a^b |\dot\gamma(t)| \, dt$$

because $\mathrm{Var}(\gamma, [a, a]) = 0$. ∎

1.4 Corollary For $\gamma = (\gamma_1, \ldots, \gamma_n) \in C^1(I, \mathbb{R}^n)$, we have

$$L(\gamma) = \int_a^b \sqrt{\big(\dot\gamma_1(t)\big)^2 + \cdots + \big(\dot\gamma_n(t)\big)^2} \, dt \ .$$

Differentiable curves

The image of a path is a set of points in E which does not depend on the particular function used to describe it. In other words, it has a geometric meaning independent of any special parametrization. To understand this precisely, we must specify which changes of parametrization leave this image invariant.

Suppose J_1 and J_2 are intervals and $q \in \mathbb{N} \cup \{\infty\}$. The map $\varphi : J_1 \to J_2$ is said to be an (**orientation-preserving**) C^q **change of parameters** if φ belongs[2] to $\mathrm{Diff}^q(J_1, J_2)$ and is strictly increasing. If $\gamma_j \in C^q(J_j, E)$ for $j = 1, 2$, then γ_1 is said to be an (**orientation-preserving**) C^q **reparametrization** of γ_2 if there is a C^q change of parameters φ such that $\gamma_1 = \gamma_2 \circ \varphi$.

1.5 Remarks (a) When $\varphi \in \mathrm{Diff}^q(J_1, J_2)$ is strictly decreasing, we then say φ is an **orientation-reversing** C^q **change of parameters**. That said, in what follows, we will always take changes of parameters to be orientation-preserving.

(b) A map $\varphi : J_1 \to J_2$ is a C^q change of parameters if and only if φ belongs to $C^q(J_1, J_2)$, is surjective, and satisfies $\dot\varphi(t) > 0$ for $t \in J_1$.

Proof This follows from Theorems III.5.7 and IV.2.8. ∎

(c) Let I_1 and I_2 be compact intervals, and suppose $\gamma_1 \in C(I_1, E)$ is a continuous reparametrization of $\gamma_2 \in C(I_2, E)$. Then

$$\mathrm{Var}(\gamma_1, I_1) = \mathrm{Var}(\gamma_2, I_2) \ .$$

Proof Suppose $\varphi \in \mathrm{Diff}^0(I_1, I_2)$ is a change of parameters satisfying $\gamma_1 = \gamma_2 \circ \varphi$. If $\mathfrak{Z} = (t_0, \ldots, t_n)$ is a partition of I_1, then $\varphi(\mathfrak{Z}) := \big(\varphi(t_0), \ldots, \varphi(t_n)\big)$ is a partition of I_2, and we have

$$L_{\mathfrak{Z}}(\gamma_1, I_1) = L_{\mathfrak{Z}}(\gamma_2 \circ \varphi, I_1) = L_{\varphi(\mathfrak{Z})}(\gamma_2, I_2) \leq \mathrm{Var}(\gamma_2, I_2) \ .$$

---

[2]If in addition J_1 and J_2 are not open, then this means $\varphi : J_1 \to J_2$ is bijective, and φ and φ^{-1} belong to the class C^q. In particular, $\mathrm{Diff}^0(J_1, J_2)$ is the set of all topological maps (homeomorphisms) of J_1 to J_2.

Thus we have the relation

$$\mathrm{Var}(\gamma_1, I_1) = \mathrm{Var}(\gamma_2 \circ \varphi, I_1) \leq \mathrm{Var}(\gamma_2, I_2) \ . \tag{1.4}$$

Noting that $\gamma_2 = (\gamma_2 \circ \varphi) \circ \varphi^{-1}$, we see from (1.4) (if we replace γ_2 by $\gamma_2 \circ \varphi$ and φ by φ^{-1}) that

$$\mathrm{Var}(\gamma_2, I_2) = \mathrm{Var}\big((\gamma_2 \circ \varphi) \circ \varphi^{-1}, I_2\big) \leq \mathrm{Var}(\gamma_2 \circ \varphi, I_1) = \mathrm{Var}(\gamma_1, I_1) \ . \ \blacksquare$$

On the set of all C^q paths in E, we define the relation $\sim$ by

$$\gamma_1 \sim \gamma_2 :\Longleftrightarrow \gamma_1 \text{ is a } C^q \text{ reparametrization of } \gamma_2 \ .$$

It is not hard to see that $\sim$ is an equivalence relation (see Exercise 5). The associated equivalence classes are called C^q **curves** in E. Every representative of a C^q curve Γ is a C^q **parametrization** of Γ. We may say a C^0 curve is **continuous**, a C^1 curve is **continuously differentiable**, and a C^∞ curve is **smooth**. If the curve Γ has a parametrization on a compact domain (or a compact **parameter interval**), we say the curve is **compact**, and then, due to Theorem III.3.6, the image of Γ is also compact. We say a parametrization γ of Γ is **regular** if $\dot{\gamma}(t) \neq 0$ for $t \in \mathrm{dom}(\gamma)$. When Γ has a regular parametrization, we call Γ a **regular curve**. Sometimes we write $\Gamma = [\gamma]$ to emphasize that Γ is an equivalence class of parametrizations that contains γ.

1.6 Remarks (a) If Γ is a compact curve, every parametrization of it has a compact domain of definition. If Γ is a regular C^1 curve, every parametrization of it is regular.

Proof Suppose $\gamma \in C^q(J, E)$ is a parametrization of Γ, and let $\gamma_1 \in C^q(J_1, E)$ be a reparametrization of γ. Then there is a $\varphi \in \mathrm{Diff}^q(J_1, J)$ such that $\gamma_1 = \gamma \circ \varphi$. When J is compact, this holds also for $J_1 = \varphi^{-1}(J)$ because continuous images of compact sets are also compact (Theorem III.3.6). From the chain rule, we find for $q \geq 1$ that

$$\dot{\gamma}_1(t) = \dot{\gamma}\big(\varphi(t)\big)\dot{\varphi}(t) \quad \text{for } t \in J_1 \ .$$

Therefore, because $\dot{\gamma}(s) \neq 0$ for $s \in J$ and $\dot{\varphi}(t) \neq 0$ for $t \in J_1$, we also have $\dot{\gamma}_1(t) \neq 0$ for $t \in J_1$. $\blacksquare$

(b) A regular curve can have a nonregular parametrization, which is, however, not equivalent to its regular parametrization.

Proof We consider the regular smooth curve Γ that is parametrized by the function

$$[-1, 1] \to \mathbb{R}^2 \ , \quad t \mapsto \gamma(t) := (t, t) \ .$$

We also consider the smooth path $\widetilde{\gamma} \colon [-1, 1] \to \mathbb{R}^2$ with $\widetilde{\gamma}(t) := (t^3, t^3)$. Then $\mathrm{im}(\gamma) = \mathrm{im}(\widetilde{\gamma})$ but $\dot{\widetilde{\gamma}}(0) = (0, 0)$. Therefore $\widetilde{\gamma}$ is not a C^1 reparametrization of γ. $\blacksquare$

(c) Suppose $I \to E$, $t \mapsto \gamma(t)$ is a regular C^1 parametrization of a curve Γ. Then

$$\dot{\gamma}(t) = \lim_{s \to t} \frac{\gamma(s) - \gamma(t)}{s - t} \quad \text{for } t \in I$$

can be interpreted the "instantaneous velocity" of the curve Γ at the point $\gamma(t)$. ■

Suppose Γ is a continuous curve in E, and suppose γ and γ_1 are (equivalent) parametrizations of Γ. Then clearly, $\text{im}(\gamma) = \text{im}(\gamma_1)$. Therefore the **image** of Γ is well defined through

$$\text{im}(\Gamma) := \text{im}(\gamma) .$$

If Γ is compact, $\text{dom}(\gamma) = [a, b]$, and $\text{dom}(\gamma_1) = [a_1, b_1]$, we have the relations

$$\gamma(a) = \gamma_1(a_1) \quad \text{and} \quad \gamma(b) = \gamma_1(b_1) .$$

Thus, for a compact curve, the **initial point** $A_\Gamma := \gamma(a)$ and the **end point** $E_\Gamma := \gamma(b)$ are well defined.

If A_Γ and E_Γ coincide, Γ is **closed** or a **loop**. Finally, we often simply write $p \in \Gamma$ for $p \in \text{im}(\Gamma)$.

Rectifiable curves

Suppose Γ is a continuous curve in E and $\gamma \in C(J, E)$ is a parametrization of Γ. Also let $\alpha := \inf J$ and $\beta := \sup J$. Then $\text{Var}(\gamma, [a, b])$ is defined for $\alpha < a < b < \beta$. Then Lemma 1.1 implies for every $c \in (\alpha, \beta)$ that the function

$$[c, \beta) \to \overline{\mathbb{R}} , \quad b \mapsto \text{Var}(\gamma, [c, b])$$

is increasing and that the map

$$(\alpha, c] \to \overline{\mathbb{R}} , \quad a \mapsto \text{Var}(\gamma, [a, c])$$

is decreasing. Thus from (1.1) the limit

$$\text{Var}(\gamma, J) := \lim_{\substack{a \downarrow \alpha \\ b \uparrow \beta}} \text{Var}(\gamma, [a, b]) \qquad (1.5)$$

exists in $\overline{\mathbb{R}}$, and we call it **(total) variation** of γ over J. If $\gamma_1 \in C(J_1, E)$ is a reparametrization of γ, it follows from Remark 1.5(c) that $\text{Var}(\gamma, J) = \text{Var}(\gamma_1, J_1)$. Therefore the **total variation** or **length** (or the **arc length**) of Γ is well defined through

$$L(\Gamma) := \text{Var}(\Gamma) := \text{Var}(\gamma, J) .$$

The curve Γ is said to be **rectifiable** if it has a finite length, that is, if $L(\Gamma) < \infty$.

1.7 Theorem *Suppose $\gamma \in C^1(J, E)$ is a parametrization of a C^1 curve Γ. Then*

$$L(\Gamma) = \int_J |\dot{\gamma}(t)| \, dt \ . \tag{1.6}$$

If Γ is compact, Γ is rectifiable.

Proof This follows immediately from (1.5), Theorem 1.3, and the definition of an improper integral. ∎

1.8 Remarks **(a)** Theorem 1.7 says in particular that $L(\Gamma)$, and thus also the integral in (1.6), is independent of the special parametrization γ.

(b) Suppose $\widetilde{\Gamma}$ is a C^q curve imbedded in $\mathbb{R}^n$ as set out in Section VII.8. Also suppose (φ, U) is a chart of $\widetilde{\Gamma}$ and (g, V) is its associated parametrization. Then[3] $\Gamma := \widetilde{\Gamma} \cap U$ is a regular C^q curve, and g is a regular C^q parametrization of Γ.

(c) Regular C^q curves are generally not imbedded C^q curves.

Proof This follows from Example VII.9.6(c). ∎

(d) Because we admit only orientation-preserving changes of parameters, our curves are all **oriented**, that is, they "pass through" in a fixed direction. ∎

1.9 Examples **(a)** (graphs of real-valued functions) Suppose $f \in C^q(J, \mathbb{R})$ and $\Gamma := \text{graph}(f)$. Then Γ is a regular C^q curve in $\mathbb{R}^2$, and the map $J \to \mathbb{R}^2$, $t \mapsto (t, f(t))$ is a regular C^q parametrization of Γ. Also

$$L(\Gamma) = \int_J \sqrt{1 + [f'(t)]^2} \, dt \ .$$

(b) (plane curves in polar coordinates) Suppose $r, \varphi \in C^q(J, \mathbb{R})$ and $r(t) \geq 0$ for $t \in J$. Also let

$$\gamma(t) := r(t)\big(\cos(\varphi(t)), \sin(\varphi(t))\big) \quad \text{for } t \in J \ .$$

Identifying $\mathbb{R}^2$ with $\mathbb{C}$ gives γ the representation

$$\gamma(t) = r(t)e^{i\varphi(t)} \quad \text{for } t \in J \ .$$

If $[\dot{r}(t)]^2 + [r(t)\dot{\varphi}(t)]^2 > 0$, then γ is a regular C^q parametrization of a curve Γ with

$$L(\Gamma) = \int_J \sqrt{[\dot{r}(t)]^2 + [r(t)\dot{\varphi}(t)]^2} \, dt \ .$$

Proof We have

$$\dot{\gamma}(t) := \big(\dot{r}(t) + i r(t)\dot{\varphi}(t)\big)e^{i\varphi(t)} \ ,$$

---

[3]Here and in other cases where there is little risk of confusion, we simply write Γ for $\text{im}(\Gamma)$.

and thus $|\dot{\gamma}(t)|^2 = [\dot{r}(t)]^2 + [r(t)\dot{\varphi}(t)]^2$, from which the claim follows. ∎

(c) Suppose $0 < b \le 2\pi$. For the circular arc parametrized by

$$\gamma : [0,b] \to \mathbb{R}^2 , \quad t \mapsto R(\cos t, \sin t)$$

or the equivalent one obtained through

$$\gamma : [0,b] \to \mathbb{C} , \quad t \mapsto Re^{it} ,$$

identifying $\mathbb{R}^2$ with $\mathbb{C}$ gives[4] $L(\Gamma) = bR$.

Proof This follows from (b). ∎

(d) (the logarithmic spiral) For $\lambda < 0$ and $a \in \mathbb{R}$,

$$\gamma_{a,\infty} : [a,\infty) \to \mathbb{R}^2 , \quad t \mapsto e^{\lambda t}(\cos t, \sin t)$$

is a smooth regular parametrization of the curve $\Gamma_{a,\infty} := [\gamma_{a,\infty}]$. It has finite length $e^{\lambda a}\sqrt{1+\lambda^2}/|\lambda|$.

When $\lambda > 0$, we have analogously that

$$\gamma_{-\infty,a} : (-\infty, a] \to \mathbb{R}^2 , \quad t \mapsto e^{\lambda t}(\cos t, \sin t)$$

is a smooth regular parametrization of the curve $\Gamma_{-\infty,a} := [\gamma_{-\infty,a}]$ with finite length $e^{\lambda a}\sqrt{1+\lambda^2}/\lambda$. For $\lambda \ne 0$, the map

$$\mathbb{R} \to \mathbb{R}^2 , \quad t \mapsto e^{\lambda t}(\cos t, \sin t)$$

is a smooth regular parametrization of a **logarithmic spiral**. Its length is infinite. When $\lambda > 0$ [or $\lambda < 0$], it spirals outward [or inward]. Identifying $\mathbb{R}^2$ with $\mathbb{C}$ gives the logarithmic spiral the simple parametrization $t \mapsto e^{(\lambda+i)t}$.

Proof Suppose $\lambda < 0$. We set $r(t) = e^{\lambda t}$ and $\varphi(t) = t$ for $t \in [a,\infty)$. According to (b), we then have

$$L(\Gamma_{a,\infty}) = \lim_{b\to\infty} \int_a^b \sqrt{1+\lambda^2}\, e^{\lambda t}\, dt = \frac{\sqrt{1+\lambda^2}}{\lambda} \lim_{b\to\infty} (e^{\lambda b} - e^{\lambda a}) = \frac{\sqrt{1+\lambda^2}}{|\lambda|} e^{\lambda a} .$$

The case $\lambda > 0$ works out analogously. ∎

---

[4]See Exercise III.6.12.

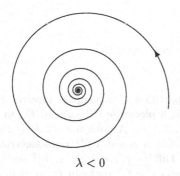

$\lambda < 0$

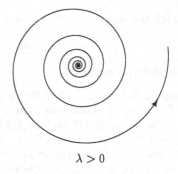

$\lambda > 0$

(e) For $R > 0$ and $h > 0$, the regular smooth curve Γ parametrized by

$$\gamma \colon \mathbb{R} \to \mathbb{R}^3 , \quad t \mapsto (R\cos t, R\sin t, ht) \quad (1.7)$$

is the **helix** with radius R and pitch $2\pi h$. By identifying[5] $\mathbb{R}^2$ with $\mathbb{C}$ and therefore $\mathbb{R}^3$ with $\mathbb{C} \times \mathbb{R}$, we give (1.7) the form $t \mapsto (Re^{it}, ht)$. We have

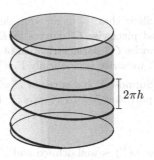

$2\pi h$

$$L(\gamma \,|\, [a,b]) = (b-a)\sqrt{R^2 + h^2}$$

for $-\infty < a < b < \infty$. The curve Γ lies on the cylinder with radius R whose axis is the z-axis. In one revolution, it rises coaxially by $2\pi h$.

Proof Because $|\dot{\gamma}(t)|^2 = R^2 + h^2$, the claim follows from Theorem 1.7. ∎

Exercises

1 A disc rolls on a line without slipping. Find a parametrization for the path traced by an arbitrary point fixed on the disc (or one "outside" the disc). Determine the arc length of this path as the disc turns once.

2 Suppose $-\infty < \alpha < \beta < \infty$. Calculate the length of these paths:

$$[0, \pi] \to \mathbb{R}^2 , \quad t \mapsto (\cos t + t \sin t, \sin t - t \cos t) \, ;$$
$$[1, \infty) \to \mathbb{R}^2 , \quad t \mapsto t^{-2}(\cos t, \sin t) \, ;$$
$$[\alpha, \beta] \to \mathbb{R}^2 , \quad t \mapsto (t^3, 3t^2/2) \, ;$$
$$[\alpha, \beta] \to \mathbb{R}^2 , \quad t \mapsto (t, t^2/2) \, .$$

3 Approximate (for example, with the trapezoid rule) the length of the limaçon of Pascal (see Example VII.9.1(c)).

---

[5]as a metric space

4 Sketch the space curve $\Gamma = [\gamma]$ with

$$\gamma : [-3\pi, 3\pi] \to \mathbb{R}^3 , \quad t \mapsto t(\cos t, \sin t, 1) ,$$

and calculate its arc length.

5 Suppose $3 = (\alpha_0, \ldots, \alpha_m)$ for $m \in \mathbb{N}^\times$ is a partition of a compact interval I and $q \in \mathbb{N}^\times \cup \{\infty\}$. A continuous path $\gamma \in C(I, E)$ is a **piecewise C^q path** in E (on 3) if $\gamma_j := \gamma\,|\,[\alpha_{j-1}, \alpha_j] \in C^q([\alpha_{j-1}, \alpha_j], E)$ for $j = 1, \ldots, m$. A path $\eta \in C(J, E)$ that is piecewise C^q on the partition $3' = (\beta_0, \ldots, \beta_m)$ of J is called a C^q **reparametrization** of γ if there is a C^q change of parameters $\varphi_j \in \mathrm{Diff}^q([\alpha_{j-1}, \alpha_j], [\beta_{j-1}, \beta_j])$ such that $\gamma_j := \eta_j \circ \varphi_j$ for $j = 1, \ldots, m$. On the set of all piecewise C^q paths in E, we define $\sim$ through

$$\gamma \sim \eta :\Longleftrightarrow \eta \text{ is a reparametrization of } \gamma .$$

(a) Show that $\sim$ is an equivalence relation. The corresponding equivalence classes are called **piecewise C^q curves** in E. Every representative of a piecewise C^q curve Γ is a **piecewise C^q parametrization** of Γ. When $\gamma \in C(I, E)$ is a piecewise C^q parametrization of Γ, we write symbolically $\Gamma = \sum_{j=1}^m \Gamma_j$, where $\Gamma_j := [\gamma_j]$ for $j = 1, \ldots, m$.

(b) Suppose Γ is a curve in E with parametrization $\gamma \in C(I, E)$ that is piecewise C^q on the partition $3 = (\alpha_0, \ldots, \alpha_m)$. Define the **length** (or the **arc length**) of Γ through

$$L(\Gamma) := \mathrm{Var}(\gamma, I) .$$

Show $L(\Gamma)$ is well defined and

$$L(\Gamma) = \sum_{j=1}^m L(\Gamma_j) = \sum_{j=1}^m \int_{\alpha_{j-1}}^{\alpha_j} |\dot{\gamma}_j(t)| \, dt .$$

(Hint: Remark 1.5(c) and Lemma 1.1.)

6 If $\Gamma = [\gamma]$ is a plane closed piecewise C^q curve and $(\alpha_0, \ldots, \alpha_m)$ is a partition for γ,

$$A(\Gamma) := \frac{1}{2} \sum_{j=1}^m \int_{\alpha_{j-1}}^{\alpha_j} \det[\gamma_j(t), \dot{\gamma}_j(t)] \, dt$$

is the **oriented area contained in Γ**.

(a) Show that $A(\Gamma)$ is well defined.

(b) Let $-\infty < \alpha < \beta < \infty$, and suppose $f \in C^1([\alpha, \beta], \mathbb{R})$ satisfies $f(\alpha) = f(\beta) = 0$. Let $a := \alpha$ and $b := 2\beta - \alpha$, and define $\gamma : [a, b] \to \mathbb{R}^2$ by

$$\gamma(t) := \begin{cases} (\alpha + \beta - t, f(\alpha + \beta - t)) , & t \in [a, \beta] , \\ (\alpha - \beta + t, 0) , & t \in [\beta, b] . \end{cases}$$

Show that

(α) $\Gamma := [\gamma]$ is a closed piecewise C^q curve (sketch it).

(β) $A(\Gamma) = \int_\alpha^\beta f(t) \, dt$, that is, the oriented area $A(\Gamma)$ equals the oriented area between the graph of f and $[\alpha, \beta]$ (see Remark VI.3.3(a)).

(c) Find the oriented area of the ellipse with semiaxes $a, b > 0$. Use the parametrization $[0, 2\pi] \to \mathbb{R}^2$, $t \mapsto (a\cos t, b\sin t)$.

(d) The path $\gamma : [-\pi/4, 3\pi/4] \to \mathbb{R}^2$ with

$$t \mapsto \begin{cases} \sqrt{\cos(2t)}\,(\cos t, \sin t) \,, & t \in [-\pi/4, \pi/4] \,, \\ \sqrt{|\cos(2t)|}\,(-\sin t, -\cos t) \,, & t \in [\pi/4, 3\pi/4] \end{cases}$$

parametrizes the **lemniscate**. Verify that $\Gamma = [\gamma]$ is a plane piecewise C^∞ curve, and find $A(\Gamma)$.

7 Find $A(\Gamma)$ if Γ is the boundary of a square.

8 Suppose $f : \mathbb{R} \to [0, 1]$ is continuous and 2-periodic with

$$f(t) = \begin{cases} 0 \,, & t \in [0, 1/3] \,, \\ 1 \,, & t \in [2/3, 1] \,. \end{cases}$$

Also let

$$r(t) := \sum_{k=1}^{\infty} 2^{-k} f(3^{2k} t) \quad \text{and} \quad \alpha(t) := 2\pi \sum_{k=1}^{\infty} 2^{-k} f(3^{2k-1} t) \quad \text{for } t \in \mathbb{R} \,.$$

Show

(a) $\gamma : \mathbb{R} \to \mathbb{R}^2$, $t \mapsto r(t)\big(\cos\alpha(t), \sin\alpha(t)\big)$ is continuous;

(b) $\gamma([0, 1]) = \overline{\mathbb{B}}$.

(Hint: Suppose $(r_0, \alpha_0) \in [0, 1] \times [0, 2\pi]$ are polar coordinates for $(x_0, y_0) \in \overline{\mathbb{B}} \backslash \{0, 0\}$. Also define $\sum_{k=1}^{\infty} g_k 2^{-k}$ and $\sum_{k=1}^{\infty} h_k 2^{-k}$ to be the respective expansions of r_0 and $\alpha_0/2\pi$ as binary fractions. For

$$a_n := \begin{cases} g_k \,, & n = 2k \,, \\ h_k \,, & n = 2k - 1 \,, \end{cases}$$

show $t_0 := 2\sum_{n=1}^{\infty} a_n 3^{-n-1} \in [0, 1]$, $f(3^n t_0) = a_n$ for $n \in \mathbb{N}$, and $\gamma(t_0) = (x_0, y_0)$.)

2   Curves in $\mathbb{R}^n$

In Section VII.6, we allowed curves to pass through infinite-dimensional vector spaces. However, the finite-dimensional case is certainly more familiar from classical (differential geometry) and gains additional structure from the Euclidean structure of $\mathbb{R}^n$. We now examine local properties of curves in $\mathbb{R}^n$.

In the following, suppose

- $n \geq 2$ and $\gamma \in C^1(J, \mathbb{R}^n)$ is a regular parametrization of a curve Γ.

Unit tangent vectors

For $t \in J$,

$$\mathfrak{t}(t) := \mathfrak{t}_\gamma(t) := \big(\gamma(t), \dot{\gamma}(t)/|\dot{\gamma}(t)|\big) \in T_{\gamma(t)}\mathbb{R}^n$$

is a tangential vector in $\mathbb{R}^n$ of length 1 and is called the **unit tangent vector** of Γ at the point $\gamma(t)$. Any element of

$$\mathbb{R}\mathfrak{t}(t) := \big(\gamma(t), \mathbb{R}\dot{\gamma}(t)\big) \subset T_{\gamma(t)}\mathbb{R}^n$$

is a **tangent** of Γ at $\gamma(t)$.

2.1 Remarks   (a) The unit tangent vector $\mathfrak{t}$ is invariant under change of parameters. Precisely, if $\zeta = \gamma \circ \varphi$ is a reparametrization of γ, then $\mathfrak{t}_\gamma(t) = \mathfrak{t}_\zeta(s)$ for $t = \varphi(s)$. Therefore is it meaningful to speak of the unit tangent vectors of Γ.

Proof   This follows immediately from the chain rule and the positivity of $\dot{\varphi}(s)$. ∎

(b) According to Corollary VII.9.8, every $t_0 \in \mathring{J}$ has an open subinterval J_0 of J around it such that $\Gamma_0 := \mathrm{im}(\gamma_0)$ with $\gamma_0 := \gamma \,|\, J_0$ is a one-dimensional C^1 submanifold of $\mathbb{R}^n$. By shrinking J_0, we can assume that Γ_0 is described by a single chart (U_0, φ_0), where $\gamma_0 = i_{\Gamma_0} \circ \varphi_0^{-1}$. Then clearly the tangent to the curve Γ_0 at $p := \gamma(t_0)$ clearly agrees with the tangential space of the *manifold* Γ_0 at the point p, that is, $T_p\Gamma_0 = \mathbb{R}\mathfrak{t}(t_0)$.

Now it can happen that there is a $t_1 \neq t_0$ in $\mathring{J}$ such that $\gamma(t_1) = p = \gamma(t_0)$. Then there is an open interval J_1 around t_1 such that $\Gamma_1 := \mathrm{im}(\gamma_1)$ with $\gamma_1 := \gamma \,|\, J_1$ is a C^1 submanifold of $\mathbb{R}^n$ for which $T_p\Gamma_1 = \mathbb{R}\mathfrak{t}(t_1)$. Thus it is generally *not* true that $T_p\Gamma_0 = T_p\Gamma_1$, that is, in general, $\mathfrak{t}(t_0) \neq \pm\mathfrak{t}(t_1)$. This shows that Γ can have several such "double point" tangents. In this case, Γ is not a submanifold of $\mathbb{R}^n$, although sufficiently small "pieces" of it are, where "sufficiently small" is measured in the parameter interval.

The limaçon of Pascal from Example VII.9.6(c) gives
a concrete example. Suppose Γ is a compact regular curve
in $\mathbb{R}^2$ parametrized by

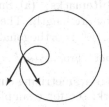

$$\gamma \colon [-\pi, \pi] \to (1 + 2\cos t)(\cos t, \sin t) \ .$$

Letting $t_0 := \arccos(-1/2)$ and $t_1 := -t_0$, we have $\gamma(t_0) = \gamma(t_1) = (0,0) \in \Gamma$. For the corresponding unit tangent
vectors, we find

$$\mathbf{t}(t_j) = \big((0,0), \big((-1)^j, -\sqrt{3}\big)/2\big) \quad \text{for } j = 0, 1 \ .$$

Therefore $\mathbf{t}(t_0) \neq \mathbf{t}(t_1)$. ∎

Parametrization by arc length

If $\gamma \colon J \to \mathbb{R}^n$ is a parametrization of Γ such that $|\dot{\gamma}(t)| = 1$ for every $t \in J$, then
clearly $L(\Gamma) = \int_J dt$, and the length of the interval J equals the length of Γ. We
say that γ **parametrizes Γ by arc length.**

2.2 Proposition *Every regular C^1 curve in $\mathbb{R}^n$ can be parametrized by arc length.
This parametrization is unique up to a reparametrization of the form $s \mapsto s + \text{const}$.
If $\eta \colon I \to \mathbb{R}^n$ is a parametrization by arc length, then $\mathbf{t}_\eta(s) = \big(\eta(s), \dot{\eta}(s)\big)$ for $s \in I$.*

Proof Suppose $\gamma \in C^1(J, \mathbb{R}^n)$ is a regular parametrization of Γ. We fix an $a \in J$
and set

$$\varphi(t) := \int_a^t |\dot{\gamma}(\tau)| \, d\tau \quad \text{for } t \in J \quad \text{and} \quad I := \varphi(J) \ .$$

The regularity of γ and $|\cdot| \in C(\mathbb{R}^n \setminus \{0\}, \mathbb{R})$ show that φ belongs to $C^1(J, \mathbb{R}^n)$. In
addition, $\dot{\varphi}(t) = |\dot{\gamma}(t)| > 0$. Due to Remark 1.5(b), φ is a C^1 change of parameters
from J onto $I := \varphi(J)$. Letting $\eta := \gamma \circ \varphi^{-1}$, this implies

$$|\dot{\eta}(s)| = \big|\dot{\gamma}\big(\varphi^{-1}(s)\big)(\varphi^{-1})^{\boldsymbol{\cdot}}(s)\big| = \frac{\big|\dot{\gamma}\big(\varphi^{-1}(s)\big)\big|}{\big|\dot{\varphi}\big(\varphi^{-1}(s)\big)\big|} = \frac{\big|\dot{\gamma}\big(\varphi^{-1}(s)\big)\big|}{\big|\dot{\gamma}\big(\varphi^{-1}(s)\big)\big|} = 1$$

for $s \in I$. Thus η parametrizes Γ by arc length.

Suppose $\widetilde{\eta} \in C^1\big(\widetilde{I}, \mathbb{R}^n\big)$ also parametrizes Γ by arc length and $\psi \in \text{Diff}^1\big(I, \widetilde{I}\big)$
satisfies $\eta = \widetilde{\eta} \circ \psi$. Then

$$1 = |\dot{\eta}(s)| = \big|\dot{\widetilde{\eta}}\big(\psi(s)\big)\dot{\psi}(s)\big| = \dot{\psi}(s) \quad \text{for } s \in I \ ,$$

and we find $\psi(s) = s + \text{const}$. ∎

2.3 Remarks   (a) Suppose Γ is a regular C^2 curve and $\gamma: I \to \mathbb{R}^n$ parametrizes it by arc length. Then $\big(\dot\gamma(s)\,\big|\,\ddot\gamma(s)\big) = 0$ for $s \in I$. Thus the vector $\big(\gamma(s), \ddot\gamma(s)\big) \subset T_{\gamma(s)}\mathbb{R}^n$ is orthogonal to the tangent $\mathbb{R}t(s) \subset T_{\gamma(s)}\mathbb{R}^n$.

Proof   From $|\dot\gamma(s)|^2 = 1$, we find $(|\dot\gamma|^2)^{\cdot}(s) = 2\big(\dot\gamma(s)\,\big|\,\ddot\gamma(s)\big) = 0$ for $s \in I$. ∎

(b) Parametrizing curves by arc length is a convenient tactic for much theoretical work; see for example the proof of Proposition 2.12. However, for any concrete example, performing such a parametrization comes at the cost of doing the arc length integral, and the resulting new parametrization may not be an elementary function.

Proof   We consider the regular parametrization

$$\gamma: [0, 2\pi] \to \mathbb{R}^2 \,, \quad t \mapsto (a\cos t, b\sin t).$$

Its image is the ellipse $(x/a)^2 + (y/b)^2 = 1$ with $a, b > 0$. The proof of Proposition 2.2 shows that every change of parametrization of γ to one by arc length has the form

$$\varphi(t) := \int_0^t \sqrt{a^2 \sin^2 s + b^2 \cos^2 s}\, ds + \text{const} \quad \text{for } 0 \le t \le 2\pi \,.$$

One can show that φ is not an elementary function.[1] ∎

Oriented bases

Suppose $\mathcal{B} = (b_1, \ldots, b_n)$ and $\mathcal{C} = (c_1, \ldots, c_n)$ are ordered bases of a real vector space E of dimension n. Also denote by $T_{\mathcal{B},\mathcal{C}} = [t_{jk}]$ the transformation matrix from $\mathcal{B}$ to $\mathcal{C}$, that is, $c_j = \sum_{k=1}^n t_{jk} b_k$ for $1 \le j \le n$. We say $\mathcal{B}$ and $\mathcal{C}$ are **identically** [or **oppositely**] **oriented** if $\det T_{\mathcal{B},\mathcal{C}}$ is positive [or negative].

2.4 Remarks   (a) On the set of ordered bases of E, the assignment

$$\mathcal{B} \sim \mathcal{C} :\Longleftrightarrow \mathcal{B} \text{ and } \mathcal{C} \text{ are identically oriented}$$

defines an equivalence relation. There are exactly two equivalent classes, the two **orientations** of E. By selecting the first and calling it $\mathcal{O}r$, we say $(E, \mathcal{O}r)$ is **oriented**. Every basis in this orientation is **positively oriented**, and every basis in the opposite orientation, which we call $-\mathcal{O}r$, is **negatively oriented**.

(b) Two ordered orthonormal bases $\mathcal{B}$ and $\mathcal{C}$ of an n-dimensional inner product space are identically oriented if and only if $T_{\mathcal{B},\mathcal{C}}$ belongs to $SO(n)$.[2]

(c) The elements of $SO(2)$ are also called **rotation matrices**, because, for every $T \in SO(2)$, there is a unique **rotation angle** $\alpha \in [0, 2\pi)$ such that T has the representation

$$T = \begin{bmatrix} \cos\alpha & -\sin\alpha \\ \sin\alpha & \cos\alpha \end{bmatrix}$$

---

[1]This φ is in fact given by an **elliptic integral**. For the related theory, see for example [FB95].
[2]See Exercises VII.9.2 and VII.9.10.

(see Exercise 10). Therefore any two identically oriented bases of a two-dimensional inner product space are related by a rotation.

(d) If E is an inner product space and $\mathcal{B}$ is an ONB, then $T_{\mathcal{B},\mathcal{C}} = \left[(c_j \mid b_k) \right]$.

Proof This follows immediately from the definition of $T_{\mathcal{B},\mathcal{C}}$. ∎

(e) An oriented basis of $\mathbb{R}^n$ is **positively oriented** if its orientation is the same as that of the canonical basis. ∎

The Frenet n-frame

In the following, we will set somewhat stronger differentiability assumptions on Γ and always assume that Γ is a regular C^n curve in $\mathbb{R}^n$ with $n \geq 2$. As usual, let $\gamma \in C^n(J, \mathbb{R}^n)$ be a parametrization of Γ. We say Γ is **complete** if the vectors $\dot{\gamma}(t), \ldots, \gamma^{(n-1)}(t)$ are linearly independent for every $t \in J$.

Suppose Γ is a complete curve in $\mathbb{R}^n$. An n-tuple $e = (e_1, \ldots, e_n)$ of functions $J \to \mathbb{R}^n$ is the **moving n-frame** or **Frenet n-frame** for Γ if

(FB$_1$) $e_j \in C^1(J, \mathbb{R}^n)$ for $1 \leq j \leq n$;

(FB$_2$) $e(t)$ is a positive orthonormal basis of $\mathbb{R}^n$ for $t \in J$;

(FB$_3$) $\gamma^{(k)}(t) \in \operatorname{span}\{e_1(t), \ldots, e_k(t)\}$ for $1 \leq k \leq n-1$ and $t \in J$;

(FB$_4$) $\left(\dot{\gamma}(t), \ldots, \gamma^{(k)}(t) \right)$ and $\left(e_1(t), \ldots, e_k(t) \right)$ for $t \in J$ and $k \in \{1, \ldots, n-1\}$
have the same orientation (as bases of $\operatorname{span}\{e_1(t), \ldots, e_k(t)\}$).

2.5 Remarks **(a)** From the chain rule, it follows easily that this concept is well defined, that is, independent of the special parametrization of Γ, in the following sense: If $\tilde{\gamma} := \gamma \circ \varphi$ is a reparametrization of γ and $\tilde{e}$ is a moving n-frame for Γ that satisfies (FB$_3$) and (FB$_4$) when γ is replaced by $\tilde{\gamma}$, then $\tilde{e}(s) = e(t)$ for $t = \varphi(s)$.

(b) A C^2 curve in $\mathbb{R}^2$ is complete if and only if it is regular.

(c) Suppose $\Gamma = [\gamma]$ is a regular C^2 curve in $\mathbb{R}^2$. Also let $e_1 := \dot{\gamma}/|\dot{\gamma}| =: (e_1^1, e_1^2)$ and $e_2 := (-e_1^2, e_1^1)$. Then (e_1, e_2) is a Frenet two-frame of Γ.

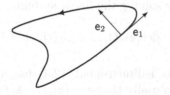 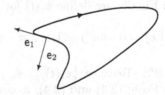

(In these and similar maps we identify $e_j(t) \in \mathbb{R}^2$ with the tangent part of $\left(\gamma(t), e_j(t) \right) \in T_{\gamma(t)}\Gamma$.)

Proof Obviously $(e_1(t), e_2(t))$ is an ONB of $\mathbb{R}^2$ and is positive because

$$\det[e_1, e_2] = (e_1^1)^2 + (e_1^2)^2 = |e_1|^2 = 1 . \quad \blacksquare$$

2.6 Theorem *Every complete C^n curve in $\mathbb{R}^n$ has a unique moving n-frame.*

Proof (i) Suppose $\Gamma = [\gamma]$ is a complete C^n curve in $\mathbb{R}^n$. One can establish the existence of a Frenet n-frame using the Gram–Schmidt orthonormalization process (see for example [Art93, VII.1.22]). Indeed, as a complete curve, Γ is regular. Therefore $e_1(t) := \dot{\gamma}(t)/|\dot{\gamma}(t)|$ is defined for $t \in I$. For $k \in \{2, \ldots, n-1\}$, we define e_k recursively. Using the already constructed $e_1, \ldots, e_{k-1}$, we set

$$\widetilde{e}_k := \gamma^{(k)} - \sum_{j=1}^{k-1}(\gamma^{(k)}\,|\,e_j)e_j \quad \text{and} \quad e_k := \widetilde{e}_k/|\widetilde{e}_k| \,. \tag{2.1}$$

Because

$$\mathrm{span}\big(\dot{\gamma}(t), \ldots, \gamma^{(k-1)}(t)\big) = \mathrm{span}\big(e_1(t), \ldots, e_{k-1}(t)\big) \tag{2.2}$$

and $\gamma^{(k)}(t) \notin \mathrm{span}\big(\dot{\gamma}(t), \ldots, \gamma^{(k-1)}(t)\big)$, we have $\widetilde{e}_k(t) \neq 0$ for $t \in J$, and thus e_k is well defined. In addition, we have

$$\big(e_j(t)\,\big|\,e_k(t)\big) = \delta_{jk} \quad \text{and} \quad \gamma^{(k)}(t) \in \mathrm{span}\big(e_1(t), \ldots, e_k(t)\big) \tag{2.3}$$

for $1 \le j, k \le n-1$ and $t \in J$.

(ii) Now suppose $T_k(t)$ is the transformation matrix from $\big(\dot{\gamma}(t), \ldots, \gamma^{(k)}(t)\big)$ to $\big(e_1(t), \ldots, e_k(t)\big)$. From (2.1), we have for $k \in \{2, \ldots, n-1\}$ the recursion formula

$$T_1(t) = |\dot{\gamma}(t)|^{-1} \quad \text{and} \quad T_k(t) = \begin{bmatrix} T_{k-1}(t) & 0 \\ * & |\widetilde{e}_k(t)|^{-1} \end{bmatrix} .$$

Thus $\det T_k(t) > 0$, which shows that $\big(\dot{\gamma}(t), \ldots, \gamma^{(k)}(t)\big)$ and $\big(e_1(t), \ldots, e_k(t)\big)$ have the same orientation.

(iii) Finally, we define $e_n(t)$ for $t \in J$ by solving the linear system

$$\big(e_j(t)\,\big|\,x\big) = 0 \quad \text{for } j = 1, \ldots, n-1 \quad \text{and} \quad \det\big[e_1(t), \ldots e_{n-1}(t), x\big] = 1 \tag{2.4}$$

for $x \in \mathbb{R}^n$. Because $\{e_1(t), \ldots e_{n-1}(t)\}^{\perp}$ is 1-dimensional, (2.4) has a unique solution. From (2.3) and (2.4), it now follows easily that $e = (e_1, \ldots, e_n)$ satisfies the postulates (FB$_2$)–(FB$_4$).

(iv) It remains to verify (FB$_1$). From (2.1), it follows easily that $e_k \in C^1(J, \mathbb{R}^n)$ for $k = 1, \ldots, n-1$. Thus (2.4) and Cramer's rule (see [Art93, § I.5]) shows that e_n also belongs to $C^1(J, \mathbb{R}^n)$. Altogether we have shown that e is a Frenet n-frame of Γ. ∎

(v) To check the uniqueness of e, suppose $(\delta_1, \ldots, \delta_n)$ is another Frenet n-frame of Γ. Then clearly we have $e_1 = \delta_1 = \dot\gamma/|\dot\gamma|$. Also, from (FB$_2$) and (FB$_3$), it follows that $\gamma^{(k)}$ has the expansion

$$\gamma^{(k)} = \sum_{j=1}^{k} (\gamma^{(k)} \,|\, \delta_j)\delta_j$$

for every $k \in \{1, \ldots, n-1\}$. Thus $e_j = \delta_j$ for $1 \le j \le k-1$ and $k \in \{2, \ldots, n-1\}$, so (2.1) implies $(\gamma^{(k)} \,|\, \delta_k)\delta_k = e_k \,|\widetilde{e}_k|$. Hence, there is for every $t \in J$ a real number $\alpha(t)$ such that $|\alpha(t)| = 1$ and $\delta_k(t) = \alpha(t)e_k(t)$. Since $\big(\delta_1(t), \ldots, \delta_k(t)\big)$ and $\big(e_1(t), \ldots, e_k(t)\big)$ are identically oriented, it follows that $\alpha(t) = 1$. Thus $(e_1, \ldots, e_{n-1})$ and therefore also e_n are uniquely fixed. ∎

2.7 Corollary *Suppose $\Gamma = [\gamma]$ is a complete C^n curve in $\mathbb{R}^n$ and $e = (e_1, \ldots, e_n)$ is a Frenet n-frame of Γ. Then we have the* **Frenet derivative formula**

$$\dot{e}_j = \sum_{k=1}^{n} (\dot{e}_j \,|\, e_k)e_k \quad \text{for } j = 1, \ldots, n \,. \tag{2.5}$$

We also have the relations

$$(\dot{e}_j \,|\, e_k) = -(e_j \,|\, \dot{e}_k) \quad \text{for } 1 \le j, k \le n \quad \text{and} \quad (\dot{e}_j \,|\, e_k) = 0 \quad \text{for } |k - j| > 1 \,.$$

Proof Because $\big(e_1(t), \ldots, e_n(t)\big)$ is an ONB of $\mathbb{R}^n$ for every $t \in J$, we get (2.5). Differentiation of $(e_j \,|\, e_k) = \delta_{jk}$ implies $(\dot{e}_j \,|\, e_k) = -(e_j \,|\, \dot{e}_k)$. Finally (2.1) and (2.2) show that $e_j(t)$ belongs to $\text{span}\big(\dot\gamma(t), \ldots, \gamma^{(j)}(t)\big)$ for $1 \le j < n$. Consequently,

$$\dot{e}_j(t) \in \text{span}\big(\dot\gamma(t), \ldots, \gamma^{(j+1)}(t)\big) = \text{span}\big(e_1(t), \ldots, e_{j+1}(t)\big) \quad \text{for } t \in J$$

and $1 \le j < n-1$, and we find $(\dot{e}_j \,|\, e_k) = 0$ for $k > j+1$ and therefore for $|k - j| > 1$. ∎

2.8 Remarks Suppose $\Gamma = [\gamma]$ is a complete C^n curve in $\mathbb{R}^n$.

(a) The curvature

$$\kappa_j := (\dot{e}_j \,|\, e_{j+1})/|\dot\gamma| \in C(J) \quad \text{with} \quad 1 \le j \le n-1 \,,$$

is well defined, that is, it is independent of the parametrization γ.

Proof This is a simple consequence of Remark 2.5(a) and the chain rule. ∎

(b) Letting $\omega_{jk} := (\dot{e}_j \,|\, e_k)$, the Frenet derivative formula reads

$$\dot{e}_j = \sum_{k=1}^{n} \omega_{jk}e_k \quad \text{for } 1 \le j \le n \,,$$

where the matrix $[\omega_{jk}]$ is

$$
|\dot\gamma|
\begin{bmatrix}
0 & \kappa_1 & & & & & \\
-\kappa_1 & 0 & \kappa_2 & & & & \\
& -\kappa_2 & 0 & \kappa_3 & & \text{\Large 0} & \\
& & \ddots & \ddots & \ddots & & \\
& \text{\Large 0} & & & & & \\
& & & & -\kappa_{n-2} & 0 & \kappa_{n-1} \\
& & & & & -\kappa_{n-1} & 0
\end{bmatrix} .
$$

Proof This follows from Corollary 2.7 and (a). ∎

In the remaining part of this section, we will discuss in more detail the implications of Theorem 2.6 and Corollary 2.7 for plane and space curves.

Curvature of plane curves

In the following, $\Gamma = [\gamma]$ is a plane regular C^2 curve and (e_1, e_2) is the associated moving two-frame. Then e_1 coincides with the tangent part of the unit tangential vector $\mathfrak{t} = (\gamma, \dot\gamma/|\dot\gamma|)$. For every $t \in J$,

$$ \mathfrak{n}(t) := \big(\gamma(t), e_2(t)\big) \in T_{\gamma(t)}\mathbb{R}^2 $$

is the **unit normal vector** of Γ at the point $\gamma(t)$. In the plane case, only κ_1 is defined, and it is called the **curvature** κ of the curve Γ. Therefore

$$ \kappa = (\dot{e}_1 \,|\, e_2)/|\dot\gamma| . \tag{2.6} $$

2.9 Remarks (a) The Frenet derivative formula reads[3]

$$ \dot{\mathfrak{t}} = |\dot\gamma|\,\kappa\mathfrak{n} \quad\text{and}\quad \dot{\mathfrak{n}} = -|\dot\gamma|\,\kappa\mathfrak{t} . \tag{2.7} $$

If $\eta \in C^2(I, \mathbb{R}^2)$ parametrizes Γ by arc length, equation (2.7) assumes the simple form

$$ \dot{\mathfrak{t}} = \kappa\mathfrak{n} \quad\text{and}\quad \dot{\mathfrak{n}} = -\kappa\mathfrak{t} . \tag{2.8} $$

In addition, we then have

$$ \kappa(s) = \big(\dot{e}_1(s) \,\big|\, e_2(s)\big) = \big(\dot{\mathfrak{t}}(s) \,\big|\, \mathfrak{n}(s)\big)_{\eta(s)} \quad\text{for } s \in I . $$

From (2.8) follows $\ddot\eta(s) = \kappa(s)e_2(s)$, and thus $|\ddot\eta(s)| = |\kappa(s)|$ for $s \in I$. On these grounds, $\kappa(s)\mathfrak{n}(s) \in T_{\eta(s)}\mathbb{R}^2$ is called the **curvature vector** of Γ at the point $\eta(s)$.

---

[3] Naturally, $\dot{\mathfrak{t}}(t)$ is understood to be the vector $\big(\gamma(t), \dot{e}_1(t)\big) \in T_{\gamma(t)}\mathbb{R}^2$, etc.

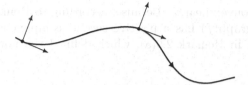

In the case $\kappa(s) > 0$ [or $\kappa(s) < 0$], the unit tangent vector $\mathfrak{t}(s)$ spins in the positive [or negative] direction (because $\big(\dot{\mathfrak{t}}(s)\,|\,\mathfrak{n}(s)\big)_{\eta(s)}$ is the projection of the instantaneous change of $\mathfrak{t}(s)$ onto the normal vector). One says that the unit normal vector $\mathfrak{n}(s)$ points to the **convex** [or **concave**] **side** of Γ.

(b) When $\eta\colon I \to \mathbb{R}^2$, $s \mapsto \big(x(s), y(s)\big)$ parametrizes Γ by arc length, $\mathfrak{e}_2 = (-\dot{y}, \dot{x})$ and

$$\kappa = \dot{x}\ddot{y} - \dot{y}\ddot{x} = \det[\dot{\eta}, \ddot{\eta}] \ .$$

Proof This follows from Remark 2.5(c) and from (2.6). ∎

For concrete calculations, it is useful to know how to find the curvature when given an arbitrary regular parametrization.

2.10 Proposition When $\gamma\colon J \to \mathbb{R}^2$, $t \mapsto \big(x(t), y(t)\big)$ *is a regular parametrization of a plane C^2 curve,*

$$\kappa = \frac{\dot{x}\ddot{y} - \dot{y}\ddot{x}}{\big((\dot{x})^2 + (\dot{y})^2\big)^{3/2}} = \frac{\det[\dot{\gamma}, \ddot{\gamma}]}{|\dot{\gamma}|^3} \ .$$

Proof From $\mathfrak{e}_1 = \dot{\gamma}/|\dot{\gamma}|$, it follows that

$$\dot{\mathfrak{e}}_1 = \frac{\ddot{\gamma}}{|\dot{\gamma}|} - \frac{\dot{x} + \dot{y}}{|\dot{\gamma}|^2}\frac{\dot{\gamma}}{|\dot{\gamma}|} = \frac{\ddot{\gamma}}{|\dot{\gamma}|} - \frac{\dot{x} + \dot{y}}{|\dot{\gamma}|^2}\mathfrak{e}_1 \ .$$

According to Remark 2.5(c), we have $\mathfrak{e}_2 = (-\dot{y}, \dot{x})/|\dot{\gamma}|$, and the theorem then follows from (2.6). ∎

2.11 Example (curvature of a graph) Suppose $f \in C^2(J, \mathbb{R})$. Then for the curve parametrized as $J \to \mathbb{R}^2$, $x \mapsto \big(x, f(x)\big)$, the curvature is

$$\kappa = \frac{f''}{\big(1 + (f')^2\big)^{3/2}} \ .$$

Proof This follows immediately from Proposition 2.10. ∎

From the characterization of convex functions of Corollary IV.2.13 and the formula above, we learn that graph(f) at the point $\big(x, f(x)\big)$ is positively curved if and only if f is convex near x. Because, according to Remark 2.9(b), the unit normal vector on graph(f) has a positive second component, this explains the convention adopted in Remark 2.9(a), which defined the notions of convex and concave.

Identifying lines and circles

Next, we apply the Frenet formula to show that line segments and circular arcs are characterized by their curvature.

2.12 Proposition *Suppose* $\Gamma = [\gamma]$ *is a plane regular* C^2 *curve. Then*

(i) Γ *is a line segment if and only if* $\kappa(t) = 0$ *for* $t \in J$;

(ii) Γ *is a circular arc with radius* r *if and only if* $|\kappa(t)| = 1/r$ *for* $t \in J$.

Proof (i) It is clear that the curvature vanishes on every line segment (see Proposition 2.10).

Conversely, suppose $\kappa = 0$. We can assume without losing generality that Γ has an arc length parametrization η. It follows from Remark 2.9(a) that $\ddot{\eta}(s) = 0$ for $s \in J$. The Taylor formula of Theorem IV.3.2 then implies that there are $a, b \in \mathbb{R}^2$ such that $\eta(s) = a + bs$.

(ii) Suppose Γ is a circular arc with radius r. Then there is an $m \in \mathbb{R}^2 = \mathbb{C}$ and an interval J such that Γ is parametrized by $J \to \mathbb{C}$, $t \mapsto re^{it} + m$ or $t \mapsto re^{-it} + m$. The claim now follows from Proposition 2.10.

Conversely, suppose $\kappa = \delta/r$ with $\delta = 1$ or $\delta = -1$. We can again assume that η parametrizes Γ by arc length. Denoting by (e_1, e_2) the moving two-frame of Γ, we find from the Frenet formula (2.8) that

$$\dot{\eta} = e_1 \,, \quad \dot{e}_1 = (\delta/r)e_2 \,, \quad \dot{e}_2 = -(\delta/r)e_1 \,.$$

From this follows

$$\big(\eta(s) + (r/\delta)e_2\big)^{\boldsymbol{\cdot}} = \dot{\eta}(s) + (r/\delta)\dot{e}_2(s) = 0 \quad \text{for } s \in I \,.$$

Hence there is an $m \in \mathbb{R}^2$ such that $\eta(s) + (r/\delta)e_2(s) = m$ for $s \in I$. Thus we get

$$|\eta(s) - m| = |re_2(s)| = r \quad \text{for } s \in I \,,$$

which shows that η parametrizes a circular arc with radius r. $\blacksquare$

Instantaneous circles along curves

Suppose $\Gamma = [\gamma]$ is a plane regular C^2 curve. If $\kappa(t_0) \neq 0$ at $t_0 \in J$, we call

$$r(t_0) := 1/|\kappa(t_0)|$$

the **instantaneous radius** and

$$m(t_0) := \gamma(t_0) + r(t_0)e_2(t_0) \in \mathbb{R}^2$$

the **instantaneous center** of Γ at the point $\gamma(t_0)$. The circle in $\mathbb{R}^2$ with center $m(t_0)$ and radius $r(t_0)$ is the **instantaneously osculating circle** of Γ at $\gamma(t_0)$.

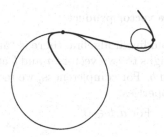

2.13 Remarks **(a)** The osculating circle has at $\gamma(t_0)$ the parametrization

$$[0, 2\pi] \to \mathbb{R}^2 , \quad t \mapsto m(t_0) + |\kappa(t_0)|^{-1} (\cos t, \sin t) .$$

(b) The osculating circle is the unique circle that touches Γ at $\gamma(t_0)$ to at least second order.

Proof We can assume that η is a parametrization of Γ by arc length with $\eta(s_0) = \gamma(t_0)$. Also assume (a, b) is a positive ONB of $\mathbb{R}^2$. Finally, let

$$\widetilde{\gamma} : [s_0, s_0 + 2\pi r] \to \mathbb{R}^2 , \quad s \mapsto \widetilde{\gamma}(s)$$

with

$$\widetilde{\gamma}(s) := m + r \cos((s - s_0)/r)a + r \sin((s - s_0)/r)b$$

parametrize by arc length the circle K with center m and radius r. Then Γ and K touch at $\eta(s_0)$ to at least second order if $\eta^{(j)}(s_0) = \widetilde{\gamma}^{(j)}(s_0)$ for $0 \le j \le 2$ or, equivalently, if the equations

$$\eta(s_0) = m + ra , \quad \dot{\eta}(s_0) = b , \quad \ddot{\eta}(s_0) = -a/r$$

hold. Because $e_1 = \dot{\eta}$, the first Frenet formula implies $\kappa(s_0)e_2(s_0) = -a/r$. Because the orthonormal basis $(-e_2(s_0), e_1(s_0))$ is positively oriented, we find that $|\kappa(s_0)| = 1/r$ and $(a, b) = (-e_2(s_0), e_1(s_0))$ and $m = \eta(s_0) + re_2(s_0) = \gamma(t_0) + [\kappa(t_0)]^{-1}e_2(t_0)$. ∎

(c) When $\kappa(t) \neq 0$ for $t \in J$,

$$t \mapsto m(t) := \gamma(t) + e_2(t)/\kappa(t)$$

is a continuous parametrization of a plane curve, the **evolute** of Γ. ∎

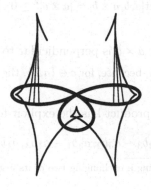

The vector product

From linear algebra, there is "another product" in $\mathbb{R}^3$ besides the inner product. It assigns to two vectors a and b another vector $a \times b$, the vector or cross product of a and b. For completeness, we recall how it is defined and collect its most important properties.

For $a, b \in \mathbb{R}^3$,

$$\mathbb{R}^3 \to \mathbb{R} , \quad c \mapsto \det[a, b, c]$$

is a (continuous) linear form on $\mathbb{R}^3$. Therefore, by the Riesz representation theorem, there is exactly one vector, called $a \times b$, such that

$$(a \times b \,|\, c) = \det[a, b, c] \quad \text{for } c \in \mathbb{R}^3 . \tag{2.9}$$

Through this, we define the **vector** or **cross product**.

$$\times : \mathbb{R}^3 \times \mathbb{R}^3 \to \mathbb{R}^3 , \quad (a, b) \mapsto a \times b .$$

2.14 Remarks (a) The cross product is an **alternating** bilinear map, that is, $\times$ is bilinear and skew-symmetric:

$$a \times b = -b \times a \quad \text{for } a, b \in \mathbb{R}^3 .$$

Proof This follows easily from (2.9), because the determinant is an alternating[4] trilinear form. ∎

(b) For $a, b \in \mathbb{R}^3$, we have $a \times b \neq 0$ if and only if a and b are linearly independent. If they are, $(a, b, a \times b)$ is a positively oriented basis of $\mathbb{R}^3$.

Proof From (2.9), we have $a \times b = 0$ if and only if $\det[a, b, c]$ vanishes for every choice of $c \in \mathbb{R}^3$. Were a and b linearly independent, we would have $\det[a, b, c] \neq 0$ by choosing c such that a, b and c are linearly independent, which, because of (2.9), contradicts the assumption $a \times b = 0$. This implies the first claim. The second claim follows because (2.9) gives

$$\det[a, b, a \times b] = |a \times b|^2 \geq 0 . ∎$$

(c) For $a, b \in \mathbb{R}^3$, the vector $a \times b$ is perpendicular to a and b.

Proof This follows from (2.9), because, for $c \in \{a, b\}$, the determinant has two identical columns. ∎

(d) For $a, b \in \mathbb{R}^3$, the cross product has the explicit form

$$a \times b = (a_2 b_3 - a_3 b_2, a_3 b_1 - a_1 b_3, a_1 b_2 - a_2 b_1) .$$

---

[4]An m-linear map is **alternating** if exchanging two of its arguments multiplies it by negative one.

Proof This follows again from (2.9) by expanding the determinant in its last column. ∎

(e) For $a, b \in \mathbb{R}^3 \setminus \{0\}$, the (unoriented) **angle** $\varphi := \sphericalangle(a, b) \in [0, \pi]$ **between a and b** is defined by $\cos\varphi := (a \mid b)/|a| \, |b|$.[5] Then we have

$$|a \times b| = \sqrt{|a|^2 \, |b|^2 - (a \mid b)^2} = |a| \, |b| \sin\varphi .$$

This means that $|a \times b|$ is precisely the (unoriented) area of the parallelogram spanned by a and b.

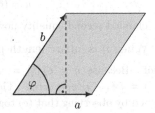

Proof Because of the Cauchy–Schwarz inequality, $\sphericalangle(a, b) \in [0, \pi]$ is well defined. The first equality can be verified directly from (d), and the second follows from $\cos^2\varphi + \sin^2\varphi = 1$ and $\sin\varphi \geq 0$. ∎

(f) When a and b are orthonormal vectors in $\mathbb{R}^3$, the (ordered) set $(a, b, a \times b)$ is a positive ONB on $\mathbb{R}^3$.

Proof This follows from (b) and (e). ∎

The curvature and torsion of space curves

Working now in three dimensions, suppose $\Gamma = [\gamma]$ is a complete C^3 curve in $\mathbb{R}^3$. Then Γ has two curvatures κ_1 and κ_2. We call $\kappa := \kappa_1$ the **curvature** and $\tau := \kappa_2$ the **torsion**. Therefore

$$\kappa = (\dot{e}_1 \mid e_2)/|\dot\gamma| \quad \text{and} \quad \tau = (\dot{e}_2 \mid e_3)/|\dot\gamma| . \tag{2.10}$$

Here we also call

$$\mathfrak{n}(t) := \big(\gamma(t), e_2(t)\big) \in T_{\gamma(t)}\mathbb{R}^3$$

the **unit normal vector**, while

$$\mathfrak{b}(t) := \big(\gamma(t), e_3(t)\big) \in T_{\gamma(t)}\mathbb{R}^3$$

is the **unit binormal vector** of Γ at the point $\gamma(t)$. The vectors $\mathfrak{t}(t)$ and $\mathfrak{n}(t)$ span the **osculating plane** of Γ at the point $\gamma(t)$. The plane spanned by $\mathfrak{n}(t)$ and $\mathfrak{b}(t)$ is the **normal plane** of Γ at $\gamma(t)$.

2.15 Remarks **(a)** We have $e_3 = e_1 \times e_2$, that is, $\mathfrak{b} = \mathfrak{t} \times \mathfrak{n}$.

(b) When Γ is parametrized by arc length, the Frenet formula gives

$$\dot{e}_1 = \kappa e_2 , \quad \dot{e}_2 = -\kappa e_1 + \tau e_3 , \quad \dot{e}_3 = -\tau e_2 ,$$

and therefore

$$\dot{\mathfrak{t}} = \kappa\mathfrak{n} , \quad \dot{\mathfrak{n}} = -\kappa\mathfrak{t} + \tau\mathfrak{b} , \quad \dot{\mathfrak{b}} = -\tau\mathfrak{n} .$$

---

[5]This angle can be defined in any inner product space, not just $\mathbb{R}^3$. Also $\sphericalangle(a, b) = \sphericalangle(b, a)$.

(c) The curvature of a complete curve is always positive. When η is an arc length parametrization,

$$\kappa = |\ddot{\eta}| = |\dot{\eta} \times \ddot{\eta}| \ .$$

Proof Because $e_1 = \dot{\eta}$ and because (2.1) and Remark 2.3(a) imply that $e_2 = \ddot{\eta}/|\ddot{\eta}|$, we get

$$\kappa = (\dot{e}_1 \,|\, e_2) = (\ddot{\eta} \,|\, \ddot{\eta}/|\ddot{\eta}|) = |\ddot{\eta}| > 0 \ .$$

The claimed second equality now follows from Remark 2.14(e). ∎

(d) When η is an arc length parametrization of Γ, we have $\tau = \det[\dot{\eta}, \ddot{\eta}, \dddot{\eta}]/\kappa^2$.

Proof Because $\ddot{\eta} = e_2 |\ddot{\eta}| = \kappa e_2$ and from the Frenet derivative formula for e_2, we get $\dddot{\eta} = \dot{\kappa} e_2 - \kappa^2 e_1 + \kappa \tau e_3$. Thus $\det[\dot{\eta}, \ddot{\eta}, \dddot{\eta}] = \det[e_1, \kappa e_2, \kappa \tau e_3] = \kappa^2 \tau$. The proof is finished by observing that (c) together with the completeness of Γ implies that κ vanishes nowhere. ∎

The torsion $\tau(t)$ is a measure of how, in the vicinity of $\gamma(t)$, the curve $\Gamma = [\gamma]$ "winds" out of the osculating plane, shown in lighter gray at right. Indeed, a plane curve is characterized by vanishing torsion, as the following theorem shows.

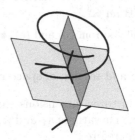

2.16 Proposition *A complete curve lies in a plane if and only if its torsion vanishes everywhere.*

Proof We can assume that η parametrizes Γ by arc length. If $\eta(s)$ lies in a plane for every $s \in I$, so does $\dot{\eta}(s)$, $\ddot{\eta}(s)$, and $\dddot{\eta}(s)$. With Remark 2.15(d), this implies $\tau = \det[\dot{\eta}, \ddot{\eta}, \dddot{\eta}]/\kappa^2 = 0$. Conversely, if $\tau = 0$, the third Frenet derivative formula gives that $e_3(s) = e_3(\alpha)$ for every $s \in I$ and an $\alpha \in I$. Because $e_1(s) \perp e_3(s)$,

$$\left(\eta(s) \,\middle|\, e_3(\alpha) \right)^{\cdot} = \left(\dot{\eta}(s) \,\middle|\, e_3(\alpha) \right) = \left(e_1(s) \,\middle|\, e_3(s) \right) = 0 \quad \text{for } s \in I \ .$$

Therefore $\left(\eta(\cdot) \,\middle|\, e_3(\alpha) \right)$ is constant on I, which means that the orthogonal projection of $\eta(s)$ onto the line $\mathbb{R}e_3(\alpha)$ is independent of $s \in I$. Therefore Γ lies in the plane orthogonal to $e_3(\alpha)$. ∎

Exercises

1 Suppose $r \in C^2(J, \mathbb{R}^+)$ has $[r(t)]^2 + [\dot{r}(t)]^2 > 0$ for $t \in J$ and $\gamma(t) := r(t)(\cos t, \sin t)$. Show the curve parametrized by γ has the curvature

$$\kappa = \frac{2[\dot{r}]^2 - r\ddot{r} + r^2}{\left(r^2 + [\dot{r}]^2 \right)^{3/2}} \ .$$

2 Calculate the curvature of

the limaçon of Pascal $\quad [-\pi, \pi] \to \mathbb{R}^2 \ , \quad t \mapsto (1 + 2\cos t)(\cos t, \sin t)$;

the logarithmic spiral $\quad \mathbb{R} \to \mathbb{R}^2 \ , \quad t \mapsto e^{\lambda t}(\cos t, \sin t)$;

the cycloid $\quad [0, 2\pi] \to \mathbb{R}^2 \ , \quad t \mapsto R(t - \sin t, 1 - \cos t)$.

3 Suppose $\gamma : J \to \mathbb{R}^3, \ t \mapsto (x(t), y(t))$ is a regular C^2 parametrization of a plane curve Γ, and $\kappa(t) \neq 0$ for $t \in J$. Show that the evolute of Γ is parametrized by $t \mapsto m(t)$, where

$$ m := \left(x - \frac{\dot{x}^2 + \dot{y}^2}{\dot{x}\ddot{y} - \ddot{x}\dot{y}} \dot{y}, y + \frac{\dot{x}^2 + \dot{y}^2}{\dot{x}\ddot{y} - \ddot{x}\dot{x}} \dot{x} \right) . $$

4 Prove that

(a) the evolute of the parabola $y = x^2/2$ is the Neil, or semicubical, parabola $\mathbb{R} \to \mathbb{R}^2$, $t \mapsto (-t^3, 1 + 3t^2/2)$;

(b) the evolute of the logarithmic spiral $\mathbb{R} \to \mathbb{R}^2, \ t \mapsto e^{\lambda t}(\cos t, \sin t)$ is the logarithmic spiral $\mathbb{R} \to \mathbb{R}^2, \ t \mapsto \lambda e^{\lambda(t - \pi/2)}(\cos t, \sin t)$;

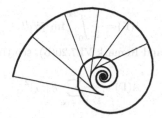

(c) the evolute of the cycloid $\mathbb{R} \to \mathbb{R}^2, \ t \mapsto (t - \sin t, 1 - \cos t)$ is the cycloid $\mathbb{R} \to \mathbb{R}^2$, $t \mapsto (t + \sin t, \cos t - 1)$.

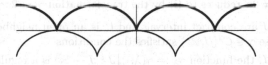

5 Let $\gamma \in C^3(J, \mathbb{R}^2)$ be a regular parametrization of a plane curve Γ, and suppose $\kappa(t)\dot{\kappa}(t) \neq 0$ for $t \in J$. Also denote by m the parametrization given in Remark 2.13(c) for the evolute M of Γ. Show

(a) M is a regular C^1 curve;

(b) the tangent to Γ at $\gamma(t)$ is perpendicular to the tangent to M at $m(t)$;

(c) the arc length of M is given by

$$ L(M) = \int_J |\dot{\kappa}_\gamma| \kappa_\gamma^{-2} \, dt . $$

(Hint for (b) and (c): the Frenet derivative formulas.)

6 Show that the logarithmic spiral Γ slices through straight lines from the origin at a constant angle, that is, at any point along the spiral, the angle between the tangent to Γ and the line from the origin is the same.

7 Calculate the curvature and the torsion

(a) for the **elliptic helix** $\mathbb{R} \to \mathbb{R}^3$, $t \mapsto (a\cos t, b\sin t, ct)$ with $ab \neq 0$ for $a, b, c \in \mathbb{R}$;

(b) for the curve $\mathbb{R} \to \mathbb{R}^3$, $t \mapsto t(\cos t, \sin t, 1)$.

8 Suppose Γ is a regular closed C^1 curve in the plane. Prove the **isoperimetric inequality**

$$4\pi A(\Gamma) \leq \big[L(\Gamma)\big]^2$$

(see Exercise 1.6).

(Hints: (i) Identify $\mathbb{R}^2$ with $\mathbb{C}$ and consider first the case $L(\Gamma) = 2\pi$. Then, without loss of generality, let $\gamma \in C^1([0, 2\pi], \mathbb{C})$ parametrize Γ by arc length. Using Parseval's equation (Corollary VI.7.16) and Remark VI.7.20(b), show

$$1 = \frac{1}{2\pi} \int_0^{2\pi} |\dot\gamma|^2\, dt = \sum_{k=-\infty}^{\infty} k^2 \,|\widehat{\gamma}_k|^2 \;.$$

Further show

$$A(\Gamma) = \frac{1}{2} \int_0^{2\pi} \mathrm{Im}\, \overline{\gamma}\dot\gamma\, dt \;.$$

Thus from Exercise VI.7.11 and Remark VI.7.20(b), see that

$$A(\Gamma) = \pi \sum_{k=-\infty}^{\infty} k \,|\widehat{\gamma}_k|^2 \;,$$

and therefore

$$A(\Gamma) \leq \pi \sum_{k=-\infty}^{\infty} k^2 \,|\widehat{\gamma}_k|^2 = \pi = \big[L(\Gamma)\big]^2/4\pi \;.$$

(ii) The case $L(\Gamma) \neq 2\pi$ reduces to (i) by the transformation $\gamma \to 2\pi\gamma/L(\Gamma)$.)

9 Suppose I and J are compact intervals and U is an open neighborhood of $I \times J$ in $\mathbb{R}^2$. Further, suppose $\gamma \in C^3(U, \mathbb{R}^2)$ satisfies the conditions

(i) for every $\lambda \in I$, the function $\gamma_\lambda := \gamma(\lambda, \cdot) \,|\, J : J \to \mathbb{R}^2$ is a regular parametrization of a C^3 curve Γ_λ;

(ii) $\partial_1\gamma(\lambda, t) = \kappa_{\Gamma_\lambda}(t)\mathfrak{n}_{\Gamma_\lambda}(t)$ for $(\lambda, t) \in I \times J$.

Show that the function

$$\ell : I \to \mathbb{R} \;, \quad \lambda \mapsto L(\Gamma_\lambda)$$

has continuous derivatives and that

$$\ell'(\lambda) = -\int_0^{\ell(\lambda)} \big[\kappa\big(s(\lambda)\big)\big]^2\, ds(\lambda) \quad \text{for } \lambda \in I \;,$$

where $s(\lambda)$ is the arc length paramater of Γ_λ for $\lambda \in I$.

(Hint: Define $v : I \times J \to \mathbb{R}$, $(\lambda, t) \mapsto |\partial_1\gamma(\lambda, t)|$. Using the Frenet formulas and (ii), conclude $\partial_1 v = -\kappa^2 v$.)

10 Show that for every $T \in SO(2)$, there is a unique $\alpha \in [0, 2\pi)$ such that

$$T = \begin{bmatrix} \cos\alpha & -\sin\alpha \\ \sin\alpha & \cos\alpha \end{bmatrix}.$$

(Hint: Exercise VII.9.2.)

3   Pfaff forms

In Section VI.5, we introduced differentials in an ad hoc way, so that we could represent integration rules more easily. We will now develop a calculus for using these previously formal objects. This calculus will naturally admit the simple calculation rules from before and will also add much clarity and understanding. Simultaneously, this development will expand the concept of an integral from something done on intervals to something that can be done on curves.

In the following let

- X be open in $\mathbb{R}^n$ and not empty; also suppose $q \in \mathbb{N} \cup \{\infty\}$.

Vector fields and Pfaff forms

A map $\boldsymbol{v} \colon X \to TX$ in which $\boldsymbol{v}(p) \in T_pX$ for $p \in X$ is a **vector field** on X. To every vector field $\boldsymbol{v}$, there corresponds a unique function $v \colon X \to \mathbb{R}^n$, the **tangent part** of $\boldsymbol{v}$, such that $\boldsymbol{v}(p) = \big(p, v(p)\big)$. With the natural projection

$$\mathrm{pr}_2 \colon TX = X \times \mathbb{R}^n \to \mathbb{R}^n \ , \quad (p, \xi) \mapsto \xi \ ,$$

we have $v = \mathrm{pr}_2 \circ \boldsymbol{v}$.

A vector field $\boldsymbol{v}$ belongs to the class C^q if $v \in C^q(X, \mathbb{R}^n)$. We denote the set of all vector field of class C^q on X by $\mathcal{V}^q(X)$.

Suppose E is a Banach space over $\mathbb{K}$ and $E' := \mathcal{L}(E, \mathbb{K})$ is its dual space. The map

$$\langle \cdot, \cdot \rangle \colon E' \times E \to \mathbb{K} \ , \quad (e', e) \mapsto \langle e', e \rangle := e'(e) \ ,$$

assigns every pair $(e', e) \in E' \times E$ the value of e' at the location e and is called the **dual pairing** between E and E'.

We will now think about the spaces dual to the tangential spaces of X. For $p \in X$, the space dual to T_pX is the **cotangent space** of X at the point p and is denoted T_p^*X. Also

$$T^*X := \bigcup_{p \in X} T_p^*X$$

is the **cotangential bundle** of X. The elements of T_p^*X, the **cotangential vectors** at p, are therefore linear forms on the tangential space of X at p. The corresponding dual pairing is denoted by

$$\langle \cdot, \cdot \rangle_p \colon T_p^*X \times T_pX \to \mathbb{R} \ .$$

Every map $\boldsymbol{\alpha} \colon X \to T^*X$ in which $\boldsymbol{\alpha}(p) \in T_p^*X$ for $p \in X$ is called a **Pfaff form** on X. Sometimes we call α a **differential form of degree 1** or a **1-form**.

3.1 Remarks   (a) If E is a Banach space over $\mathbb{K}$, we have $\langle \cdot, \cdot \rangle \in \mathcal{L}(E', E; \mathbb{K})$.

Proof   It is clear that the dual pairing is bilinear. Further, Conclusion VI.2.4(a) gives the estimate

$$|\langle e', e \rangle| = |e'(e)| \leq \|e'\|_{E'} \|e\|_E \quad \text{for } (e', e) \in E' \times E .$$

Therefore Proposition VII.4.1 shows that $\langle \cdot, \cdot \rangle$ belongs to $\mathcal{L}(E', E; \mathbb{K})$. ∎

(b) For $(p, e') \in \{p\} \times (\mathbb{R}^n)'$, define $J(p, e') \in T_p^* X$ through

$$\langle J(p, e'), (p, e) \rangle_p = \langle e', e \rangle \quad \text{for } e \in \mathbb{R}^n , \tag{3.1}$$

where $\langle \cdot, \cdot \rangle$ denotes the dual pairing between $\mathbb{R}^n$ and $(\mathbb{R}^n)'$. Then J is an isometric isomorphism of $\{p\} \times (\mathbb{R}^n)'$ onto $T_p^* X$. *Using this isomorphism, we identify $T_p^* X$ with $\{p\} \times (\mathbb{R}^n)'$.*

Proof   Obviously J is linear. If $J(p, e') = 0$ for some $(p, e') \in \{p\} \times (\mathbb{R}^n)'$, then $e' = 0$ follows from (3.1). Therefore J is injective. Because

$$\dim\big(\{p\} \times (\mathbb{R}^n)'\big) = \dim\big(\{p\} \times \mathbb{R}^n\big) = \dim(T_p X) = \dim(T_p^* X) ,$$

J is also surjective. The claim then follows from Theorem VII.1.6. ∎

Because $T_p^* X = \{p\} \times (\mathbb{R}^n)'$, there is for every Pfaff form $\boldsymbol{\alpha}$ a unique map $\alpha \colon X \to (\mathbb{R}^n)'$, the **cotangent part** of $\boldsymbol{\alpha}$, such that $\boldsymbol{\alpha}(p) = \big(p, \alpha(p)\big)$. A 1-form $\boldsymbol{\alpha}$ belongs to the **class** C^q if $\alpha \in C^q\big(X, (\mathbb{R}^n)'\big)$. We denote the set of all 1-forms of class C^q on X by $\Omega_{(q)}(X)$.

On $\mathcal{V}^q(X)$, we define multiplication by functions in $C^q(X)$ by

$$C^q(X) \times \mathcal{V}^q(X) \to \mathcal{V}^q(X) , \qquad (a, \boldsymbol{v}) \mapsto a\boldsymbol{v}$$

using the pointwise multiplication $(a\boldsymbol{v})(p) := a(p)\boldsymbol{v}(p)$ for $p \in X$. Likewise, on $\Omega_{(q)}(X)$, we define multiplication by functions in $C^q(X)$ by

$$C^q(X) \times \Omega_{(q)}(X) \to \Omega_{(q)}(X) , \qquad (a, \boldsymbol{\alpha}) \mapsto a\boldsymbol{\alpha}$$

using the pointwise multiplication $(a\boldsymbol{\alpha})(p) := a(p)\boldsymbol{\alpha}(p)$ for $p \in X$.

Then one easily verifies that $\mathcal{V}^q(X)$ and $\Omega_{(q)}(X)$ are modules[1] over the (commutative) ring $C^q(X)$. Because $C^q(X)$ naturally contains $\mathbb{R}$ as a subring (using the identification of $\lambda \in \mathbb{R}$ and $\lambda \mathbf{1} \in C^q(X)$), we have in particular that $\mathcal{V}^q(X)$ and $\Omega_{(q)}(X)$ are real vector spaces.

3.2 Remarks   (a) For $\boldsymbol{\alpha} \in \Omega_{(q)}(X)$ and $\boldsymbol{v} \in \mathcal{V}^q(X)$,

$$\Big(p \mapsto \langle \boldsymbol{\alpha}(p), \boldsymbol{v}(p) \rangle_p = \langle \alpha(p), v(p) \rangle\Big) \in C^q(X) .$$

---

[1]For completeness, we have put the relevant facts about modules at the end of this section.

(b) When no confusion is expected, we identify $\alpha \in \Omega_{(q)}(X)$ with its cotangent part $\alpha = \mathrm{pr}_2 \circ \boldsymbol{\alpha}$ and identify $\boldsymbol{v} \in \mathcal{V}^q(X)$ with its tangent part $v = \mathrm{pr}_2 \circ \boldsymbol{v}$. Then

$$\mathcal{V}^q(X) = C^q(X, \mathbb{R}^n) \quad \text{and} \quad \Omega_{(q)}(X) = C^q\big(X, (\mathbb{R}^n)'\big) .$$

(c) Suppose $f \in C^{q+1}(X)$, and

$$d_p f = \mathrm{pr}_2 \circ T_p f \quad \text{for } p \in X ,$$

is the differential of f (see Section VII.10). Then the map $df := (p \mapsto d_p f)$ belongs to $\Omega_{(q)}(X)$, and we have $d_p f = \partial f(p)$. From now on, we write $df(p)$ for $d_p f$. $\blacksquare$

The canonical basis

In the following, let $(e_1, \dots, e_n)$ be the standard basis of $\mathbb{R}^n$, and let $(\varepsilon^1, \dots, \varepsilon^n)$ be the corresponding dual basis[2] of $(\mathbb{R}^n)'$, that is,[3]

$$\langle \varepsilon^j, e_k \rangle = \delta_k^j \quad \text{for } j, k \in \{1, \dots, n\} .$$

We also set

$$dx^j := d(\mathrm{pr}_j) \in \Omega_{(\infty)}(X) \quad \text{for } j = 1, \dots, n ,$$

using the j-th projection $\mathrm{pr}_j : \mathbb{R}^n \to \mathbb{R}$, $x = (x_1, \dots, x_n) \mapsto x_j$.

3.3 Remarks **(a)** According to Remark 3.1(b), $(\varepsilon^j)_p = (p, \varepsilon^j)$ belongs to $T_p^* X$ for every $j = 1, \dots, n$, and $\big((\varepsilon^1)_p, \dots, (\varepsilon^n)_p\big)$ is the basis that is dual to the canonical basis $\big((e_1)_p, \dots, (e_n)_p\big)$ of $T_p X$.

(b) We have
$$dx^j(p) = (p, \varepsilon^j) \quad \text{for } p \in X \text{ and } j = 1, \dots, n .$$

Proof The claim follows from Proposition VII.2.8, which gives

$$\big\langle dx^j(p), (e_k)_p \big\rangle_p = \langle \partial(\mathrm{pr}_j)(p), e_k \rangle = \partial_k \, \mathrm{pr}_j(p) = \delta_{jk} . \ \blacksquare$$

(c) We set $e_j(p) := (p, e_j)$ for $1 \le j \le n$ and $p \in X$. Then $(e_1, \dots, e_n)$ is a module basis of $\mathcal{V}^q(X)$, and $(dx^1, \dots, dx^n)$ is a module basis of $\Omega_{(q)}(X)$. Also, we have

$$\big\langle dx^j(p), e_k(p) \big\rangle_p = \delta_k^j \quad \text{for } 1 \le j, k \le n \text{ and } p \in X , \tag{3.2}$$

and $(e_1, \dots, e_n)$ is the **canonical basis** of $\mathcal{V}^q(X)$. Likewise, $(dx^1, \dots, dx^n)$ is the canonical basis of $\Omega_{(q)}(X)$.[4]

---

[2]The existence of dual bases is a standard result of linear algebra.
[3]$\delta_k^j := \delta^{jk} := \delta_{jk}$ is the Kronecker symbol.
[4]These basis are dual to each other in the sense of module duality (see [SS88, § 70]).

Finally, we define

$$\langle \boldsymbol{\alpha}, \boldsymbol{v} \rangle : \Omega_{(q)}(X) \times \mathcal{V}^q(X) \to C^q(X)$$

through

$$\langle \boldsymbol{\alpha}, \boldsymbol{v} \rangle(p) := \langle \boldsymbol{\alpha}(p), \boldsymbol{v}(p) \rangle_p \quad \text{for } p \in X \ .$$

Then (3.2) takes the form

$$\langle dx^j, e_k \rangle = \delta_k^j \quad \text{for } 1 \le j, k \le n \ .$$

Proof Suppose $\boldsymbol{\alpha} \in \Omega_{(q)}(X)$. We set

$$a_j(p) := \langle \boldsymbol{\alpha}(p), e_j(p) \rangle_p \quad \text{for } p \in X \ .$$

Then, due to Remark 3.2(a), a_j belongs to $C^q(X)$, and therefore $\beta := \sum a_j \, dx^j$ belongs to $\Omega_{(q)}(X)$. Also,

$$\langle \beta(p), e_k(p) \rangle_p = \sum_{j=1}^n a_j(p) \langle dx^j(p), e_k(p) \rangle_p = a_k(p) = \langle \boldsymbol{\alpha}(p), e_k(p) \rangle_p \quad \text{for } p \in X$$

and $k = 1, \ldots, n$. Therefore $\boldsymbol{\alpha} = \beta$, which shows that $(e_1, \ldots e_n)$ and $(dx^1, \ldots, dx^n)$ are module bases and that (3.2) holds. ∎

(d) Every $\boldsymbol{\alpha} \in \Omega_{(q)}(X)$ has the **canonical basis representation**

$$\boldsymbol{\alpha} = \sum_{j=1}^n \langle \boldsymbol{\alpha}, e_j \rangle \, dx^j \ .$$

(e) For $f \in C^{q+1}(X)$,

$$df = \partial_1 f \, dx^1 + \cdots + \partial_n f \, dx^n \ .$$

The map $d : C^{q+1}(X) \to \Omega_{(q)}(X)$ is $\mathbb{R}$-linear.

Proof Because

$$\langle df(p), e_j(p) \rangle_p = \partial_j f(p) \quad \text{for } p \in X \ ,$$

the first claim follows from (d). The second is obvious. ∎

(f) $\mathcal{V}^q(X)$ and $\Omega_{(q)}(X)$ are infinite-dimensional vectors spaces over $\mathbb{R}$.

Proof We consider $X = \mathbb{R}$ and leave the general case to you. From the fundamental theorem of algebra, it follows easily that the monomials $\{ X^m \ ; \ m \in \mathbb{N} \}$ in $\mathcal{V}^q(\mathbb{R}) = C^q(\mathbb{R})$ are linearly independent over $\mathbb{R}$. Therefore $\mathcal{V}^q(\mathbb{R})$ is an infinite-dimensional vector space. We already know that $\Omega_{(q)}(\mathbb{R})$ is an $\mathbb{R}$-vector space. As in the case of $\mathcal{V}^q(\mathbb{R})$, we see that $\{ X^m \, dx \ ; \ m \in \mathbb{N} \} \subset \Omega_{(q)}(\mathbb{R})$ is a linearly independent set over $\mathbb{R}$. Therefore $\Omega_{(q)}(\mathbb{R})$ is also an infinite-dimensional $\mathbb{R}$-vector space. ∎

(g) The $C^q(X)$-modules $\mathcal{V}^q(X)$ and $\Omega_{(q)}(X)$ are isomorphic. We define a module isomorphism, the **canonical isomorphism**, through

$$\Theta : \mathcal{V}^q(X) \to \Omega_{(q)}(X) , \quad \sum_{j=1}^{n} a_j \boldsymbol{e}_j \mapsto \sum_{j=1}^{n} a_j \, dx^j .$$

Proof This follows from (c) and (d). ∎

(h) For $f \in C^{q+1}(X)$, we have $\Theta^{-1} df = \operatorname{grad} f$, that is, the diagram[5]

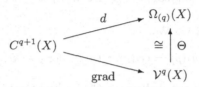

is commutative.

Proof Due to (e), we have

$$\Theta^{-1} df = \Theta^{-1} \Big(\sum_{j=1}^{n} \partial_j f \, dx^j \Big) = \sum_{j=1}^{n} \partial_j f \, \boldsymbol{e}_j = \operatorname{grad} f \quad \text{for } f \in C^{q+1}(X) . ∎$$

Exact forms and gradient fields

We have seen that every function $f \in C^{q+1}(X)$ induces the Pfaff form df. Now we ask whether every Pfaff form is so induced.

Suppose $\alpha \in \Omega_{(q)}(X)$, where in the following we apply the identification rule of Remark 3.2(b) If there is a $f \in C^{q+1}(X)$ such that $df = \alpha$, we say α is **exact**, and f is an **antiderivative** of α.

3.4 Remarks **(a)** Suppose X is a domain and $\alpha \in \Omega_{(0)}(X)$ is exact. If f and g are antiderivatives of α, then $f - g$ is constant.

Proof First, $\alpha = df = dg$ implies $d(f - g) = 0$. Then apply Remark VII.3.11(c). ∎

(b) Suppose $\alpha = \sum_{j=1}^{n} a_j \, dx^j \in \Omega_{(1)}(X)$ is exact. Then it satisfies **integrability conditions**

$$\partial_k a_j = \partial_j a_k \quad \text{for } 1 \le j, k \le n .$$

Proof Suppose $f \in C^2(X)$ has $df = \alpha$. From Remarks 3.3(d) and (e), we have $a_j = \partial_j f$. Thus it follows from the Schwarz theorem (Corollary VII.5.5(ii)) that

$$\partial_k a_j = \partial_k \partial_j f = \partial_j \partial_k f = \partial_j a_k \quad \text{for } 1 \le j, k \le n . ∎$$

---

[5]Analogously to the notation df, we write $\operatorname{grad} f$, or ∇f, for the map $p \mapsto \nabla_p f$ (see Section VII.10). In every concrete case, it will be clear from context whether $\operatorname{grad} f(p)$ means the value of this function at the point p or the tangent part of $\nabla_p f$.

(c) If $X \subset \mathbb{R}$, every form $\alpha \in \Omega_{(q)}(X)$ is exact.

Proof Because X is a disjoint union of open intervals (see Exercise III.4.6), it suffices to consider the case $X = (a,b)$ with $a < b$. According to Remark 3.3(c), there is for every $\alpha \in \Omega_{(q)}(X)$ an $a \in C^q(X)$ such that $\alpha = a\,dx$. Suppose $p_0 \in (a,b)$ and

$$f(p) := \int_{p_0}^p a(x)\,dx \quad \text{for } p \in (a,b) \ .$$

Then f belongs to $C^{q+1}(X)$ with $f' = a$. Therefore $df = a\,dx = \alpha$ follows from Remark 3.3(e). $\blacksquare$

With the help of the canonical isomorphism $\Theta : \mathcal{V}^q(X) \to \Omega_{(q)}(X)$, the properties of Pfaff forms can be transferred to vector fields. We thus introduce the following notation: a vector field $v \in \mathcal{V}^q(X)$ is a **gradient field** if there is an $f \in C^{q+1}(X)$ such that $v = \nabla f$. In this context, one calls f a **potential** for v.

Remarks 3.3 and 3.4 imply immediately the facts below about gradient fields.

3.5 Remarks **(a)** If X is a domain, then two potentials of any gradient field on it differ by at most a constant.

(b) If $v = (v_1, \ldots, v_n) \in \mathcal{V}^1(X)$ is a gradient field, the integrability conditions $\partial_k v_j = \partial_j v_k$ hold for $1 \leq j, k \leq n$.

(c) A Pfaff form $\alpha \in \Omega_{(q)}(X)$ is exact if and only if $\Theta^{-1}\alpha \in \mathcal{V}^q(X)$ is a gradient field. $\blacksquare$

3.6 Examples **(a)** (central fields) Suppose $X := \mathbb{R}^n \setminus \{0\}$ and the components of the Pfaff form

$$\alpha = \sum_{j=1}^n a_j\,dx^j \in \Omega_{(q)}(X)$$

have the representation $a_j(x) := x^j \varphi(|x|)$ for $x = (x^1, \ldots, x^n)$ and $1 \leq j \leq n$, where $\varphi \in C^q((0,\infty), \mathbb{R})$. Then α is exact. An antiderivative f is given by $f(x) := \Phi(|x|)$ with

$$\Phi(r) := \int_{r_0}^r t\varphi(t)\,dt \quad \text{for } r > 0 \ ,$$

where r_0 is a strictly positive number. Therefore the vector field $v := \sum_{j=1}^n a_j e_j = \Theta^{-1}\alpha$ is a gradient field in which the vector $v(x) \in T_x(X)$ lies on the line through x and 0 for every $x \in X$. The **equipotential surfaces**, that is, the level sets of f, are the spheres rS^{n-1} for $r > 0$, these spheres are perpendicular to the gradient fields (see Example VII.3.6(c)).

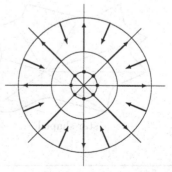

Proof It is clear that Φ belongs to $C^{q+1}((0, \infty), \mathbb{R})$ and that $\Phi'(r) = r\varphi(r)$. It then follows from the chain rule that

$$\partial_j f(x) = \Phi'(|x|) x^j / |x| = x^j \varphi(|x|) = a_j(x) \quad \text{for } j = 1, \ldots, n \text{ and } x \in X .$$

The remaining claims are obvious. $\blacksquare$

(b) The central field $x \mapsto cx/|x|^n$ with $c \in \mathbb{R}^\times$ and $n \geq 2$ is a gradient field. A potential U is given by

$$U(x) := \begin{cases} c \log |x| , & n = 2 , \\ c \, |x|^{2-n} / (2 - n), & n > 2 , \end{cases}$$

for $x \neq 0$. This potential plays an important role in physics.[6] Depending on the physical setting, it may be either the **Newtonian** or the **Coulomb potential**.

Proof This is a special case of (a). $\blacksquare$

The Poincaré lemma

The exact 1-forms are especially important class of differential forms. We already know that every exact 1-form satisfies the integrability conditions. On the other hand, there are 1-forms that are not exact but still satisfy the integrability conditions. It is therefore useful to seek conditions that, in addition to the already necessary integrability conditions, will secure the existence of an antiderivative. We will now see, perhaps surprisingly, that the existence depends on the topological properties of the underlying domain.

A continuously differentiable Pfaff form is a **closed** if it satisfies the integrability conditions. A subset of M of $\mathbb{R}^n$ is **star shaped** (with respect to $x_0 \in M$) if there is an $x_0 \in M$ such that for every $x \in M$ the straight line path $[\![x_0, x]\!]$ from x_0 to x lies in M.

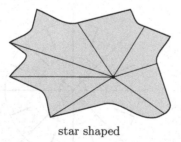

star shaped

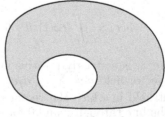

not star shaped

---

[6]The name "potential" comes from physics.

3.7 Remarks (a) Every exact, continuously differentiable 1-form is closed.

(b) If M is star shaped, M is connected.

Proof This follows from Proposition III.4.8. ∎

(c) Every convex set is star shaped with respect to any of its points. ∎

Suppose X is star shaped with respect to 0 and $f \in C^1(X)$ is an antiderivative of $\alpha = \sum_{j=1}^{n} a_j \, dx^j$. Then $a_j = \partial_j f$, and Remark 3.3(c) gives

$$\langle \alpha(tx), e_j \rangle = a_j(tx) = \partial_j f(tx) = \left(\nabla f(tx) \,\middle|\, e_j \right) \quad \text{for } j = 1, \dots, n \, .$$

Therefore $\langle \alpha(tx), x \rangle = \left(\nabla f(tx) \,\middle|\, x \right)$ for $x \in X$, and the mean value theorem in integral form (Theorem VII.3.10) gives the representation

$$f(x) - f(0) = \int_0^1 \left(\nabla f(tx) \,\middle|\, x \right) dt = \int_0^1 \langle \alpha(tx), x \rangle \, dt \, , \tag{3.3}$$

which is the key to the next theorem.

3.8 Theorem (the Poincaré lemma) *Suppose X is star shaped and $q \geq 1$. When $\alpha \in \Omega_{(q)}(X)$ is closed, α is exact.*

Proof Suppose X is star shaped with respect to 0 and $\alpha = \sum_{j=1}^{n} a_j \, dx^j$. Because, for $x \in X$, the segment $[\![0, x]\!]$ lies in X,

$$f(x) := \int_0^1 \langle \alpha(tx), x \rangle \, dt = \sum_{k=1}^{n} \int_0^1 a_k(tx) x^k \, dt \quad \text{for } x \in X \tag{3.4}$$

is well defined. Also

$$\left[(t, x) \mapsto \langle \alpha(tx), x \rangle \right] \in C^q\left([0, 1] \times X \right) \, ,$$

and, from the integrability conditions $\partial_j a_k = \partial_k a_j$, we get

$$\frac{\partial}{\partial x^j} \langle \alpha(tx), x \rangle = a_j(tx) + \sum_{k=1}^{n} t x^k \partial_k a_j(tx) \, .$$

Thus Proposition VII.6.7 implies that f belongs to $C^q(X)$ and that

$$\partial_j f(x) = \int_0^1 a_j(tx) \, dt + \int_0^1 t \left(\sum_{k=1}^{n} x^k \partial_k a_j(tx) \right) dt \, .$$

Integrating the the second integral by parts with $u(t) := t$ and $v_j(t) := a_j(tx)$ gives

$$\partial_j f(x) = \int_0^1 a_j(tx) \, dt + t a_j(tx) \Big|_0^1 - \int_0^1 a_j(tx) \, dt = a_j(x) \quad \text{for } x \in X \, .$$

Finally, since $\partial_j f = a_j$ belongs to $C^q(X)$ for $j = 1, \ldots, n$, we find $f \in C^{q+1}(X)$ and $df = \alpha$ (see Remark 3.3(e)).

The case in which X is star shaped with respect to the point $x_0 \in X$ can clearly be reduced to the situation above by the translation $x \mapsto x - x_0$. ∎

3.9 Remarks   (a) The central fields of Example 3.6(a) show that a domain need not be star shaped for a potential (or an antiderivative) to exist.

(b) The proof of Theorem 3.8 is constructive, that is, the formula (3.4) gives an antiderivative. ∎

Until now, we exclusively used in $\Omega_{(q)}(X)$ the canonical module basis consisting of the 1-forms $dx^1, \ldots, dx^n$. This choice is natural if we are using in $V^q(X)$ the canonical basis $(e_1, \ldots, e_n)$ (see Remark 3.3(c)). In other words, if we understand X as a submanifold of $\mathbb{R}^n$, the module basis $(dx^1, \ldots, dx^n)$ is adapted to the trivial chart $\varphi = \mathrm{id}_X$. If we describe X using another chart ψ, we anticipate that $\alpha \in \Omega_{(q)}(X)$ will be better represented better in a module basis adapted to ψ.[7] To pass from one representation to another, we must study how the Pfaff forms transform under change of charts.

Dual operators

We first introduce — in a generalization of the transpose of a matrix — the concept of the "transpose" or "dual" of a linear operator.

Suppose E and F are Banach spaces. Then, to each $A \in \mathcal{L}(E, F)$, we define a **dual** or **transposed operator** through

$$A^\top : F' \to E' , \quad f' \mapsto f' \circ A .$$

3.10 Remarks   (a) The operator dual to A is linear and continuous, that is, $A^\top \in \mathcal{L}(F', E')$. Also[8]

$$\langle A^\top f', e \rangle = \langle f', Ae \rangle \quad \text{for } e \in E \text{ and } f' \in F' , \tag{3.5}$$

and

$$\|A^\top\|_{\mathcal{L}(F', E')} \leq \|A\|_{\mathcal{L}(E, F)} \quad \text{for } A \in \mathcal{L}(E, F) . \tag{3.6}$$

Proof   From $A \in \mathcal{L}(E, F)$ it follows that $A^\top f'$ belongs to E' for $f' \in F'$. It is obvious that $A^\top : F' \to E'$ is linear and that (3.5) holds. The latter implies

$$|\langle A^\top f', e \rangle| = |\langle f', Ae \rangle| \leq \|f'\| \|Ae\| \leq \|f'\| \|A\| \|e\| \quad \text{for } e \in E \text{ and } f' \in F'.$$

---

[7]We will gain a better understanding of this issue in Volume III when we study differential forms on manifolds.

[8]In functional analysis, it is shown that $A^\top$ is characterized by (3.5) and that (3.6) holds with equality.

From this we read off (see Remark VII.2.13(a)) the inequality

$$\| A^\top f' \| \le \| A \| \, \| f' \| \quad \text{for } f' \in F' \ .$$

Therefore we get (3.6) from the definition of the operator norm (see Section VI.2). Now Theorem VI.2.5 guarantees that $A^\top$ is continuous. ∎

(b) If $\mathcal{E} = (e_1, \dots, e_n)$ and $\mathcal{F} = (f_1, \dots, f_m)$ are bases of the finite-dimensional Banach spaces E and F, then $[A^\top]_{\mathcal{E}',\mathcal{F}'} = \big([A]_{\mathcal{E},\mathcal{F}}\big)^\top$, where $M^\top \in \mathbb{K}^{n \times m}$ denotes the matrix dual to $M \in \mathbb{K}^{m \times n}$ and $\mathcal{E}'$ and $\mathcal{F}'$ are the bases dual to $\mathcal{E}$ and $\mathcal{F}$.

Proof This follows easily from the definition of the representation matrix (see Section VII.1) and by expanding f' in the basis dual to $\mathcal{F}$. ∎

(c) If G is another Banach space,

$$(BA)^\top = A^\top B^\top \quad \text{for } A \in \mathcal{L}(E, F) \text{ and } B \in \mathcal{L}(F, G) \ ,$$

and $(1_E)^\top = 1_{E'}$.

(d) For $A \in \mathcal{L}\mathrm{is}(E, F)$, $A^\top$ belongs to $\mathcal{L}\mathrm{is}(F', E')$, and $(A^\top)^{-1} = (A^{-1})^\top$.

Proof This follows immediately from (c). ∎

Transformation rules

In the following, suppose X is open in $\mathbb{R}^n$ and Y is open in $\mathbb{R}^m$. In the special cases $n = 1$ or $m = 1$, we also allow X or Y (as the case may be) to be a perfect interval. Also suppose $\varphi \in C^q(X, Y)$ with $q \in \mathbb{N}^\times \cup \{\infty\}$.

Using φ, we define the map

$$\varphi^* \colon \mathbb{R}^Y \to \mathbb{R}^X \ ,$$

the **pull back**, through

$$\varphi^* f := f \circ \varphi \quad \text{for } f \in \mathbb{R}^Y \ .$$

More precisely, φ^* is the **pull back of functions**.[9]

For Pfaff forms, we define the **pull back** by φ through

$$\varphi^* \colon \Omega_{(q-1)}(Y) \to \Omega_{(q-1)}(X) \ , \quad \alpha \mapsto \varphi^* \alpha$$

with

$$\varphi^* \alpha(p) := (T_p \varphi)^\top \alpha\big(\varphi(p)\big) \ .$$

From its operand, it will always be obvious whether φ^* means the pull back of functions or of 1-forms. The pull back take functions (or 1-forms) "living" on Y to ones on X, whereas the original φ goes the other way.

---

[9]This definition is obviously meaningful for $f \in Z^Y$ and any sets X, Y, and Z.

3.11 Remarks **(a)** The pull backs of $f \in C^q(Y)$ and $\alpha \in \Omega_{(q-1)}(Y)$ are defined through the commutativity of the diagrams

$$
\begin{array}{ccc}
X & \xrightarrow{\;\;\varphi\;\;} & Y \\
& \varphi^* f \searrow \quad \swarrow f & \\
& \mathbb{R} &
\end{array}
\qquad\qquad
\begin{array}{ccc}
TX & \xrightarrow{\;\;T\varphi\;\;} & TY \\
& \varphi^* \alpha \searrow \quad \swarrow \alpha & \\
& \mathbb{R} &
\end{array}
$$

where $(T\varphi)v \in TY$ for $v \in TX$ is determined through

$$\big((T\varphi)v\big)\big(\varphi(p)\big) := (T_p\varphi)v(p) \quad \text{for } p \in X \ .$$

(b) For $\alpha \in \Omega_{(q-1)}(Y)$,

$$\varphi^*\alpha(p) = (T_p\varphi)^\top \alpha\big(\varphi(p)\big) = \big(p, \big(\partial\varphi(p)\big)^\top \alpha\big(\varphi(p)\big)\big) \in T_p^* X \quad \text{for } p \in X \ .$$

Proof From the definition of $\varphi^*\alpha$, we have

$$
\begin{aligned}
\big\langle \varphi^*\alpha(p), v(p) \big\rangle_p &= \big\langle \alpha\big(\varphi(p)\big), (T_p\varphi)v(p) \big\rangle_{\varphi(p)} \\
&= \big\langle \alpha\big(\varphi(p)\big), \partial\varphi(p)v(p) \big\rangle = \big\langle \big(\partial\varphi(p)\big)^\top \alpha\big(\varphi(p)\big), v(p) \big\rangle
\end{aligned}
$$

for $p \in X$ and $v \in \mathcal{V}^q(X)$, and the claim follows. ∎

(c) The maps

$$\varphi^* : C^q(Y) \to C^q(X) \tag{3.7}$$

and

$$\varphi^* : \Omega_{(q-1)}(Y) \to \Omega_{(q-1)}(X) \tag{3.8}$$

are well defined and $\mathbb{R}$-linear. It should be pointed out specifically that in (3.8) the "order of regularity" of the considered Pfaff forms is $q - 1$ and not q, whereas q is the correct order of regularity in (3.7). Because the definition involves the tangential map, the one derivative order is "lost".

Proof The statements about the map (3.7) follow immediately from the chain rule. The properties of (3.8) are likewise implied by the chain rule and (b). ∎

In the next theorem, we gather the rules for pull backs.

3.12 Proposition *Suppose Z is open in $\mathbb{R}^\ell$ or (in the case $\ell = 1$) is a perfect interval. Then*

(i) $(\mathrm{id}_X)^* = \mathrm{id}_{\mathcal{F}(X)}$ *for $\mathcal{F}(X) := C^q(X)$ or $\mathcal{F}(X) := \Omega_{(q-1)}(X)$,*
and
$(\psi \circ \varphi)^* = \varphi^*\psi^*$ *for $\varphi \in C^q(X, Y)$ and $\psi \in C^q(Y, Z)$;*

(ii) $\varphi^*(fg) = (\varphi^*f)(\varphi^*g)$ for $f, g \in C^q(Y)$,
and
$\varphi^*(h\boldsymbol{\alpha}) = (\varphi^*h)\varphi^*\boldsymbol{\alpha}$ for $h \in C^{(q-1)}(Y)$ and $\boldsymbol{\alpha} \in \Omega_{(q-1)}(Y)$;

(iii) $\varphi^*(df) = d(\varphi^*f)$ for $f \in C^q(Y)$, that is, the diagram

$$
\begin{array}{ccc}
C^q(Y) & \xrightarrow{\ \ d\ \ } & \Omega_{(q-1)}(Y) \\
\varphi^* \downarrow & & \downarrow \varphi^* \\
C^q(X) & \xrightarrow{\ \ d\ \ } & \Omega_{(q-1)}(X)
\end{array}
$$

commutes.

Proof (i) The first claim is clear. To prove the second formula, we must show

$$(\psi \circ \varphi)^* f = (\varphi^* \circ \psi^*)f \quad \text{for } f \in C^q(Z) \ ,$$

and

$$(\psi \circ \varphi)^* \boldsymbol{\alpha} = (\varphi^* \circ \psi^*)\boldsymbol{\alpha} \quad \text{for } \boldsymbol{\alpha} \in \Omega_{(q-1)}(Z) \ .$$

For $f \in C^q(Z)$, we have

$$(\psi \circ \varphi)^* f = f \circ (\psi \circ \varphi) = (f \circ \psi) \circ \varphi = (\psi^* f) \circ \varphi = (\varphi^* \circ \psi^*)f \ .$$

For $\boldsymbol{\alpha} \in \Omega_{(q-1)}(Y)$, it follows from the chain rule of Remark VII.10.2(b) and from Remark 3.10(c) that, for $p \in X$,

$$
\begin{aligned}
(\psi \circ \varphi)^* \boldsymbol{\alpha}(p) &= \big(T_p(\psi \circ \varphi)\big)^{\top} \boldsymbol{\alpha}\big(\psi \circ \varphi(p)\big) = \big((T_{\varphi(p)}\psi) \circ T_p\varphi\big)^{\top} \boldsymbol{\alpha}\big(\psi(\varphi(p))\big) \\
&= (T_p\varphi)^{\top} (T_{\varphi(p)}\psi)^{\top} \boldsymbol{\alpha}\big(\psi(\varphi(p))\big) = (T_p\varphi)^{\top} \psi^* \boldsymbol{\alpha}\big(\varphi(p)\big) \\
&= \varphi^* \psi^* \boldsymbol{\alpha}(p) \ .
\end{aligned}
$$

(ii) The first statement is clear. The second follows, for $p \in X$, from

$$
\begin{aligned}
\varphi^*(h\boldsymbol{\alpha})(p) &= (T_p\varphi)^{\top} (h\boldsymbol{\alpha})\big(\varphi(p)\big) = (T_p\varphi)^{\top} h\big(\varphi(p)\big)\boldsymbol{\alpha}\big(\varphi(p)\big) \\
&= h\big(\varphi(p)\big)(T_p\varphi)^{\top}\boldsymbol{\alpha}\big(\varphi(p)\big) = \big((\varphi^*h)\varphi^*\boldsymbol{\alpha}\big)(p) \ .
\end{aligned}
$$

(iii) From the chain rule for the tangential map and the definition of differentials, we get

$$
\begin{aligned}
\varphi^*(df)(p) &= (T_p\varphi)^{\top} df\big(\varphi(p)\big) = df\big(\varphi(p)\big) \circ T_p\varphi \\
&= \mathrm{pr}_2 \circ T_{\varphi(p)}f \circ T_p\varphi = \mathrm{pr}_2 \circ T_p(f \circ \varphi) = d(\varphi^*f)(p) \ ,
\end{aligned}
$$

and therefore the claim. ∎

3.13 Corollary *For $\varphi \in \mathrm{Diff}^q(X, Y)$, the maps*

$$\varphi^* : C^q(Y) \to C^q(X)$$

and

$$\varphi^* : \Omega_{(q-1)}(Y) \to \Omega_{(q-1)}(X)$$

are bijective with $(\varphi^*)^{-1} = (\varphi^{-1})^*$.

Proof This follows from Proposition 3.12(i). ∎

3.14 Examples (a) Denote points in X and Y by $x = (x^1, \ldots, x^n) \in X$ and $y = (y^1, \ldots, y^m) \in Y$, respectively. Then

$$\varphi^* \, dy^j = d\varphi^j = \sum_{k=1}^n \partial_k \varphi^j \, dx^k \quad \text{for } 1 \le j \le m \ .$$

Proof From Proposition 3.12(iii), we have

$$\varphi^* \, dy^j = \varphi^* \, d(\mathrm{pr}_j) = d(\varphi^* \, \mathrm{pr}_j) = d\varphi^j = \sum_{k=1}^n \partial_k \varphi^j \, dx^k \ ,$$

where the last equality is implied by Remark 3.3(e). ∎

(b) For $\alpha = \sum_{j=1}^m a_j \, dy^j \in \Omega_{(0)}(Y)$, we have

$$\varphi^* \alpha = \sum_{j=1}^m \sum_{k=1}^n (a_j \circ \varphi) \partial_k \varphi^j \, dx^k \ .$$

Proof Proposition 3.12(ii) and the linearity of φ^* give

$$\varphi^* \Big(\sum_{j=1}^m a_j \, dy^j \Big) = \sum_{j=1}^m \varphi^*(a_j \, dy^j) = \sum_{j=1}^m (\varphi^* a_j)(\varphi^* \, dy^j) \ .$$

The claim then follows from (a). ∎

(c) Suppose X is open in $\mathbb{R}$ and $\alpha \in \Omega_{(0)}(Y)$. Then

$$\varphi^* \alpha(t) = \big\langle \alpha\big(\varphi(t)\big), \dot\varphi(t) \big\rangle \, dt \quad \text{for } t \in X \ .$$

Proof This is a special case of (b). ∎

(d) Suppose $Y \subset \mathbb{R}$ is a compact interval, $f \in C(Y)$, and $\alpha := f \, dy \in \Omega_{(0)}(Y)$. Also define $X := [a, b]$, and let $\varphi \in C^1(X, Y)$. Then, according to (b),

$$\varphi^* \alpha = \varphi^*(f \, dy) = (f \circ \varphi) \, d\varphi = (f \circ \varphi)\varphi' \, dx \ .$$

Thus, the substitution rule for integrals (Theorem VI.5.1),

$$\int_a^b (f \circ \varphi)\varphi'\, dx = \int_{\varphi(a)}^{\varphi(b)} f\, dy$$

can be written formally with using the Pfaff forms $\alpha = f\, dy$ and $\varphi^*\alpha$ as

$$\int_X \varphi^*\alpha = \int_{\varphi(X)} \alpha \ . \tag{3.9}$$

Because $\Omega_{(0)}(X)$ is a one-dimensional $C(X)$-module, every continuous 1-form β on X can be represented uniquely in the form $\beta = b\, dx$ for $b \in C(X)$. Motivated by (3.9), we define the **integral of** $\beta = b\, dx \in \Omega_{(0)}(X)$ **over** X by

$$\int_X \beta := \int_X b \ , \tag{3.10}$$

where the right side is interpreted as a Cauchy–Riemann integral. Thus we have replaced the previously formal definition of the "differential", which we gave in Remarks VI.5.2, with a mathematically precise definition (at least in the case of real-valued functions). ∎

Modules

For completeness, we collect here the most relevant facts from the theory of modules. In Volume III, we will further study the modules $\mathcal{V}^q(X)$ and $\Omega_{(q)}(X)$ through some examples.

Suppose $R = (R, +, \cdot)$ is a commutative ring with unity. A **module over the ring** R, an R-**module**, is a triple $(M, +, \cdot)$ consisting of a nonempty set M and two operations: an "inner" operation $+$, **addition**, and an "outer" operation $\cdot$, **multiplication**, with elements of R. These must satisfy

(M_1) $(M, +)$ is an Abelian group;

(M_2) the distribute property, that is,

$$\lambda \cdot (v + w) = \lambda \cdot v + \lambda \cdot w\ , \quad (\lambda + \mu) \cdot v = \lambda \cdot v + \mu \cdot v \quad \text{for } \lambda, \mu \in R \text{ and } v, w \in M\ ;$$

(M_3) $\lambda \cdot (\mu \cdot v) = (\lambda \cdot \mu) \cdot v\ , \quad 1 \cdot v = v \quad \text{for } \lambda, \mu \in R \text{ and } v \in M.$

3.15 Remarks Suppose $M = (M, +, \cdot)$ is an R-module.

(a) As in the case of vector spaces, we stipulate that multiplication binds stronger than addition, and we usually write λv for $\lambda \cdot v$.

(b) The axiom (M_3) shows that the ring R operates on the set M from the left (see Exercise I.7.6).

(c) The addition operations in R and in M will both be denoted by $+$. Likewise, we denote the ring multiplication in R and the multiplication product of R on M by the same $\cdot$. Finally, we write 0 for the zero elements of $(R,+)$ and of $(M,+)$. Experience shows that this simplified notation does not lead to misunderstanding.

(d) If R is a field, M is a vector space over R. ∎

3.16 Examples (a) Every commutative ring R with unity is an R-module.

(b) Suppose $G = (G,+)$ is an Abelian group. For $(z,g) \in \mathbb{Z} \times G$, let[10]

$$z \cdot g := \begin{cases} \sum_{k=1}^{z} g \,, & z > 0 \,, \\ 0 \,, & z = 0 \,, \\ -\sum_{k=1}^{-z} g \,, & z < 0 \,. \end{cases}$$

Then is $G = (G,+,\cdot)$ is a $\mathbb{Z}$-module.

(c) Suppose G is an Abelian group and R is a commutative subring of the ring of all endomorphisms of G. Then G together with the operation

$$R \times G \to G \,, \quad (T,g) \mapsto Tg \,,$$

is an R-module.

(d) Suppose M is an R-module. A nonempty subset U of M is a **submodule** of M if

(UM$_1$) U is a subgroup of M,

(UM$_2$) $R \cdot U \subseteq U$.

It is not hard to verify that a nonempty subset U of M is a submodule of M if and only if U is a module with the operation induced by M.

If $\{U_\alpha \,;\, \alpha \in \mathsf{A}\}$ is a family of submodules of M, then $\bigcap_{\alpha \in \mathsf{A}} U_\alpha$ is also a submodule of M. ∎

Suppose M and N are modules over R. The map $T : M \to N$ is a **module homomorphism** if

$$T(\lambda v + \mu w) = \lambda T(v) + \mu T(w) \quad \text{for } \lambda, \mu \in R \text{ and } v, w \in M \,.$$

A bijective module homomorphism is a **module isomorphism**. Then the inverse map $T^{-1} : N \to M$ is also a module homomorphism. When there is a module isomorphism from M to N, we say M and N are isomorphic, and we write $M \cong N$.

Suppose M is an R-module and $A \subset M$. Then

$$\mathrm{span}(A) := \bigcap \{ U \,;\, U \text{ is a submodule of } M \text{ with } U \supseteq A \}$$

is called the **span** of A in M. Clearly $\mathrm{span}(A)$ is the smallest submodule of M containing A. When $\mathrm{span}(A)$ equals M, we call A a **generating system** of M.

The elements $v_0, \ldots, v_k \in M$ are said to be **linearly independent** over R if $\lambda_j \in R$ and $\sum_{j=0}^{k} \lambda_j v_j = 0$ imply $\lambda_0 = \cdots = \lambda_k = 0$. A subset B of M is **free** over R if every finite subset of B is linearly independent. If $\mathrm{span}(B) = M$ for some subset B over R, then B forms a **basis** of M. An R-module that has a basis is called a **free R-module**.

---

[10]See Example I.5.12.

3.17 Remarks (a) These statements are equivalent:

(i) M is a free R-module with basis B.

(ii) Every $v \in M$ is of the form $v = \sum_{j=0}^{k} \lambda_j v_j$ for unique $\lambda_j \in R^\times$ and $v_j \in B$.

Proof Refer to texts on algebra (for example [SS88, §§ 19 and 22]). ■

(b) Suppose M and N are modules over R and M is free with basis B. Further suppose S and T are module homomorphisms from M to N with $S\,|\,B = T\,|\,B$. Then $S = T$, that is, *a homomorphism over a free module is uniquely determined by how it maps its basis*.

(c) Suppose B and B' are bases of a free R-module M. Then one can show that $\mathrm{Num}(B) = \mathrm{Num}(B')$ if R is not the null ring (for example [Art93, § XII.2]). Thus the **dimension** of M is defined by $\dim(M) := \mathrm{Num}(B)$.

(d) $\mathbb{Z}_2$ is *not* a free $\mathbb{Z}$-module.

Proof From Exercise I.9.2, we know that $\mathbb{Z}_2$ is a ring. Therefore $(\mathbb{Z}_2, +)$ is an Abelian group, and Example 3.16(b) shows that $\mathbb{Z}_2$ is a $\mathbb{Z}$-module. Suppose B is a basis of $\mathbb{Z}_2$ over $\mathbb{Z}$. Then the coset $[1]$ belongs to B. On the other hand, $[0] = 2 \cdot [1]$ shows that $[1]$ is not linearly independent over $\mathbb{Z}$, that is, $[1] \notin B$. ■

(e) Suppose M is an R-module and $M = \mathrm{span}(B)$ for $B \subset M$. Then[11] B does not (generally) contain a basis of M.

Proof We consider in the $\mathbb{Z}$-module[12] $\mathbb{Z}$ the set $B = \{2,3\}$. From $\mathbb{Z} = \mathbb{Z}(3 - 2)$, we have $\mathrm{span}(B) = \mathbb{Z}$. Further, it follows from

$$3 \cdot 2 - 2 \cdot 3 = 0 \text{ and } z \cdot 2 \neq 5 \neq z \cdot 3 \quad \text{for } z \in \mathbb{Z}$$

that neither $\{2,3\}$ nor $\{2\}$ nor $\{3\}$ is a basis of $\mathbb{Z}$. ■

(f) Suppose M is a free R-module and U is a free submodule of M with basis B'. Then B' does not (generally) generate a basis of M.

Proof $\mathbb{Z}$ is a free module over $\mathbb{Z}$. The even integers $2\mathbb{Z}$ form a free submodule of $\mathbb{Z}$. However, neither $\{2\}$ nor $\{-2\}$ generates a basis of $\mathbb{Z}$. Also, $\dim(\mathbb{Z}) = \dim(2\mathbb{Z}) = 1$. ■

(g) If R is a field, the above notions of "span", "linear independence", "bases", and "dimension" agree with those of Section I.12. ■

Exercises

1 Suppose E, F and G are Banach spaces; also let $A \in \mathcal{L}(E, F)$ and $B \in \mathcal{L}(G, E)$. Prove

(a) $(AB)^\top = B^\top A^\top$;

(b) For $A \in \mathcal{L}\mathrm{is}(E, F)$, we have $A^\top \in \mathcal{L}\mathrm{is}(F', E')$ and $[A^\top]^{-1} = [A^{-1}]^\top$.

2 Which of these 1-forms are closed?

(a) $2xy^3\, dx + 3x^2y^2\, dy \in \Omega_{(\infty)}(\mathbb{R}^2)$.

---

[11] Suppose V is a K vector space and $V = \mathrm{span}(B)$. The it is proved in linear algebra that B contains a basis of V.

[12] Note that $\mathbb{Z}$ is a free $\mathbb{Z}$-module.

(b) $\dfrac{2(x^2 - y^2 - 1)\, dy - 4xy\, dx}{(x^2 + y^2 - 1)^2 + 4y^2} \in \Omega_{(\infty)}(X)$, where $X := \mathbb{R}^2 \setminus \{(-1,0),(1,0)\}$.

3 Suppose $X := \mathbb{R}^2 \setminus \{(-1,0),(1,0)\}$,

$$\varphi : X \to \mathbb{R}^2 \setminus \{(0,0)\} , \quad (x,y) \mapsto (x^2 + y^2 - 1, 2y)$$

and

$$\alpha := \frac{u\, dv - v\, du}{u^2 + v^2} \in \Omega(\mathbb{R}^2 \setminus \{(0,0)\}) .$$

Show that $\varphi^* \alpha$ coincides with the 1-form of Exercise 2(b).

4 Suppose X is open in $\mathbb{R}^n$ and $\alpha \in \Omega_{(1)}(X)$. Denote by $a \in C^1(X, \mathbb{R}^n)$ the cotangent part of $\Theta^{-1}\alpha$. Prove that α is closed if and only if $\partial a(x)$ is symmetric for every $x \in X$.

5 Suppose

$$\alpha := x\, dy - y\, dx \in \Omega_{(\infty)}(\mathbb{R}^2) \quad \text{and} \quad \beta := \alpha/(x^2 + y^2) \in \Omega_{(\infty)}(\mathbb{R}^2 \setminus \{(0,0)\})$$

and

$$\varphi : (0, \infty) \times (0, 2\pi) \to \mathbb{R}^2 \setminus \{(0,0)\} , \quad (r, \theta) \mapsto (r\cos\theta, r\sin\theta) .$$

Calculate $\varphi^* \alpha$ and $\varphi^* \beta$.

6 Determine $\varphi^* \alpha$ for $\alpha := x\, dy - y\, dx - (x + y)\, dz \in \Omega_{(\infty)}(\mathbb{R}^3)$ and

$$\varphi : (0, 2\pi) \times (0, \pi) \to \mathbb{R}^3 , \quad (\theta_1, \theta_2) \mapsto (\cos\theta_1 \sin\theta_2, \sin\theta_1 \sin\theta_2, \cos\theta_2) .$$

7 Suppose X and Y are open in $\mathbb{R}^n$, $\varphi \in \mathrm{Diff}^q(X, Y)$, and $\alpha \in \Omega_{(q-1)}(Y)$ for some $q \geq 1$. Prove

(a) α is exact if and only if $\varphi^* \alpha$ is exact;

(b) for $q \geq 2$, α is closed if and only if $\varphi^* \alpha$ is closed.

8 Suppose X is open in $\mathbb{R}^2$ and $\alpha \in \Omega_{(1)}(X)$. A function $h \in C^1(X)$ is an **Euler multiplier** (or **integrating factor**) for α if $h\alpha$ is closed and $h(x,y) \neq 0$ for $(x,y) \in X$.

(a) Show that if $h \in C^1(X)$ satisfies $h(x,y) \neq 0$ for $(x,y) \in X$, then h is an Euler multiplier for $\alpha = a\, dx + b\, dy$ if and only if

$$ah_y - bh_x + (a_y - b_x)h = 0 . \tag{3.11}$$

(b) Suppose $a, b, c, e > 0$, and let

$$\beta := (c - ex)y\, dx + (a - by)x\, dy .$$

Show that β has an integrating factor of the form $h(x,y) = m(xy)$ on $X = (0, \infty)^2$. (Hint: Apply (3.11) to find a differential equation for m, and guess the solution.)

9 Let J and D be open intervals, $f \in C(J \times D, \mathbb{R})$ and $\alpha := dx - f\, dt \in \Omega_{(0)}(J \times D)$. Further suppose $u \in C^1(J, D)$, and define $\varphi_u : J \to \mathbb{R}^2$, $t \mapsto (t, u(t))$. Prove that $\varphi_u^* \alpha = 0$ if and only if u solves $\dot{x} = f(t, x)$.

10 Verify that $\mathcal{V}^q(X)$ and $\Omega_{(q)}(X)$ are $C^q(X)$-modules and that $\mathcal{V}^{q+1}(X)$ and $\Omega_{(q+1)}(X)$ are respectively submodules of $\mathcal{V}^q(X)$ and $\Omega_{(q)}(X)$.

11 Suppose R is a commutative ring with unity and $a \in R$. Prove

(a) aR is a submodule of R;

(b) $aR \neq 0 \Longleftrightarrow a \neq 0$;

(c) $aR \neq R \Longleftrightarrow a$ has no inverse element.

A submodule U of an R-module M is said to be **nontrivial** if $U \neq \{0\}$ and $U \neq M$. What are the nontrivial submodules of $\mathbb{Z}_2$ and of $\mathbb{Z}_6$?

12 Prove the statements of Example 3.16(d).

13 Suppose M and N are R-modules and $T : M \to N$ is a module homomorphism. Verify that $\ker(T) := \{v \in M \ ; \ Tv = 0\}$ and $\mathrm{im}(T)$ are submodules of M and N, respectively.

4 Line integrals

In Chapter VI, we developed theory for integrating functions of a real variable over intervals. If we regard these intervals as especially simple curves, we may then wonder whether the concept of integration can be extended to arbitrary curves. This section will do just that. Of course, we can already integrate functions of paths, so to develop a new way to integrate that depends only on the curve, we will have to ensure that it is independent of parametrization. It will turn out that we must integrate Pfaff forms, not functions.

In these sections, suppose

- X is open in $\mathbb{R}^n$;
  I and I_1 are compact perfect intervals.

In addition, we always identify vector fields and Pfaff forms with their tangent and cotangent parts, respectively.

The definition

We now take up the substitution rule for integrals. Suppose I and J are compact intervals and $\varphi \in C^1(I, J)$. Further suppose $a \in C(J)$ and $\alpha = a\, dy$ is a continuous 1-form on J. Then according to 3.14(d) and Definition (3.10), the substitution rule has the form

$$\int_{\varphi(I)} \alpha = \int_I \varphi^* \alpha = \int_I (a \circ \varphi) \dot{\varphi}\, dt \ . \tag{4.1}$$

Because $\Omega_{(0)}(J)$ is a one-dimensional $C(J)$-module, every $\alpha \in \Omega_{(0)}(J)$ has a unique representation $\alpha = a\, dy$ such that $a \in C(J)$. Therefore (4.1) is defined for every $\alpha \in \Omega_{(0)}(J)$. The last integrand can also be expressed in the form

$$\langle \alpha(\varphi(t)), \dot{\varphi}(t) \rangle\, dt = \langle \alpha(\varphi(t)), \dot{\varphi}(t)1 \rangle\, dt \ .$$

This observation is the basic result needed for defining line integrals of 1-forms.

For $\alpha \in \Omega_{(0)}(X)$ and $\gamma \in C^1(I, X)$

$$\int_\gamma \alpha := \int_I \gamma^* \alpha = \int_I \langle \alpha(\gamma(t)), \dot{\gamma}(t) \rangle\, dt$$

is the **integral of α along the path γ**.

4.1 Remarks   (a) In a basis representation $\alpha = \sum_{j=1}^n a_j\, dx^j$, we have

$$\int_\gamma \alpha = \sum_{j=1}^n \int_I (a_j \circ \gamma) \dot{\gamma}^j\, dt \ .$$

Proof   This follows directly from Example 3.14(b). ∎

(b) Suppose $\varphi \in C^1(I_1, I)$ and $\gamma_1 := \gamma \circ \varphi$. Then

$$\int_\gamma \alpha = \int_{\gamma_1} \alpha \quad \text{for } \alpha \in \Omega_{(0)}(X) .$$

Proof   Proposition 3.12(i) and the substitution rule (3.9) give

$$\int_{\gamma_1} \alpha = \int_{I_1} (\gamma \circ \varphi)^* \alpha = \int_{I_1} \varphi^*(\gamma^* \alpha) = \int_I \gamma^* \alpha = \int_\gamma \alpha \quad \text{for } \alpha \in \Omega_{(0)}(X) . \blacksquare$$

This result shows that the integral of α along γ does not change under a change of parametrization. Thus, for every compact C^1 curve Γ in X and every $\alpha \in \Omega_{(0)}(X)$, the **line integral of α along Γ**

$$\int_\Gamma \alpha := \int_\gamma \alpha$$

is well defined, where γ is an arbitrary C^1 parametrization of Γ.

4.2 Examples   **(a)** Let Γ be the circle parametrized by $\gamma \colon [0, 2\pi] \to \mathbb{R}^2$, $t \mapsto R(\cos t, \sin t)$. Then

$$\int_\Gamma X\,dy - Y\,dx = 2\pi R^2 .$$

Proof   From

$$\gamma^* dx = d\gamma^1 = -(R\sin)\,dt \text{ and } \gamma^* dy = d\gamma^2 = (R\cos)\,dt$$

it follows that $\gamma^*(X\,dy - Y\,dx) = R^2\,dt$, and the claim follows. $\blacksquare$

(b) Suppose $\alpha \in \Omega_{(0)}(X)$ is exact. If $f \in C^1(X)$ is an antiderivative,[1]

$$\int_\Gamma \alpha = \int_\Gamma df = f(E_\Gamma) - f(A_\Gamma) ,$$

where A_Γ and E_Γ are the initial and final points of Γ.

Proof   We fix a C^1 parametrization $\gamma \colon [a, b] \to X$ of Γ. Then it follows from Proposition 3.12(iii) that

$$\gamma^* df = d(\gamma^* f) = d(f \circ \gamma) = (f \circ \gamma)^{\cdot}\,dt .$$

Therefore we find

$$\int_\Gamma \alpha = \int_\Gamma df = \int_a^b \gamma^* df = \int_a^b (f \circ \gamma)^{\cdot}\,dt = f \circ \gamma\big|_a^b = f(E_\Gamma) - f(A_\Gamma)$$

because $E_\Gamma = \gamma(b)$ and $A_\Gamma = \gamma(a)$. $\blacksquare$

---

[1] This statement is a generalization of Corollary VI.4.14.

(c) The integral of an exact Pfaff form along a C^1 curve Γ depends on the initial and final points, but not on the path between them. If Γ is closed, the integral evaluates to 0.

Proof This follows directly from (b). ∎

(d) There are closed 1-forms that are not exact.

Proof Suppose $X := \mathbb{R}^2 \setminus \{(0,0)\}$ and $\alpha \in \Omega_{(\infty)}(X)$ with

$$\alpha(x,y) := \frac{x\,dy - y\,dx}{x^2 + y^2} \quad \text{for } (x,y) \in X .$$

One easily checks that α is closed. The parametrization $\gamma : [0, 2\pi] \to X$, $t \mapsto (\cos t, \sin t)$ of the circle Γ gives $\gamma^* \alpha = dt$ and therefore

$$\int_\Gamma \alpha = \int_0^{2\pi} dt = 2\pi \neq 0 .$$

Then (c) shows α is not exact. ∎

(e) Suppose $x_0 \in X$ and $\gamma_{x_0} : I \to X$, $t \mapsto x_0$. This means that γ_{x_0} parametrizes the **point curve** Γ, whose image is simply $\{x_0\}$. Then

$$\int_{\gamma_{x_0}} \alpha = 0 \quad \text{for } \alpha \in \Omega_{(0)}(X) .$$

Proof This is obvious because $\dot{\gamma}_{x_0} = 0$. ∎

Elementary properties

Suppose $I = [a,b]$ and $\gamma \in C(I, X)$. Then

$$\gamma^- : I \to X , \quad t \mapsto \gamma(a + b - t)$$

is the **path inverse to** γ, and $-\Gamma := [\gamma^-]$ is the **curve inverse to** $\Gamma := [\gamma]$. (Note that Γ and $-\Gamma$ have the same image but the opposite orientation).

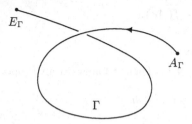

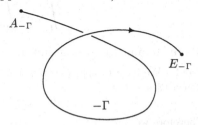

Now let $q \in \mathbb{N}^\times \cup \{\infty\}$. Further suppose $\gamma \in C(I, X)$ and $(t_0, \ldots, t_m)$ is a partition of I. If[2]

$$\gamma_j := \gamma \,|\, [t_{j-1}, t_j] \in C^q\big([t_{j-1}, t_j], X\big) \quad \text{for } 1 \le j \le m ,$$

[2]See Exercise 1.5 and the definition of a piecewise continuously differentiable function in Section VI.7. Here we also assume that γ is continuous.

then we say γ is **piecewise-C^q path** in X or a **sum** of the C^q paths γ_j. The curve $\Gamma = [\gamma]$ parametrized by γ is said to be a **piecewise C^q curve** in X, and we write $\Gamma := \Gamma_1 + \cdots + \Gamma_m$, where $\Gamma_j := [\gamma_j]$. Given a piecewise C^1 curve $\Gamma = \Gamma_1 + \cdots + \Gamma_m$ in X and $\alpha \in \Omega_{(0)}(X)$ we define the **line integral of α along Γ** by

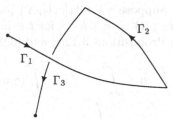

$$\int_\Gamma \alpha := \sum_{j=1}^m \int_{\Gamma_j} \alpha .$$

Finally, we can also join piecewise C^q curves. So let $\Gamma := \sum_{j=1}^m \Gamma_j$ and $\widetilde{\Gamma} := \sum_{j=1}^{\widetilde{m}} \widetilde{\Gamma}_j$ be piecewise C^q curves with $E_\Gamma = A_{\widetilde{\Gamma}}$. Then $\Sigma = \Sigma_1 + \cdots + \Sigma_{m+\widetilde{m}}$ with

$$\Sigma_j := \begin{cases} \Gamma_j , & 1 \leq j \leq m , \\ \widetilde{\Gamma}_{j-m} , & m + 1 \leq j \leq m + \widetilde{m} , \end{cases}$$

is a piecewise C^q curve, the **sum**[3] of Γ and $\widetilde{\Gamma}$. In this case, we write $\Gamma + \widetilde{\Gamma} := \Sigma$ and set

$$\int_{\Gamma+\widetilde{\Gamma}} \alpha := \int_\Sigma \alpha \quad \text{for } \alpha \in \Omega_{(0)}(X) .$$

Clearly, these notations are consistent with the prior notation $\Gamma = \Gamma_1 + \cdots + \Gamma_m$ for a piecewise C^q curve and the notation $\int_\Gamma \alpha = \int_{\Gamma_1 + \cdots + \Gamma_m} \alpha$ for line integrals.

The next theorem lists some basic properties of line integrals.

4.3 Proposition *Suppose* Γ, Γ_1 *and* Γ_2 *are piecewise* C^1 *curves and* $\alpha, \alpha_1, \alpha_2 \in \Omega_{(0)}(X)$. *Then*

(i) $\int_\Gamma (\lambda_1 \alpha_1 + \lambda_2 \alpha_2) = \lambda_1 \int_\Gamma \alpha_1 + \lambda_2 \int_\Gamma \alpha_2, \quad \lambda_1, \lambda_2 \in \mathbb{R},$
   that is, $\int_\Gamma : \Omega_{(0)}(X) \to \mathbb{R}$ *is a vector space homomorphism;*

(ii) $\int_{-\Gamma} \alpha = -\int_\Gamma \alpha,$
   that is, the line integral is oriented;

(iii) *if* $\Gamma_1 + \Gamma_2$ *is defined, we have*

$$\int_{\Gamma_1 + \Gamma_2} \alpha = \int_{\Gamma_1} \alpha + \int_{\Gamma_2} \alpha ,$$

   that is, the line integral is additive with respect to its integrated curves;

(iv) *for* $\alpha = \sum_{j=1}^n a_j \, dx^j$ *and* $a := \Theta^{-1} \alpha = \sum_{j=1}^n a_j e_j = (a_1, \ldots, a_n) \in \mathcal{V}^0(X)$, *we have*

$$\left| \int_\Gamma \alpha \right| \leq \max_{x \in \Gamma} |a(x)| \, L(\Gamma) .$$

---

[3] Note that for two piecewise C^q curves Γ and $\widetilde{\Gamma}$, the sum $\Gamma + \widetilde{\Gamma}$ is only defined when the final point of Γ is the initial point of $\widetilde{\Gamma}$.

Proof   The statements (i) and (iii) are obvious.

(ii) Due to (iii), it is sufficient to consider the case where Γ is a C^q curve.

Suppose $\gamma \in C^1([a,b], X)$ parametrizes Γ. Then we write γ^- as $\gamma^- = \gamma \circ \varphi$, where $\varphi(t) := a + b - t$ for $t \in [a,b]$. Because $\varphi(a) = b$ and $\varphi(b) = a$, it follows from Propositions 3.12(i) and (3.9) that

$$\int_{-\Gamma} \alpha = \int_a^b (\gamma^-)^* \alpha = \int_a^b (\gamma \circ \varphi)^* \alpha = \int_a^b \varphi^*(\gamma^*\alpha) = \int_b^a \gamma^*\alpha = -\int_a^b \gamma^*\alpha$$

$$= -\int_\Gamma \alpha \ .$$

(iv) The length of the piecewise C^1 curve $\Gamma = \Gamma_1 + \cdots + \Gamma_m$ is clearly equal to the sum of the lengths of its pieces[4] Γ_j. Thus it suffices to prove the statement for one C^1 curve $\Gamma := [\gamma]$. With the help of the Cauchy–Schwarz inequality and Proposition VI.4.3, it follows that

$$\left| \int_\Gamma \alpha \right| = \left| \int_I \gamma^*\alpha \right| = \left| \int_I \langle \alpha(\gamma(t)), \dot\gamma(t) \rangle \, dt \right| = \left| \int_I \big(a(\gamma(t)) \, |\dot\gamma(t)|\big) \, dt \right|$$

$$\leq \int_I |a(\gamma(t))| \, |\dot\gamma(t)| \, dt \leq \max_{x \in \Gamma} |a(x)| \int_I |\dot\gamma(t)| \, dt$$

$$= \max_{x \in \Gamma} |a(x)| \, L(\Gamma) \ ,$$

where we identify $(\mathbb{R}^n)'$ with $\mathbb{R}^n$, and in the last step, we have applied Theorem 1.3. ■

The fundamental theorem of line integrals

4.4 Theorem   *Suppose $X \subset \mathbb{R}^n$ is a domain and $\alpha \in \Omega_{(0)}(X)$. Then these statements are equivalent:*

(i) *α is exact;*

(ii) *$\int_\Gamma \alpha = 0$ for every closed piecewise C^1 curve in X.*

Proof   The case "(i)$\Rightarrow$(ii)" follows from Example 4.2(b) and Proposition 4.3(iii).

"(ii)$\Rightarrow$(i)" Suppose $x_0 \in X$. According to Theorem III.4.10, there is for every $x \in X$ a continuous, piecewise straight path in X that leads to x from x_0. Thus there is for every $x \in X$ a piecewise C^1 curve Γ_x in X with initial point x_0 and final point x. We set

$$f : X \to \mathbb{R} \ , \quad x \mapsto \int_{\Gamma_x} \alpha \ .$$

---

[4]See Exercise 1.5.

To verify that f is well defined, that
is, independent of the special curve
Γ_x, we choose a second piecewise C^1
curve $\widetilde{\Gamma}_x$ in X with the same end
points. Then $\Sigma := \Gamma_x + (-\widetilde{\Gamma}_x)$ is a
closed piecewise C^1 curve in X. By
assumption we have $\int_\Sigma \alpha = 0$, and
we deduce from Proposition 4.3 that
$0 = \int_{\Gamma_x} \alpha - \int_{\widetilde{\Gamma}_x} \alpha$. Therefore f is well
defined.

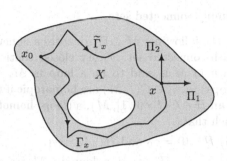

Suppose now $h \in \mathbb{R}^+$ with $\overline{\mathbb{B}}(x,h) \subset X$ and $\Pi_j := [\pi_j]$ with

$$\pi_j : [0,1] \to X , \quad t \mapsto x + t\,h\,e_j \quad \text{for } j = 1, \dots, n .$$

Then $\Gamma_x + \Pi_j$ and $\Pi_j = (-\Gamma_x) + (\Gamma_x + \Pi_j)$ are curves X. Because Γ_x and $\Gamma_x + \Pi_j$
have the same initial point x_0,

$$f(x + he_j) - f(x) = \int_{\Gamma_x + \Pi_j} \alpha - \int_{\Gamma_x} \alpha = \int_{\Pi_j} \alpha .$$

Letting $a_j := \langle \alpha, e_j \rangle$, we find

$$\int_{\Pi_j} \alpha = \int_0^1 \pi_j^* \alpha = \int_0^1 \langle \alpha(x + the_j), he_j \rangle \, dt = h \int_0^1 a_j(x + the_j) \, dt .$$

Therefore

$$f(x + he_j) - f(x) - a_j(x)h = h \int_0^1 \left[a_j(x + the_j) - a_j(x) \right] dt = o(h)$$

as $h \to 0$. Thus f has continuous partial derivatives $\partial_j f = a_j$ for $j = 1, \dots, n$.
Now Corollary VII.2.11 shows that f belongs to $C^1(X)$, and Remark 3.3(e) gives
$df = \alpha$. ∎

4.5 Corollary *Suppose X is open in $\mathbb{R}^n$ and star shaped, and let $x_0 \in X$. Also
suppose $q \in \mathbb{N}^\times \cup \{\infty\}$ and $\alpha \in \Omega_{(q)}(X)$ is closed. Let*

$$f(x) := \int_{\Gamma_x} \alpha \quad \text{for } x \in X ,$$

*where Γ_x is a piecewise C^1 curve in X with initial point x_0 and final point x. This
function satisfies $f \in C^{q+1}(X)$ and $df = \alpha$.*

Proof The proof of the Poincaré lemma (Theorem 3.8) and of Theorem 4.4 guar-
antee that $f \in C^1(X)$ and $df = \alpha$. Because $\partial_j f = a_j \in C^q(X)$ for $j = 1, \dots, n$, it
follows from Theorem VII.5.4 that f belongs to $C^{q+1}(X)$. ∎

This corollary gives a prescription for constructing the antiderivative of a
closed Pfaff form on a star shaped domain. In a concrete calculation, one chooses
the curve Γ_x that makes the resulting integration the easiest to do (see Exercise 7).

Simply connected sets

In the following, $M \subset \mathbb{R}^n$ denotes a nonempty
path-connected set. Every closed continuous
path in M is said to be a **loop** in M. Two
loops[5] $\gamma_0, \gamma_1 \in C(I, M)$ are **homotopic** if there
is an $H \in C(I \times [0,1], M)$, a **(loop) homotopy**,
such that

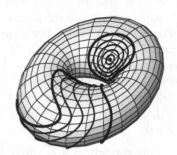

(i) $H(\cdot, 0) = \gamma_0$ and $H(\cdot, 1) = \gamma_1$;

(ii) $\gamma_s := H(\cdot, s)$ is a loop in M for every
$s \in (0, 1)$.

When two loops γ_0 and γ_1 are homotopic, we write $\gamma_0 \sim \gamma_1$. We denote by γ_{x_0}
the **point loop** $[t \mapsto x_0]$ with $x_0 \in M$. Every loop in M that is homotopic to a
point loop is **null homotopic**. Finally, M is said to be **simply connected** if every
loop in M is null homotopic.

4.6 Remarks **(a)** On the set of all loops in M, $\sim$ is an equivalence relation.

Proof (i) It is clear the every loop is homotopic to itself, that is, the relation $\sim$ is
reflexive.

(ii) Suppose H is a homotopy from γ_0 to γ_1 and

$$H^-(t, s) := H(t, 1 - s) \quad \text{for } (t, s) \in I \times [0, 1] \ .$$

Then H^- is a homotopy from γ_1 to γ_0. Therefore the relation $\sim$ is symmetric.

(iii) Finally, let H_0 be a homotopy from γ_0 to γ_1, and let H_1 be a homotopy from
γ_1 to γ_2. We set

$$H(t, s) := \begin{cases} H_0(t, 2s) \ , & (t, s) \in I \times [0, 1/2] \ , \\ H_1(t, 2s - 1) \ , & (t, s) \in I \times [1/2, 1] \ . \end{cases}$$

It is not hard to check that H belongs to $C(I \times [0, 1], M)$; see Exercise III.2.13. Also,
$H(\cdot, 0) = \gamma_0$ and $H(\cdot, 1) = \gamma_2$, and every $H(\cdot, s)$ is a loop in M. ∎

(b) Suppose γ is a loop in M. These statements are equivalent:

(i) γ is null homotopic;

(ii) $\gamma \sim \gamma_{x_0}$ for some $x_0 \in M$;

(iii) $\gamma \sim \gamma_x$ for every $x \in M$.

Proof It suffices to verify the implication "(ii)⇒(iii)". For that, let $\gamma \in C(I, M)$ be a
loop with $\gamma \sim \gamma_{x_0}$ for some $x_0 \in M$. Also suppose $x \in M$. Because M is path connected,
there is a continuous path $w \in C(I, M)$ that connects x_0 to x. We set

$$H(t, s) := w(s) \quad \text{for } (t, s) \in I \times [0, 1] \ ,$$

and we thus see that γ_{x_0} and γ_x are homotopic. The claim now follows using (a). ∎

---

[5]It is easy to see that, without loss of generality, we can define γ_0 and γ_1 over the same
parameter interval. In particular, we can choose $I = [0, 1]$.

(c) Every star shaped set is simply connected.

Proof Suppose M is star shaped with respect to x_0 and $\gamma \in C(I, M)$ is a loop in M. We set
$$H(t, s) := x_0 + s\big(\gamma(t) - x_0\big) \quad \text{for } (t, s) \in I \times [0, 1] \ .$$
Then H is a homotopy from γ_{x_0} to γ. ∎

(d) The set $Q := \mathbb{B}_2 \backslash T$ with $T := \big((-1/2, 1/2) \times \{0\}\big) \cup \big(\{0\} \times (-1, 0)\big)$ is simply connected, but not star shaped.

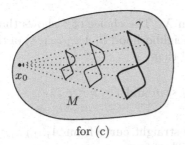
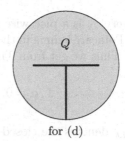

for (c)   for (d)

The homotopy invariance of line integrals

The line integral of closed Pfaff forms is invariant under loop homotopies:

4.7 Proposition *Suppose $\alpha \in \Omega_{(1)}(X)$ is closed. Let γ_0 and γ_1 be homotopic piecewise C^1 loops in X. Then $\int_{\gamma_0} \alpha = \int_{\gamma_1} \alpha$.*

Proof (i) Suppose $H \in C\big(I \times [0, 1], X\big)$ is a homotopy from γ_0 to γ_1. Because $I \times [0, 1]$ is compact, it follows from Theorem III.3.6 that $K := H\big(I \times [0, 1]\big)$ is compact. Because X^c is closed and $X^c \cap K = \emptyset$, there is, according to Example III.3.9(c), an $\varepsilon > 0$ such that
$$|H(t, s) - y| \geq \varepsilon \quad \text{for } (t, s) \in I \times [0, 1] \text{ and } y \in X^c \ .$$
Theorem III.3.13 guarantees that H is uniformly continuous. Hence there is a $\delta > 0$ such that
$$|H(t, s) - H(\tau, \sigma)| < \varepsilon \quad \text{for } |t - \tau| < \delta \ , \quad |s - \sigma| < \delta \ .$$

(ii) Now we choose a partition $(t_0, \ldots, t_m)$ of I and a partition $(s_0, \ldots, s_\ell)$ of $[0, 1]$, both with mesh $< \delta$. Letting $A_{j,k} := H(t_j, s_k)$, we set
$$\widetilde{\gamma}_k(t) := A_{j-1,k} + \frac{t - t_{j-1}}{t_j - t_{j-1}} \big(A_{j,k} - A_{j-1,k}\big) \quad \text{for } t_{j-1} \leq t \leq t_j \ ,$$
$1 \leq j \leq m$, and $0 \leq k \leq \ell$.

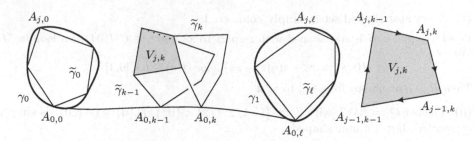

Clearly every $\widetilde{\gamma}_k$ is a piecewise C^1 loop in X. The choice of δ shows that we can apply the Poincaré lemma in the convex neighborhood $\mathbb{B}(A_{j-1,k-1}, \varepsilon)$ of the points $A_{j-1,k-1}$. Thus we get from Theorem 4.4 that

$$\int_{\partial V_{j,k}} \alpha = 0 \quad \text{for } 1 \leq j \leq m \text{ and } 1 \leq k \leq \ell,$$

where $\partial V_{j,k}$ denotes the closed piecewise straight curve from $A_{j-1,k-1}$ to $A_{j,k-1}$ to $A_{j,k}$ to $A_{j-1,k}$ and back to $A_{j-1,k-1}$. Therefore

$$\int_{\widetilde{\gamma}_{k-1}} \alpha = \int_{\widetilde{\gamma}_k} \alpha \quad \text{for } 1 \leq k \leq \ell,$$

because the integral cancels itself over the "connecting pieces" between $\widetilde{\gamma}_{k-1}$ and $\widetilde{\gamma}_k$. Likewise, using the Poincaré lemma we conclude that $\int_{\widetilde{\gamma}_0} \alpha = \int_{\gamma_0} \alpha$ and $\int_{\widetilde{\gamma}_\ell} \alpha = \int_{\gamma_\ell} \alpha$, as the claim requires. ∎

As an application of homotopy invariance theorem, we now get a wide-reaching generalization of the Poincaré lemma.

4.8 Theorem *Suppose X is open in $\mathbb{R}^n$ and simply connected. If $\alpha \in \Omega_{(1)}(X)$ is closed, α is exact.*

Proof Suppose γ is a piecewise C^1 loop in X and $x_0 \in X$. According to Remark 4.6(b), γ and γ_{x_0} are homotopic. Then Proposition 4.7 and Example 4.2(e) give $\int_\gamma \alpha = \int_{\gamma_{x_0}} \alpha = 0$, and the claim follows from Theorem 4.4. ∎

4.9 Examples (a) The "punctured plane" $\mathbb{R}^2 \setminus \{(0,0)\}$ is not simply connected.

Proof This follows from Theorem 4.8 and Example 4.2(d). ∎

(b) For $n \geq 3$, the set $\mathbb{R}^n \setminus \{0\}$ is simply connected.

Proof See Exercise 12. ∎

4.10 Remarks (a) Suppose X is open in $\mathbb{R}^n$ and simply connected. Also, suppose the vector field $v = (v_1, \ldots, v_n) \in \mathcal{V}^1(X)$ satisfies the integrability conditions

$\partial_j v_k = \partial_k v_j$ for $1 \leq j, k \leq n$. Then v has a potential U, that is, v is a gradient field. If x_0 is any point in X, then U can be calculated through

$$U(x) := \int_0^1 \left(v\big(\gamma_x(t)\big) \,\big|\, \dot{\gamma}_x(t) \right) dt \quad \text{for } x \in X ,$$

where $\gamma_x : [0,1] \to X$ is a piecewise C^1 path in X for which $\gamma_x(0) = x_0$ and $\gamma_x(1) = x$.

Proof The canonical isomorphism Θ defined in Remark 3.3(g) assigns to v the Pfaff form $\alpha := \Theta v \in \Omega_{(1)}(X)$. Because v satisfies the integrability conditions, α is closed and therefore exact by Theorem 4.8. Because, in the proof of Corollary 4.5, we can replace the Poincaré lemma by Theorem 4.8, it follows that, with $\Gamma_x := [\gamma_x]$,

$$U(x) := \int_{\Gamma_x} \alpha \quad \text{for } x \in X ,$$

defines a potential for α. ∎

(b) Suppose $\alpha = \sum_{j=1}^n a_j \, dx^j \in \Omega_{(0)}(X)$, and let $a := \Theta^{-1}\alpha = (\alpha_1, \ldots, \alpha_n) \in \mathcal{V}^0(X)$ be the associated vector field. It is conventional to write α symbolically as a scalar product

$$\alpha = a \cdot ds$$

using the **vector line element** $ds := (dx^1, \ldots, dx^n)$. Then, due to Remark 3.3(g), the line integral

$$\int_\Gamma a \cdot ds := \int_\Gamma \Theta a$$

is well defined for every vector field $a \in \mathcal{V}(X)$ and for every piecewise C^1 curve Γ in X.

(c) Suppose F is a continuous force field (for example, the electrostatic or gravitational field) in an open subset X of (three dimensional Euclidean) space. If a (small) test particle is guided along the piecewise C^1 curve $\Gamma = [\gamma]$, then the **work** done by F along Γ is given by the line integral

$$\int_\Gamma F \cdot ds .$$

In particular, if γ is a piecewise C^1 parametrization of Γ, then

$$A := \int_\Gamma F \cdot ds = \int_I \left(F\big(\gamma(t)\big) \,\big|\, \dot{\gamma}(t) \right) dt$$

is a Cauchy–Riemann integral. Therefore it is approximated by the Riemann sum

$$\sum_{j=1}^n F\big(\gamma(t_{j-1})\big) \cdot \dot{\gamma}(t_{j-1})(t_j - t_{j-1}) ,$$

where $\mathfrak{Z} := (t_0, \ldots, t_n)$ is a suitable partition of I. Because

$$\gamma(t_j) - \gamma(t_{j-1}) = \dot{\gamma}(t_{j-1})(t_j - t_{j-1}) + o(\triangle_{\mathfrak{Z}}) \ ,$$

where $\triangle_{\mathfrak{Z}}$ is the mesh of $\mathfrak{Z}$, the sum

$$\sum_{j=1}^{n} F\big(\gamma(t_{j-1})\big) \cdot \big(\gamma(t_j) - \gamma(t_{j-1})\big)$$

represents, for a partition with sufficiently small mesh, a good approximation for A which, by passing the limit $\triangle_{\mathfrak{Z}} \to 0$, converges to A (see Theorem VI.3.4). The scalar product

$$F\big(\gamma(t_{j-1})\big) \cdot \big(\gamma(t_j) - \gamma(t_{j-1})\big)$$

is the work of a (constant) force $F\big(\gamma(t_{j-1})\big)$ done on a particle displaced on a line segment from $\gamma(t_{j-1})$ to $\gamma(t_j)$ ("work = force × distance" or, more precisely, "work = force × displacement in the direction of the force").

A force field F is said to be **conservative** if it has a potential. In that case, according to Example 4.2(c), the work is "path independent" and only depends on the initial and final position. If X is simply connected (for example, X is the whole space) and F has continuous derivatives, then F, according to (a) and Theorem 4.8, is conservative if and only if the integrability conditions are satisfied. If the latter is the case, one says that the force field is **irrotational**.[6] ∎

Exercises

1   Calculate

$$\int_{\Gamma} (x^2 - y^2) \, dx + 3z \, dy + 4xy \, dz$$

along one winding of a helix.

2   Suppose $\varphi \in C^1\big((0, \infty), \mathbb{R}\big)$ and

$$\alpha(x, y) := x\varphi\big(\sqrt{x^2 + y^2}\big) \, dy - y\varphi\big(\sqrt{x^2 + y^2}\big) \, dx \quad \text{for } (x, y) \in \mathbb{R}^2 \ .$$

What is the value of $\int_{\Gamma} \alpha$ when Γ is the positively oriented[7] circle centered at 0 with radius $R > 0$?

3   Calculate

$$\int_{\Gamma} 2xy^3 \, dx + 3x^2 y^2 \, dy \ ,$$

where Γ is parametrized by $[\alpha, \beta] \to \mathbb{R}^2$, $t \mapsto (t, t^2/2)$.

---

[6]We will clarify this language in Volume III.

[7]A circle is by convention positively oriented when it is traversed in the counterclockwise direction (see Remark 5.8).

4 Suppose Γ_+ are Γ_- are positively oriented circles of radius 1 centered at $(1,0)$ and $(-1,0)$, respectively. Also suppose

$$\alpha := \frac{2(x^2 - y^2 - 1)\,dy - 4xy\,dx}{(x^2 + y^2 - 1)^2 + 4y^2} \in \Omega_{(\infty)}(\mathbb{R}^2 \setminus \{(-1,0),(1,0)\}) \ .$$

Prove

$$\frac{1}{2\pi} \int_{\Gamma_\pm} \alpha = \pm 1 \ .$$

(Hint: Exercise 3.3.)

5 Suppose X is open in $\mathbb{R}^n$ and Γ is a C^1 curve in X. For $q \geq 0$, prove or disprove that

$$\int_\Gamma : \Omega_{(q)}(X) \to \mathbb{R} \ , \quad \alpha \mapsto \int_\Gamma \alpha$$

is a $C^q(X)$ module homomorphism.

6 Suppose X and Y are open in $\mathbb{R}^n$ and $\mathbb{R}^m$, respectively, and $\alpha \in \Omega_{(q)}(X \times Y)$. Further suppose Γ is a C^1 curve in X. Show that

$$f : Y \to \mathbb{R} \ , \quad y \mapsto \int_\Gamma \alpha(\cdot, y)$$

belongs to $C^q(Y, \mathbb{R})$. Calculate ∇f for $q \geq 1$.

7 Suppose $-\infty \leq \alpha_j < \beta_j \leq \infty$ for $j = 1, \ldots, n$ and $X := \prod_{j=1}^n (\alpha_j, \beta_j)$. Also suppose $\alpha = \sum_{j=1}^n a_j \, dx^j \in \Omega_{(1)}(X)$ is closed. Determine (in a simple way) all antiderivatives of α.

8 Suppose $X := (0, \infty)^2$ and $a, b, c, e > 0$. According to Exercise 3.8(b),

$$\alpha := (c - ex)y \, dx + (a - by)x \, dy$$

has an integrating factor h of the form $h(x, y) = m(xy)$. Determine all antiderivatives of $h\alpha$. (Hint: Exercise 7.)

9 Suppose X is open in $\mathbb{R}^n$ and Γ is a compact piecewise C^1 curve in X with the parametrization γ. Also suppose $f, f_k \in C(X, \mathbb{R}^n)$ for $k \in \mathbb{N}$. Prove

(a) if $(\gamma^* f_k)$ converges uniformly to $\gamma^* f$,

$$\lim_k \int_\Gamma \sum_{j=1}^n f_k^j \, dx^j = \int_\Gamma \sum_{j=1}^n f^j \, dx^j \ ;$$

(b) if $\sum_k \gamma^* f_k$ converges uniformly to $\gamma^* f$,

$$\sum_k \int_\Gamma \sum_{j=1}^n f_k^j \, dx^j = \int_\Gamma \sum_{j=1}^n f^j \, dx^j \ .$$

10 Suppose $a_j \in C^1(\mathbb{R}^n, \mathbb{R})$ is positively homogeneous of degree $\lambda \neq -1$ and the 1-form $\alpha := \sum_{j=1}^n a_j \, dx^j$ is closed. Show that

$$f(x) := \frac{1}{\lambda + 1} \sum_{j=1}^n x^j a_j(x) \quad \text{for } x = (x^1, \ldots, x^n) \in \mathbb{R}^n$$

defines an antiderivative for α. (Hint: Apply the Euler homogeneity relation (see Exercise VII.3.2).)

11 Show that if $\gamma \in C(I, X)$ is a loop, there is a change of parameter $\varphi : [0, 1] \to I$ such that $\gamma \circ \varphi$ is a loop.

12 (a) Suppose X is open in $\mathbb{R}^n$ and γ is a loop in X. Show that γ is homotopic to a polygonal loop in X.

(b) Show that $\mathbb{R}^n \backslash \mathbb{R}^+ a$ is simply connected for $a \in \mathbb{R}^n \backslash \{0\}$.

(c) Suppose γ is a polygonal loop in $\mathbb{R}^n \backslash \{0\}$ for $n \geq 3$. Show that there is a half ray $\mathbb{R}^+ a$ with $a \in \mathbb{R}^n \backslash \{0\}$ that does not intersect the image of γ.

(d) Prove that $\mathbb{R}^n \backslash \{0\}$ is simply connected for $n \geq 3$.

13 Prove or disprove that every closed 1-form in $\mathbb{R}^n \backslash \{0\}$ is exact for $n \geq 3$.

14 Suppose $X \subset \mathbb{R}^n$ is nonempty and path connected. For $\gamma_1, \gamma_2 \in C([0, 1], X)$ with $\gamma_1(1) = \gamma_2(0)$,

$$\gamma_2 \oplus \gamma_1 : [0, 1] \to X , \quad t \mapsto \begin{cases} \gamma_1(2t) , & 0 \leq t \leq 1/2 , \\ \gamma_2(2t - 1) , & 1/2 \leq t \leq 1 \end{cases}$$

is the **path joining** γ_2 to γ_1. For $x_0 \in X$, define

$$S_{x_0} := \left\{ \gamma \in C([0, 1], X) \; ; \; \gamma(0) = \gamma(1) = x_0 \right\} .$$

Also denote by $\sim$ the equivalence relation induced by the loop homotopies of S_{x_0} (see Example I.4.2(d)).

Verify that

(a) the map

$$(S_{x_0}/\sim) \times (S_{x_0}/\sim) \to S_{x_0}/\sim , \quad ([\gamma_1], [\gamma_2]) \mapsto [\gamma_2 \oplus \gamma_1]$$

is a well defined operation on $S_{x_0}/\sim$, and $\Pi_1(X, x_0) := (S_{x_0}/\sim, \oplus)$ is a group, the **fundamental group**, or the **first homotopy group**, of X with respect to x_0;

(b) for $x_0, x_1 \in X$, the groups $\Pi_1(X, x_0)$ and $\Pi_1(X, x_1)$ are isomorphic.

Remark Part (b) justifies speaking of the "fundamental group" $\Pi_1(X)$ of X; see the construction in Example I.7.8(g). The set X is simply connected if and only if $\Pi_1(X)$ is trivial, that is, if it consists only of the identity element.

(Hint for (b)): Let $w \in C([0, 1], X)$ be a path in X with $w(0) = x_0$ and $w(1) = x_1$. Then the map $[\gamma] \mapsto [w \oplus \gamma \oplus w^-]$ is a group isomorphism from $\Pi_1(X, x_0)$ to $\Pi_1(X, x_1)$.)

5 Holomorphic functions

Line integrals becomes particularly useful in complex analysis, the theory of complex functions. With the results from the previous section, we may deduce almost effortlessly the (global) Cauchy integral theorem and the Cauchy integral formula. These theorems form the core of complex analysis and have wide-reaching consequences, some of which we will see in this section and the next.

Here, suppose

- U is in $\mathbb{C}$ and $f : U \to \mathbb{C}$ is continuous.

Also in this section, we often decompose $z \in U$ and f into real and imaginary parts, that is, we write

$$z = x + iy \in \mathbb{R} + i\mathbb{R} \text{ and } f(z) = u(x,y) + iv(x,y) \in \mathbb{R} + i\mathbb{R} .$$

Complex line integrals

Suppose $I \subset \mathbb{R}$ is a compact interval, and suppose Γ is a piecewise C^1 curve in U parametrized by

$$I \to U , \quad t \mapsto z(t) = x(t) + iy(t) .$$

Then

$$\int_\Gamma f \, dz := \int_\Gamma f(z) \, dz := \int_\Gamma u \, dx - v \, dy + i \int_\Gamma u \, dy + v \, dx$$

is the **complex line integral** of f along Γ.[1]

5.1 Remarks (a) We denote by $\Omega(U, \mathbb{C})$ the space of **continuous complex 1-forms** and define it as follows:

On the product group $\big(\Omega_{(0)}(U) \times \Omega_{(0)}(U), +\big)$, we define an outer multiplication

$$C(U, \mathbb{C}) \times \big[\Omega_{(0)}(U) \times \Omega_{(0)}(U)\big] \to \Omega_{(0)}(U) \times \Omega_{(0)}(U) , \quad (f, \alpha) \mapsto f\alpha ,$$

where

$$f\alpha := \big((ua_1 - vb_1) \, dx + (ua_2 - vb_2) \, dy, (ub_1 + va_1) \, dx + (va_2 + ub_2) \, dy\big)$$

for $\alpha = (a_1 \, dx + a_2 \, dy, b_1 \, dx + b_2 \, dy)$. Then one immediately verifies that

$$\Omega(U, \mathbb{C}) := \big(\Omega_{(0)}(U) \times \Omega_{(0)}(U), +, \cdot\big)$$

---

[1]In this context, the curve Γ may also be called the **contour**.

is a free module over $C(U, \mathbb{C})$. In addition, we have

$$1(a_1 \, dx + a_2 \, dy, 0) = (a_1 \, dx + a_2 \, dy, 0) \ ,$$
$$i \, (a_1 \, dx + a_2 \, dy, 0) = (0, a_1 \, dx + a_2 \, dy)$$

for $a_1 \, dx + a_2 \, dy \in \Omega_{(0)}(U)$. Therefore we can identify $\Omega_{(0)}(U)$ with $\Omega_{(0)}(U) \times \{0\}$ in $\Omega(U, \mathbb{C})$ and represent $(a_1 \, dx + a_2 \, dy, b_1 \, dx + b_2 \, dy) \in \Omega(U, \mathbb{C})$ uniquely in the form

$$a_1 \, dx + a_2 \, dy + i \, (b_1 \, dx + b_2 \, dy) \ ,$$

which we express through the notation

$$\Omega(U, \mathbb{C}) = \Omega_{(0)}(U) + i \Omega_{(0)}(U) \ .$$

Finally, we have

$$(a_1 + i \, b_1)(dx, 0) = (a_1 \, dx, b_1 \, dx) \ ,$$
$$(a_2 + i \, b_2)(dy, 0) = (a_2 \, dy, b_2 \, dy) \ ,$$

and thus

$$(a_1 + i \, b_1)(dx, 0) + (a_2 + i \, b_2)(dy, 0) = (a_1 \, dx + a_2 \, dy, b_1 \, dx + b_2 \, dy) \ .$$

This means we can also write $(a_1 \, dx + a_2 \, dy, b_1 \, dx + b_2 \, dy) \in \Omega(U, \mathbb{C})$ uniquely as

$$(a_1 + i b_1) \, dx + (a_2 + i b_2) \, dy \ ,$$

that is,

$$\Omega(U, \mathbb{C}) = \left\{ a \, dx + b \, dy \ ; \ a, b \in C(U, \mathbb{C}) \right\} \ .$$

(b) With $u_x := \partial_1 u$ etc, we call

$$df := (u_x + i v_x) \, dx + (u_y + i v_y) \, dy \in \Omega(U, \mathbb{C})$$

the **complex differential** of f. Clearly, $dz = dx + i \, dy$, and we get[2]

$$f \, dz = (u + iv)(dx + i \, dy) = u \, dx - v \, dy + i \, (u \, dy + v \, dx) \qquad (5.1)$$

for $f = u + iv \in C(U, \mathbb{C})$. ∎

5.2 Proposition   *Suppose Γ is a piecewise C^1 curve parametrized by $I \to U$, $t \mapsto z(t)$. Then*

(i) $\int_\Gamma f(z) \, dz = \int_I f(z(t)) \dot{z}(t) \, dt$;

(ii) $\left| \int_\Gamma f(z) \, dz \right| \leq \max_{z \in \Gamma} |f(z)| \, L(\Gamma)$.

---

[2]Compare (5.1) with the definition of the complex line integral of f.

Proof (i) For the piecewise C^1 path $\gamma : I \to \mathbb{R}^2$, $t \mapsto (x(t), y(t))$, we have

$$\int_\Gamma f \, dz = \int_I \gamma^*(u \, dx - v \, dy) + i \int_I \gamma^*(u \, dy + v \, dx)$$

$$= \int_I [(u \circ \gamma)\dot{x} - (v \circ \gamma)\dot{y}] \, dt + i \int_I [(u \circ \gamma)\dot{y} + (v \circ \gamma)\dot{x}] \, dt$$

$$= \int_I (u \circ \gamma + iv \circ \gamma)(\dot{x} + i\dot{y}) \, dt$$

$$= \int_I f(z(t)) \dot{z}(t) \, dt .$$

(ii) This statement follows from (i) by estimating the last integral using Theorem 1.3 and Proposition VI.4.3. ∎

5.3 Examples (a) Suppose $z_0 \in \mathbb{C}$ and $r > 0$, and let $z(t) := z_0 + re^{it}$ for $t \in [0, 2\pi]$. Then, for $\Gamma := [z]$, we have

$$\int_\Gamma (z - z_0)^m \, dz = \begin{cases} 0 , & m \in \mathbb{Z}\backslash\{-1\} , \\ 2\pi i , & m = -1 . \end{cases}$$

Proof Because

$$\int_\Gamma (z - z_0)^m \, dz = \int_0^{2\pi} r^m e^{itm} ire^{it} \, dt = ir^{m+1} \int_0^{2\pi} e^{i(m+1)t} \, dt ,$$

the claim follows from the $2\pi i$-periodicity of the exponential function. ∎

(b) Suppose Γ is as in (a) with $z_0 = 0$. Then for $z \in \mathbb{C}$, we have

$$\frac{1}{2\pi i} \int_\Gamma \lambda^{-k-1} e^{\lambda z} \, d\lambda = \frac{z^k}{k!} \quad \text{for } k \in \mathbb{N} .$$

Proof From Exercise 4.9, it follows that

$$\int_\Gamma \lambda^{-k-1} e^{\lambda z} \, d\lambda = \sum_{n=0}^\infty \frac{z^n}{n!} \int_\Gamma \lambda^{n-k-1} \, d\lambda ,$$

and the claim is implied by (a). ∎

(c) Suppose Γ is the curve in $\mathbb{R}$ parametrized by $I \to \mathbb{C}$, $t \mapsto t$. Then

$$\int_\Gamma f(z) \, dz = \int_I f(t) \, dt ,$$

that is, the complex line integral and the Cauchy–Riemann integral from Chapter VI agree in this case.

Proof This follows from Proposition 5.2(i). ∎

(d) For the curves Γ_1 and Γ_2 parametrized respectively by

$$[0,\pi] \to \mathbb{C} , \quad t \mapsto e^{i(\pi-t)} ,$$

and

$$[-1,1] \to \mathbb{C} , \quad t \mapsto t ,$$

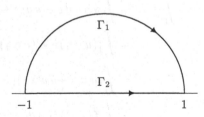

we have

$$\int_{\Gamma_1} |z| \, dz = -i \int_0^\pi e^{i(\pi-t)} \, dt = e^{i(\pi-t)} \Big|_0^\pi = 2$$

and

$$\int_{\Gamma_2} |z| \, dz = \int_{-1}^1 |t| \, dt = 1 .$$

Thus complex line integrals generally depend on the integration path and not only on the initial and final points. ∎

Holomorphism

A **holomorphic** function f is a function that has continuous complex derivatives, that is, $f \in C^1(U,\mathbb{C})$. We call f **continuously real differentiable** if

$$U \to \mathbb{R}^2 , \quad (x,y) \mapsto \big(u(x,y), v(x,y) \big)$$

belongs to $C^1(U,\mathbb{R}^2)$ (see Section VII.2).

5.4 Remarks **(a)** At the end of this section we will show that the assumption of continuous complex derivatives is unnecessary, that is, every complex differentiable function has a continuous derivative and is therefore holomorphic. We apply here the stronger assumption of *continuous* complex differentiability because it brings us more quickly to the fundamental theorem of **complex analysis**, that is, of the theory of complex-valued functions of a complex variable.

(b) A function f is holomorphic if and only if it is continuously real differentiable and satisfies the Cauchy–Riemann equations

$$u_x = v_y \text{ and } u_y = -v_x .$$

In this case, $f' = u_x + i v_x$.

Proof This a reformulation of Theorem VII.2.18. ∎

(c) According to Remark 5.1(b), we have

$$f \, dz = u \, dx - v \, dy + i \, (u \, dy + v \, dx) \ .$$

If f is holomorphic, the Cauchy–Riemann equations say that the 1-forms $u \, dx - v \, dy$ and $u \, dy + v \, dx$ are closed.

(d) Suppose U is a domain and f is holomorphic. Then f is constant if and only if these conditions are satisfied: [3]

  (i) $u = \text{const}$;

  (ii) $v = \text{const}$;

  (iii) $\bar{f}$ is holomorphic;

  (iv) $|f| = \text{const}$.

Proof If f is constant, (i)–(iv) clearly hold.

   Because $f' = u_x + i v_x$, it follows from the Cauchy–Riemann equations and from Remark VII.3.11(c) that either (i) or (ii) implies f is constant.

   (iii) If f and $\bar{f}$ are holomorphic, so is $u = (f + \bar{f})/2$. Since $\text{Im}(u) = 0$, it follows from (ii) that u is constant and therefore, by (i), so is f.

   (iv) It suffices to consider the case $f \neq 0$. Because $|f|$ is constant, f vanishes nowhere. Therefore $1/f$ is defined and holomorphic on U. Thus $\bar{f} = |f|^2/f$ is also holomorphic in U, and the claim follows from (iii). ∎

The Cauchy integral theorem

We have seen in Example 5.3(d) that complex line integrals generally depend on more than the initial and final points of the integration path. However, in the case of holomorphic integrands, the theorems of the previous sections give the following important results about path independence.

5.5 Theorem (the Cauchy integral theorem) *Suppose U is simply connected and f is holomorphic. Then, for every closed piecewise C^1 curve Γ in U,*

$$\int_\Gamma f \, dz = 0 \ .$$

Proof According to Remark 5.4(c) the 1-forms $\alpha_1 := u \, dx - v \, dy$ and $\alpha_2 := u \, dy + v \, dx$ are both closed. Because U is simply connected, it follows from Theorem 4.8 that α_1 and α_2 are exact. Now Theorem 4.4 implies

$$\int_\Gamma f \, dz = \int_\Gamma \alpha_1 + i \int_\Gamma \alpha_2 = 0$$

for every closed, piecewise C^1 curve Γ in U. ∎

----
[3]See Exercise V.3.5.

5.6 Theorem *Suppose U is simply connected and f is holomorphic. Then f has a holomorphic antiderivative. Every antiderivative φ of f satisfies*

$$\int_\Gamma f\, dz = \varphi(E_\Gamma) - \varphi(A_\Gamma)$$

for every piecewise C^1 curve Γ in U.

Proof We use the notation of the previous proofs. Because α_1 and α_2 are exact, there are $h_1, h_2 \in C^2(U, \mathbb{R})$ such that $dh_1 = \alpha_1$ and $dh_2 = \alpha_2$. From this, we read off

$$(h_1)_x = u \ , \quad (h_1)_y = -v \ , \quad (h_2)_x = v \ , \quad (h_2)_y = u \ .$$

Therefore $\varphi := h_1 + i h_2$ satisfies the Cauchy–Riemann equations. Thus φ is holomorphic, and

$$\varphi' = (h_1)_x + i(h_2)_x = u + iv = f \ .$$

This shows that φ is an antiderivative of f. The second statement follows from Example 4.2(b). ∎

5.7 Proposition *Suppose f is holomorphic and γ_1 and γ_2 are homotopic piecewise C^1 loops in U. Then*

$$\int_{\gamma_1} f\, dz = \int_{\gamma_2} f\, dz \ .$$

Proof This follows from Remark 5.4(c) and Proposition 4.7 (see also the proof of Theorem 5.5). ∎

The orientation of circles

5.8 Remark We recall the notation $\mathbb{D}(a, r) = a + r\mathbb{D}$ for an open disc in $\mathbb{C}$ with center $a \in \mathbb{C}$ and radius $r > 0$. In the following, we understand by $\partial\mathbb{D}(a, r) = a + r\partial\mathbb{D}$ the **positively oriented circle** with center a and radius r. This means that $\partial\mathbb{D}(a, r)$ is the curve $\Gamma = [\gamma]$ with $\gamma \colon [0, 2\pi] \to \mathbb{C}$, $t \mapsto a + re^{it}$. With this orientation, the circle is traversed so that the disc $\mathbb{D}(a, r)$ stays on the left side, or it is traversed counterclockwise. This is equivalent to saying that the negative unit normal vector $-\mathfrak{n}$ always points outward.

Proof From Remark 2.5(c) and the canonical identification of $\mathbb{C}$ with $\mathbb{R}^2$, it follows that the Frenet two-frame is given by $e_1 = (-\sin, \cos)$ and $e_2 = (-\cos, -\sin)$. Letting $x(t) = a + r(\cos t, \sin t) \in \partial\mathbb{D}(a, r)$, we have the relation

$$x(t) + re_2(t) = a \quad \text{for } 0 \le t \le 2\pi \ .$$

Thus $e_2(t)$ points inward along $\partial\mathbb{D}(a, r)$, and its negative points outward. ∎

The Cauchy integral formula

Holomorphic functions have a remarkable integral representation, which we derive in the next theorem. The implications of this formula will be explored in later applications.

5.9 Theorem (the Cauchy integral formula) *Suppose f is holomorphic and* $\overline{\mathbb{D}}(z_0, r) \subset U$. *Then*

$$ f(z) = \frac{1}{2\pi i} \int_{\partial \mathbb{D}(z_0, r)} \frac{f(\zeta)}{\zeta - z} \, d\zeta \quad \text{for } z \in \mathbb{D}(z_0, r) \, . $$

Proof Suppose $z \in \mathbb{D}(z_0, r)$ and $\varepsilon > 0$. Then there is an $\delta > 0$ such that $\overline{\mathbb{D}}(z, \delta) \subset U$ and

$$ |f(\zeta) - f(z)| \leq \varepsilon \quad \text{for } \zeta \in \overline{\mathbb{D}}(z, \delta) \, . \tag{5.2} $$

We set $\Gamma_\delta := \partial \mathbb{D}(z, \delta)$ and $\Gamma := \partial \mathbb{D}(z_0, r)$. Exercise 6 shows that Γ_δ and Γ are homotopic in $U \backslash \{z\}$. In addition,

$$ U \backslash \{z\} \to \mathbb{C} \, , \quad \zeta \mapsto 1/(\zeta - z) $$

is holomorphic. Therefore it follows from Proposition 5.7 and Example 5.3(a) that

$$ \int_\Gamma \frac{d\zeta}{\zeta - z} = \int_{\Gamma_\delta} \frac{d\zeta}{\zeta - z} = 2\pi i \, . \tag{5.3} $$

Because

$$ U \backslash \{z\} \to \mathbb{C} \, , \quad \zeta \mapsto \big(f(\zeta) - f(z)\big)/(\zeta - z) $$

is also holomorphic, we know from Proposition 5.7 that

$$ \int_\Gamma \frac{f(\zeta) - f(z)}{\zeta - z} \, d\zeta = \int_{\Gamma_\delta} \frac{f(\zeta) - f(z)}{\zeta - z} \, d\zeta \, . \tag{5.4} $$

Combining (5.3) with (5.4), we get

$$ \frac{1}{2\pi i} \int_\Gamma \frac{f(\zeta)}{\zeta - z} \, d\zeta = \frac{1}{2\pi i} \int_\Gamma \frac{f(z)}{\zeta - z} \, d\zeta + \frac{1}{2\pi i} \int_\Gamma \frac{f(\zeta) - f(z)}{\zeta - z} \, d\zeta $$

$$ = f(z) + \frac{1}{2\pi i} \int_{\Gamma_\delta} \frac{f(\zeta) - f(z)}{\zeta - z} \, d\zeta \, . $$

Then Proposition 5.2(ii) and (5.2) imply the estimate

$$ \left| \frac{1}{2\pi i} \int_{\Gamma_\delta} \frac{f(\zeta) - f(z)}{\zeta - z} \, d\zeta \right| \leq \frac{\varepsilon}{2\pi \delta} 2\pi \delta = \varepsilon \, , $$

and we find

$$ \left| \frac{1}{2\pi i} \int_\Gamma \frac{f(\zeta)}{\zeta - z} \, d\zeta - f(z) \right| \leq \varepsilon \, . $$

Because $\varepsilon > 0$ was arbitrary, the claim is proved. ∎

5.10 Remarks   **(a)** Under the assumptions of Theorem 5.9, we have

$$f(z) = \frac{r}{2\pi} \int_0^{2\pi} \frac{f(z_0 + re^{it})}{z_0 + re^{it} - z} e^{it} \, dt \quad \text{for } z \in \mathbb{D}(z_0, r)$$

and, in particular,

$$f(z_0) = \frac{1}{2\pi} \int_0^{2\pi} f(z_0 + re^{it}) \, dt \ .$$

(b) A function f has the **mean value property** if, for every $z_0 \in U$, there is an $r_0 > 0$ such that

$$f(z_0) = \frac{1}{2\pi} \int_0^{2\pi} f(z_0 + re^{it}) \, dt \quad \text{for } r \in [0, r_0] \ .$$

It follows from (a) that *holomorphic functions have the mean value property.*

(c) If f has the mean value property, then so does $\operatorname{Re} f$, $\operatorname{Im} f$, $\bar{f}$, and λf for $\lambda \in \mathbb{C}$. If $g \in C(U, \mathbb{C})$ also has the mean value property, then so does $f + g$.

Proof   This follows from Theorems III.1.5 and III.1.10 and Corollary VI.4.10. ∎

(d) The Cauchy integral formula remains true for essentially general curves, as we will show in the next section. ∎

Analytic functions

As another consequence of the Cauchy integral formula, we now prove the fundamental theorem, which says that holomorphic functions are analytic. We proceed as follows: Suppose $z_0 \in U$ and $r > 0$ with $\overline{\mathbb{D}}(z_0, r) \subset U$. Then the function

$$\mathbb{D}(z_0, r) \to \mathbb{C} , \quad z \mapsto f(\zeta)/(\zeta - z)$$

is analytic for every $\zeta \in \partial\mathbb{D}(z_0, r)$ and admits a power series expansion. With the help of the Cauchy integral formula we can transfer this expansion to f.

5.11 Theorem   *A function f is holomorphic if and only if it is analytic. Therefore*

$$C^1(U, \mathbb{C}) = C^\omega(U, \mathbb{C}) \ .$$

Proof   (i) Suppose f is holomorphic. Let $z_0 \in U$ and $r > 0$ with $\overline{\mathbb{D}}(z_0, r) \subset U$. We choose $z \in \mathbb{D}(z_0, r)$ and set $r_0 := |z - z_0|$. Then $r_0 < r$, and

$$\frac{|z - z_0|}{|\zeta - z_0|} = \frac{r_0}{r} < 1 \quad \text{for } \zeta \in \Gamma := \partial\mathbb{D}(z_0, r) \ .$$

Therefore the geometric series $\sum a^k$ with $a := (z - z_0)/(\zeta - z_0)$ converges and has the value

$$\sum_{k=0}^{\infty} \left(\frac{z - z_0}{\zeta - z_0} \right)^k = \frac{1}{1 - (z - z_0)/(\zeta - z_0)} = \frac{\zeta - z_0}{\zeta - z} .$$

Then

$$\frac{f(\zeta)}{\zeta - z} = \frac{f(\zeta)}{\zeta - z_0} \sum_{k=0}^{\infty} \left(\frac{z - z_0}{\zeta - z_0} \right)^k = \sum_{k=0}^{\infty} \frac{f(\zeta)}{(\zeta - z_0)^{k+1}} (z - z_0)^k . \qquad (5.5)$$

Because Γ is compact, there is an $M \geq 0$ such that $|f(\zeta)| \leq M$ for $\zeta \in \Gamma$. It follows that

$$\left| \frac{f(\zeta)}{(\zeta - z_0)^{k+1}} (z - z_0)^k \right| \leq \frac{M}{r^{k+1}} r_0^k = \frac{M}{r} \left(\frac{r_0}{r} \right)^k \quad \text{for } \zeta \in \Gamma .$$

Because $r_0/r < 1$, the Weierstrass majorant criterion (Theorem V.1.6) says that the series in (5.5) converges uniformly with respect to $\zeta \in \Gamma$. Setting

$$a_k := \frac{1}{2\pi i} \int_{\Gamma} \frac{f(\zeta)}{(\zeta - z_0)^{k+1}} \, d\zeta \quad \text{for } k \in \mathbb{N} , \qquad (5.6)$$

it follows[4] from Proposition VI.4.1 that

$$\begin{aligned}
f(z) &= \frac{1}{2\pi i} \int_{\Gamma} \frac{f(\zeta)}{(\zeta - z)} \, d\zeta \\
&= \frac{1}{2\pi i} \int_{\Gamma} \sum_{k=0}^{\infty} \frac{f(\zeta)}{(\zeta - z_0)^{k+1}} (z - z_0)^k \, d\zeta \\
&= \sum_{k=0}^{\infty} \left[\frac{1}{2\pi i} \int_{\Gamma} \frac{f(\zeta)}{(\zeta - z_0)^{k+1}} \, d\zeta \right] (z - z_0)^k \\
&= \sum_{k=0}^{\infty} a_k (z - z_0)^k .
\end{aligned}$$

Because this is true for every $z_0 \in U$, we find f can be expanded in the neighborhood of every point in U in a convergent power series. Therefore f is analytic.

(ii) When f is analytic, it has continuous complex derivatives and is thus holomorphic. ∎

5.12 Corollary (Cauchy's derivative formula) *Suppose f is holomorphic, $z \in U$, and $r > 0$ with $\overline{\mathbb{D}}(z, r) \subset U$. Then*

$$f^{(n)}(z) = \frac{n!}{2\pi i} \int_{\partial \mathbb{D}(z,r)} \frac{f(\zeta)}{(\zeta - z)^{n+1}} \, d\zeta \quad \text{for } n \in \mathbb{N} .$$

---

[4]See also Exercise 4.9.

Proof From Theorem 5.11, Remark V.3.4(b), and the identity theorem for analytic functions, it follows that f is represented in $\mathbb{D}(z, r)$ by its Taylor series, that is, $f = \mathcal{T}(f, z)$. The claim now follows from (5.6) and the uniqueness theorem for power series. ∎

5.13 Remarks **(a)** Suppose f is holomorphic and $z \in U$. Then the above proof shows that the Taylor series $\mathcal{T}(f, z)$ replicates the function f in at least the largest disc that lies entirely in U.

(b) If f is holomorphic, u and v belong to $C^\infty(U, \mathbb{R})$.

Proof This follows from Theorem 5.11. ∎

(c) If f is analytic, then so is $1/f$ (in $U \backslash f^{-1}(0)$).

Proof From the quotient rule, $1/f$ has continuous derivatives in $U \backslash f^{-1}(0)$ and is therefore holomorphic. Then the claim follows from Theorem 5.11. ∎

(d) Suppose V is open in $\mathbb{C}$. If $f \colon U \to \mathbb{C}$ and $g \colon V \to \mathbb{C}$ are analytic with $f(U) \subset V$, the composition $g \circ f \colon U \to \mathbb{C}$ is also analytic.

Proof This follows from the chain rule and Theorem 5.11. ∎

Liouville's theorem

A function holomorphic on all of $\mathbb{C}$ said to be **entire**.

5.14 Theorem (Liouville) *Every bounded entire function is constant.*

Proof According to Remark 5.13(a), we have

$$f(z) = \sum_{k=0}^{\infty} \frac{f^{(k)}(0)}{k!} z^k \quad \text{for } z \in \mathbb{C} .$$

By assumption, there is an $M < \infty$ such that $|f(z)| \leq M$ for $z \in \mathbb{C}$. Thus it follows from Proposition 5.2(ii) and Corollary 5.12 that

$$\left| \frac{f^{(k)}(0)}{k!} \right| \leq \frac{M}{r^k} \quad \text{for } r > 0 .$$

For $k \geq 1$, the limit $r \to \infty$ shows that $f^{(k)}(0) = 0$. Therefore $f(z)$ equals $f(0)$ for every $z \in \mathbb{C}$. ∎

5.15 Application Liouville's theorem helps with an easy proof of the fundamental theorem of algebra, that is, *every nonconstant polynomial on $\mathbb{C}[X]$ has a zero.*

Proof We write $p \in \mathbb{C}[X]$ in the form $p(z) = \sum_{k=0}^{n} a_k z^k$ with $n \geq 1$ and $a_n \neq 0$. Because

$$p(z) = z^n \left(a_n + \frac{a_{n-1}}{z} + \cdots + \frac{a_0}{z^n} \right) ,$$

we have $|p(z)| \to \infty$ for $|z| \to \infty$. Thus there is an $R > 0$ such that $|p(z)| \geq 1$ for $z \notin R\mathbb{D}$. We assume that p has no zeros in $R\overline{\mathbb{D}}$. Because $R\overline{\mathbb{D}}$ is compact, it follows from the extreme value theorem (Corollary III.3.8) that there is a positive number ε such that $|p(z)| \geq \varepsilon$. Therefore $1/p$ is entire and satisfies $|1/p(z)| \leq \max\{1, 1/\varepsilon\}$ for $z \in \mathbb{C}$. Liouville's theorem now implies that $1/p$ is constant and thus so is p, a contradiction. ∎

The Fresnel integral

The Cauchy integral theorem also can be used to calculate real integrals whose integrands are related to holomorphic functions. We can regard integrals with real integration limits as complex line integrals on a real path. Then the greater freedom to choose the integration curve, which is guaranteed by the Cauchy integral theorem, will allow us to calculate many new integrals.

The next theorem demonstrates this method, and we will generalize the techniques in the subsequent sections. Still more can be learned from the exercises.

5.16 Proposition *The following improper Fresnel integrals converge, and*

$$\int_0^\infty \cos(t^2)\, dt = \int_0^\infty \sin(t^2)\, dt = \frac{\sqrt{2\pi}}{4} .$$

Proof The convergence of these integrals follows from Exercise VI.8.7.

Consider the entire function $z \mapsto e^{-z^2}$ along the closed piecewise C^1 curve $\Gamma = \Gamma_1 + \Gamma_2 + \Gamma_3$ that follows straight line segments from 0 to $\alpha > 0$ to $\alpha + i\alpha$ and finally back 0. From the Cauchy integral theorem, we have

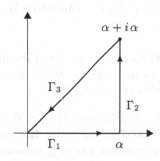

$$-\int_{\Gamma_3} e^{-z^2}\, dz = \int_{\Gamma_1} e^{-z^2}\, dz + \int_{\Gamma_2} e^{-z^2}\, dz .$$

Application VI.9.7 shows

$$\int_{\Gamma_1} e^{-z^2}\, dz = \int_0^\alpha e^{-t^2}\, dt \to \int_0^\infty e^{-t^2}\, dt = \frac{\sqrt{\pi}}{2} \quad (\alpha \to \infty) .$$

The integral over Γ_2 can be estimated as

$$\left| \int_{\Gamma_2} e^{-z^2}\, dz \right| = \left| \int_0^\alpha e^{-(\alpha+it)^2} i\, dt \right| \leq \int_0^\alpha e^{-\operatorname{Re}(\alpha+it)^2}\, dt = e^{-\alpha^2} \int_0^\alpha e^{t^2}\, dt .$$

Noting

$$\int_0^\alpha e^{t^2}\, dt \leq \int_0^\alpha e^{\alpha t}\, dt = \frac{1}{\alpha}(e^{\alpha^2} - 1) ,$$

we find

$$\left| \int_{\Gamma_2} e^{-z^2} \, dz \right| \le \frac{1}{\alpha} (1 - e^{-\alpha^2}) \to 0 \quad (\alpha \to \infty) .$$

Thus we get

$$\lim_{\alpha \to \infty} \left(- \int_{\Gamma_3} e^{-z^2} \, dz \right) = \frac{\sqrt{\pi}}{2} . \tag{5.7}$$

With the parametrization $t \mapsto t + it$ of $-\Gamma_3$, we have

$$-\int_{\Gamma_3} e^{-z^2} \, dz = \int_0^\alpha e^{-(1+i)^2 t^2} (1+i) \, dt = (1+i) \int_0^\alpha e^{-2it^2} \, dt$$

$$= (1+i) \left(\int_0^\alpha \cos(2t^2) \, dt - i \int_0^\alpha \sin(2t^2) \, dt \right) .$$

By the substitution $\sqrt{2}t = \tau$ and the limit $\alpha \to \infty$, we find using (5.7) that

$$\int_0^\infty \cos(\tau^2) \, d\tau - i \int_0^\infty \sin(\tau^2) \, d\tau = \frac{\sqrt{2}\sqrt{\pi}}{2(1+i)} = \frac{\sqrt{2\pi}}{4} (1-i) . \ \blacksquare$$

The maximum principle

We have already seen in Remark 5.10(b) that holomorphic functions have the mean value property. In particular, the absolute value of a holomorphic function f at the center of a disc cannot be larger than the maximum absolute value of f on the boundary.

5.17 Theorem (generalized maximum principle) *Suppose the function f has the mean value property. If $|f|$ has a local maximum at $z_0 \in U$, then f is constant in a neighborhood of z_0.*

Proof (i) The case $f(z_0) = 0$ is clear. Because $f(z_0) \ne 0$, there is a c such that $|c| = 1$ and $cf(z_0) > 0$. Because cf also has the mean value property, we can assume without loss of generality that $f(z_0)$ is real and positive. By assumption, there is an $r_0 > 0$ such that $\overline{\mathbb{D}}(z_0, r_0) \subset U$ and $|f(z)| \le f(z_0)$ for $z \in \overline{\mathbb{D}}(z_0, r_0)$. Also

$$f(z_0) = \frac{1}{2\pi} \int_0^{2\pi} f(z_0 + re^{it}) \, dt \quad \text{for } r \in [0, r_0] .$$

(ii) The function $h : U \to \mathbb{R}, \ z \mapsto \operatorname{Re} f(z) - f(z_0)$ satisfies $h(z_0) = 0$ and

$$h(z) \le |f(z)| - f(z_0) \le 0 \quad \text{for } z \in \mathbb{D}(z_0, r_0) .$$

According to Remark 5.10(c), h also has the mean value property. Therefore

$$0 = h(z_0) = \frac{1}{2\pi} \int_0^{2\pi} h(z_0 + re^{it}) \, dt \quad \text{for } 0 \le r \le r_0 . \tag{5.8}$$

Because $h(z_0 + re^{it}) \leq 0$ for $r \in [0, r_0]$ and $t \in [0, 2\pi]$, Proposition VI.4.8 and (5.8) imply that h vanishes identically on $\mathbb{D}(z_0, r_0)$. Therefore $\operatorname{Re} f(z) = f(z_0)$ for $z \in \mathbb{D}(z_0, r_0)$. Now it follows from $|f(z)| \leq |f(z_0)| = \operatorname{Re} f(z_0)$ that $\operatorname{Im}(f(z)) = 0$, and therefore $f(z)$ equals $f(z_0)$ for every $z \in \mathbb{D}(z_0, r_0)$. ∎

5.18 Corollary (maximum principle) *Suppose U is connected and f has the mean value property.*

  (i) *If $|f|$ has a local maximum at $z_0 \in U$, then f is constant.*

  (ii) *If U is bounded and $f \in C(\overline{U}, \mathbb{C})$, then $|f|$ assumes its maximum on ∂U, that is, there is a $z_0 \in \partial U$ such that $|f(z_0)| = \max_{z \in \overline{U}} |f(z)|$.*

Proof   (i) Suppose $f(z_0) = w_0$ and $M := f^{-1}(w_0)$. The continuity of f shows that M is closed in U (see Example III.2.22(a)). According to Theorem 5.17, every $z_1 \in M$ has a neighborhood V such that $f(z) = f(z_0) = w_0$ for $z \in V$. Therefore M is open in U. Thus it follows from Remark III.4.3 that M coincides with U. Therefore $f(z) = w_0$ for every $z \in U$.

(ii) Because f is continuous on the compact set $\overline{U}$, we know $|f|$ assumes its maximum at some point $z_0 \in \overline{U}$. If z_0 belongs to ∂U, there is nothing to prove. If z_0 lies inside U, the claim follows from (i). ∎

Harmonic functions

Suppose X is open in $\mathbb{R}^n$ and not empty. The linear map

$$\Delta : C^2(X, \mathbb{K}) \to C(X, \mathbb{K}) , \quad f \mapsto \sum_{j=1}^{n} \partial_j^2 f$$

is called the **Laplace operator** or **Laplacian** (on X). A function $g \in C^2(X, \mathbb{K})$ is **harmonic** on X if $\Delta g = 0$. We denote the set of all functions harmonic on X by $\mathcal{H}\mathrm{arm}(X, \mathbb{K})$.

5.19 Remarks   (a) We have $\Delta \in \operatorname{Hom}(C^2(X, \mathbb{K}), C(X, \mathbb{K}))$, and

$$\mathcal{H}\mathrm{arm}(X, \mathbb{K}) = \Delta^{-1}(0) .$$

Thus the harmonic functions form a vector subspace of $C^2(X, \mathbb{K})$.

(b) For $f \in C^2(X, \mathbb{C})$, we have

$$f \in \mathcal{H}\mathrm{arm}(X, \mathbb{C}) \iff \operatorname{Re} f, \operatorname{Im} f \in \mathcal{H}\mathrm{arm}(X, \mathbb{R}) .$$

(c) Every holomorphic function in U is harmonic, that is, $C^\omega(U, \mathbb{C}) \subset \mathcal{H}\mathrm{arm}(U, \mathbb{C})$.

Proof   If f is holomorphic, the Cauchy–Riemann equations imply

$$\partial_x^2 f = \partial_x \partial_y v - i \partial_x \partial_y u , \quad \partial_y^2 f = -\partial_y \partial_x v + i \partial_y \partial_x u .$$

Therefore we find $\Delta f = 0$ because of Corollary VII.5.5(ii). ∎

(d) $\mathcal{H}arm(U, \mathbb{C}) \neq C^\omega(U, \mathbb{C})$.

Proof The function $U \to \mathbb{C}$, $x + iy \mapsto x$ is harmonic but not holomorphic (see Remark 5.4(b)). ∎

The previous considerations imply that the real part of a holomorphic function is harmonic. The next theorem shows that on simply connected domains every harmonic real-valued function is the real part of a holomorphic function.

5.20 Proposition *Suppose* $u: U \to \mathbb{R}$ *is harmonic. Then we have these:*

(i) *Suppose* $V := \mathbb{D}(z_0, r) \subset U$ *for some* $(z_0, r) \in U \times (0, \infty)$. *Then there is a holomorphic function* g *in* V *such that* $u = \operatorname{Re} g$.

(ii) *If* U *is simply connected, there is a* $g \in C^\omega(U, \mathbb{C})$ *such that* $u = \operatorname{Re} g$.

Proof Because u is harmonic, the 1-form $\alpha := -u_y \, dx + u_x \, dy$ satisfies the integrability conditions. Therefore α is closed.

(i) Because V is simply connected, Theorem 4.8 says there is a $v \in C^1(V, \mathbb{R})$ such that $dv = \alpha | V$. Therefore $v_x = -u_y | V$ and $v_y = u_x | V$. Setting $g(z) := u(x, y) + i v(x, y)$, we find from Remark 5.4(b) and Theorem 5.11 that g belongs to $C^\omega(V, \mathbb{C})$.

(ii) In this case, we can replace the disc V in the proof of (i) by U. ∎

5.21 Corollary *Suppose* $u: U \to \mathbb{R}$ *is harmonic. Then*

(i) $u \in C^\infty(U, \mathbb{R})$;

(ii) u *has the mean value property;*

(iii) *if* U *is a domain and there is a nonempty subset* V *of* U *such that* $u|V = 0$, *then* $u = 0$.

Proof (i) and (ii) Suppose $V := \mathbb{D}(z_0, r) \subset U$ for some $r > 0$. Because differentiability and the mean value property are local properties, it suffices to consider u on V. Using Proposition 5.20, we find a $g \in C^\omega(V, \mathbb{C})$ such that $\operatorname{Re} g = u|V$, and the claims follow from Remark 5.13(b) and Remarks 5.10(b) and (c).

(iii) Suppose M is the set of all $z \in U$ for which there is a neighborhood V such that $u|V = 0$. Then M is open and, by assumption, not empty. Suppose $z_0 \in U$ is a cluster point of M. By Proposition 5.20, there is an $r > 0$ and a $g \in C^\omega(\mathbb{D}(z_0, r), \mathbb{C})$ such that $\operatorname{Re} g = u|\mathbb{D}(z_0, r)$. Further, $M \cap \mathbb{D}(z_0, r)$ is not empty because z_0 is a cluster point of M. For $z_1 \in M \cap \mathbb{D}(z_0, r)$, there is a neighborhood V of z_1 in U such that $u|V = 0$. Therefore it follows from Remark 5.4(d) that g is constant on $V \cap \mathbb{D}(z_0, r)$. The identity theorem for analytic functions (Theorem V.3.13) then shows that g is constant on $\mathbb{D}(z_0, r)$. Therefore $u = 0$ on $\mathbb{D}(z_0, r)$, that is, z_0 belongs to M. Consequently M is closed in U, and Remark III.4.3 implies $M = U$. ∎

5.22 Corollary (maximum and minimum principle for harmonic functions)
Suppose U is a domain and $u : U \to \mathbb{R}$ is harmonic. Then

(i) *if u has a local extremum in U, then u is constant;*

(ii) *if U is bounded and $u \in C(\overline{U}, \mathbb{R})$, then u assumes its maximum and its minimum on ∂U.*

Proof According to Corollary 5.21(ii), u has the mean value property.

(i) Suppose $z_0 \in U$ is a local extremal point of u. When $u(z_0)$ is a positive maximum of u, the claim follows from Theorem 5.17 and Corollaries 5.18 and 5.21(iii). If $u(z_0)$ is a positive minimum of u, then $z \mapsto \big[2u(z_0) - u(z)\big]$ has a positive maximum at z_0, and the claim follows as in the first case. The remaining cases can be treated similarly.

(ii) This follows from (i). ∎

5.23 Remarks (a) The set of zeros of a holomorphic function is discrete.[5] However, the set of zeros of a real-valued harmonic function is not generally discrete.

Proof The first statement follows from Theorem 5.11 and the identity theorem analytic functions (Theorem V.3.13). The second statement follows by considering the harmonic function $\mathbb{C} \to \mathbb{R}$, $x + iy \mapsto x$. ∎

(b) One can show that a function is harmonic if and only if has the mean value property (see for example [Con78, Theorem X.2.11]). ∎

Goursat's theorem

We will now prove, as promised in Remark 5.4(a), that every complex differentiable function has continuous derivatives and is therefore holomorphic. To prepare, we first prove Morera's theorem, which gives criterion for proving a function is holomorphic. This criterion is useful elsewhere, as we shall see.

Suppose $X \subset \mathbb{C}$. Every closed path Δ of line segments in X that has exactly three sides is called a **triangular path in X** if the closure of the triangle bounded by Δ lies in X.

5.24 Theorem (Morera) *Suppose the function f satisfies $\int_\Delta f \, dz = 0$ for every triangular path Δ in U. Then f is analytic.*

Proof Suppose $a \in U$ and $r > 0$ with $\mathbb{D}(a, r) \subset U$. It suffices to verify that $f \,|\, \mathbb{D}(a, r)$ is analytic. So suppose $z_0 \in \mathbb{D}(a, r)$ and

$$ F : \mathbb{D}(a, r) \to \mathbb{C} , \quad z \mapsto \int_{[\![a, z]\!]} f(w) \, dw . $$

---

[5]A subset D of a metric space X is said to be **discrete** if every $d \in D$ has a neighborhood U in X such that $U \cap D = \{d\}$.

Our assumptions imply the identity

$$F(z) = \int_{[\![a,z_0]\!]} f(w)\, dw + \int_{[\![z_0,z]\!]} f(w)\, dw = F(z_0) + \int_{[\![z_0,z]\!]} f(w)\, dw \ .$$

Therefore

$$\frac{F(z) - F(z_0)}{z - z_0} - f(z_0) = \frac{1}{z - z_0} \int_{[\![z_0,z]\!]} \big(f(w) - f(z_0)\big)\, dw \quad \text{for } z \neq z_0 \ . \qquad (5.9)$$

Suppose now $\varepsilon > 0$. Then there is a $\delta \in (0, r - |z_0 - a|)$ such that $|f(w) - f(z_0)| < \varepsilon$ for $w \in \mathbb{D}(z_0, \delta)$. Thus it follows from (5.9) that

$$\left| \frac{F(z) - F(z_0)}{z - z_0} - f(z_0) \right| < \varepsilon \quad \text{for } 0 < |z - z_0| < \delta \ .$$

Therefore $F'(z_0) = f(z_0)$, which shows that F has continuous derivatives. Then Theorem 5.11 shows F belongs to $C^\omega\big(\mathbb{D}(a,r), \mathbb{C}\big)$, and we find that $F' = f\,|\,\mathbb{D}(a,r)$ belongs to $C^\omega\big(\mathbb{D}(a,r), \mathbb{C}\big)$. ∎

5.25 Theorem (Goursat) *Suppose f is differentiable. Then $\int_\Delta f\, dz = 0$ for every triangular path Δ in U.*

Proof (i) Suppose Δ is a triangular path in U. Without loss of generality, we can assume that Δ bounds a triangle with positive area. Then Δ has the vertices z_0, z_1, and z_2, and $z_2 \in [\![z_0, z_1]\!]$. Thus $\int_\Delta f(z)\, dz = 0$, as is easily verified. We denote by K the closure of the triangle bounded by Δ.

By connecting the midpoints of the three sides of Δ, we get four congruent closed subtriangles $K_1, \ldots, K_4$ of K. We orient the (topological) boundaries of the K_j so that shared sides are oppositely oriented and denote the resulting triangular paths by $\Delta_1, \ldots, \Delta_4$. Then

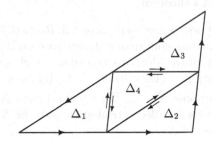

$$\left| \int_\Delta f(z)\, dz \right| = \left| \sum_{j=1}^{4} \int_{\Delta_j} f(z)\, dz \right| \leq 4 \max_{1 \leq j \leq 4} \left| \int_{\Delta_j} f(z)\, dz \right| \ .$$

Of the four triangular paths $\Delta_1, \ldots, \Delta_4$, there is one, which we call Δ^1, that satisfies

$$\left| \int_{\Delta^1} f(z)\, dz \right| = \max_{1 \leq j \leq 4} \left| \int_{\Delta_j} f(z)\, dz \right| \ .$$

Then

$$\left| \int_\Delta f(z)\, dz \right| \leq 4 \left| \int_{\Delta^1} f(z)\, dz \right| \ .$$

(ii) To Δ^1, we apply the same decomposition and selection procedure as we did for Δ. Thus we inductively obtain a sequence (Δ^n) of triangular paths and corresponding closed triangles K^n such that

$$K^1 \supset K^2 \supset \cdots \supset K^n \supset \cdots \text{ and } \left| \int_{\Delta^n} f(z)\, dz \right| \le 4 \left| \int_{\Delta^{n+1}} f(z)\, dz \right| \qquad (5.10)$$

for $n \in \mathbb{N}^\times$. Clearly $\{\, K^n \;;\; n \in \mathbb{N}^\times \,\}$ has the finite intersection property. Therefore it follows from the compactness of K^1 that $\bigcap_n K^n$ is not empty (see Exercise III.3.5). We fix a z_0 in $\bigcap K^n$.

(iii) The inequality in (5.10) implies

$$\left| \int_\Delta f(z)\, dz \right| \le 4^n \left| \int_{\Delta^n} f(z)\, dz \right| . \qquad (5.11)$$

In addition, we have the elementary geometric relations

$$L(\Delta^{n+1}) = L(\Delta^n)/2 \text{ and } \mathrm{diam}(K^{n+1}) = \mathrm{diam}(K^n)/2 \quad \text{for } n \in \mathbb{N}^\times .$$

From this follows

$$L(\Delta^n) = \ell/2^n \text{ and } \mathrm{diam}(K^n) = d/2^n \quad \text{for } n \in \mathbb{N}^\times ,$$

where $\ell := L(\Delta)$ and $d := \mathrm{diam}(K)$.

(iv) Suppose $\varepsilon > 0$. The differentiability of f at z_0 implies the existence of a $\delta > 0$ such that $\mathbb{D}(z_0, \delta) \subset U$ and

$$|f(z) - f(z_0) - f'(z_0)(z - z_0)| \le \frac{\varepsilon}{d\ell} |z - z_0| \quad \text{for } z \in \mathbb{D}(z_0, \delta) .$$

We now choose an $n \in \mathbb{N}^\times$ with $\mathrm{diam}(K^n) = d/2^n < \delta$. Because $z_0 \in K^n$, we then have $\Delta^n \subset \mathbb{D}(z_0, \delta)$. The Cauchy integral theorem implies

$$0 = \int_{\Delta^n} dz = \int_{\Delta^n} z\, dz .$$

Thus

$$\left| \int_{\Delta^n} f(z)\, dz \right| = \left| \int_{\Delta^n} \left(f(z) - f(z_0) - f'(z_0)(z - z_0) \right) dz \right|$$

$$\le \frac{\varepsilon}{d\ell} \max_{z \in \Delta^n} |z - z_0|\, L(\Delta^n) \le \frac{\varepsilon}{d\ell} \mathrm{diam}(K^n) L(\Delta^n) = \frac{\varepsilon}{4^n} .$$

Now (5.11) finishes the proof. ∎

5.26 Corollary *If f is differentiable, f is holomorphic.*

Proof This follows from Theorems 5.24 and 5.25. ∎

The Weierstrass convergence theorem

As a further application of Morera's theorem, we prove a theorem due to Weierstrass concerning the limit of a locally uniformly convergent sequence of holomorphic functions. We have already applied this result — in combination with Theorem 5.11 — in Application VI.7.23(b) to prove the product representation of the sine.

5.27 Theorem (Weierstrass convergence theorem) *Suppose* (g_n) *is a locally uniformly convergent sequence of holomorphic functions in* U. *Then* $g := \lim g_n$ *is holomorphic in* U.

Proof According to Theorem V.2.1, g is continuous. From Remark V.2.3(c) we know that (g_n) is uniformly convergent on every compact subset of U. Therefore (g_n) to g converges uniformly on every triangular path Δ in U. Therefore Proposition VI.4.1(i) implies

$$\int_\Delta g\,dz = \lim_{n \to \infty} \int_\Delta g_n\,dz = 0 \,, \tag{5.12}$$

where the last equality follows from the Cauchy integral theorem applied to g_n. Because (5.12) holds for every triangular path in U, Morera's theorem finishes the proof. ∎

Exercises

1 If $f : U \to \mathbb{C}$ is real differentiable,

$$\partial_W f := \frac{1}{2}(\partial_x f - i\partial_y f) \quad \text{and} \quad \overline{\partial}_W f := \frac{1}{2}(\partial_x f + i\partial_y f)$$

are called the **Wirtinger derivatives**[6] of f.
Show

(a) $\overline{\partial}_W f = \overline{\partial_W \overline{f}}$, and $\overline{\partial}_W \overline{f} = \overline{\partial_W f}$;

(b) f is holomorphic $\iff \overline{\partial}_W f = 0$;

(c) $4\partial_W \overline{\partial}_W f = 4\overline{\partial}_W \partial_W f = \Delta f$ if f is twice real differentiable;

(d) $\det \begin{bmatrix} u_x & u_y \\ v_x & v_y \end{bmatrix} = \det \begin{bmatrix} \partial_W f & \overline{\partial}_W f \\ \overline{\partial_W f} & \overline{\overline{\partial}_W f} \end{bmatrix} = |\partial_W f|^2 - |\overline{\partial}_W f|^2.$

2 Suppose $d\bar{z} := d(z \mapsto \bar{z})$. Then $d\bar{z} = dx - i\,dy$ and

$$df = \partial_x f\,dx + \partial_y f\,dy = f'\,dz = \partial_W f\,dz + \overline{\partial}_W f\,d\bar{z} \quad \text{for } f \in C^1(U, \mathbb{C}) \,.$$

---

[6]Usually, the Wirtinger derivative will be denoted by ∂ (or $\overline{\partial}$). However, when this notation might collide with our notation ∂ for the derivative operator, we will write ∂_W (or respectively $\overline{\partial}_W$).

3 The 1-form $\alpha \in \Omega_{(1)}(U, \mathbb{C})$ is said to be **holomorphic** if every $z_0 \in U$ has a neighborhood V such that there is an $f \in C^1(V, \mathbb{C})$ with $df = \alpha|V$.

(a) Show these statements are equivalent:

  (i) α is holomorphic.

 (ii) There is a real differentiable function $a \in C(U, \mathbb{C})$ such that $\alpha = a \, dz$ and α is closed.

(iii) There is an $a \in C^1(U, \mathbb{C})$ such that $\alpha = a \, dz$.

(b) Show that
$$\alpha := \frac{x \, dx + y \, dy}{x^2 + y^2} + i \frac{x \, dy - y \, dx}{x^2 + y^2}$$
is a holomorphic 1-form in $\mathbb{C}^\times$. Is α is globally exact, that is, is there a holomorphic function f in $\mathbb{C}^\times$ such that $\alpha = f \, dz$?

4 Suppose U is connected and $u : U \to \mathbb{R}$ is harmonic. If $v \in C^1(U, \mathbb{R})$ satisfies the relations $v_x = -u_y$ and $v_y = u_x$, we say that v is **conjugate** to u. Prove

(a) if v is conjugate to u, then v is harmonic;

(b) if U is simply connected, then to every function harmonic in U there is a harmonic conjugate. (Hint: Consider $u_x - i u_y$.)

5 Prove the Weierstrass convergence theorem using the Cauchy integral formula. Also show that, under the assumptions of Theorem 5.27, for every $k \in \mathbb{N}$ the sequence $(f_n^{(k)})_{n \in \mathbb{N}}$ of k-th derivatives converges locally uniformly on U to $f^{(k)}$. (Hint: Proposition VII.6.7).

6 Suppose $z_0 \in U$ and $r > 0$ with $\overline{\mathbb{D}}(z_0, r) \subset U$. Further suppose $z \in \mathbb{D}(z_0, r)$ and $\delta > 0$ with $\mathbb{D}(z, \delta) \subset U$. Verify that $\partial \mathbb{D}(z, \delta)$ and $\partial \mathbb{D}(z_0, r)$ are homotopic in $U \backslash \{z\}$.

7 With the help of the Cauchy integral theorem, show $\int_{-\infty}^{\infty} dx/(1 + x^2) = \pi$.

8 Suppose $p = \sum_{k=0}^{n} a_k X^k \in \mathbb{C}[X]$ is such that $a_n = 1$. Let $R := \max\{1, 2 \sum_{k=0}^{n-1} |a_k|\}$. Show
$$|z|^n / 2 \le |p(z)| \le 2 |z|^n \quad \text{for } z \in R\mathbb{D}^c .$$

9 Suppose $0 \le r_0 < r < r_1 < \infty$ and f is holomorphic in $\mathbb{D}(z_0, r_1)$. Prove that
$$|f^{(n)}(z)| \le \frac{r n!}{(r - r_0)^{n+1}} \max_{w \in \partial \mathbb{D}(z_0, r)} |f(w)| \quad \text{for } z \in \mathbb{D}(z_0, r_0) \text{ and } n \in \mathbb{N} .$$

10 Given an entire function f, suppose there are $M, R > 0$ and an $n \in \mathbb{N}$ such that $|f(z)| \le M |z|^n$ for $z \in R\mathbb{D}^c$. Show f is a polynomial with $\deg(f) \le n$. (Hint: Exercise 9.)

11 Suppose Γ_1 and Γ_2 are compact piecewise C^1 curves in U and $f \in C(U \times U, \mathbb{C})$. Show
$$\int_{\Gamma_2} \left(\int_{\Gamma_1} f(w, z) \, dw \right) dz = \int_{\Gamma_1} \left(\int_{\Gamma_2} f(w, z) \, dz \right) dw .$$
(Hint: Exercise VII.6.2).

12 Suppose Γ is a compact piecewise C^1 curve in U and $f \in C(U \times U, \mathbb{C})$. Also suppose $f(w, \cdot)$ is holomorphic for every $w \in U$. Show that
$$F : U \to \mathbb{C} , \quad z \mapsto \int_{\Gamma} f(w, z) \, dw$$

is holomorphic and that $F' = \int_\Gamma \partial_2 f(w, \cdot)\, dw$.
(Hint: Exercise 11, Morera's theorem, and Proposition VII.6.7.)

13 Suppose $L = a + \mathbb{R}b$, $a, b \in \mathbb{C}$ is a straight line in $\mathbb{C}$, and $f \in C(U, \mathbb{C})$ is holomorphic in $U \backslash L$. Show f is holomorphic in all of U. (Hint: Morera's theorem.)

14 Suppose U is connected and f is holomorphic in U. Prove the following.

(a) If $|f|$ has a global minimum at z_0, then either f is constant or $f(z_0) = 0$.

(b) Suppose U is bounded and $f \in C(\overline{U}, \mathbb{C})$. Show either that f has a zero in U or that $|f|$ assumes its minimum on ∂U.

15 For $R > 0$,

$$P_R : R\partial\mathbb{D} \times R\mathbb{D} \to \mathbb{C}, \quad (\zeta, z) \mapsto \frac{R^2 - |z|^2}{|\zeta - z|^2}$$

is called the **Poisson kernel** for $R\mathbb{D}$.

Prove

(a) $P_R(\zeta, z) = \mathrm{Re}((\zeta + z)/(\zeta - z))$ for $(\zeta, z) \in R\partial\mathbb{D} \times R\mathbb{D}$;

(b) for every $\zeta \in R\partial\mathbb{D}$, the function $P_R(\zeta, \cdot)$ is harmonic in $R\mathbb{D}$;

(c) for $r \in [0, R]$ and $t, \theta \in [0, 2\pi)$,

$$P_R(Re^{i\theta}, re^{it}) = \frac{R^2 - r^2}{R^2 - 2Rr\cos(\theta - t) + r^2} \;;$$

(d) $P_1(1, re^{it}) = \sum_{n=-\infty}^{\infty} r^{|n|} e^{int}$ for $r \in [0, 1)$, and $t \in \mathbb{R}$;

(e) $\int_0^{2\pi} P_R(Re^{i\theta}, z)\, d\theta = 2\pi$ for $z \in R\mathbb{D}$.

(Hint for (d): $\sum_{k=0}^{\infty}(re^{it})^k = 1/(1 - re^{it})$.)

16 Suppose $\rho > 1$ and f is holomorphic in $\rho\mathbb{D}$. Show that

$$f(z) = \frac{1}{2\pi} \int_0^{2\pi} P_1(e^{i\theta}, z) f(e^{i\theta})\, d\theta \quad \text{for } z \in \mathbb{D} .$$

(Hints: (i) For $g \in C^1(\rho_0\mathbb{D}, \mathbb{C})$ with $\rho_0 > 1$, show

$$g(z) = \frac{1}{2\pi} \int_0^{2\pi} \frac{g(e^{i\theta})}{1 - e^{-i\theta}z}\, d\theta \quad \text{for } z \in \mathbb{D} .$$

(ii) For $z \in \mathbb{D}$, $\rho_0 := \min(\rho, 1/|z|)$, $g : \rho_0\mathbb{D} \to \mathbb{C}$, show $w \mapsto f(w)/(1 - w\bar{z})$ is holomorphic.)

17 Show these:

(a) For $g \in C(\partial\mathbb{D}, \mathbb{R})$,

$$\mathbb{D} \to \mathbb{C}, \quad z \mapsto \int_0^{2\pi} P_1(e^{i\theta}, z) g(e^{i\theta})\, d\theta$$

is harmonic.

(b) If $f \in C(\overline{\mathbb{D}}, \mathbb{R})$ is harmonic in $\mathbb{D}$, then

$$f(z) = \frac{1}{2\pi} \int_0^{2\pi} P_1(e^{i\theta}, z) f(e^{i\theta}) \, d\theta \quad \text{for } z \in \mathbb{D} .$$

(Hints: (a) Exercise 15(b) and Proposition VII.6.7. (b) Suppose $0 < r_k < 1$ with $\lim r_k = 1$ and $f_k(z) := f(r_k z)$ for $z \in r_k^{-1}\mathbb{D}$. Exercise 16 gives

$$f_k(z) = \frac{1}{2\pi} \int_0^{2\pi} P_1(e^{i\theta}, z) f_k(e^{i\theta}) \, dz \quad \text{for } z \in \mathbb{D} .$$

Now consider the limit $k \to \infty$.)

18 Suppose $a \in \mathbb{C}$ and $\alpha \neq \operatorname{Re} a$. Show that for $\gamma_\alpha : \mathbb{R} \to \mathbb{C}$, $s \mapsto \alpha + is$, we have

$$e^{ta} = \frac{1}{2\pi i} \int_{\gamma_\alpha} e^{\lambda t} (\lambda - a)^{-1} \, d\lambda \quad \text{for } t > 0 .$$

(Hint: The Cauchy integral formula gives

$$e^{ta} = \frac{1}{2\pi i} \int_{\partial \mathbb{D}(a,r)} e^{\lambda t} (\lambda - a)^{-1} \, d\lambda \quad \text{for } t \in \mathbb{R} \text{ and } r > 0 .$$

Now apply the Cauchy integral theorem.)

6   Meromorphic functions

In central sections of this chapter, we explored complex functions that were holomorphic except at isolated points. Typical examples are

$$\mathbb{C}^\times \to \mathbb{C}\,, \quad z \mapsto e^{1/z}\,, \qquad \mathbb{C}\backslash\{\pm i\} \to \mathbb{C}\,, \quad z \mapsto 1/(1+z^2)\,,$$
$$\mathbb{C}^\times \to \mathbb{C}\,, \quad z \mapsto \sin(z)/z\,, \qquad \mathbb{C}\backslash\pi\mathbb{Z} \to \mathbb{C}\,, \quad z \mapsto \cot z\,.$$

It turns out that these "exceptional points" lead to an amazingly simple classification. This classification is aided by the Laurent series, which expands a function as a power series in both positive and negative coefficients. It thus generalizes the Taylor series to one that can expand holomorphic functions.

In this context, we will also extend the Cauchy integral theorem and prove the residue theorem, which has many important applications, of which we give only a few.

The Laurent expansion

For $c := (c_n) \in \mathbb{C}^\mathbb{Z}$, we consider the power series

$$nc := \sum_{n \geq 0} c_n X^n \text{ and } hc := \sum_{n \geq 1} c_{-n} X^n \,.$$

Suppose their convergence radii are ρ_1 and $1/\rho_0$, and $0 \leq \rho_0 < \rho_1 \leq \infty$. Then the function $\underline{nc}$ [or $\underline{hc}$] represented by nc [or hc] in $\rho_1 \mathbb{D}$ [or $(1/\rho_0)\mathbb{D}$] is holomorphic by Theorem V.3.1. Because $z \mapsto 1/z$ is holomorphic in $\mathbb{C}^\times$ and $|1/z| < 1/\rho_0$ for $|z| > \rho_0$, Remark 5.13(d) guarantees that the function

$$z \mapsto \underline{hc}(1/z) = \sum_{n=1}^{\infty} c_{-n} z^{-n}$$

is holomorphic at $|z| > \rho_0$. Therefore

$$z \mapsto \sum_{n=-\infty}^{\infty} c_n z^n := \sum_{n=0}^{\infty} c_n z^n + \sum_{n=1}^{\infty} c_{-n} z^{-n}$$

is a holomorphic function in the annulus around 0 given by

$$\Omega(\rho_0, \rho_1) := \rho_1 \mathbb{D} \backslash \rho_0 \overline{\mathbb{D}} = \{\, z \in \mathbb{C} \,;\, \rho_0 < |z| < \rho_1 \,\}\,.$$

Suppose now $z_0 \in \mathbb{C}$. Then

$$\sum_{n \in \mathbb{Z}} c_n (z - z_0)^n := \sum_{n \geq 0} c_n (z - z_0)^n + \sum_{n \geq 1} c_{-n} (z - z_0)^{-n} \qquad (6.1)$$

is called the **Laurent series** about the **expansion point** z_0, and (c_n) is the sequence of coefficients. In (6.1), $\sum_{n\geq 1} c_{-n}(z-z_0)^{-n}$ is the **principal part** and $\sum_{n\geq 0} c_n(z-z_0)^n$ is the **auxiliary part**. The Laurent series (6.1) is **convergent** [or **norm convergent**] in $M \subset \mathbb{C}$ if both the principal part and the auxiliary part are convergent [or norm convergent] in M. Its **value** at $z \in M$ is by definition the sum of the value of the principal and auxiliary parts in z, that is,

$$\sum_{n=-\infty}^{\infty} c_n(z-z_0)^n := \sum_{n=0}^{\infty} c_n(z-z_0)^n + \sum_{n=1}^{\infty} c_{-n}(z-z_0)^{-n} .$$

From these preliminaries and Theorem V.1.8, it follows that the Laurent series $\sum_{n\in\mathbb{Z}} c_n(z-z_0)^n$ converges normally in every compact subset of the annulus around z_0 given by

$$z_0 + \Omega(\rho_0, \rho_1) = \{ z \in \mathbb{C} \, ; \, \rho_0 < |z-z_0| < \rho_1 \}$$

and that the function

$$z_0 + \Omega(\rho_0, \rho_1) \to \mathbb{C} , \quad z \mapsto \sum_{n=-\infty}^{\infty} c_n(z-z_0)^n$$

is holomorphic. The considerations above show that, conversely, every function holomorphic in an annulus can be represented by a Laurent series.

6.1 Lemma *For $\rho > 0$ and $a \in \mathbb{C}$,*

$$\int_{\rho\partial\mathbb{D}} \frac{dz}{z-a} = \begin{cases} 2\pi i \, , & |a| < \rho \, , \\ 0 \, , & |a| > \rho \, . \end{cases}$$

Proof (i) Suppose $|a| < \rho$ and $\delta > 0$ with $\mathbb{D}(a,\delta) \subset \rho\mathbb{D}$. Because $\partial\mathbb{D}(a,\delta)$ and $\rho\partial\mathbb{D}$ are homotopic in $\mathbb{C}\backslash\{a\}$ (see Exercise 5.6), the claim follows from Proposition 5.7 and Example 5.3(a).

(ii) If $|a| > \rho$, then $\rho\partial\mathbb{D}$ is null homotopic in $\mathbb{C}\backslash\{a\}$, and the claim again follows from Proposition 5.7. ∎

6.2 Lemma *Suppose $f : \Omega(r_0, r_1) \to \mathbb{C}$ is holomorphic.*

(i) *For $r, s \in (r_0, r_1)$, we have*

$$\int_{r\partial\mathbb{D}} f(z) \, dz = \int_{s\partial\mathbb{D}} f(z) \, dz .$$

(ii) *Suppose $a \in \Omega(\rho_0, \rho_1)$ with $r_0 < \rho_0 < \rho_1 < r_1$. Then*

$$f(a) = \frac{1}{2\pi i} \int_{\rho_1\partial\mathbb{D}} \frac{f(z)}{z-a} \, dz - \frac{1}{2\pi i} \int_{\rho_0\partial\mathbb{D}} \frac{f(z)}{z-a} \, dz .$$

Proof (i) Because $r\partial\mathbb{D}$ and $s\partial\mathbb{D}$ are homotopic in $\Omega := \Omega(r_0, r_1)$, the claim follows from Proposition 5.7.

(ii) Suppose $g : \Omega \to \mathbb{C}$ is defined through

$$g(z) := \begin{cases} \big(f(z) - f(a)\big)/(z - a) , & z \in \Omega \backslash \{a\} , \\ f'(a) , & z = a . \end{cases}$$

Obviously g is holomorphic in $\Omega \backslash \{a\}$ with

$$g'(z) = \frac{f'(z)(z - a) - f(z) + f(a)}{(z - a)^2} \quad \text{for } z \in \Omega \backslash \{a\} . \tag{6.2}$$

With Taylor's formula (Corollary IV.3.3)

$$f(z) = f(a) + f'(a)(z - a) + \frac{1}{2} f''(a)(z - a)^2 + o\big(|z - a|^2\big) \tag{6.3}$$

as $z \to a$, we find

$$\frac{g(z) - g(a)}{z - a} = \frac{1}{z - a} \Big(\frac{f(z) - f(a)}{z - a} - f'(a) \Big) = \frac{1}{2} f''(a) + \frac{o\big(|z - a|^2\big)}{(z - a)^2}$$

as $z \to a$ in $\Omega \backslash \{a\}$. Thus g is differentiable at a with $g'(a) = f''(a)/2$. Therefore g is holomorphic in Ω. The claim follows by applying Lemma 6.1 and (i) to g. ∎

Now we can prove the aforementioned expansion theorem.

6.3 Theorem (Laurent expansion) *Every function f holomorphic in $\Omega := \Omega(r_0, r_1)$ has a unique Laurent expansion*

$$f(z) = \sum_{n=-\infty}^{\infty} c_n z^n \quad \text{for } z \in \Omega . \tag{6.4}$$

The Laurent series converges normally on every compact subset of Ω, and its coefficients are given by

$$c_n = \frac{1}{2\pi i} \int_{r\partial\mathbb{D}} \frac{f(\zeta)}{\zeta^{n+1}} \, d\zeta \quad \text{for } n \in \mathbb{Z} \text{ and } r_0 < r < r_1 . \tag{6.5}$$

Proof (i) We first verify that f is represented by the Laurent series with the coefficients given in (6.5). From Lemma 6.2(i) it follows that c_n is well defined for $n \in \mathbb{Z}$, that is, it is independent of r.

Suppose $r_0 < s_0 < s_1 < r_1$ and $z \in \Omega(s_0, s_1)$. For $\zeta \in \mathbb{C}$ with $|\zeta| = s_1$, we have $|z/\zeta| < 1$, and thus

$$\frac{1}{\zeta - z} = \frac{1}{\zeta} \cdot \frac{1}{1 - z/\zeta} = \sum_{n=0}^{\infty} \frac{z^n}{\zeta^{n+1}} ,$$

with normal convergence on $s_1 \partial \mathbb{D}$. Therefore we have

$$\frac{1}{2\pi i} \int_{s_1 \partial \mathbb{D}} \frac{f(\zeta)}{\zeta - z} \, d\zeta = \sum_{n=0}^{\infty} c_n z^n$$

(see Exercise 4.9). For $\zeta \in \mathbb{C}$ with $|\zeta| = s_0$, we have $|\zeta/z| < 1$, and therefore

$$\frac{1}{\zeta - z} = -\frac{1}{z} \cdot \frac{1}{1 - \zeta/z} = -\sum_{m=0}^{\infty} \frac{\zeta^m}{z^{m+1}} \ ,$$

with normal convergence on $s_0 \partial \mathbb{D}$. Therefore

$$\frac{1}{2\pi i} \int_{s_0 \partial \mathbb{D}} \frac{f(\zeta)}{\zeta - z} \, d\zeta = -\sum_{m=0}^{\infty} \left(\frac{1}{2\pi i} \int_{s_0 \partial \mathbb{D}} f(\zeta) \zeta^m \, d\zeta \right) z^{-m-1}$$

$$= -\sum_{n=1}^{\infty} c_{-n} z^{-n} \ .$$

Thus we get from Lemma 6.2(ii) the representation (6.4).

(ii) Now we prove the coefficients in (6.4) are unique. Suppose $f(z) = \sum_{n=-\infty}^{\infty} a_n z^n$ converges normally on compact subsets of Ω. For $r \in (r_0, r_1)$ and $m \in \mathbb{Z}$, we have (see Example 5.3(a))

$$\frac{1}{2\pi i} \int_{r \partial \mathbb{D}} f(z) z^{-m-1} \, dz = \frac{1}{2\pi i} \sum_{n=-\infty}^{\infty} a_n \int_{r \partial \mathbb{D}} z^{n-m-1} \, dz = a_m \ .$$

This shows $a_m = c_m$ for $m \in \mathbb{Z}$.

(iii) Because we already know that the Laurent series converges normally on compact subsets of Ω, the theorem is proved. ∎

As a simple consequence of this theorem, we obtain the Laurent expansion of a function holomorphic in the open **punctured disc**

$$\mathbb{D}^{\bullet}(z_0, r) := z_0 + \Omega(0, r) = \{ z \in \mathbb{C} \ ; \ 0 < |z - z_0| < r \} \ .$$

6.4 Corollary *Suppose f is holomorphic in $\mathbb{D}^{\bullet}(z_0, r)$. Then f has a unique Laurent expansion*

$$f(z) = \sum_{n=-\infty}^{\infty} c_n (z - z_0)^n \quad \text{for } z \in \mathbb{D}^{\bullet}(z_0, r) \ ,$$

where

$$c_n := \frac{1}{2\pi i} \int_{\partial \mathbb{D}(z_0, \rho)} \frac{f(z)}{(z - z_0)^{n+1}} \, dz \quad \text{for } n \in \mathbb{Z} \text{ and } \rho \in (0, r) \ .$$

The series converges normally on every compact subset of $\mathbb{D}^\bullet(z_0, r)$, *and*

$$|c_n| \leq \rho^{-n} \max_{z \in \partial\mathbb{D}(z_0, \rho)} |f(z)| \quad \text{for } n \in \mathbb{Z} \text{ and } \rho \in (0, r) \ . \tag{6.6}$$

Proof With the exception of (6.6), all the statements follow from Theorem 6.3 applied to $z \mapsto f(z + z_0)$. The estimate (6.6) is implied by (6.5) and Proposition 5.2(ii). ∎

Removable singularities

In the following suppose

- U is an open subset of $\mathbb{C}$ and z_0 is a point in U.

Given a holomorphic function $f : U \backslash \{z_0\} \to \mathbb{C}$, the point z_0 is a **removable singularity** if f has a holomorphic extension $F : U \to \mathbb{C}$. When there is no fear of confusion, we reuse the symbol f for this extension.

6.5 Example Let $f : U \to \mathbb{C}$ be holomorphic. Then z_0 is removable singularity of

$$g : U \backslash \{z_0\} \to \mathbb{C} \ , \quad z \mapsto \big(f(z) - f(z_0)\big)/(z - z_0) \ .$$

In particular, 0 is a removable singularity of

$$z \mapsto \sin(z)/z \ , \quad z \mapsto (\cos z - 1)/z \ , \quad z \mapsto \log(z + 1)/z \ .$$

Proof This follows from the proof of Lemma 6.2(ii). ∎

Removable singularities of a function f can be characterized by the local boundedness of f:

6.6 Theorem (Riemann's removability theorem) *Suppose* $f : U \backslash \{z_0\} \to \mathbb{C}$ *is holomorphic. Then the point* z_0 *is a removable singularity of* f *if and only if* f *is bounded in a neighborhood of* z_0.

Proof Suppose $r > 0$ with $\overline{\mathbb{D}}(z_0, r) \subset U$.

If z_0 is a removable singularity of f, there is an $F \in C^\omega(U)$ such that $F \supset f$. From the compactness of $\overline{\mathbb{D}}(z_0, r)$, it follows that

$$\sup_{z \in \mathbb{D}^\bullet(z_0, r)} |f(z)| = \sup_{z \in \mathbb{D}^\bullet(z_0, r)} |F(z)| \leq \max_{z \in \overline{\mathbb{D}}(z_0, r)} |F(z)| < \infty \ .$$

Therefore f is bounded in the neighborhood $\mathbb{D}^\bullet(z_0, r)$ of z_0 in $U \backslash \{z_0\}$.

To prove the converse, we set

$$M(\rho) := \max_{z \in \partial\mathbb{D}(z_0, \rho)} |f(z)| \quad \text{for } \rho \in (0, r) \ .$$

By assumption, there is an $M \geq 0$ such that $M(\rho) \leq M$ for $\rho \in (0, r)$ (because $|f|$ is continuous on $\overline{\mathbb{D}}(z_0, r) \backslash \{z_0\}$ and thus is bounded for every $0 < r_0 < r$ on the compact set $\overline{\mathbb{D}}(z_0, r) \backslash \mathbb{D}(z_0, r_0)$). Thus it follows from (6.6) that

$$|c_n| \leq M(\rho)\rho^{-n} \leq M\rho^{-n} \quad \text{for } n \in \mathbb{Z} \text{ and } \rho \in (0, r) .$$

Therefore the principal part of Laurent expansion of f vanishes, and from Corollary 6.4 it follows that

$$f(z) = \sum_{n=0}^{\infty} c_n(z - z_0)^n \quad \text{for } z \in \mathbb{D}^{\bullet}(z_0, r) .$$

The function defined through

$$z \mapsto \sum_{n=0}^{\infty} c_n(z - z_0)^n \quad \text{for } z \in \mathbb{D}(z_0, r) ,$$

is holomorphic on $\mathbb{D}(z_0, r)$ and agrees with f on $\mathbb{D}^{\bullet}(z_0, r)$. Therefore z_0 is a removable singularity of f. ∎

Isolated singularities

Suppose $f \colon U \backslash \{z_0\} \to \mathbb{C}$ is holomorphic and $r > 0$ with $\overline{\mathbb{D}}(z_0, r) \subset U$. Further suppose

$$f(z) = \sum_{n=-\infty}^{\infty} c_n(z - z_0)^n \quad \text{for } z \in \mathbb{D}^{\bullet}(z_0, r)$$

is the Laurent expansion of f in $\mathbb{D}^{\bullet}(z_0, r)$. If z_0 is a singularity of f, it is **isolated** if it is not removable. Due to (the proof of) Riemann's removability theorem, this is the case if and only if the principal part of the Laurent expansion of f does not vanish identically. If z_0 is an isolated singularity of f, we say z_0 is a **pole** of f if there is an $m \in \mathbb{N}^{\times}$ such that $c_{-m} \neq 0$ and $c_{-n} = 0$ for $n > m$. In this case, m is the **order** of the pole. If infinitely many coefficients of the principal part of the Laurent series are different from zero, z_0 is an **essential singularity** of f. Finally, we define the **residue** of f at z_0 through

$$\text{Res}(f, z_0) := c_{-1} .$$

A function g is said to be **meromorphic** in U if there is a closed subset $P(g)$ of U on which g is holomorphic in $U \backslash P(g)$ and every $z \in P(g)$ is a pole of g.[1] Then $P(g)$ is the **set of poles** of g.

---

[1] $P(g)$ can also be empty. Therefore every holomorphic function on U is also meromorphic.

6.7 Remarks   **(a)** Suppose $f : U \setminus \{z_0\} \to \mathbb{C}$ is holomorphic. Then we have

$$\operatorname{Res}(f, z_0) = \frac{1}{2\pi i} \int_{\partial \mathbb{D}(z_0, r)} f(z) \, dz$$

for every $r > 0$ such that $\overline{\mathbb{D}}(z_0, r) \subset U$. The residue of f at z_0 is therefore (up to the factor $1/2\pi i$) what is "left over" (or residual) after integrating f over the path $\partial \mathbb{D}(z_0, r)$.

Proof   This follows from Corollary 6.4 and Example 5.3(a). ∎

(b) The set of poles $P(f)$ of a function f meromorphic in U is discrete and countable, and this set has no cluster point in U.[2]

Proof   (i) Suppose $z_0 \in P(f)$. Then there is an $r > 0$ such that $\overline{\mathbb{D}}(z_0, r) \subset U$. Hence f is holomorphic in $\mathbb{D}^\bullet(z_0, r)$. Therefore $P(f) \cap \mathbb{D}(z_0, r) = \{z_0\}$, which shows that $P(f)$ is discrete.

(ii) Assume $P(f)$ has a cluster point z_0 in U. Because $P(f)$ is discrete, z_0 does not belong to $P(f)$. Therefore z_0 lies in an open set $U \setminus P(f)$, and we find an $r > 0$ with $\mathbb{D}(z_0, r) \subset U \setminus P(f)$. Therefore z_0 is not a cluster point of $P(f)$, which is a contradiction.

(iii) For every $z \in P(f)$, there is an $r_z > 0$ with $\mathbb{D}^\bullet(z, r_z) \cap P(f) = \emptyset$. If K is a compact subset of U, then $K \cap P(f)$ is also compact. Therefore we find $z_0, \ldots, z_m \in P(f)$ such that

$$K \cap P(f) \subset \bigcup_{j=0}^{m} \mathbb{D}(z_j, r_{z_j}) \ .$$

Consequently $K \cap P(f)$ is a finite set.

(iv) To verify that $P(f)$ is countable, we set

$$K_j := \{ x \in U \ ; \ d(x, \partial U) \geq 1/j, \ |x| \leq j \} \quad \text{for } j \in \mathbb{N}^\times \ .$$

Due to Examples III.1.3(l) and III.2.22(c) and because of the Heine–Borel theorem, every K_j is compact. Also $\bigcup_j K_j = U$, and $K_j \cap P(f)$ is finite for every $j \in \mathbb{N}^\times$. It then follows from Proposition I.6.8 that $P(f) = \bigcup_j (K_j \cap P(f))$ is countable. ∎

The next theorem shows that a function is meromorphic if and only if it is locally the quotient of two holomorphic functions.[3]

6.8 Proposition   *The function f is meromorphic in U if and only if there is a closed subset A of U and these conditions are satisfied:*

(i) *f is holomorphic in $U \setminus A$.*

(ii) *To every $a \in A$, there is an $r > 0$ such that $\mathbb{D}(a, r) \subset U$, $g, h \in C^\omega\big(\mathbb{D}(a, r)\big)$ with $h \neq 0$, and $f = g/h$ in $\mathbb{D}^\bullet(a, r)$.*

---

[2]However, it is entirely possible that the f's poles may accumulate on the boundary of U.
[3]This explains the name "meromorphic", which means "fractional form", while "holomorphic" can be translated as "whole form".

Proof (a) Suppose f is meromorphic in U. Then, for $a \in P(f) =: A$, there is an $r > 0$ with $\mathbb{D}(a,r) \subset U$ such that f has the Laurent expansion

$$f(z) = \sum_{n=-m}^{\infty} c_n(z-a)^n \quad \text{for } z \in \mathbb{D}^\bullet(a,r)$$

in $\mathbb{D}^\bullet(a,r)$ for a suitable $m \in \mathbb{N}^\times$. The holomorphic function

$$\mathbb{D}^\bullet(a,r) \to \mathbb{C}, \quad z \mapsto (z-a)^m f(z) = \sum_{k=0}^{\infty} c_{k-m}(z-a)^k \qquad (6.7)$$

has a removable singularity at a. Thus there is a $g \in C^\omega(\mathbb{D}(a,r))$ such that

$$g(z) = (z-a)^m f(z) \quad \text{for } 0 < |z-a| < r \ .$$

Therefore f has the representation $f = g/h$ in $\mathbb{D}^\bullet(a,r)$ with $h := (X-a)^m \in C^\omega(\mathbb{C})$.

(b) Suppose the given conditions are satisfied. If $h(a) \neq 0$, we can assume (by shrinking r) that $h(z) \neq 0$ for $z \in \mathbb{D}(a,r)$. Then $f = g/h$ is holomorphic in $\mathbb{D}(a,r)$. In the case $h(a) = 0$, we can assume that there is an $m \in \mathbb{N}^\times$ such that h has the power series expansion

$$h(z) = \sum_{k=m}^{\infty} c_k(z-a)^k = (z-a)^m \sum_{n=0}^{\infty} c_{n+m}(z-a)^n$$

in $\mathbb{D}(a,r)$, where c_m is distinct from zero because $h \neq 0$. Therefore the function defined through

$$\varphi(z) := \sum_{n=0}^{\infty} c_{n+m}(z-a)^n \quad \text{for } z \in \mathbb{D}(a,r)$$

is holomorphic on $\mathbb{D}(a,r)$ (see Proposition V.3.5) with $\varphi(a) = c_m \neq 0$. Thus there is a $\rho \in (0,r)$ such that $\varphi(z) \neq 0$ for $|z-a| \leq \rho$, which implies $g/\varphi \in C^\omega(\mathbb{D}(a,\rho))$. Denoting by $\sum_{n \geq 0} b_n(z-a)^n$ the Taylor series for g/φ, we find

$$f(z) = \frac{g(z)}{h(z)} = \frac{1}{(z-a)^m}\frac{g}{\varphi}(z) = \sum_{n=0}^{\infty} b_n(z-a)^{n-m} \quad \text{for } z \in \mathbb{D}^\bullet(a,\rho) \ .$$

From this expression and from the uniqueness of the Laurent expansion, we learn that a is a pole of f. ∎

6.9 Examples   (a) Every rational function is meromorphic in $\mathbb{C}$ and has finitely many poles.

(b) The tangent and the cotangent are meromorphic in $\mathbb{C}$. Their sets of poles are respectively $\pi(\mathbb{Z} + 1/2)$ and $\pi\mathbb{Z}$. The Laurent expansion of the cotangent in $\pi\mathbb{D}^{\bullet}$ reads

$$\cot z = \frac{1}{z} - 2\sum_{k=1}^{\infty} \frac{\zeta(2k)}{\pi^{2k}} z^{2k-1} \quad \text{for } z \in \pi\mathbb{D}^{\bullet}\ .$$

Proof   Because $\tan = \sin/\cos$ and $\cot = 1/\tan$, the first two claims follow from Proposition 6.8. The given representation of the cotangent is implied by (VI.7.18), Theorem VI.6.15, and the uniqueness of the Laurent expansion. ∎

(c) The gamma function is meromorphic in $\mathbb{C}$. Its set of poles is $-\mathbb{N}$.

Proof   The Weierstrass product representation of Proposition VI.9.5 and the Weierstrass convergence theorem (Theorem 5.27) implies that Γ is the reciprocal of an entire function whose set of zeros is $-\mathbb{N}$. The claim then follows from Proposition 6.8. ∎

(d) The Riemann ζ function is meromorphic in $\mathbb{C}$.

Proof   This follows from Theorem VI.6.15. ∎

(e) The function $z \mapsto e^{1/z}$ is not meromorphic in $\mathbb{C}$.

Proof   Because $e^{1/z} = \exp(1/z)$, we have

$$e^{1/z} = \sum_{k=0}^{\infty} \frac{1}{k!\,z^k} = 1 + \frac{1}{z} + \frac{1}{2z^2} + \cdots \quad \text{for } z \in \mathbb{C}^{\times}\ .$$

Therefore 0 is an essential singularity of $z \mapsto e^{1/z}$. ∎

Simple poles

As we shall see, the residues of meromorphic functions play a particularly important role. Therefore, we should seek ways to determine residues without explicitly carrying out the Laurent expansion. This is especially easy for first order poles, the **simple poles**, as we show next.

6.10 Proposition   *The holomorphic function $f : U\backslash\{z_0\} \to \mathbb{C}$ has a simple pole at z_0 if and only if $z \mapsto g(z) := (z - z_0)f(z)$ has a removable singularity at z_0 with $g(z_0) \neq 0$. Then*

$$\mathrm{Res}(f, z_0) = \lim_{z \to z_0} (z - z_0)f(z)\ .$$

Proof   Suppose g is holomorphic in U with $g(z_0) \neq 0$. Then there is an $r > 0$ with $\mathbb{D}(z_0, r) \subset U$ and a sequence (b_n) in $\mathbb{C}$ such that

$$g(z) = \sum_{n=0}^{\infty} b_n(z - z_0)^n \text{ for } z \in \mathbb{D}(z_0, r) \text{ and } g(z_0) = b_0 \neq 0\ .$$

Because

$$f(z) = \frac{g(z)}{z - z_0} = \sum_{n=-1}^{\infty} c_n(z - z_0)^n \quad \text{for } z \in \mathbb{D}^{\bullet}(z_0, r) \, ,$$

where $c_n := b_{n+1}$ and $c_{-1} := b_0 \neq 0$, the point z_0 is a simple pole, and

$$\mathrm{Res}(f, z_0) = c_{-1} = b_0 = g(z_0) = \lim_{z \to z_0}(z - z_0)f(z) \, .$$

Conversely, if z_0 is a simple pole of f, there is an $r > 0$ such that $\mathbb{D}(z_0, r) \subset U$ and

$$f(z) = \sum_{n=-1}^{\infty} c_n(z - z_0)^n \quad \text{for } z \in \mathbb{D}^{\bullet}(z_0, r) \text{ and } c_{-1} \neq 0 \, .$$

From this follows

$$(z - z_0)f(z) = \sum_{n=0}^{\infty} c_{n-1}(z - z_0)^n \quad \text{for } z \in \mathbb{D}^{\bullet}(z_0, r) \, .$$

Now the claim is implied by the Riemann removability theorem and the identity theorem for analytic functions. ∎

6.11 Examples (a) Suppose g and h are holomorphic in U and h has a **simple zero** at z_0, that is,[4] $h(z_0) = 0$ and $h'(z_0) \neq 0$. Then $f := g/h$ is meromorphic in U, and z_0 is a simple pole of f with $\mathrm{Res}(f, z_0) = g(z_0)/h'(z_0)$ if $g(z_0) \neq 0$.

Proof The Taylor formula gives

$$h(z) = (z - z_0)h'(z_0) + o(|z - z_0|) \quad (z \to z_0) \, ,$$

and therefore $h(z)(z - z_0)^{-1} \to h'(z_0)$ for $z \to z_0$. This implies

$$\lim_{z \to z_0}(z - z_0)f(z) = \lim_{z \to z_0} g(z)/\big(h(z)(z - z_0)^{-1}\big) = g(z_0)/h'(z_0) \, ,$$

and the claim then follows from Theorems 6.8 and 6.10 and the Riemann removability theorem. ∎

(b) The tangent and the cotangent have only simple pole poles. Their residues are given by

$$\mathrm{Res}\big(\tan, \pi(k + 1/2)\big) = -\mathrm{Res}(\cot, k\pi) = -1 \quad \text{for } k \in \mathbb{Z} \, .$$

Proof This follows immediately from (a). ∎

---

[4] A simple zero is a zero of order 1 (see Exercise IV.3.10).

(c) The gamma function has only first order poles, and

$$\operatorname{Res}(\Gamma, -n) = (-1)^n/n! \quad \text{for } n \in \mathbb{N} .$$

Proof From (VI.9.2) we obtain for $z \in \mathbb{C}\backslash(-\mathbb{N})$ with $\operatorname{Re} z > -n - 1$ the representation

$$(z + n)\Gamma(z) = \frac{\Gamma(z + n + 1)}{z(z + 1) \cdot \cdots \cdot (z + n - 1)} .$$

Therefore $(z + n)\Gamma(z)$ converges for $z \to -n$ to $(-1)^n \Gamma(1)/n!$. Because $\Gamma(1) = 1$, the claim is implied by Example 6.9(c) and Proposition 6.10. ∎

(d) The Riemann ζ function has a simple pole at 1 with residue 1.

Proof This follows from Theorem VI.6.15. ∎

(e) Suppose $p \in \mathbb{R}$ and $a > 0$. Then the function defined by $f(z) := e^{-ipz}/(z^2 + a^2)$ is meromorphic in $\mathbb{C}$ and has simple poles at $\pm ia$ with

$$\operatorname{Res}(f, \pm ia) = \pm e^{\pm pa}/2ia .$$

Proof Obviously $(z \mp ia)f(z) = e^{-ipz}/(z \pm ia)$ is holomorphic at $\pm ia$, and

$$\lim_{z \to \pm ia} (z \mp ia)f(z) = \pm e^{\pm pa}/2ia .$$

The claim therefore follows from Proposition 6.10. ∎

(f) For $p \in \mathbb{R}$ and $f(z) := e^{-ipz}/(z^4 + 1)$, the function f is meromorphic in $\mathbb{C}$ and

$$P(f) = \{ z_j := e^{i(\pi/4 + j\pi/2)} ; \ j = 0, \ldots, 3 \} .$$

Every pole is simple, and every residue has

$$\operatorname{Res}(f, z_j) = e^{-ipz_j}/4z_j^3 \quad \text{for } 0 \le j \le 3 .$$

Proof An elementary calculation gives

$$\prod_{\substack{k=0 \\ k \ne j}}^{3} (z_j - z_k) = 4z_j^3 .$$

Therefore this claim also follows from Proposition 6.10. ∎

The winding number

We have already seen in the last section that the homotopy invariance of line integrals of holomorphic functions (Proposition 5.7) can be used to effectively calculate other integrals. We will now derive a like result for meromorphic functions. For this we must know more about the position of the poles relative to the integration curve.

In the following

- Γ is always a closed compact piecewise C^1 curve in $\mathbb{C}$.

For $a \in \mathbb{C}\backslash\Gamma$,

$$w(\Gamma, a) := \frac{1}{2\pi i} \int_\Gamma \frac{dz}{z - a}$$

is called the **winding number**, the **circulation number**, or the **index** of Γ about a.

6.12 Examples (a) Suppose $m \in \mathbb{Z}^\times$ and $r > 0$. Then for $z_0 \in \mathbb{C}$, the parametrization $\gamma_m : [0, 2\pi] \to z_0 + re^{imt}$ is that of a smooth curve $\Gamma_m := \Gamma_m(z_0, r)$, whose image is the oriented circle $\partial \mathbb{D}(z_0, r)$. If m is positive [or negative], then Γ_m is oriented the same as [or opposite to] $\partial \mathbb{D}(z_0, r)$. Therefore $\partial \mathbb{D}(z_0, r)$ will circulate $|m|$ times in the positive [or negative] direction as t goes from 0 to 2π. On these grounds, we call Γ_m the **m-times traversed circle** with center z_0 and radius r. We have

$$w(\Gamma_m, a) = \begin{cases} m, & |a - z_0| < r, \\ 0, & |a - z_0| > r. \end{cases}$$

Proof As in the proof of Lemma 6.1, it follows in the case $|a - z_0| < r$ that

$$\int_{\Gamma_m} \frac{dz}{z - a} = \int_{\Gamma_m} \frac{dz}{z - z_0} = \int_0^{2\pi} \frac{imre^{imt}}{re^{imt}} \, dt = 2\pi i m .$$

When $|a - z_0| > r$, the curve Γ_m is null homotopic in $\mathbb{C}\backslash\{a\}$. $\blacksquare$

(b) For $z_0 \in \mathbb{C}$, suppose Γ_{z_0} is a point curve with $\text{im}(\Gamma_{z_0}) = \{z_0\}$. Then $w(\Gamma_{z_0}, a) = 0$ for $a \in \mathbb{C}\backslash\{z_0\}$.

(c) If γ_1 and γ_2 are homotopic piecewise C^1 loops in U and $a \in U^c$, then

$$w(\Gamma_1, a) = w(\Gamma_2, a)$$

for $\Gamma_1 = [\gamma_1]$ and $\Gamma_2 = [\gamma_2]$.

Proof For $a \in U^c$, the function $z \mapsto 1/(z - a)$ is holomorphic in U. Therefore the claim follows from Proposition 5.7. $\blacksquare$

(d) If U is simply connected and $\Gamma \subset U$, then $w(\Gamma, a) = 0$ for $a \in U^c$.

Proof This follows from (b) and (c). $\blacksquare$

According to Example 6.12(a), the winding number $w(\Gamma_m, a)$ of the m-times traversed circle $\Gamma_m := \Gamma_m(z_0, r)$ is the (signed) number of times Γ_m "winds around" the point a. We next show that this geometric interpretation of the winding is valid for every closed piecewise C^1 curve, but we must first prove a technical result.

6.13 Lemma *Suppose I is a perfect compact interval and $\gamma : I \to \mathbb{C}^\times$ has piecewise continuous derivatives. Then there is a continuous and piecewise continuously differentiable function $\varphi : I \to \mathbb{C}$ such that $\exp \circ \varphi = \gamma$. Here, γ and φ have continuous derivatives on the same interval I.*

Proof (i) We define $\log_\alpha := \mathrm{Log} \, | \, (\mathbb{C} \backslash \mathbb{R}^+ e^{i\alpha})$ for $\alpha \in \mathbb{R}$. Then from Proposition III.6.19 and Exercise III.6.9, $\log_\alpha$ is a topological map of $\mathbb{C} \backslash \mathbb{R}^+ e^{i\alpha}$ onto $\mathbb{R} + i(\alpha, \alpha + 2\pi)$ satisfying

$$\exp(\log_\alpha z) = z \quad \text{for } z \in \mathbb{C} \backslash \mathbb{R}^+ e^{i\alpha} \,. \tag{6.8}$$

From this we find (see Example IV.1.13(e)) that $\log_\alpha$ is a C^1 diffeomorphism of $\mathbb{C} \backslash \mathbb{R}^+ e^{i\alpha}$ onto $\mathbb{R} + i(\alpha, \alpha + 2\pi)$ satisfying

$$(\log_\alpha)'(z) = 1/z \quad \text{for } z \in \mathbb{C} \backslash \mathbb{R}^+ e^{i\alpha} \,. \tag{6.9}$$

(ii) Because $\gamma(I)$ is compact, there exists an $r > 0$ with $r\overline{\mathbb{D}} \cap \gamma(I) = \emptyset$. From the uniform continuity of γ, there is a partition $(t_0, \ldots, t_m)$ of I such that $\mathrm{diam}\big(\gamma \, | \, [t_{j-1}, t_j]\big) < r/2$ for $1 \leq j \leq m$. Because γ has piecewise continuous derivatives, we can choose this partition so that $\gamma \, | \, [t_{j-1}, t_j]$ is continuously differentiable for $1 \leq j \leq m$. Because the disc $D_j := \mathbb{D}\big(\gamma(t_j), r\big)$ has no zeros, we can fix an $\alpha_j \in (-\pi, \pi)$ such that $\mathbb{R}^+ e^{i\alpha_j} \cap \overline{D}_j = \emptyset$ for $1 \leq j \leq m$. Then we set $\log_j := \log_{\alpha_j}$ and $\varphi_j := \log_j \circ \gamma \, | \, [t_{j-1}, t_j]$. From (i), we know φ_j belongs to $C^1\big([t_{j-1}, t_j]\big)$, and because $\gamma(t) \in D_j \cap D_{j+1}$ for $t \in [t_{j-1}, t_j]$, we find using (6.8) that

$$\exp\big(\varphi_j(t_j)\big) = \gamma(t_j) = \exp\big(\varphi_{j+1}(t_j)\big) \quad \text{for } 1 \leq j \leq m - 1 \,.$$

Consequently, Proposition III.6.13 and the addition theorem for the exponential function guarantee the existence of $k_j \in \mathbb{Z}$ such that

$$\varphi_j(t_j) - \varphi_{j+1}(t_j) = 2\pi i k_j \quad \text{for } 1 \leq j \leq m - 1 \,. \tag{6.10}$$

Now we define $\varphi : I \to \mathbb{C}$ by

$$\varphi(t) := \varphi_j(t) + 2\pi i \sum_{n=0}^{j-1} k_n \quad \text{for } t_{j-1} \leq t \leq t_j \text{ and } 1 \leq j \leq m \,, \tag{6.11}$$

where $k_0 := 0$. Then it follows from (6.10) and the continuous differentiability of $\log_j$ and $\gamma \, | \, [t_{j-1}, t_j]$ that φ has piecewise continuous derivatives. Finally, we get $\exp \circ \varphi = \gamma$ from (6.8), the definition of φ_j, and the $2\pi i$-periodicity of the exponential function. ∎

6.14 Theorem *For every $a \in \mathbb{C} \backslash \Gamma$, the index $w(\Gamma, a)$ is a whole number.*

Proof Suppose γ is a piecewise C^1 parametrization of Γ and $(t_0, \ldots, t_m)$ is a partition of the parameter interval I such that $\gamma \, | \, [t_{j-1}, t_j]$ is continuously differentiable for $1 \leq j \leq m$. Then, according to Lemma 6.13, to $a \in \mathbb{C} \backslash \Gamma$ there is a $\varphi \in C(I)$ such that $\varphi | [t_{j-1}, t_j] \in C^1[t_{j-1}, t_j]$ for $1 \le j \le m$, and $e^\varphi = \gamma - a$. From this it follows that $\dot{\gamma}(t) = \dot{\varphi}(t)(\gamma(t) - a)$ for $t_{j-1} \le t \le t_j$ and $1 \le j \le m$. Therefore we get

$$\int_\Gamma \frac{dz}{z-a} = \sum_{j=1}^m \int_{t_{j-1}}^{t_j} \frac{\dot{\gamma}(t)\,dt}{\gamma(t)-a} = \sum_{j=1}^m \int_{t_{j-1}}^{t_j} \dot{\varphi}(t)\,dt = \varphi(t_m) - \varphi(t_0) . \qquad (6.12)$$

Because Γ is closed, we have

$$\exp(\varphi(t_m)) = E_\Gamma - a = A_\Gamma - a = \exp(\varphi(t_0)) ,$$

and therefore $\varphi(t_m) - \varphi(t_0) \in 2\pi i \mathbb{Z}$, according to Proposition III.6.13(i). So $w(\Gamma, a)$ is whole number. ∎

6.15 Remarks (a) Suppose the assumptions of Lemma 6.13 are satisfied. With the notation of its proof, we denote by $\arg_j(\gamma(t))$ for $1 \le j \le m$ and $t \in [t_{j-1}, t_j]$ the unique number $\eta \in (\alpha_j, \alpha_j + 2\pi)$ such that $\operatorname{Im}(\varphi_j(t)) = \eta$. Then it follows from

$$e^{\log |\gamma(t)|} = |\gamma(t)| = |e^{\varphi(t)}| = e^{\operatorname{Re} \varphi(t)}$$

that

$$\varphi_j(t) = \log_j \gamma(t) = \log |\gamma(t)| + i \arg_j(\gamma(t)) \qquad (6.13)$$

for $t_{j-1} \le t \le t_j$ and $1 \le j \le m$. Now we set

$$\arg_{c,\gamma}(t) := \arg_j(\gamma(t)) + 2\pi \sum_{n=0}^{j-1} k_n$$

for $t_{j-1} \le t \le t_j$ and $1 \le j \le m$. Then (6.11) and (6.13) imply

$$\varphi = \log \circ |\gamma| + i \arg_{c,\gamma} . \qquad (6.14)$$

Because φ is piecewise continuously differentiable, (6.14) shows that this also holds for $\arg_{c,\gamma}$, where $\arg_{c,\gamma}$ and γ are continuously differentiable on the same subintervals of I. In other words, $\arg_{c,\gamma}$ is a piecewise continuously differentiable function that "selects from" the set-valued function $\operatorname{Arg} \circ \gamma$, that is,

$$\arg_{c,\gamma} \in \operatorname{Arg}(\gamma(t)) \quad \text{for } t \in I .$$

Likewise, φ is a piecewise continuously differentiable "selection from" the set-valued logarithm $\operatorname{Log} \circ \gamma$.

(b) Suppose γ is a piecewise C^1 parametrization of Γ and $a \in \mathbb{C} \backslash \Gamma$. Also suppose

$$\varphi = \log \circ |\gamma - a| + i \arg_{c, \gamma - a}$$

is a piecewise C^1 selection from $\mathrm{Log} \circ (\gamma - a)$, where φ belongs to $C^1[t_{j-1}, t_j]$ for $1 \leq j \leq m$. Then we have

$$\int_{t_{j-1}}^{t_j} \dot{\varphi}(t)\, dt = \log \circ |\gamma - a|\big|_{t_{j-1}}^{t_j} + i \arg_{c,\gamma-a}\big|_{t_{j-1}}^{t_j} = \varphi(t_j) - \varphi(t_{j-1}) \ .$$

Because $\log |\gamma(t_m) - a| = \log |\gamma(t_0) - a|$, it thus follows from (6.12) that

$$w(\Gamma, a) = \frac{1}{2\pi i} \int_\Gamma \frac{dz}{z - a} = \frac{1}{2\pi}\big(\arg_{c,\gamma-a}(t_m) - \arg_{c,\gamma-a}(t_0)\big) \ .$$

This shows that 2π times the winding number of Γ about a is the "accumulated change" of the argument $\arg_{c,\gamma-a}$ of $\gamma - a$ as Γ passes from $\gamma(t_0)$ to $\gamma(t_m)$. Therefore $w(\Gamma, a)$ specifies how many times Γ winds around the point a, where $w(\Gamma, a) > 0$ means it winds counterclockwise and $w(\Gamma, a) < 0$ means it winds clockwise. ∎

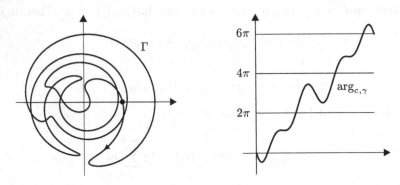

A curve with $w(\Gamma, 0) = 3$

The continuity of the winding number

We first prove a simple lemma for maps in discrete spaces.

6.16 Lemma *Suppose X and Y are metric spaces and Y is discrete. If $f : X \to Y$ is continuous, then f is constant on every connected component of X.*

Proof Suppose Z is a connected component of X. By Theorem III.4.5, $f(Z)$ is connected. Because Y is discrete, every connected component of Y consists of exactly one point. Therefore f is constant on Z. ∎

Because Γ is compact, there is an $R > 0$ such that $\Gamma \subset R\mathbb{D}$. Therefore $\mathbb{C}\backslash\Gamma$ contains the set $R\mathbb{D}^c$, which shows that $\mathbb{C}\backslash\Gamma$ has exactly one unbounded connected component.

6.17 Corollary *The map $w(\Gamma, \cdot) \colon \mathbb{C}\backslash\Gamma \to \mathbb{Z}$ is constant on every connected component. If a belongs to the unbounded connected component of $\mathbb{C}\backslash\Gamma$, then $w(\Gamma, a) = 0$.*

Proof Suppose $a \in \mathbb{C}\backslash\Gamma$, and define $d := d(a, \Gamma)$. We then choose an $\varepsilon > 0$ and define $\delta := \min\{\varepsilon\pi d^2 \backslash L(\Gamma), d/2\}$. Then we have

$$|z - a| \geq d > 0 \quad \text{and} \quad |z - b| \geq d/2 \quad \text{for } z \in \Gamma \text{ and } b \in \mathbb{D}(a, \delta) \,.$$

From this follows

$$|w(\Gamma, a) - w(\Gamma, b)| \leq \frac{1}{2\pi} \int_\Gamma \left| \frac{a - b}{(z - a)(z - b)} \right| dz < \frac{1}{2\pi} L(\Gamma) \frac{2}{d^2} \delta \leq \varepsilon$$

for $b \in \mathbb{D}(a, \delta)$. Because ε was arbitrary, $w(\Gamma, \cdot)$ is continuous in $a \in \mathbb{C}\backslash\Gamma$. The first statement now follows from Theorem 6.14 and Lemma 6.16.

Denote by Z the unbounded connected component of $\mathbb{C}\backslash\Gamma$. Also suppose $R > 0$ is chosen so that Z contains the set $R\mathbb{D}^c$. Finally, we choose $a \in Z$ with $|a| > R$ and $|a| > L(\Gamma)/\pi + \max\{|z| \,;\, z \in \Gamma\}$. Then we have

$$|w(\Gamma, a)| \leq \frac{1}{2\pi} \int_\Gamma \frac{dz}{|z - a|} < \frac{1}{2} \,.$$

Because $w(\Gamma, a) \in \mathbb{Z}$, we have $w(\Gamma, a) = 0$, and therefore $w(\Gamma, b) = 0$ for every $b \in Z$. ∎

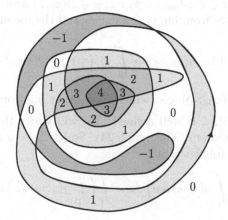

6.18 Corollary *Suppose f is meromorphic in U and $w(\Gamma, a) = 0$ for $a \in U^c$. Then $\{z \in P(f)\backslash\Gamma \,;\, w(\Gamma, z) \neq 0\}$ is a finite set.*

Proof By Corollary 6.17, $B := \{z \in U\backslash\Gamma \,;\, w(\Gamma, z) \neq 0\}$ is bounded. When $B \cap P(f)$ is not finite, this set has a cluster point $z_0 \in \bar{B}$. According to Remark 6.7(b), $P(f)$ has a cluster point in U, and therefore z_0 belongs to U^c. Thus, by assumption, $w(\Gamma, z_0) = 0$. On the other hand, it follows from the continuity of $w(\Gamma, \cdot)$ and from $w(\Gamma, B) \subset \mathbb{Z}^\times$ that $w(\Gamma, z_0)$ is distinct from zero. Therefore $B \cap P(f)$ is finite. ∎

The generalized Cauchy integral theorem

The concept of winding number allows us to expand the domain of the Cauchy integral theorem and the Cauchy integral formula. We first present a lemma.

6.19 Lemma Suppose $f : U \to \mathbb{C}$ is holomorphic and $\Gamma \subset U$.

(i) *The function*

$$
g : U \times U \to \mathbb{C} , \quad (z, w) \mapsto
\begin{cases}
\big(f(w) - f(z)\big)/(w - z) , & z \neq w , \\
f'(z) , & z = w ,
\end{cases}
$$

is continuous.

(ii) *The map*

$$
h : U \to \mathbb{C} , \quad z \mapsto \int_\Gamma g(z, w) \, dw
$$

is analytic.

Proof (i) Obviously g is continuous at every point (z, w) with $z \neq w$.

Suppose $z_0 \in U$ and $\varepsilon > 0$. Then there is an $r > 0$ such that $\mathbb{D}(z_0, r) \subset U$ and $|f'(\zeta) - f'(z_0)| < \varepsilon$ for $\zeta \in \mathbb{D}(z_0, r)$. For $z, w \in \mathbb{D}(z_0, r)$ and $\gamma(t) := (1 - t)z + tw$ with $t \in [0, 1]$, it follows from $\operatorname{im}(\gamma) \subset \mathbb{D}(z_0, r)$ and the mean value theorem that

$$
g(z, w) - g(z_0, z_0) = \int_0^1 \big[f'(\gamma(t)) - f'(z_0) \big] \, dt .
$$

Therefore $|g(z, w) - g(z_0, z_0)| < \varepsilon$, which shows that g is continuous at (z_0, z_0).

(ii) Because of (i), h is well defined. We will verify that h is analytic with help from Morera's theorem (Theorem 5.24). So let Δ be a triangular path in U. From Exercise 5.11 it follows that[5]

$$
\int_\Delta \left(\int_\Gamma g(z, w) \, dw \right) dz = \int_\Gamma \left(\int_\Delta g(z, w) \, dz \right) dw . \tag{6.15}
$$

Also, the proof of Lemma 6.2 shows that $g(\cdot, w)$ belongs to $C^\omega(U)$ for every $w \in U$. Therefore it follows from Theorem 5.25 that $\int_\Delta g(z, w) \, dz = 0$ for $w \in \Gamma$, and we get $\int_\Delta h(z) \, dz = 0$ from (6.15). Therefore h is analytic. ∎

A curve Γ in U is said to be **null homologous in U** if it is closed and piecewise continuously differentiable and if $w(\Gamma, a) = 0$ for $a \in U^c$.[6]

---

[5]This is an elementary version of the Fubini's theorem, which will be proved in full generality in Volume III. See also Exercise 5.12.

[6]When $U = \mathbb{C}$, every closed piecewise C^1 curve is null homologous.

6.20 Theorem (homology version of the Cauchy integral theorem and formula)
Suppose U is open in $\mathbb{C}$ and f is holomorphic in U. Then, if the curve Γ is null homologous in U,

$$\frac{1}{2\pi i} \int_\Gamma \frac{f(\zeta)}{\zeta - z}\, d\zeta = w(\Gamma, z) f(z) \quad \text{for } z \in U \backslash \Gamma, \tag{6.16}$$

and

$$\int_\Gamma f(z)\, dz = 0. \tag{6.17}$$

Proof   We use the notations of Lemma 6.19.

(i) Clearly (6.16) is equivalent to the statement

$$h(z) = 0 \quad \text{for } z \in U \backslash \Gamma. \tag{6.18}$$

To verify this, suppose $U_0 := \{\, z \in \mathbb{C} \backslash \Gamma \;;\; w(\Gamma, z) = 0 \,\}$ and

$$h_0(z) := \frac{1}{2\pi i} \int_\Gamma \frac{f(\zeta)}{\zeta - z}\, d\zeta \quad \text{for } z \in U_0.$$

According to Theorem 6.14 and because $w(\Gamma, \cdot)$ is continuous, U_0 is open. We also have

$$h_0(z) = \frac{1}{2\pi i} \int_\Gamma \frac{f(\zeta)}{\zeta - z}\, d\zeta - f(z) w(\Gamma, z) = h(z) \text{ for } z \in U \cap U_0.$$

Because h_0 is holomorphic, the uniqueness theorem for analytic functions (Theorem V.3.13) says there is a function H holomorphic on $U \cup U_0$ such that $H \supset h_0$ and $H \supset h$. By assumption, we have $w(\Gamma, a) = 0$ for $a \in U^c$. Thus U^c belongs to U_0, and we learn that H is an entire function.

(ii) Suppose $R > 0$ with $\Gamma \subset R\mathbb{D}$. Because $R\mathbb{D}^c$ lies in the unbounded connected component of $\mathbb{C} \backslash \Gamma$, we have $R\mathbb{D}^c \subset U_0$. Suppose now $\varepsilon > 0$. We set $M := \max_{\zeta \in \Gamma} |f(\zeta)|$ and $R' := R + L(\Gamma) M / 2\pi\varepsilon$. For $z \in R'\mathbb{D}^c$, we then have

$$|\zeta - z| \geq |z| - |\zeta| > L(\Gamma) M / 2\pi\varepsilon \quad \text{for } \zeta \in \Gamma,$$

and we find

$$|h_0(z)| \leq \frac{1}{2\pi} \int_\Gamma \left| \frac{f(\zeta)}{\zeta - z} \right| d\zeta < \varepsilon \quad \text{for } z \in R'\mathbb{D}^c. \tag{6.19}$$

Because $h_0 \subset H$ and because H as an entire function is bounded on bounded sets, it follows that H is bounded on all of $\mathbb{C}$. Therefore we get from Liouville's theorem and (6.19) that H vanishes everywhere. Now $h \subset H$ implies (6.18).

(iii) Suppose $a \in \mathbb{C} \backslash \Gamma$ and

$$F: U \to \mathbb{C}, \quad z \mapsto (z - a) f(z).$$

Because F is holomorphic and $F(a) = 0$, (6.16) shows that

$$\frac{1}{2\pi i} \int_\Gamma f(z)\, dz = \frac{1}{2\pi i} \int_\Gamma \frac{F(z)}{z - a}\, dz = w(\Gamma, a) F(a) = 0. \quad \blacksquare$$

6.21 Remarks (a) If U is simply connected, every closed piecewise C^1 curve is null homologous in U. So Theorem 6.20 is a generalization of Theorems 5.5 and 5.9.

Proof This follows from Example 6.12(d). ∎

(b) Under the assumptions of Theorem 6.20, we have the generalized Cauchy derivative formula

$$ w(\Gamma, z) f^{(k)}(z) = \frac{k!}{2\pi i} \int_\Gamma \frac{f(\zeta)}{(\zeta - z)^{k+1}} \, d\zeta \quad \text{for } z \in U \backslash \Gamma, \quad k \in \mathbb{N} . $$

Proof This follows easily from Theorem 6.20. ∎

The residue theorem

The next theorem further generalizes the homology version of the Cauchy integral theorem to the case of meromorphic functions.

6.22 Theorem (residue theorem) *Suppose U is open in $\mathbb{C}$ and f is meromorphic in U. Further suppose Γ is a curve in $U \backslash P(f)$ that is null homologous in U. Then*

$$ \int_\Gamma f(z) \, dz = 2\pi i \sum_{p \in P(f)} \mathrm{Res}(f, p) w(\Gamma, p) , \qquad (6.20) $$

where only finitely many terms in the sum are distinct from zero.

Proof From Corollary 6.18 we know that $A := \{ a \in P(f) ; \ w(\Gamma, a) \neq 0 \}$ is a finite set. Thus the sum in (6.20) has only finitely many terms distinct from zero.

Suppose $A = \{ a_0, \ldots, a_m \}$ and f_j is the principal part of the Laurent expansion of f at a_j for $0 \leq j \leq m$. Then f_j is holomorphic in $\mathbb{C} \backslash \{ a_j \}$, and the singularities of $F := f - \sum_{j=0}^m f_j$ at $a_0, \ldots, a_m$ are removable (because F locally has the form

$$ F = g_j - \sum_{\substack{k=0 \\ k \neq j}}^m f_j $$

at a_j, where g_j is the auxiliary part of f at a_j). Therefore, by the Riemann removability theorem, F has a holomorphic continuation (also denoted by F) on

$$ U_0 := U \backslash \big(P(f) \backslash A \big) = A \cup \big(U \backslash P(f) \big) . $$

Because Γ lies in $U \backslash P(f)$ and is null homologous in U, Γ lies in U_0 and is null homologous there. Thus it follows from the generalized Cauchy integral theorem that $\int_\Gamma F \, dz = 0$, which implies

$$ \int_\Gamma f \, dz = \sum_{j=0}^m \int_\Gamma f_j \, dz . \qquad (6.21) $$

Because a_j is a pole of f, there are $n_j \in \mathbb{N}^\times$ and $c_{jk} \in \mathbb{C}$ for $1 \leq k \leq n_j$ and $0 \leq j \leq m$ such that

$$f_j(z) = \sum_{k=1}^{n_j} c_{jk}(z - a_j)^{-k} \quad \text{for } 0 \leq j \leq m .$$

It therefore follows from Remark 6.21(b) (with $f = 1$) that

$$\int_\Gamma f_j \, dz = \sum_{k=1}^{n_j} c_{jk} \int_\Gamma \frac{dz}{(z - a_j)^k} = 2\pi i\, c_{j1} w(\Gamma, a_j) .$$

Because $c_{j1} = \mathrm{Res}(f, a_j)$, the claim follows from (6.21) using the definition of A. ∎

Fourier integrals

The residue theorem has many important applications, which we now exemplify by calculating some improper integrals. We concentrate on a particularly significant class, namely, Fourier integrals.

Suppose $f \colon \mathbb{R} \to \mathbb{C}$ is absolutely integrable. For $p \in \mathbb{R}$, $|e^{-ipx}| = 1$ for $x \in \mathbb{R}$, and therefore the function $x \mapsto e^{-ipx} f(x)$ is absolutely integrable. Thus the **Fourier integral** of f at p, given by

$$\widehat{f}(p) := \int_{-\infty}^{\infty} e^{-ipx} f(x) \, dx \in \mathbb{C} , \tag{6.22}$$

is defined for every $p \in \mathbb{R}$. The function $\widehat{f} \colon \mathbb{R} \to \mathbb{C}$ defined through (6.22) is called the **Fourier transform** of f.

The next theorem and the subsequent examples show that the residue theorem helps calculate many Fourier transforms.

6.23 Proposition *Suppose the function f meromorphic on $\mathbb{C}$ has the properties*

(i) $P(f)$ *is finite,*

(ii) $P(f) \cap \mathbb{R} = \emptyset$,

(iii) $\lim_{|z| \to \infty} z f(z) = 0$.

Then

$$\widehat{f}(p) = \begin{cases} -2\pi i \sum_{\substack{z \in P(f) \\ \mathrm{Im}\, z < 0}} \mathrm{Res}(f e^{-ip\cdot}, z) , & p \geq 0 , \\ 2\pi i \sum_{\substack{z \in P(f) \\ \mathrm{Im}\, z > 0}} \mathrm{Res}(f e^{-ip\cdot}, z) , & p \leq 0 . \end{cases}$$

Proof Suppose $p \leq 0$. By assumption (i), there is an $r > 0$ such that $P(f) \subset r\mathbb{D}$. We choose Γ as a positively oriented boundary of $V := r\mathbb{D} \cap \{\, z \in \mathbb{C} \;;\; \mathrm{Im}\, z > 0 \,\}$.

Then we have

$$w(\Gamma, z) = \begin{cases} 1, & z \in V, \\ 0, & z \in (\overline{V})^c, \end{cases}$$

(see Exercise 13). Therefore it follows
from the residue theorem that

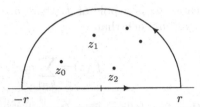

$$\int_{-r}^{r} f(x) e^{-ipx}\, dx + i \int_{0}^{\pi} f(re^{it}) e^{-ipre^{it}} re^{it}\, dt = 2\pi i \sum_{z \in V} \mathrm{Res}(fe^{-ip\cdot}, z)\ .$$

Because $p \le 0$, the integral over the semicircle can be estimated as

$$\left| \int_{0}^{\pi} f(re^{it}) e^{-ipre^{it}} re^{it}\, dt \right| \le \pi \max_{0 \le t \le \pi} \left| rf(re^{it}) e^{pr\sin t} \right| \le \pi \max_{|z|=r} |zf(z)|\ .$$

The assumption (iii) now implies the claim for $p \le 0$.

The case $p \ge 0$ is similar, but the integration path is in $\{\, z \in \mathbb{C}\ ;\ \mathrm{Im}\, z \le 0\,\}$. ∎

6.24 Examples   **(a)** For $f(x) := 1/(x^2 + a^2)$ with $a > 0$, we have

$$\widehat{f}(p) = \pi e^{-|p|\, a}/a \quad \text{for } p \in \mathbb{R}\ .$$

Proof   The function $f(z) := 1/(z^2 + a^2)$ is meromorphic in $\mathbb{C}$. Further $f\,|\,\mathbb{R}$ is absolutely integrable (see Example VI.8.4(e)), and $\lim_{|z| \to \infty} zf(z) = 0$. For the residues of $z \mapsto e^{-ipz} f(z)$ at the simple poles $\pm ia$, we found in Example 6.11(e) that

$$\mathrm{Res}(e^{-ipz} f(z), \pm ia) = \pm e^{\pm pa}/2ia\ .$$

Therefore the claim follows from Proposition 6.23. ∎

(b) We have

$$\int_{-\infty}^{\infty} \frac{dx}{x^4 + 1} = \frac{\pi}{\sqrt{2}}\ .$$

Proof   We consider the function f meromorphic in $\mathbb{C}$ defined by $f(z) := 1/(z^4 + 1)$. The poles of f in $\{\, z \in \mathbb{C}\ ;\ \mathrm{Im}\, z > 0\,\}$ are $z_0 := (1 + i)/\sqrt{2}$ and $z_1 := iz_0 = (-1 + i)/\sqrt{2}$. From Example 6.11(f), we know

$$\mathrm{Res}(fe^{-ip\cdot}, z_0) + \mathrm{Res}(fe^{-ip\cdot}, z_1) = \frac{1}{4z_0^3}(e^{-ipz_0} + ie^{-ipz_1})\ .$$

Now it follows from Proposition 6.23 that

$$\int_{-\infty}^{\infty} \frac{dx}{x^4 + 1} = \widehat{f}(0) = 2\pi i \left(\frac{1}{4z_0^3}(1 + i) \right) = \frac{\pi i}{2}(-z_0)(1 + i) = \frac{\pi}{\sqrt{2}}\ . \blacksquare$$

We can use the residue theorem to evaluate the convergent improper integral $\int_{\mathbb{R}} \sin(x)\, dx/x$, even though 0 is a removable singularity of $x \mapsto \sin(x)/x$ and the integral is not absolutely convergent (see Exercise VI.8.1).

6.25 Proposition   *For $p \in \mathbb{R}$, we have*

$$\lim_{R \to \infty} \int_{-R}^{R} \frac{\sin x}{x} e^{-ipx} \, dx = \begin{cases} \pi, & |p| < 1, \\ \pi/2, & |p| = 1, \\ 0, & |p| > 1. \end{cases}$$

Proof   (i) Suppose $R > 1$. We integrate the entire function

$$z \mapsto \frac{\sin z}{z} e^{-ipz}$$

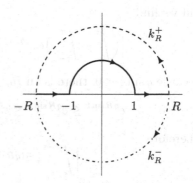

over the path γ_R that goes from $-R$ along the real axis to -1, along the upper half of the unit circle to $+1$, and finally along the real axis to R. The Cauchy integral theorem gives

$$\int_{-R}^{R} \frac{\sin x}{x} e^{-ipx} \, dx = \int_{\gamma_R} \frac{\sin z}{z} e^{-ipz} \, dz .$$

Because $\sin z = (e^{iz} - e^{-iz})/2i$, it follows that

$$\int_{-R}^{R} \frac{\sin x}{x} e^{-ipx} \, dx = \frac{1}{2i} \int_{\gamma_R} (e^{-iz(p-1)} - e^{-iz(p+1)}) \frac{dz}{z} .$$

(ii) Now we wish to calculate the integral

$$h_R(q) := \frac{1}{2\pi i} \int_{\gamma_R} \frac{e^{-izq}}{z} \, dz \quad \text{for } q \in \mathbb{R} .$$

So consider the curve $\Gamma_R^{\pm}$ parametrized by the loop $\gamma_R + k_R^{\pm}$, where

$$k_R^{\pm} : [0, \pi] \to \mathbb{C} , \quad t \mapsto R e^{\pm it} .$$

Then $w(\Gamma_R^+, 0) = 0$ and $w(\Gamma_R^-, 0) = -1$. Because $\mathrm{Res}(e^{-izq}/z, 0) = 1$, it follows from the residue theorem that

$$h_R(q) = -\frac{1}{2\pi i} \int_{k_R^+} \frac{e^{-izq}}{z} \, dz = -\frac{1}{2\pi} \int_{0}^{\pi} e^{-iqRe^{it}} \, dt$$

and

$$h_R(q) = -1 - \frac{1}{2\pi i} \int_{k_R^-} \frac{e^{-izq}}{z} \, dz = -1 + \frac{1}{2\pi} \int_{-\pi}^{0} e^{-iqRe^{it}} \, dt .$$

(iii) Next we want to show

$$\int_0^\pi e^{-iqRe^{it}} \, dt \to 0 \quad (R \to \infty) \quad \text{for } q < 0 \, . \tag{6.23}$$

So suppose $\varepsilon \in (0, \pi/2)$. Because $q \sin t \leq 0$ for $t \in [0, \pi]$ it follows that

$$|e^{-iqRe^{it}}| = e^{qR \sin t} \leq 1 \quad \text{for } q < 0 \, , \quad R > 1 \, , \quad t \in [0, \pi] \, ,$$

and we find

$$\left| \left(\int_0^\varepsilon + \int_{\pi-\varepsilon}^\pi \right) e^{-iqRe^{it}} \, dt \right| \leq 2\varepsilon \quad \text{for } q < 0 \text{ and } R > 1 \, . \tag{6.24}$$

Because $q \sin \varepsilon < 0$, there is an $R_0 > 1$ such that

$$e^{qR \sin t} \leq e^{qR \sin \varepsilon} \leq \varepsilon \quad \text{for } R \geq R_0 \text{ and } \varepsilon \leq t \leq \pi - \varepsilon \, .$$

Therefore

$$\left| \int_\varepsilon^{\pi-\varepsilon} e^{-iqRe^{it}} \, dt \right| \leq \varepsilon \pi \quad \text{for } R \geq R_0 \, ,$$

which, together with (6.24), proves (6.23). Similarly, one verifies for $q > 0$ that

$$\int_{-\pi}^0 e^{-iqRe^{it}} \, dt \to 0 \quad (R \to \infty) \, .$$

(iv) One easily checks that $h_R(0) = -1/2$ for $R > 1$. Therefore from (ii) and (iii), we have

$$\lim_{R \to \infty} h_R(q) = \begin{cases} 0 \, , & q < 0 \, , \\ -1/2 \, , & q = 0 \, , \\ -1 \, , & q > 0 \, . \end{cases}$$

Because of (i), we know

$$\lim_{R \to \infty} \int_{-R}^R \frac{\sin x}{x} e^{-ipx} \, dx = \pi \lim_{R \to \infty} \left(h_R(p-1) - h_R(p+1) \right) \, ,$$

from which the claim follows. ∎

6.26 Corollary

$$\int_{-\infty}^\infty \frac{\sin x}{x} \, dx = \pi \, .$$

Proof   The convergence of the improper integral $\int_{-\infty}^\infty (\sin x / x) \, dx$ follows from Exercises VI.4.11(ii) and VI.8.1(ii). We get its value from Proposition 6.25. ∎

6.27 Remarks (a) It does *not* follow from Proposition 6.25 that improper integral

$$\int_{-\infty}^{\infty} \frac{\sin x}{x} e^{i p x}\, dx$$

converges for $p \in \mathbb{R}^{\times}$. For that to hold we must confirm that $\lim_{R \to \infty} \int_0^R \cdots dx$ and $\lim_{R \to \infty} \int_{-R}^0 \cdots dx$ exist independently. When the limit

$$PV \int_{-\infty}^{\infty} f := \lim_{R \to \infty} \int_{-R}^{R} f(x)\, dx$$

exists for a (continuous) function $f : \mathbb{R} \to \mathbb{C}$, it is called the **Cauchy principal value**[7] of $\int_{-\infty}^{\infty} f$. If f is improperly integrable, the Cauchy principal value exists and agrees with $\int_{-\infty}^{\infty} f$. However, the converse of this statement does not hold, as shown by the example $f(t) := t$ for $t \in \mathbb{R}$.

(b) Because the function $g : \mathbb{R} \to \mathbb{R}$, $x \mapsto \sin(x)/x$ is not absolutely integrable, we cannot assign to g a Fourier transform $\widehat{g}$. Instead, one can (and must!) use a more general definition of the Fourier transformation—from the framework of the theory of distributions[8]—as we have done here. In the general theory, the Fourier transform of g *is* defined, and $\widehat{g}$ is the same piecewise constant function defined through the Cauchy principal value given in Proposition 6.25.

(c) The **Fourier transformation** $f \mapsto \widehat{f}$ is very important in many areas of math and physics. However, a more precise study of this map awaits the Lebesgue integration theory. Accordingly, we will return to the Fourier integral in Volume III. ∎

We refer you to the exercises and to the literature of complex analysis (for example, [Car66], [Con78], [FB95], [Rem92]) for a great number of further applications of the Cauchy integral theorem and the residue theorem, as well as for a more in-depth treatment of the theory of holomorphic and meromorphic functions.

Exercises

1 Suppose f is holomorphic in U and $z_0 \in U$ with $f'(z_0) \neq 0$. In addition, suppose $g : \mathbb{C} \to \mathbb{C}$ is meromorphic and has a simple pole at $w_0 := f(z_0)$. Show that $g \circ f$ has a simple pole at z_0 with $\mathrm{Res}(g \circ f, z_0) = \mathrm{Res}(g, w_0)/f'(z_0)$.

2 Suppose the function f is meromorphic in $\mathbb{C}$ and, in $\Omega(r_0, r_1)$ for $r_0 > 0$, has the Laurent expansion $\sum_{n=-\infty}^{\infty} c_n z^n$ for $n \in \mathbb{N}$ with $c_{-n} \neq 0$. Prove or disprove that f has an essential singularity. (Hint: Consider $z \mapsto 1/(z-1)$ in $\Omega(1, 2)$.)

3 Suppose $a, b \in \mathbb{C}$ with $0 < |a| < |b|$ and

$$f : \mathbb{C} \setminus \{a, b\} \to \mathbb{C}, \quad z \mapsto \frac{a - b}{(z - a)(z - b)}\ .$$

---

[7]See Exercise VI.8.9.

[8]For an introduction of the theory of distributions and many important applications, see for example [Sch65].

(a) Determine the Laurent expansion of f around 0 in $\Omega(0,|a|)$, $\Omega(|a|,|b|)$, and $\Omega(|b|,\infty)$.

(b) What is the Laurent expansion of f at a? (Hint: geometric series.)

4 Calculate

$$\int_{\partial\mathbb{D}} \frac{4\,dz}{1+4z^2} \quad \text{and} \quad \int_{\partial\mathbb{D}} \frac{e^z \sin z}{(1-e^z)^2}\,dz\ .$$

5 The holomorphic function $f: U\backslash\{z_0\} \to \mathbb{C}$ has an isolated singularity at $z_0 \in U$. Prove the equivalence of these statements:

  (i) z_0 is a pole of order n.

 (ii) $g: U\backslash\{z_0\} \to \mathbb{C}$, $z \mapsto (z-z_0)^n f(z)$ has a removable singularity at z_0 with $g(z_0) \neq 0$.

(iii) There are $\varepsilon, M_1, M_2 > 0$ such that $\mathbb{D}(z_0,\varepsilon) \subset U$ and

$$M_1\,|z-z_0|^{-n} \le |f(z)| \le M_2\,|z-z_0|^{-n} \quad \text{for } z \in \mathbb{D}^\bullet(z_0,\varepsilon)\ .$$

6 Show that $z_0 \in U$ is a pole of a function f holomorphic in $U\backslash\{z_0\}$ if and only if $|f(z)| \to \infty$ as $z \to z_0$.

7 Suppose z_0 is an isolated singularity of f, which is holomorphic in $U\backslash\{z_0\}$. Show that these statements are equivalent:

 (i) z_0 is an essential singularity.

(ii) For every $w_0 \in \mathbb{C}$, there exists a sequence (z_n) in $U\backslash\{z_0\}$ such that $\lim z_n = z_0$ and $\lim f(z_n) = w_0$.

(Hint: "(i)$\Rightarrow$(ii)" If the statement is false, there are a $w_0 \in \mathbb{C}$ and $r, s > 0$ such that $f\big(\mathbb{D}(z_0,r)\big) \cap \mathbb{D}(w_0,s) = \emptyset$. Then $g: \mathbb{D}^\bullet(z_0,r) \to \mathbb{C}$, $z \mapsto 1/\big(f(z)-w_0\big)$ is holomorphic and bounded. With help from Theorem 6.6 and Exercise 6, rule out the cases $g(z_0) = 0$ and $g(z_0) \neq 0$.)

8 Determine the singular points of

$$z \mapsto e^{1/(z-1)}/(e^z - 1)\ , \quad z \mapsto (z+1)\sin\big(1/(z-1)\big)\ .$$

Are they poles or essential singularities?

9 Verify that i is an essential singularity of $z \mapsto \sin\big(\pi/(z^2+1)\big)$.
(Hint: Exercise 7.)

10 Suppose U is simply connected and $a \in U$ is an isolated singularity of f, which is a holomorphic function in $U\backslash\{a\}$. Prove that f has an antiderivative in $U\backslash\{a\}$ if and only if $\text{Res}(f,a) = 0$.

11 Prove Remark 6.21(a).

12 For $p \in (0,1)$, prove that

$$PV \int_{-\infty}^{\infty} \frac{e^{rx}}{1+e^x}\,dx = \frac{\pi}{\sin(p\pi)}\ .$$

(Hint: Integrate $z \mapsto e^{pz}(1+e^z)^{-1}$ counterclockwise along the rectangular path with corners $\pm R$ and $\pm R + 2\pi i$.)

13 Suppose Γ_j is a piecewise C^1 curve parametrized by $\gamma_j \in C(I, \mathbb{C})$ for $j = 0, 1$. Also suppose $z \in \mathbb{C}$ with $|\gamma_0(t) - \gamma_1(t)| < |\gamma_0(t) - z|$ for $t \in I$. Show $w(\Gamma_0, z) = w(\Gamma_1, z)$. (Hint: For $\gamma := (\gamma_1 - z)/(\gamma_0 - z)$, show

$$w([\gamma], 0) = w(\Gamma_1, z) - w(\Gamma_0, z)$$

and $|1 - \gamma(t)| < 1$ for $t \in I$.)

14 Let $N(f) := \{ z \in U \backslash P(f) \; ; \; f(z) = 0 \}$ define the **set of zeros** of a meromorphic function f. Prove these:

(i) If $f \neq 0$, then $N(f)$ is a discrete subset of U.

(ii) The function

$$1/f : U \backslash N(f) \to \mathbb{C} , \quad z \mapsto 1/f(z)$$

is meromorphic in U and $P(1/f) = N(f)$ and $N(1/f) = P(f)$.

(Hints: (i) The identity theorem for analytic functions. (ii) Exercise 6.)

15 Show that the set of functions meromorphic in U is a field with respect to pointwise addition and multiplication.

16 Show that

$$\int_{-\infty}^{\infty} \frac{x^2}{1 + x^4} \, dx = \frac{\pi}{\sqrt{2}} .$$

(Hint: Proposition 6.23.)

17 Verify that

$$\int_0^{\infty} \frac{\cos x}{x^2 + a^2} \, dx = \frac{\pi e^{-a}}{2a} \quad \text{for } a > 0 .$$

(Hint: Example 6.24(a).)

18 Suppose f is meromorphic in U, $f \neq 0$, and $g := f'/f$. Prove

(i) g is meromorphic in U and has only simple poles;

(ii) if z_0 is a zero of f of order m, then $\mathrm{Res}(g, z_0) = m$;

(iii) if z_0 is a pole of f of order m, then $\mathrm{Res}(g, z_0) = -m$.

19 Suppose f is meromorphic in U and P is a curve in $U \backslash (N(f) \cup P(f))$ that is null homologous in U. For $z \in N(f) \cup P(f)$, denote by $\nu(z)$ the multiplicity of z. Then

$$\frac{1}{2\pi i} \int_\Gamma \frac{f'(z)}{f(z)} \, dz = \sum_{n \in N(f)} w(\Gamma, n)\nu(n) - \sum_{p \in P(f)} w(\Gamma, p)\nu(p) .$$

20 For $0 < a < b < 1$, calculate

$$\frac{1}{2\pi i} \int_{\partial \mathbb{D}} \frac{2z - (a + b)}{z^2 - (a + b)z + ab} \, dz .$$

21 Determine $\int_{-\infty}^{\infty} dx/(x^4 + 4x^2 + 3)$.

References

[Ama95]   H. Amann. *Gewöhnliche Differentialgleichungen.* W. de Gruyter, Berlin, 1983, 2. Aufl. 1995.

[Ape96]   A. Apelblat. *Tables of Integrals and Series.* Verlag Harri Deutsch, Frankfurt, 1996.

[Art31]   E. Artin. *Einführung in die Theorie der Gammafunktion.* Teubner, Leipzig, 1931.

[Art93]   M. Artin. *Algebra.* Birkhäuser, Basel, 1993.

[BBM86]   A.P. Brudnikov, Yu.A. Brychkov, O.M. Marichev. *Integrals and Series, I.* Gordon & Breach, New York, 1986.

[BF87]   M. Barner, F. Flohr. *Analysis I, II.* W. de Gruyter, Berlin, 1987.

[Bla91]   Ch. Blatter. *Analysis I–III.* Springer Verlag, Berlin, 1991, 1992.

[Brü95]   J. Brüdern. *Einführung in die analytische Zahlentheorie.* Springer Verlag, Berlin, 1995.

[Car66]   H. Cartan. *Elementare Theorie der analytischen Funktionen einer oder mehrerer komplexen Veränderlichen.* BI Hochschultaschenbücher, 112/112a. Bibliographisches Institut, Mannheim, 1966.

[Con78]   J.B. Conway. *Functions of One Complex Variable.* Springer Verlag, Berlin, 1978.

[FB95]   E. Freitag, R. Busam. *Funktionentheorie.* Springer Verlag, Berlin, 1995.

[Gab96]   P. Gabriel. *Matrizen, Geometrie, Lineare Algebra.* Birkhäuser, Basel, 1996.

[GR81]   I.S. Gradstein, I.M. Ryshik. *Tables of Series, Products, and Integrals.* Verlag Harri Deutsch, Frankfurt, 1981.

[Koe83]   M. Koecher. *Lineare Algebra und analytische Geometrie.* Springer Verlag, Berlin, 1983.

[Kön92]   K. Königsberger. *Analysis 1, 2.* Springer Verlag, Berlin, 1992, 1993.

[Pra78]   K. Prachar. *Primzahlverteilung.* Springer Verlag, Berlin, 1978.

[Rem92]   R. Remmert. *Funktionentheorie 1, 2.* Springer Verlag, Berlin, 1992.

[Sch65]   L. Schwartz. *Méthodes Mathématiques pour les Sciences Physiques.* Hermann, Paris, 1965.

[Sch69]   W. Schwarz. *Einführung in die Methoden und Ergebnisse der Primzahltheorie.*
            BI Hochschultaschenbücher, 278/278a. Bibliographisches Institut, Mannheim,
            1969.

[SS88]   G. Scheja, U. Storch. *Lehrbuch der Algebra.* Teubner, Stuttgart, 1988.

[Wal92]   W. Walter. *Analysis 1, 2.* Springer Verlag, Berlin, 1992.

Index